Medical Virology: From Pathogenesis to Disease Control

Series Editor
Shailendra K. Saxena
Centre for Advanced Research
King George's Medical University
Lucknow, India

This book series reviews the recent advancement in the field of medical virology including molecular epidemiology, diagnostics and therapeutic strategies for various viral infections. The individual books in this series provide a comprehensive overview of infectious diseases that are caused by emerging and re-emerging viruses including their mode of infections, immunopathology, diagnosis, treatment, epidemiology, and etiology. It also discusses the clinical recommendations in the management of infectious diseases focusing on the current practices, recent advances in diagnostic approaches and therapeutic strategies. The books also discuss progress and challenges in the development of viral vaccines and discuss the application of viruses in the translational research and human healthcare.

Mrinal K. Ghosh • Malini Basu
Editors

Pathogenesis, Management and Socio-Economic Impact of COVID-19 Pandemic

Springer

Editors
Mrinal K. Ghosh
Department of Cancer Biology and Inflammatory Disorder
CSIR-Indian Institute of Chemical Biology
Kolkata, West Bengal, India

Malini Basu
Department of Microbiology
Dhruba Chand Halder (DCH) College - University of Calcutta
Jaynagar Mazilpur, West Bengal, India

ISSN 2662-981X ISSN 2662-9828 (electronic)
Medical Virology: From Pathogenesis to Disease Control
ISBN 978-981-95-1425-0 ISBN 978-981-95-1426-7 (eBook)
https://doi.org/10.1007/978-981-95-1426-7

This Springer imprint is published by the registered company Springer Nature Singapore Pte Ltd.
The registered company address is: 152 Beach Road, #21-01/04 Gateway East, Singapore 189721, Singapore

Contents

About the Editors

Mrinal K. Ghosh is a chief scientist, In-Charge of CSIR-IICB(TRUE), Salt Lake Campus; head (2019–2025) of the cancer biology and inflammatory disorder division at CSIR- Indian Institute of Chemical Biology (IICB), Kolkata, India; chairman of management council at CSIR-IICB (TRUE); and a professor at the Academy of Scientific and Innovative Research (AcSIR), India. He completed his PhD at the University of Calcutta and postdoctoral research (1997–2005) at Cleveland Clinic Foundation, Ohio, USA. His research interests include signal transduction in cancer focusing on the regulation of various oncogenes and tumor suppressor proteins at transcriptional, and post-translational levels, and developing nanotechnology-based targeted drug delivery systems to target identified molecular vulnerabilities in brain, breast, and colon cancers. He has published several papers in peer-reviewed national and international journals, authored books, delivered numerous lectures, and is a reviewer of various national and international journals. He is a lifetime member of the Indian Society of Translational Research (ISTR), the Chemical Biology Society (CBS), the Indian Association of Cancer Research (IACR), the Society of Biological Chemistry (SBC), and the Indian Science Congress Association (ISCA), India.

Malini Basu is an assistant professor and head at the Department of Microbiology, Dhruba Chand Halder College, University of Calcutta, Kolkata, India. She completed her PhD at the University of Calcutta and postdoctoral research at Case Western Reserve University, USA, in collaboration with Rainbow Babies and Children Hospital, Ohio, USA. Her research interests include heavy metals, microbial bioremediation and bioindicators, unique virulence, pathogenicity, and host immune response of *Helicobacter pylori* in gastric cancer. She has published several papers in peer-reviewed national and international journals and some book chapters. She is a lifetime member of the Microbiologists Society India (MBSI), the Association of Microbiologists of India (AMI), and Indian Science Congress Association (ISNA), India. She is also the state vice-president of the Microbiologists Society of India.

Part I

COVID-19 Pandemic: Pathogenesis, Inflammation, and Biology

1 The COVID-19 Outbreak

Mrinal K. Ghosh

Abstract

The COVID-19 outbreak, a systemic event that began in late 2019, has left an indelible mark on global health, economy, and the very fabric of daily life. Emerging from the bustling city of Wuhan in China, this novel coronavirus swiftly traversed the globe, sparking a pandemic that has tested the resilience and adaptability of humanity. The virus, officially named SARS-CoV-2, has been responsible for millions of deaths and countless more illnesses, challenging medical systems, governments, and societies in unprecedented ways. This chapter delves into the origins, spread, and multifaceted impacts of the COVID-19 pandemic. It examines the early detection and response to the virus, the international efforts to contain its spread, and the subsequent development and distribution of vaccines. Additionally, it explores the profound economic, social, and psychological ramifications that continue to ripple through the societies worldwide. The narrative of COVID-19 is not just a tale of disease but a comprehensive saga of scientific triumph, international collaboration, and the remarkable resilience of human communities. As we unravel the complexities of this outbreak, we aim to glean crucial lessons and insights that will fortify our defenses against future global health threats. Through a detailed examination of the events and responses surrounding COVID-19, this chapter seeks to contribute to a deeper understanding of how such a microscopic entity could wield such colossal power over our world.

M. K. Ghosh (✉)
Cancer Biology and Inflammatory Disorder Division, Council of Scientific and Industrial Research-Indian Institute of Chemical Biology (CSIR-IICB), TRUE Campus, CN-6, Sector–V, Salt Lake, Kolkata, India
e-mail: mrinalghosh@iicb.res.in

M. K. Ghosh, M. Basu (eds.), *Pathogenesis, Management and Socio-Economic Impact of COVID-19 Pandemic*, Medical Virology: From Pathogenesis to Disease Control, https://doi.org/10.1007/978-981-95-1426-7_1

Keywords

COVID-19 · SARS-CoV-2 · Variants · Genome · Pathology

Introduction

The COVID-19 outbreak, originating in late 2019 in Wuhan, China, stands as one of the most significant global health crises in modern history. Rapidly spreading worldwide, the novel coronavirus (severe acute respiratory syndrome (SARS)-CoV-2) swiftly escalated into a pandemic, challenging medical and scientific communities and exposing existing vulnerabilities across social, economic, and political spheres. This respiratory illness, caused by SARS-CoV-2, spread primarily through respiratory droplets, aided by global travel and dense urban populations. The World Health Organization (WHO) declared it a public health emergency of international concern and later a pandemic, initiating intense international cooperation and conflict as nations struggled to ombat the virus while preserving economic stability. Healthcare systems worldwide faced unprecedented strain, leading to shortages of medical supplies, personal protective equipment (PPE), and hospital beds. Frontline healthcare workers worked tirelessly amidst high morbidity and mortality rates, particularly affecting vulnerable populations. Governments implemented various public health measures, including lockdowns and social distancing mandates, with profound social and economic consequences such as business closures and job losses. Existing inequalities were exacerbated, underscoring the need for more inclusive public health strategies. The pandemic spurred rapid scientific innovation, resulting in diagnostic tests, treatments, and vaccines at unprecedented speed. Mental health challenges, exacerbated by uncertainty and isolation, prompted broader conversations about accessible mental healthcare. Education saw significant disruption with school closures, emphasizing the digital divide and the importance of adaptable educational systems. Economically, the pandemic caused widespread disruption, prompting stimulus packages and unemployment benefits, while environmental implications prompted discussions on sustainable practices. Geopolitically, the crisis affected international relations and global governance, exposing vulnerabilities in global supply chains and prompting calls for self-sufficiency. The pandemic accelerated the adoption of digital technologies, reshaping work, education, and social interaction [1, 2].

As the world moves forward, lessons learned from the pandemic underscore the importance of preparedness, robust healthcare systems, and global cooperation. Investments in public health infrastructure, research, and education are crucial, highlighting the interconnectedness of global challenges. In conclusion, the COVID-19 outbreak has been transformative, testing societal resilience, emphasizing cooperation, and highlighting the need for inclusive approaches to public health and socioeconomic policies. As recovery efforts continue, the experiences and lessons of the pandemic will shape a more updated and prepared future.

SARS-CoV-2: The Deadly Virus

SARS-CoV-2 is the deadly strain of virus that has contributed to the spread of COVID-19 pandemic. Biologically, it is an enveloped, positive-sense single-stranded RNA virus. The virus exhibits an expanding range of host adaptability and can endanger the lives of humans and animals by causing respiratory and gastrointestinal sickness. Its mode of transmission is through aerosols and direct/indirect contact. It belongs to the Coronavirus (CoV) family. The CoVs belong to the order Nidovirales, family Coronaviridae, and the subfamily Coronaviridae. Genetically, they are categorized into four critical genera: the Alpha-coronavirus, Beta-coronavirus, Gamma-coronavirus, and Delta-coronavirus [3, 4].

The diagnosis of the viral load is done by identifying specific viral gene regions through nucleic acid amplification techniques (RT-PCR) and isothermal nucleic acid amplification. It is also detected by observing the presence of serum antibodies produced in response to the viral infection (serology/IgG/IgM tests), and lateral flow assays for virus antigen testing under specified experimental and biological safety level conditions. Despite the unprecedented efforts by researchers and clinicians, the virus in its full-fledged stage does not get adequately diagnosed across the staggeringly large number of infected populations. Therefore, during the pandemic, the government bodies across the world announced total lockdown, mask use, and social distancing for effectively preventing the spread of the disease [5, 6].

Box 1.1
The NIH classifies COVID-19 into five types based on symptoms and severity, radiographic results, and organ function.

Asymptomatic/Pre-symptomatic: Positive SARS-CoV-2 test with no symptoms.
Mild: Symptoms like fever, cough, sore throat, headache, muscle pain, vomiting, diarrhea, but no shortness of breath or abnormal chest imaging.
Moderate: Symptoms with SpO2 $\geq$94% on room air.
Severe: SpO2 $\leq$94% on room air and marked tachypnea.
Critical: Acute respiratory failure, septic shock, and/or multiple organ dysfunction.

Symptoms

The symptoms are characterized as mild, moderate, severe, critical, and asymptomatic. Mild clinical symptoms range from cough, nasal congestion, headache, malaise, fever, anorexia, fatigue, muscle pain, sore throat, and dyspnea. They do not show any abnormal chest imaging findings. Moderate clinical features are characterized by the above symptoms along with chest imaging showing mild pneumonia.

Severe symptoms display suspected severe respiratory infection, shortness of breath, levels of oxygen saturation drop below 93%, and distinct lesions are observed in chest images. Finally, critical symptoms show a drastic and quick weakening of body conditions and aggravation of symptoms along with rapid disease progress, respiratory failure, septic shock, multiorgan failure, and immediate need for ventilator support and ICU monitoring [3].

Taxonomy

According to the International Committee on Taxonomy of Viruses, the taxonomic position of Coronavirus reveals that it belongs to the order Nidovirales, family Coronaviridae, and subfamily Coronavirinae. Furthermore, they can be classified into four genera: Alpha, Beta, Gamma, and Delta. Mammals are affected by alpha- and beta-coronaviruses, avians by gamma-coronaviruses, and delta-coronaviruses affect both. Human coronavirus NL63 (HCoV-NL63), porcine transmissible gastroenteritis coronavirus, and porcine respiratory coronavirus are encompassed under the genus Alpha-coronavirus. Similarly, SARS-CoV, Middle East respiratory syndrome (MERS)-CoV, HKU4 (bat coronavirus), bovine coronavirus, mouse hepatitis coronavirus, and OC43 (human coronavirus) are represented as Beta-coronavirus. Finally, examples of representative gamma-and delta-coronaviruses are avian infectious bronchitis coronavirus (IBV) and porcine delta-coronavirus, respectively [1, 2] (Fig. 1.1).

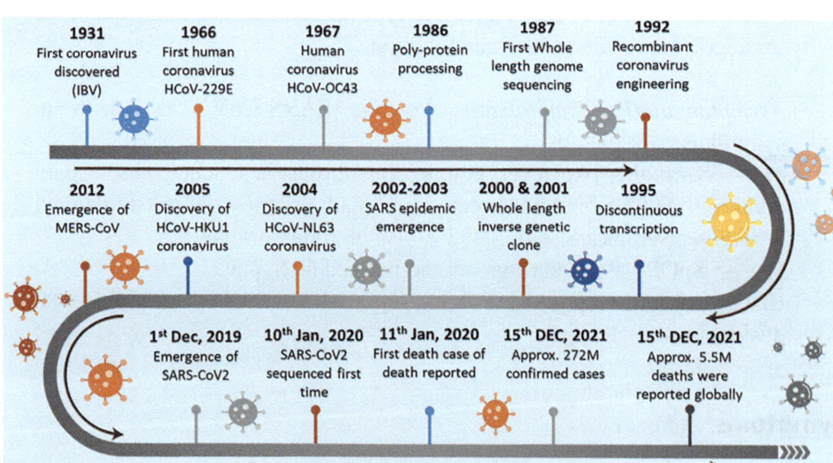

Fig. 1.1 Outbreak of SARS-CoV-2 infection: the worldwide scenario. The figure depicts the total number of cases and deaths reported per day in the top five countries. The left-hand side panel from top to bottom represents the wavy pattern in the number of cases reported per day in top five countries viz., United States, India, Brazil, United Kingdom, and Russia, respectively. And right-hand side panel represents the wavy pattern in the number of deaths reported per day in the same top five countries

Genomic Structure

The term Coronavirus was derivative of the Latin term corona, meaning crown or halo. This was due to the crown-like spikes visibly seen on the surface of this virus under an electron microscope. The shape of the virus is spherical or pleomorphic and its diameter is about 60–140 nm. The SARS-CoV-2 virus is also called human 2019-nCoV HKU-SZ-005b. Its GenBank accession number is MN975262. The genome is composed of a single-stranded RNA (29,891 nucleotides) that encodes for 9860 amino acids. The genomic structure is composed of a 5′-untranslated region (UTR), an open reading frame (ORF) 1a/b, and a 3′-UTR. 16 nonstructural proteins (nsps) are encoded by the ORF that becomes part of the replication-transcription complex (RTC) in the form of double-membrane vesicles (DMVs). Structural proteins like Spike (S), Envelope (E), Membrane (M), and Nucleoprotein (N) arranged sequentially are also encoded by the ORF. Along with the above two categories, many accessory proteins like ORF 3, 6, 7a, 7b, 8, and 9b are also encompassed under it. RTC, through discontinuous transcription, synthesizes negative subgenomic RNAs (sgRNAs); all of which have a leader sequence at 5′-end and terminal sequences at 3′-end, and serve as effective templates for subgenomic mRNAs. The most important structure that facilitates host cell binding is the SARS-CoV-2 spike protein. These are type-I transmembrane, clove-shaped proteins having 3 segments: an ectodomain, a single-pass transmembrane, and an intracellular tail. The ectodomain comprises two subunits: S1 and S2 (highly conserved). Structurally, the S1 subunit is further composed of a signal peptide (SP), an N-terminal domain (NTD), and a receptor-binding domain (RBD). The RBD facilitates binding to the host angiotensin-converting enzyme 2 (ACE2) receptor and transmembrane serine protease 2 (TMPRSS2) that aids in its entry into target host cells [7–10]. The ACE2 protein resides in various mammalian body tissues, such as the heart, lungs, gastrointestinal tract, kidneys, liver, and blood vessels. They are critical players to regulate the renin-angiotensin-aldosterone system pathway.

The structural and accessory proteins are synthesized after the entry of the virus into the host cell begins with transcription and translation processes. The synthesis occurs with the aid of RNA-dependent RNA polymerase, which employs the negative strand as a template. The structural membrane protein M is a glycoprotein and assists in transporting nutrients across the cell membrane thereby providing shape to the virus. The S or spike protein belongs to type I membrane glycoprotein (virus peplomers). The nucleocapsid (N) protein maintains RNA stability and also aids in viral RNA genome binding. The envelope (E) protein is involved in the release of viral particles and their assembly during pathogenesis [11–13].

Pathophysiology and Immune Response

SARS-CoV-2 binds to the ACE2 receptors present on the host endothelial cells through its receptor binding domain (RBD). Within host cells, the virus is primarily recognized by toll-like receptors that activate NF-κB, subsequently leading to ACE2 receptor activation on cells in the respiratory tract, gastrointestinal tract, and kidney.

ACE2 is a type I transmembrane metallo-carboxypeptidase. Following this S1-RBD attachment, TMPRSS2 (transmembrane serine protease 2), a surface protease present on the host cell surface, acts upon a cleavage site on S2 and aids in membrane fusion and viral infection. S-proteins can adversely affect tropism, infectivity, and viral lethality. After entry, translation of uncoated genomic RNA takes place, and the polyproteins (pp1a and pp1ab) form replication/transcription complexes with virus-induced double-membrane vesicles (DMVs). Subsequently, the synthesis of nested subgenomic RNAs and encoding of structural and accessory proteins take place. The assembly of new virus particles is mediated by the endoplasmic reticulum and the Golgi complex. The virus then buds off into the extracellular milieu compartment and thereafter its replication and progression commence. The symptoms in COVID-19 patients range from fatigue, headache, fever, cold, cough, loss of smell or taste, etc. Gastrointestinal symptoms like nausea, diarrhea, and stomachache are also seen. Also, ACE2 converts angiotensin II to angiotensin, thereby preventing atherosclerosis. The binding of S-protein to ACE2 robs it of its protective function causing vascular occlusion in patients [3, 12, 14].

Immunologically, the entry of the virus initiates a typical "cytokine storm," a hallmark of acute respiratory distress syndrome which is seen during severe infection of SARS-CoV-2 that leads to multiorgan failure and septic shock. After its entry, the RNA genome triggers an immune response via pathogen-associated molecular patterns (PAMPs) and also leads to the release of proinflammatory damage-associates molecular patterns. Further, various cytokines like IL6 are released to trigger a local immunological response. They also chemotactically attract other immune cells like macrophages, monocytes, and T cells, leading to an inflammatory response and therefore a "cytokine storm" [15].

On the other hand, the immune evasion mechanism of the virus inside the host cell is described as follows. Pathogen-associated molecular patterns (PAMPs) are absent on DMVs arising from the first step of SARS-CoV infection. Therefore, they evade recognition by pattern recognition receptors of the host immune system. Moreover, the nsps play a diverse role in effectively impeding immune responses. For instance, Nsp1 hampers the interferon (IFN-I) responses by silencing the translation mechanism in the host or degrading host mRNA. By blocking interferon regulatory factor 3 (IRF3) phosphorylation and disturbing NF-KB pathway, Nsp3 antagonizes cytokine and interferon production. Accessory proteins like ORF3b and ORF6 inhibit IRF3 and NF-KB-dependent IFN-β expression and impede the JAK-STAT signaling pathway, respectively. Also, structural proteins M and N inhibit IFN signaling that disrupts TANK-binding kinase 1 (TBK1)/IKB kinase ε and TRAF/NF-KB signaling [12, 16].

Variants

Genome sequencing of SARS-CoV-2 has revealed that it has the potential to undergo extensive mutation. Its nucleotide substitution rate is 1×10^{-3} substitutions per year. These mutations and variations give rise to altered pathophysiology and the birth of new variants of the disease-causing organism. WHO has designated the genetically similar group of variants into different categories, namely, variant being monitored, variant of concern (VOC), variant of interest (VOI), and variant of high consequence due to shared attributes. VOCs show enhanced transmissibility or virulence, reduced antibody neutralization through natural infection or vaccination, mechanisms to evade detection or a decrease in therapeutics or vaccination effectiveness. Variants of concern (VOCs) are viral strains associated with increased transmissibility. According to the collaborative efforts of the U.S. SARS-CoV-2 Interagency Group (SIG) with Centres for Disease Control and Prevention (CDC), National Institutes of Health, Food and Drug Administration, and WHO, there are different rounds of monitoring and designating different variants of COVID-19. According to the WHO (December 11, 2021), there were five known SARS-CoV-2 VOCs; Alpha (B.1.1.7, United Kingdom), Beta (B.1.351, South Africa), Gamma (P.1, Brazil), Delta (B.1.617.2, India), and Omicron (B.1.1.529, South Africa). These variants have mutations in their RBD or NTD. N501Y mutation (RBD) is shared by all variants except the Delta variant. Currently, there have been several additions to the list and also changes to different designations depending upon the severity and transmission abilities of the different strains. All the present variants of concern affect the function of the spike glycoprotein and other proteins that can influence the interaction with receptor hACE2. Tracking these variants is necessary to analyze and frequently upgrade the vaccine development according to the prevalent mutated version of the virus [17–19] (Table 1.1).

Table 1.1 Major SARS-CoV-2 variants

	WHO label	Pango lineage	Gisaid clade	Next strain clade	Earliest documented samples	Date of designation
VOC	Beta	B.1.351	GH/501Y.V2	20H (V2)	South Africa May-2020	VOC: 18-Dec-2020 VOC: 09-Mar-2022
	Gamma	P.1	GR/501Y.V3	20J (V3)	Brazil Nov-2020	VOC: 11-Jan-2021 VOC: 09-Mar-2022
	Delta	B.1.617.2	G/478K.V1	21A, 21I, 21J	India Oct-2020	VOI: 04-Apr-2021 VOC: 11-May-2021 VOC: 07-Jun-2022
	Omicron	B.1.1.529	GR/484A	21K, 21L, 21M, 22A, 22B, 22C, 22D	Multiple countries Nov-2021	VUM: 24-Nov-2021 VOC: 26-Nov-2021
VOI	Lambda	C.37	GR/452Q.V1	20D, 21G	Peru Aug-2020	VOI: 14-June-2021
	Mu	B.1.621	GH	21H	Columbia Jan-2021	VOI: 30-Aug-2021
	Zeta	P.2	GR/484K.V2	20B/S.484K	Rio de Janeiro Apr-2020	VOI: 06-July-2021
	Theta	P.3	GR/1092K.V1	21E	Philippines Feb-2021	VOI: 06-July-2021
	Eta	B.1.525	G/484K.V3	21D	UK and Nigeria Dec-2020	VOI: 20-Sep-2021
VUM	Kappa	B.1.617.1	G/452R.V3	21B	India Dec-2020.	VUI: 21-Apr-2021
De-escalated	Alpha	B.1.1.7	GRY	20I (V1)	UK Sep-2020	VOC: 18-Dec-2020 VOC: 09-Mar-2022
	Epsilon	B.1.427 B.1.429	GH/452R.V1	21C	California, USSA Jul-2020	VOI: 06-July-2021
	Iota	B.1.526	GH/253G.V1	21F	New York, USA Nov-2020	VOI: 20-Sep-2021

Outbreak and Milestones in Coronavirus Research

SARS-CoV-2 has rapidly spread across the globe, creating a daunting global scenario. This virus has impacted the human population in waves, with varying numbers of cases and deaths. To analyze this trend, we collected data from March 2020 to September 2023 for the top five countries with the highest total cases. The United States, India, France, Germany, and Brazil have reported approximately 108 million, 44 million, 40 million, 38 million, and 37 million confirmed COVID-19 cases, respectively. In terms of fatalities, these countries have reported 1.2 million, 0.5 million, 0.15 million, 0.175 million, and 0.7 million deaths, respectively. Most countries have experienced at least the first and second waves of infection, with some also reporting a third wave. It is evident that the total number of confirmed cases and daily deaths during the second and third waves increased exponentially compared to the previous waves [20–22].

Coronavirus belongs to a family of viruses widely known to infect various mammalian and avian species. IBV (Avian infectious bronchitis virus) was the first coronavirus, identified in 1931 [23]. After another 35 years, coronavirus HCoV-229E (1966) and HCoV-OC43 (1967) were discovered [24, 25]. Since 1970, major milestones in coronavirus research were achieved by the scientific community. These milestones include discoveries like polyprotein processing of coronaviruses studied in 1986 [26], whole-length genome sequencing in 1987 [27], engineering of recombinant coronavirus in 1992 [28, 29], discontinuous transcription studied in 1995 [30] and whole-length inverse genetic clones studied in 2000 and 2001 [31, 32]. In 2002–2003, SARS epidemic was reported with approx. 8000 cases of which 10% showed severe illness [33, 34].

After this SARS epidemic, many laboratories started to explore the screening and sequencing of viral genome that lead to the characterization of two additional human specific strains, viz., HCoV-HKU1 and HCoV-NL63 [35, 36]. These coronaviruses spread every year but typically they produce only mild upper respiratory tract infection in immune-compromised individuals [35, 36]. In 2008, an electron tomography report showed that SARS-CoV virus as double-membrane vesicles [37] (Fig. 1.2).

In 2012, the second-most pathogenic virus MERS-CoV was reported with 2500 cases with 36% fatality rate [38]. On December 1, 2019, the first clinical case of a highly pathogenic SARS-CoV-2 strain was reported [27] and on January 10, 2020, the SARS-CoV-2 virus was sequenced for the first time. WHO affirmed it to be a pandemic on March 11, 2020 [39]. It targets both upper and lower respiratory tracts and spreads via human-to-human transmission [40]. Till now, no proper medication against SARS-CoV-2-induced disease COVID-19 is available. SARS-CoV-2 infection is spreading much faster than the previous epidemics of SARS and MERS [5, 41, 42].

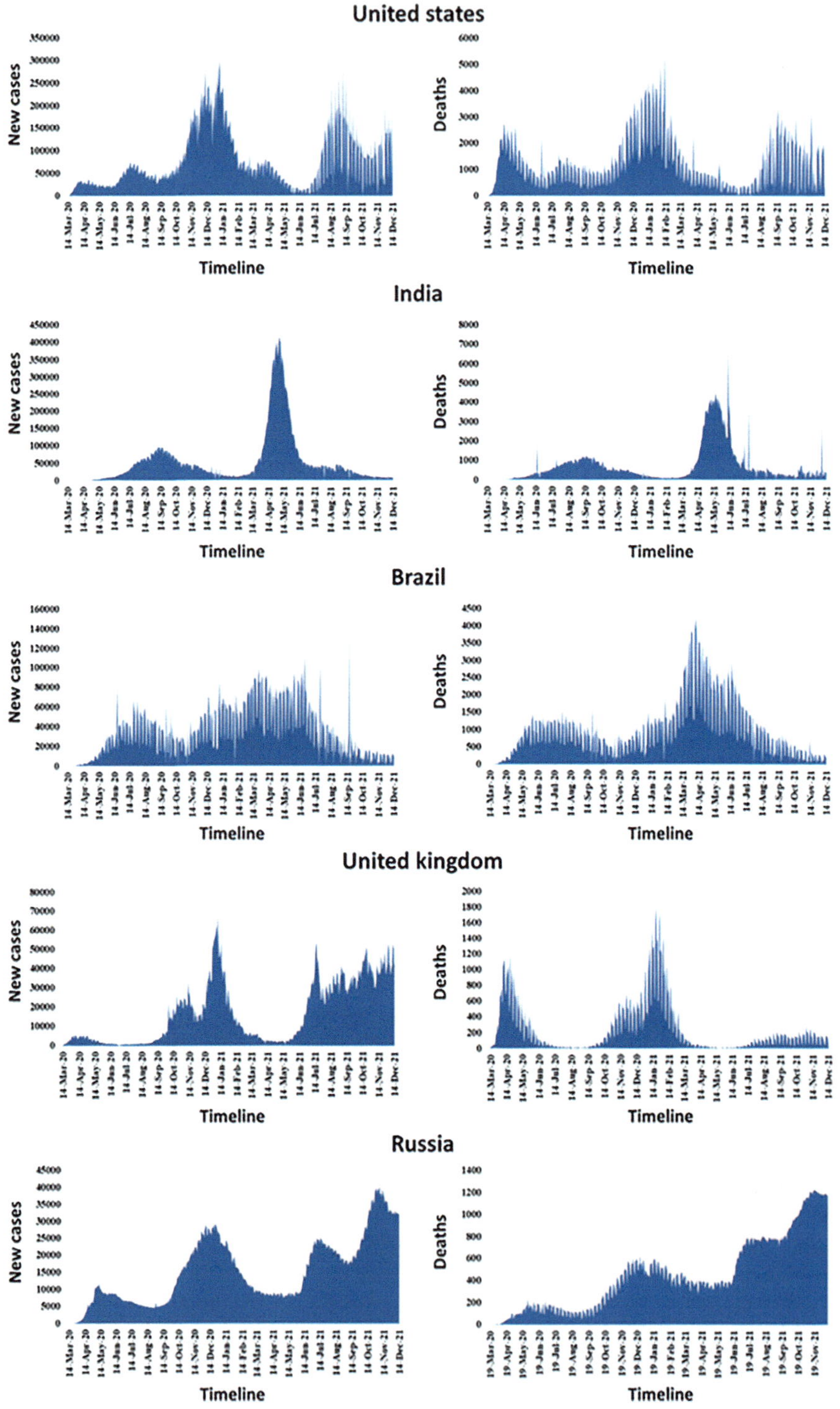

Fig. 1.2 Milestones in coronaviruses research. The figure depicts the timeline of all the major milestones in the coronavirus research

Origins of the Virus

Early Cases and Identification

The origins of the COVID-19 pandemic trace back to December 2019 in Wuhan, a major city in Hubei province, China. Medical practitioners observed an unusual cluster of pneumonia cases of unknown etiology. Patients presented with severe respiratory symptoms, which did not respond to conventional treatments. Initial epidemiological investigations linked many of these cases to the Huanan Seafood Wholesale Market, a location known for selling not only seafood but also a variety of live wild animals. On December 31, 2019, Chinese authorities reported the outbreak to the World Health Organization (WHO), marking the first step in the international awareness of what would become a global crisis [43].

Genetic Analysis and Zoonotic Transmission

Subsequent analysis identified the causative agent as a novel coronavirus, which was temporarily named 2019-nCoV and later officially designated SARS-CoV-2. Genetic sequencing revealed that this virus shares significant similarity with other coronaviruses found in bats, specifically those in the Rhinolophus genus. This finding suggested that bats were the primary reservoir for the virus. However, the exact pathway of transmission to humans remains a subject of investigation.

Coronaviruses are known for their ability to jump from animals to humans, often via an intermediate host. This zoonotic transmission mechanism had been observed in previous outbreaks, such as severe acute respiratory syndrome (SARS) and Middle East respiratory syndrome (MERS), where civet cats and dromedary camels, respectively, acted as intermediary hosts. In the case of SARS-CoV-2, while bats are considered the likely original source, it is hypothesized that another animal, potentially one sold at the Huanan market, acted as an intermediary host, facilitating the virus's transmission to humans. Several candidates, including pangolins, have been studied due to their coronavirus strains' genetic similarities to SARS-CoV-2, though definitive evidence of the intermediate host remains elusive [12] (Fig. 1.3).

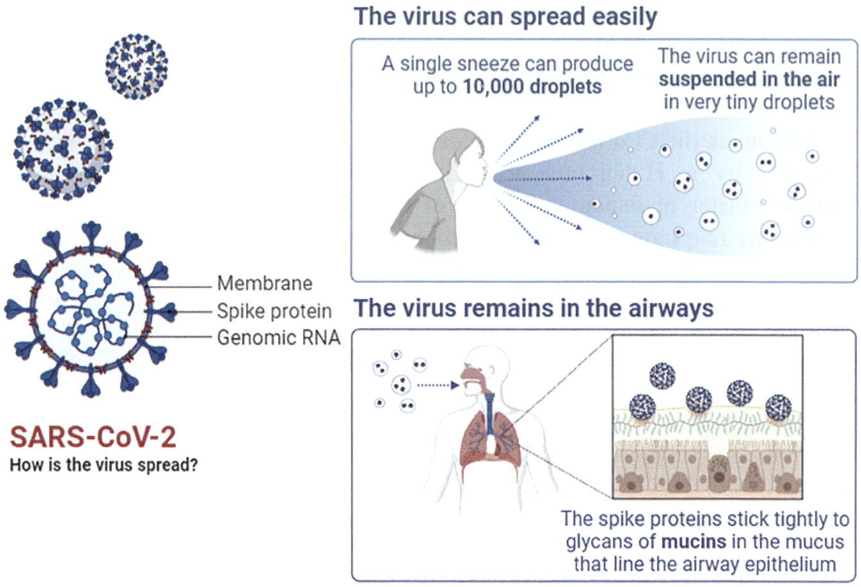

Fig. 1.3 Virus spreading through respiratory airway. Figure illustrates how the virus gets spread through the respiratory airway. (This image is adopted from Biorender (https://www.biorender.com/))

Global Spread

Initial Spread in China

The COVID-19 pandemic's initial epicenter was in Wuhan, China, where the first cases were identified in December 2019. Despite rapid and stringent measures by Chinese authorities, including the lockdown of Wuhan and other cities in Hubei province on January 23, 2020, the virus had already begun to spread. By the end of January, cases were reported in other parts of China, indicating the virus's high transmissibility. Measures such as travel bans, mass testing, and quarantine were implemented nationwide, yet the virus managed to break through containment efforts, underscoring its contagious nature and the challenges of controlling its spread in densely populated areas [44, 45].

Transmission Beyond China

The interconnected nature of modern society facilitated the rapid spread of SARS-CoV-2 beyond China's borders. International travel, particularly during the Lunar New Year, significantly contributed to the virus's dissemination. The first confirmed case outside China was reported in Thailand on January 13, 2020. Within weeks, cases appeared in numerous countries across different continents, indicating that the virus had established a global foothold.

Europe It became a significant hotspot for COVID-19 early in the pandemic. Italy was among the first European countries to be severely affected, with the northern regions experiencing high transmission rates and a dramatic surge in cases by late February 2020. The Italian government imposed a nationwide lockdown on March 9, 2020, as hospitals in Lombardy and other regions were overwhelmed. Other European countries, including Spain, France, and Germany, soon reported increasing case numbers. The rapid spread within Europe was attributed to several factors: delayed implementation of restrictive measures, high levels of intra-European travel, and initial underestimation of the virus's severity. By mid-March, the WHO declared Europe the new epicenter of the pandemic.

North America The first confirmed case in the United States was reported on January 20, 2020. Community transmission was identified in February, with significant outbreaks in states like Washington and New York. The US response involved a combination of state-level measures and federal guidelines, leading to a patchwork of policies that varied in strictness and timing. Canada also reported early cases, with its first confirmed infection on January 25, 2020. Both countries faced challenges in ramping up testing capacity and implementing consistent public health measures. By March 2020, widespread community transmission was evident, and major cities began to implement lockdowns and social distancing mandates.

Asia and the Pacific While China managed to bring the initial outbreak under control by March 2020 through strict lockdowns and extensive testing, other Asian countries quickly became battlegrounds against COVID-19. South Korea and Japan reported early cases, with South Korea in particular adopting aggressive testing and contact tracing strategies that were effective in flattening the curve. Despite these efforts, sporadic outbreaks continued, necessitating ongoing vigilance and intermittent restrictions. In Southeast Asia, countries like Singapore, Malaysia, and Indonesia faced significant challenges. Singapore initially managed the outbreak well through robust contact tracing and quarantine measures but later experienced waves of infections among migrant workers. Indonesia and the Philippines struggled with high case numbers and limited healthcare resources. Australia and New Zealand also encountered the virus, with New Zealand implementing one of the world's most stringent lockdowns, successfully eliminating community transmission for extended periods. Australia faced several waves but employed strict lockdowns and border controls to manage the spread.

Africa COVID-19's arrival in Africa was somewhat delayed compared to other continents, with the first confirmed case reported in Egypt on February 14, 2020. The virus spread to sub-Saharan Africa by early March. The continent's relatively younger population provided some demographic advantage, but healthcare infrastructure limitations posed significant challenges. African countries adopted various measures, including lockdowns, curfews, and travel restrictions, to curb the spread. South Africa became the epicenter on the continent, implementing a strict lockdown early on but still facing high transmission rates. The African Union and the Africa Centres for Disease Control and Prevention (Africa CDC) played critical roles in coordinating the response and resource allocation across the continent.

Latin America Faced a severe impact from COVID-19, with Brazil emerging as one of the hardest-hit countries. The first case in Brazil was reported on February 26, 2020. The country experienced high transmission rates and significant mortality, exacerbated by political and social factors, including resistance to lockdown measures and challenges in healthcare delivery. Other countries in the region, including Mexico, Peru, and Argentina, also faced high case numbers and strained healthcare systems. The pandemic's socioeconomic impact was profound, with increased poverty and unemployment.

Public Health Responses

National Responses

Countries worldwide implemented a variety of public health measures to combat the spread of COVID-19. The effectiveness of these measures varied widely, influenced by factors such as healthcare infrastructure, public compliance, and government efficiency.

China Response to the initial outbreak was marked by aggressive and extensive measures. Following the identification of the virus, the Chinese government swiftly imposed a lockdown in Wuhan and other cities in Hubei province. This lockdown restricted the movement of millions of people and included the closure of public transportation, businesses, and schools. Mass testing and contact tracing were implemented to identify and isolate cases. Temporary hospitals were constructed at an unprecedented speed to manage the surge in patients. These measures, though drastic, proved effective in curbing the initial spread within China. However, the early lack of transparency and delays in information sharing with the global community were criticized and highlighted the need for more prompt and open communication during health crises.

South Korea Emerged as a model for effective public health response early in the pandemic. The country implemented widespread testing and contact tracing, leveraging technology to track and isolate cases quickly. Drive-through testing centers and mobile phone alerts about potential exposure helped keep the public informed and safe. South Korea also avoided a nationwide lockdown by enforcing localized restrictions and maintaining public compliance through clear and transparent communication from health officials.

Italy One of the first European countries to experience a severe outbreak. The Italian government imposed a nationwide lockdown on March 9, 2020, following significant outbreaks in the northern regions. The lockdown included travel bans, the closure of nonessential businesses, and strict social distancing measures. Italy's healthcare system faced immense pressure, particularly in Lombardy, where hospitals were overwhelmed. The Italian experience underscored the importance of rapid response and adequate healthcare capacity.

United States The United States had a varied and decentralized response, with significant differences in measures across states. Initial responses were hampered by delays in testing and mixed messages from federal and state authorities. States like New York and California implemented strict lockdowns early on, while others adopted less stringent measures. The U.S. response highlighted the challenges of a federated system where state and federal actions were sometimes uncoordinated. Despite these challenges, the Operation Warp Speed initiative facilitated the rapid development and distribution of vaccines.

New Zealand New Zealand adopted one of the most stringent and successful responses to COVID-19. The government, led by Prime Minister Jacinda Ardern, implemented a strict nationwide lockdown early in the outbreak, aiming to eliminate the virus rather than just mitigate its spread. New Zealand's approach included robust border controls, widespread testing, and comprehensive contact tracing. The country successfully eliminated community transmission for extended periods, allowing it to lift restrictions and resume normal activities while remaining vigilant against new outbreaks [1, 44].

International Collaboration

The COVID-19 pandemic prompted unprecedented levels of international collaboration, although it also revealed gaps in global health governance and solidarity.

WHO The WHO played a central role in coordinating the global response to COVID-19. It provided guidelines on testing, treatment, and preventive measures and facilitated the sharing of information and resources among countries. The WHO declared COVID-19 a Public Health Emergency of International Concern on January 30, 2020, and later a pandemic on March 11, 2020. The organization's efforts included supporting vaccine development through the Solidarity Trials and ensuring equitable vaccine distribution via the COVAX initiative. Despite its efforts, the WHO faced criticism for its initial response, particularly regarding its handling of information from China and the timing of declaring a pandemic. These critiques underscored the need for reforms in international health governance to enhance transparency and responsiveness.

COVAX Initiative The COVAX initiative, coled by the WHO, Gavi, and the Coalition for Epidemic Preparedness Innovations (CEPI), aimed to ensure equitable access to COVID-19 vaccines worldwide. COVAX sought to distribute vaccines to low- and middle-income countries that might otherwise struggle to secure supplies. While COVAX successfully delivered millions of vaccine doses, it faced challenges including supply shortages, logistical issues, and vaccine hesitancy in some regions. The initiative highlighted both the potential and the limitations of global solidarity in addressing public health crises.

European Union The European Union (EU) coordinated a collective response among its member states, focusing on joint procurement of medical supplies and vaccines. The EU's response included establishing a recovery fund to support member states economically impacted by the pandemic. However, the EU faced criticism for delays in vaccine rollout due to supply chain issues and bureaucratic hurdles. These challenges underscored the complexities of coordinating a unified response across multiple countries with diverse healthcare systems and policies [46, 47].

Vaccine Development and Distribution

The rapid development and distribution of vaccines were pivotal in the global response to COVID-19. Several vaccines received emergency use authorization within a year of the virus's emergence, a testament to unprecedented scientific collaboration and innovation.

Development Efforts

The development of COVID-19 vaccines was accelerated through global collaboration and substantial financial investments. The mRNA vaccines developed by Pfizer-BioNTech and Moderna were among the first to receive emergency use authorization. These vaccines demonstrated high efficacy and were developed using novel mRNA technology, which allowed for faster production compared to traditional vaccine methods. Other vaccines, such as those developed by AstraZeneca, Johnson & Johnson, and Sinovac, employed different technologies and contributed to the diverse portfolio of vaccines available.

Distribution Challenges

Despite the rapid development, vaccine distribution faced significant challenges. High-income countries secured large quantities of vaccines through advance purchase agreements, leading to disparities in vaccine access. Low- and middle-income countries relied heavily on initiatives like COVAX to obtain vaccines. Logistical challenges, including cold chain requirements for mRNA vaccines, further complicated distribution efforts, particularly in regions with limited infrastructure.

Vaccine Hesitancy

Vaccine hesitancy emerged as a critical issue in the global vaccination campaign. Misinformation, distrust in government and healthcare systems, and cultural factors contributed to varying levels of vaccine acceptance. Public health campaigns aimed at educating populations about vaccine safety and efficacy were essential to addressing hesitancy and achieving higher vaccination rates [5].

Social and Economic Impact

The COVID-19 pandemic placed unprecedented strain on healthcare systems worldwide. Hospitals faced surges in patient numbers, leading to shortages of beds, ventilators, and personal protective equipment (PPE). Healthcare workers experienced high levels of stress and burnout, compounded by the risk of infection. Many routine medical services were disrupted, resulting in delayed diagnoses and treatments for non-COVID conditions.

The pandemic triggered a global economic crisis. Lockdowns and restrictions led to significant contractions in economic activity, with sectors such as travel, hospitality, and retail being particularly hard hit. Unemployment rates soared, and many businesses faced closures. Governments implemented various stimulus packages to mitigate the economic fallout, but recovery has been uneven across different regions and industries. The social and psychological impact of COVID-19 has been profound. Prolonged lockdowns, social distancing, and the loss of loved ones contributed to widespread mental health issues, including anxiety, depression, and loneliness. Educational disruptions affected millions of students, with long-term implications for learning and development. The pandemic also exacerbated existing social inequalities, disproportionately affecting marginalized communities [48–50].

Lessons Learned and Future Preparedness

The COVID-19 pandemic highlighted the need for a robust public health infrastructure. Investment in healthcare systems, including increased capacity for testing, contact tracing, and vaccination, is crucial for future preparedness. Strengthening global health governance, with enhanced coordination and resource sharing, can improve collective responses to pandemics. Effective surveillance and early warning systems are vital for detecting and responding to emerging infectious diseases. Integrating advanced technologies such as artificial intelligence and genomic sequencing can enhance the speed and accuracy of outbreak detection. International collaboration and data sharing are essential components of a comprehensive surveillance network. Ensuring equitable access to vaccines and other medical interventions is a moral and practical imperative. The unequal distribution of COVID-19 vaccines underscored the disparities between high-income and low-income countries. Mechanisms to ensure fair distribution and address logistical challenges must be prioritized to prevent similar inequities in future pandemics. The COVID-19 pandemic demonstrated that infectious diseases know no borders. Global solidarity, through cooperation and mutual support, is crucial in tackling such crises. Strengthening international institutions and fostering a spirit of collaboration can enhance global resilience against future health threats [2].

Conclusion

The COVID-19 pandemic has been a defining event of the twenty-first century, reshaping our world in profound ways. It has exposed vulnerabilities in public health systems, highlighted the importance of scientific research and innovation, and underscored the need for global cooperation. As the world moves forward, the lessons learned from this pandemic will be invaluable in preparing for and mitigating the impact of future infectious disease outbreaks. The story of COVID-19 is still unfolding, with ongoing efforts to control the virus, understand its long-term effects, and rebuild societies. By reflecting on the challenges faced and the successes achieved, humanity can emerge stronger and more resilient, ready to confront the health challenges of the future.

Author's Contributions Conceived the idea and review structure: MKG; wrote the manuscript: MKG; revised and edited the manuscript: MKG. Author has read, agreed, and approved the final draft of the manuscript.

Funding This work is jointly supported by the Department of Science and Technology (NanoMission: DST/NM/NT/2018/105(G); SERB: EMR/2017/000992) and Focused Basic Research (FBR), HCT, CSIR, Govt. of India.

Data Availability The software Biorender (https://www.biorender.com/) is used to draw some elements of a few figures.

Declarations

Conflicts of Interests There is no competing and conflict of interest.

References

1. Ciotti M, Ciccozzi M, Terrinoni A, Jiang WC, Wang CB, Bernardini S. The COVID-19 pandemic. Crit Rev Clin Lab Sci. 2020;57(6):365–88. https://doi.org/10.1080/10408363.2020.1783198.
2. Pokhrel S, Chhetri R. A literature review on impact of COVID-19 pandemic on teaching and learning. High Educ Future. 2021;8(1):133–41. https://doi.org/10.1177/2347631120983481.
3. Kumar S, Basu M, Ghosh P, Pal U, Ghosh MK. COVID-19 therapeutics: clinical application of repurposed drugs and futuristic strategies for target-based drug discovery. Genes Dis. 2023;10(4):1402–28. https://doi.org/10.1016/j.gendis.2022.12.019.

4. Ghosh MK, Kumar S, Ganguly KK, et al. COVID-19 and cancer: insights into their association and influence on genetic and epigenetic landscape. Epigenomics. 2023;15(4):227–48. https://doi.org/10.2217/epi-2023-0052.
5. Kumar S, Basu M, Ghosh P, Ansari A, Ghosh MK. COVID-19: clinical status of vaccine development to date. Br J Clin Pharmacol. 2023;89(1):114–49. Published online 2022.
6. Kumar S, Basu M, Ghosh MK. Chaperone-assisted E3 ligase CHIP: a double agent in cancer. Genes Dis. 2021;9(6). https://doi.org/10.1016/j.gendis.2021.08.003.
7. Pathania AS, Prathipati P, Abdul BA, et al. COVID-19 and cancer comorbidity: therapeutic opportunities and challenges. Theranostics. 2021;11(2):731–53. https://doi.org/10.7150/thno.51471.
8. Chen Y, Liu Q, Guo D. Emerging coronaviruses: genome structure, replication, and pathogenesis. J Med Virol. 2020;92(4):418–23. https://doi.org/10.1002/jmv.25681.
9. Chilamakuri R, Agarwal S. COVID-19: characteristics and therapeutics. Cells. 2021;10(2). https://doi.org/10.3390/cells10020206.
10. Mittal A, Manjunath K, Ranjan RK, Kaushik S, Kumar S, Verma V. COVID-19 pandemic: insights into structure, function, and hACE2 receptor recognition by SARS-CoV-2. PLoS Pathog. 2020;16(8):e1008762. https://doi.org/10.1371/journal.ppat.1008762.
11. Helmy YA, Fawzy M, Elaswad A, Sobieh A, Kenney SP, Shehata AA. The COVID-19 pandemic: a comprehensive review of taxonomy, genetics, epidemiology, diagnosis, treatment, and control. J Clin Med. 2020;9(4):1225.
12. Krishnan A, Mukherjee D, Hamilton JP, Alqahtani SA, Woreta TA. Coronavirus disease 2019: taxonomy, genomic structure, replication and pathogenesis. J Infect Dis Ther. 2021;3:2.
13. Naqvi AAT, Fatima K, Mohammad T, et al. Insights into SARS-CoV-2 genome, structure, evolution, pathogenesis and therapies: structural genomics approach. Biochimica et Biophysica Acta (BBA). 2020;1866(10):165878.
14. Kumar S, Chatterjee M, Ghosh P, Ganguly KK, Basu M, Ghosh MK. Targeting PD-1/PD-L1 in cancer immunotherapy: an effective strategy for treatment of triple-negative breast cancer (TNBC) patients. Genes Dis. 2023;10(4):1318–50.
15. Mangalmurti N, Hunter CA. Cytokine storms: understanding COVID-19. Immunity. 2020;53(1):19–25.
16. Fahmi M, Kitagawa H, Yasui G, Kubota Y, Ito M. The functional classification of ORF8 in SARS-CoV-2 replication, immune evasion, and viral pathogenesis inferred through phylogenetic profiling. Evol Bioinformatics Online. 2021;17:11769343211003080. https://doi.org/10.1177/11769343211003079.
17. Vasireddy D, Vanaparthy R, Mohan G, Malayala SV, Atluri P. Review of COVID-19 variants and COVID-19 vaccine efficacy: what the clinician should know? J Clin Med Res. 2021;13(6):317.
18. Hadj Hassine I. Covid-19 vaccines and variants of concern: a review. Rev Med Virol. 2022;32(4):e2313.
19. Duong D. What's important to know about the new COVID-19 variants? Published online 2021. Accessed 28 May 2024. https://www.cmaj.ca/content/193/4/E141.short.
20. Bok K, Sitar S, Graham BS, Mascola JR. Accelerated COVID-19 vaccine development: milestones, lessons, and prospects. Immunity. 2021;54(8):1636–51.
21. Garagiola ER, Lam Q, Wachsmuth LS, et al. Adolescent resilience during the COVID-19 pandemic: a review of the impact of the pandemic on developmental milestones. Behav Sci. 2022;12(7):220.
22. Li M, Wang H, Tian L, et al. COVID-19 vaccine development: milestones, lessons and prospects. Signal Transduct Target Ther. 2022;7(1):146.
23. Schalk AF. An apparently new respiratory disease of baby chicks. J Am Vet Med Assoc. 1931;78:413–23.
24. Hamre D, Procknow JJ. A new virus isolated from the human respiratory tract. Proc Soc Exp Biol Med. 1966;121(1):190–3.

25. McIntosh K, Dees JH, Becker WB, Kapikian AZ, Chanock RM. Recovery in tracheal organ cultures of novel viruses from patients with respiratory disease. Proc Natl Acad Sci USA. 1967;57(4):933.
26. Denison MR, Perlman S. Translation and processing of mouse hepatitis virus virion RNA in a cell-free system. J Virol. 1986;60(1):12–8.
27. Boursnell MEG, Brown TDK, Foulds IJ, Green PF, Tomley FM, Binns MM. Completion of the sequence of the genome of the coronavirus avian infectious bronchitis virus. J Gen Virol. 1987;68(1):57–77.
28. Koetzner CA, Parker MM, Ricard CS, Sturman LS, Masters PS. Repair and mutagenesis of the genome of a deletion mutant of the coronavirus mouse hepatitis virus by targeted RNA recombination. J Virol. 1992;66(4):1841–8.
29. Van Der Most RG, Heijnen L, Spaan WJ, De Groot RJ. Homologous RNA recombination allows efficient introduction of site-specific mutations into the genome of coronavirus MHV-A59 via synthetic co-replicating RNAs. Nucleic Acids Res. 1992;20(13):3375–81.
30. Sawicki SG, Sawicki DL. Coronaviruses use discontinuous extension for synthesis of subgenome-length negative strands. Corona Relat Viruses. 1995;380:499–506. Published online 1995.
31. Thiel V, Herold J, Schelle B, Siddell SG. Infectious RNA transcribed in vitro from a cDNA copy of the human coronavirus genome cloned in vaccinia virus. J Gen Virol. 2001;82(6):1273–81.
32. Engineering the largest RNA virus genome as an infectious bacterial artificial chromosome – PubMed. Accessed 6 Aug 2021. https://pubmed.ncbi.nlm.nih.gov/10805807/.
33. Zhong NS, Zheng BJ, Li YM, et al. Epidemiology and cause of severe acute respiratory syndrome (SARS) in Guangdong, People's Republic of China, in February, 2003. Lancet. 2003;362(9393):1353–8.
34. Ksiazek TG, Erdman D, Goldsmith CS, et al. A novel coronavirus associated with severe acute respiratory syndrome. N Engl J Med. 2003;348(20):1953–66.
35. Van Der Hoek L, Pyrc K, Jebbink MF, et al. Identification of a new human coronavirus. Nat Med. 2004;10(4):368–73.
36. Woo PC, Lau SK, Chung-ming Chu, et al. Characterization and complete genome sequence of a novel coronavirus, coronavirus HKU1, from patients with pneumonia. J Virol. 2005;79(2):884–95.
37. Knoops K, Kikkert M, van den Worm SH, et al. SARS-coronavirus replication is supported by a reticulovesicular network of modified endoplasmic reticulum. PLoS Biol. 2008;6(9):e226.
38. Zaki AM, Van Boheemen S, Bestebroer TM, Osterhaus AD, Fouchier RA. Isolation of a novel coronavirus from a man with pneumonia in Saudi Arabia. N Engl J Med. 2012;367(19):1814–20.
39. Jebril N. World Health Organization declared a pandemic public health menace: a systematic review of the coronavirus disease 2019 "COVID-19." Available SSRN 3566298. Published online 2020.
40. Corti C, Crimini E, Tarantino P, et al. Current perspectives: SARS-CoV-2 vaccines for cancer patients: a call to action. Eur J Cancer. 2021;148:316–27. Published online 2021.
41. Rahimi I, Chen F, Gandomi AH. A review on COVID-19 forecasting models. Neural Comput & Applic. 2023;35(33):23671–81.
42. Team CC 19 R, et al. Geographic differences in COVID-19 cases, deaths, and incidence—United States, February 12–April 7, 2020. Morb Mortal Wkly Rep. 2020;69(15):465–71.
43. Dwyer DE. The origins of severe acute respiratory syndrome-coronavirus-2. Semin Respir Crit Care Med. 2023;44(01):003–7. https://doi.org/10.1055/s-0042-1759564.
44. MacIntyre CR. Global spread of COVID-19 and pandemic potential. Glob Biosecurity. 2019;1(3).
45. Keni R, Alexander A, Nayak PG, Mudgal J, Nandakumar K. COVID-19: emergence, spread, possible treatments, and global burden. Front Public Health. 2020;8:216. https://doi.org/10.3389/fpubh.2020.00216.
46. Kim K, Cho KT. A review of global collaboration on COVID-19 research during the pandemic in 2020. Sustainability. 2021;13(14):7618.

47. Cai X, Fry CV, Wagner CS. International collaboration during the COVID-19 crisis: autumn 2020 developments. Scientometrics. 2021;126(4):3683–92.
48. Clemente-Suárez VJ, Navarro-Jiménez E, Moreno-Luna L, et al. The impact of the COVID-19 pandemic on social, health, and economy. Sustainability. 2021;13(11):6314.
49. McBride O, Murphy J, Shevlin M, et al. Monitoring the psychological, social, and economic impact of the COVID-19 pandemic in the population: context, design and conduct of the longitudinal COVID-19 Psychological Research Consortium (C19PRC) study. Int J Methods Psychiatr Res. 2021;30(1):e1861.
50. Das K, Behera RL, Paital B. Socio-economic impact of COVID-19. In: COVID-19 in the environment. Elsevier; 2022. p. 153–90. Accessed 28 May 2024. https://www.sciencedirect.com/science/article/pii/B9780323902724000142.

Pathogenesis and Biology of SARS-CoV-2

2

Sunny Kumar and Malini Basu

Abstract

This chapter delves into the intricate web of interactions between the severe acute respiratory syndrome coronavirus 2 (SARS-CoV-2) and the human host, elucidating the pathogenesis and biology that underlie the ongoing COVID-19 pandemic. Beginning with an overview of the viral structure and its entry mechanisms, we explore how SARS-CoV-2 hijacks host cellular machinery to propagate and disseminate. Emphasis is placed on the multifaceted immune responses triggered by viral invasion, including innate and adaptive immunity, and their role in disease severity and outcomes. Furthermore, we delve into the emerging evidence on viral variants, their implications for transmissibility, immune evasion, and vaccine efficacy. Special attention is given to the extrapulmonary manifestations and long-term consequences of SARS-CoV-2 infection, shedding light on the virus's systemic impact beyond the respiratory tract. Additionally, we discuss therapeutic strategies targeting viral replication, host immune responses, and inflammatory pathways, highlighting the challenges and opportunities in the development of effective treatments. By synthesizing current knowledge and ongoing research efforts, this chapter aims to provide a comprehensive understanding of the pathogenesis and biology of SARS-CoV-2, laying the groundwork for informed interventions and strategies to combat the pandemic.

S. Kumar
Cancer Biology and Inflammatory Disorder Division, Council of Scientific and Industrial Research-Indian Institute of Chemical Biology (CSIR-IICB), TRUE Campus, CN-6, Sector–V, Salt Lake, Kolkata, India

M. Basu (✉)
Department of Microbiology, Dhruba Chand Halder (DCH) College - University of Calcutta, Jaynagar Mazilpur, West Bengal, India

M. K. Ghosh, M. Basu (eds.), *Pathogenesis, Management and Socio-Economic Impact of COVID-19 Pandemic*, Medical Virology: From Pathogenesis to Disease Control, https://doi.org/10.1007/978-981-95-1426-7_2

Keywords

SARS-CoV-2 · Pathogenesis · Genomic organization · Life cycle · Biology

Introduction

The emergence of the novel coronavirus SARS-CoV-2 in late 2019 unleashed an unprecedented global health crisis, culminating in the COVID-19 pandemic. As the world grappled with the rapid spread of the virus and its devastating impact on public health, economy, and societies, scientists embarked on a quest to unravel the mysteries surrounding its pathogenesis and biology. This chapter serves as a comprehensive exploration into the intricate interplay between SARS-CoV-2 and the human host, shedding light on the underlying mechanisms driving viral infection, immune responses, disease progression, and therapeutic interventions [1, 2].

At its core, understanding the pathogenesis of SARS-CoV-2 necessitates a deep dive into the viral structure and its intricate interactions with host cells. From the spike protein-mediated viral entry to the exploitation of host cellular machinery for replication and spread, deciphering these molecular mechanisms is fundamental to elucidating how the virus establishes infection and evades immune surveillance (Fig. 2.1).

Moreover, as SARS-CoV-2 continues to evolve, with new variants emerging and circulating globally, it becomes imperative to dissect the genetic determinants driving viral diversity and their implications for transmission dynamics and vaccine efficacy. Beyond the respiratory tract, SARS-CoV-2 has revealed its systemic impact, manifesting in a spectrum of clinical manifestations ranging from mild respiratory illness to severe pneumonia and multiorgan dysfunction. Exploring the extrapulmonary manifestations and long-term consequences of COVID-19 unravels the complex interplay between viral pathogenesis, dysregulated immune responses, and host factors predisposing individuals to severe disease outcomes. In the face of the COVID-19 pandemic, therapeutic interventions have been at the forefront of scientific enquiry, ranging from antiviral drugs targeting viral replication to immunomodulatory agents mitigating cytokine storm and inflammation. Yet, the quest for effective treatments continues amidst the evolving landscape of viral variants and the imperative to balance therapeutic efficacy with safety and accessibility [3–7].

Through synthesizing current knowledge and ongoing research efforts, this chapter endeavors to provide a comprehensive understanding of the pathogenesis and biology of SARS-CoV-2. By shedding light on the molecular, immunological, and clinical aspects of COVID-19, we aim to pave the way for informed interventions and strategies to combat the pandemic and mitigate its impact on global health and society.

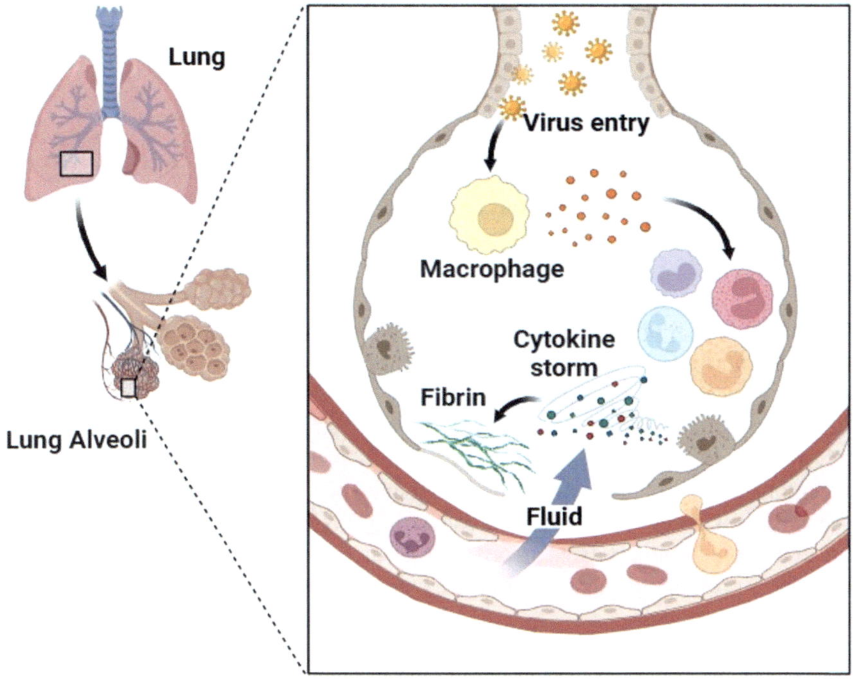

Fig. 2.1 SARS-CoV-2 pathogenesis. (This image is adopted from Biorender (https://www.biorender.com/))

Structure, Genomic Organization, and Biology of SARS-CoV-2

COVID-19 pandemic, caused by the severe acute respiratory syndrome coronavirus 2 (SARS-CoV-2), has spurred an unprecedented global health crisis. Understanding the structure and genomic organization of SARS-CoV-2 is pivotal for developing effective diagnostic tools, therapeutics, and vaccines to combat this viral threat. In this comprehensive discourse, we delve into the intricate architecture and genomic blueprint of SARS-CoV-2, exploring its structural components, genetic makeup, and implications for viral pathogenesis.

Structure

At its core, SARS-CoV-2 is a member of the Coronaviridae family, characterized by its enveloped, positive-sense single-stranded RNA genome. The virus exhibits a distinctive morphology comprising structural proteins embedded within a lipid bilayer envelope derived from the host cell membrane. Key structural proteins of SARS-CoV-2 include the spike (S), membrane (M), envelope (E), and nucleocapsid

(N) proteins, each playing pivotal roles in viral entry, assembly, and pathogenesis [8–11].

Box 2.1. Key Structural Proteins of SARS-CoV-2

- *Spike (S) Protein*: Mediates virus entry into host cells.
- *Envelope (E) Protein*: Involved in virus assembly and release.
- *Membrane (M) Protein*: Plays a key role in virus assembly.
- *Nucleocapsid (N) Protein*: Encapsulates the viral RNA genome.

Protein S The S protein stands as the hallmark feature of coronaviruses, facilitating viral entry into host cells. Structurally, the S protein forms protruding glycoprotein spikes on the viral surface, resembling a crown under electron microscopy, hence the name "corona." This trimeric protein comprises two functional subunits: S1, responsible for receptor binding, and S2, mediating membrane fusion. The receptor-binding domain within the S1 subunit specifically recognizes and binds to the host cell receptor angiotensin-converting enzyme 2 (ACE2), initiating viral attachment and entry. Given its pivotal role in viral tropism and host-cell interaction, the S protein represents a prime target for vaccine development and therapeutic intervention strategies.

Protein M The M protein serves as a central organizer in virus assembly and morphogenesis. Spanning the lipid bilayer envelope, the M protein interacts with other structural components, including the nucleocapsid protein and envelope protein, to orchestrate viral budding and assembly. Additionally, the M protein contributes to viral particle stability and membrane curvature during the budding process, influencing viral infectivity and pathogenicity.

Protein E Although relatively small in size, the E protein plays diverse roles in viral replication and pathogenesis. Embedded within the lipid bilayer envelope, the E protein contributes to viral assembly and release by interacting with other structural proteins, including the M protein. Furthermore, the E protein modulates ion channel activity, membrane permeability, and host immune responses, influencing viral pathogenesis and virulence.

Protein N The N protein encapsulates the viral RNA genome, forming the helical nucleocapsid core essential for viral replication and transcription. Beyond its structural role, the N protein interacts with viral RNA and host cell factors to facilitate RNA synthesis, packaging, and evasion of host immune surveillance. Moreover, the N protein engages in multifaceted interactions with host cell proteins, modulating cellular processes and contributing to viral pathogenesis [10–12].

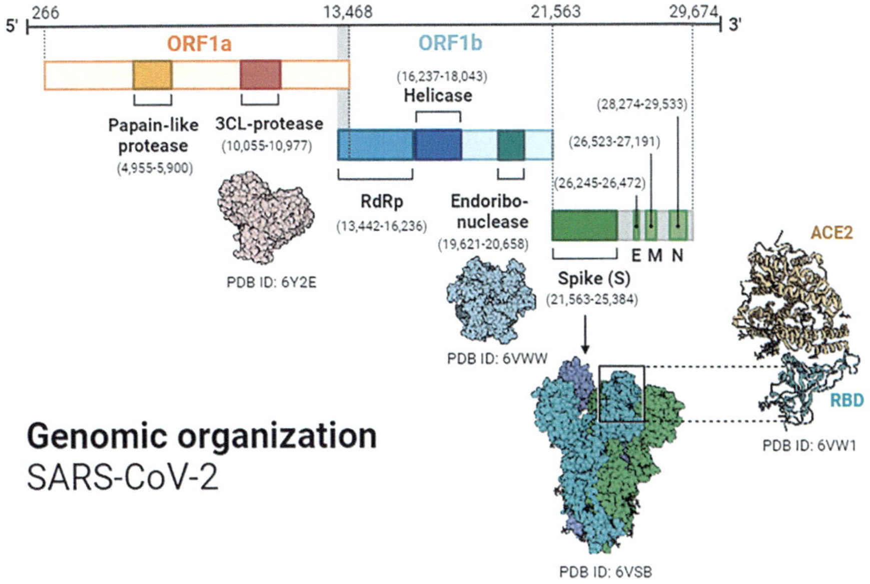

Fig. 2.2 Genomic organization of SARS-CoV-2 virus. (This image is adopted from Biorender (https://www.biorender.com/))

Genomic Organization

The genomic blueprint of SARS-CoV-2 harbors critical insights into viral replication, evolution, and pathogenesis. As a single-stranded positive-sense RNA virus, SARS-CoV-2 possesses a genome of approximately 30,000 nucleotides, encoding a repertoire of structural, nonstructural, and accessory genes. The genomic organization of SARS-CoV-2 can be delineated into distinct regions, each harboring genes essential for viral replication, transcription, and host interaction (Fig. 2.2).

Structural genes Gene S occupies the prominent position in the viral genome; the S gene encodes the spike protein, a pivotal determinant of viral entry and host-cell tropism. Genetic variations within the S gene can influence viral infectivity, transmissibility, and immune evasion, underscoring its significance as a therapeutic target and diagnostic marker. Gene E encodes the envelope protein, the E gene plays indispensable roles in viral assembly, morphogenesis, and pathogenesis. Mutations or deletions within the E gene can alter viral particle morphology, membrane permeability, and host immune recognition, modulating viral fitness and virulence. The M gene encodes the membrane protein, a fundamental constituent of the viral envelope essential for virus assembly and budding. Genetic variations within the M gene can impact viral particle assembly, stability, and host-cell interaction, influencing viral replication kinetics and pathogenicity. Gene N is essential for genome packaging and RNA synthesis; the N gene encodes the nucleocapsid protein critical for viral replication and transcription. The N protein also plays pivotal roles in modulating host immune responses, viral RNA stability, and virion assembly, highlighting its significance in viral pathogenesis and host adaptation [3, 11, 13, 14].

Nonstructural genes SARS-CoV-2 harbors a plethora of nonstructural genes encoding proteins essential for viral replication, transcription, and evasion of host immune surveillance. These include the RNA-dependent RNA polymerase (RdRp), helicase, and various proteases involved in processing viral polyproteins. Targeting nonstructural genes offers promising avenues for antiviral drug development and therapeutic intervention strategies against COVID-19 [9].

Accessory genes In addition to structural and nonstructural genes, SARS-CoV-2 encompasses accessory genes that vary in number and function across different viral strains. These accessory genes encode proteins implicated in viral pathogenesis, host-cell interaction, and immune evasion, contributing to viral fitness and virulence. Understanding the role of accessory genes in viral replication and pathogenesis is crucial for elucidating SARS-CoV-2's evolutionary dynamics and adaptive strategies [13].

The Nonstructural Proteins (NSPs)

NSPs of SARS-CoV-2 play a crucial role in the viral life cycle and pathogenesis by mediating various processes essential for viral replication and evasion of host immune responses. These NSPs are translated from the viral RNA following infection and are involved in viral RNA synthesis, processing, modification, and assembly. Here is a breakdown of the biology of NSP proteins in SARS-CoV-2.

NSP1 NSP1 is a multifunctional protein that interferes with host gene expression by binding to the ribosome and inhibiting host mRNA translation. This suppression of host protein synthesis helps the virus evade host antiviral responses and create an environment favorable for viral replication.

NSP2 NSP2 is involved in viral replication and the formation of double-membrane vesicles (DMVs), which serve as replication organelles for the virus. NSP2 also plays a role in suppressing host innate immune responses, potentially contributing to viral immune evasion.

NSP3 (papain-like protease, PLpro) NSP3 is a multifunctional protein with protease activity that cleaves the viral polyprotein, aiding in the processing of replicase proteins. Additionally, NSP3 has deubiquitinating activity, which helps the virus evade host immune surveillance by interfering with innate immune signaling pathways.

NSP4 NSP4 is a transmembrane protein that is involved in the formation of replication complexes and the modification of host membranes to support viral replication.

NSP5 (main protease, Mpro) NSP5 is a viral protease responsible for cleaving the viral polyprotein at specific sites, generating functional viral proteins necessary for viral replication and assembly. Due to its essential role in viral replication, NSP5 is considered a potential target for antiviral drug development.

NSP6 NSP6 is involved in the formation of DMVs and plays a role in viral replication and immune evasion by modulating host autophagy pathways.

NSP7, NSP8, and NSP9 These NSPs form a complex known as the RNA-dependent RNA polymerase (RdRp) complex, which is responsible for replicating the viral RNA genome. NSP12, the RNA polymerase catalytic subunit, along with NSP7 and NSP8, forms the core of this complex, while NSP9 acts as a cofactor. The RdRp complex is a primary target for antiviral drug development.

NSP10 NSP10 functions as a cofactor for the viral RdRp complex, enhancing its RNA polymerase activity.

NSP13 NSP13 is a helicase enzyme that unwinds the viral RNA during replication, facilitating RNA synthesis. It is also involved in RNA capping, a process crucial for stabilizing viral RNA and evading host immune detection.

NSP14 NSP14 possesses both exoribonuclease and N7-methyltransferase activities, which are involved in proofreading and capping viral RNA, respectively. This contributes to the fidelity of viral replication and enhances viral fitness.

Understanding the biology of NSP proteins in SARS-CoV-2 is essential for elucidating viral replication mechanisms, identifying potential targets for antiviral drug development, and devising strategies to combat COVID-19 [15–17].

Implications for Viral Pathogenesis

The intricate interplay between the structure and genomic organization of SARS-CoV-2 underpins its pathogenic potential and clinical manifestations. Variations in viral structural proteins, such as the spike protein, can influence viral tropism, transmission efficiency, and disease severity by modulating host-cell receptor binding and membrane fusion kinetics. Similarly, genetic alterations within nonstructural genes may impact viral replication fidelity, RNA synthesis kinetics, and host immune evasion strategies, shaping the course of infection and clinical outcomes [9, 18–21].

Moreover, the presence of accessory genes underscores the adaptive capacity of SARS-CoV-2 to exploit host cellular machinery and evade immune surveillance, facilitating viral replication and spread. Mutational hotspots within the viral genome may engender novel viral variants with altered phenotypic traits, including enhanced transmissibility, immune evasion, and resistance to antiviral interventions. Consequently, ongoing genomic surveillance and characterization of SARS-CoV-2 variants are imperative for informing public health strategies and pandemic response efforts [21, 22].

In conclusion, the structure and genomic organization of SARS-CoV-2 constitute pivotal determinants of viral pathogenesis, transmission dynamics, and host adaptation. Elucidating the molecular architecture of viral structural proteins, including the spike, membrane, envelope, and nucleocapsid proteins, offers valuable insights into viral entry mechanisms.

Life Cycle

Physical contacts and air droplets are the two potential modes of community spread among human population [23–25]. Immediately after its transmission, viral particles enter the host organism by binding to their epithelial cell surface receptor proteins [26–29]. After replication, the matured viral particles are released from donor host and transmitted to another recipient (Fig. 2.3).

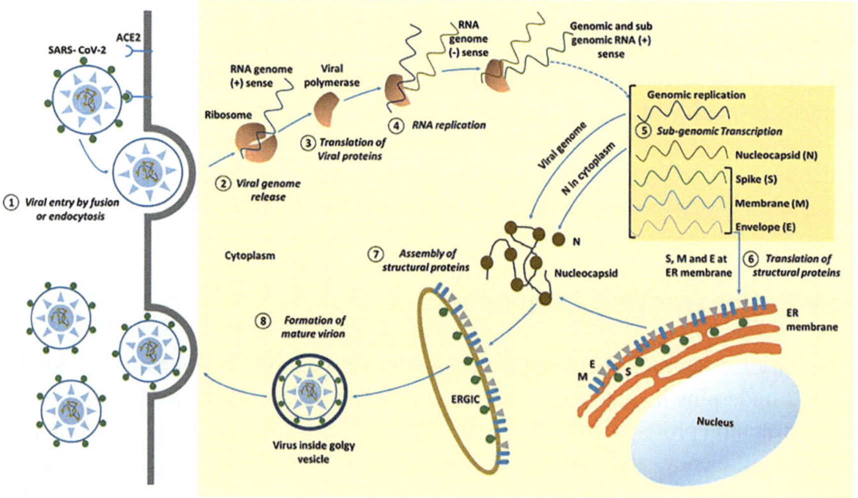

Fig. 2.3 Life cycle. The figure depicts the various steps in the life cycle of coronavirus viz., viral entry by fusion or endocytosis, release of viral genome, translation of viral proteins, RNA replication, subgenomic transcription, translation and assembly of structural proteins, formation of mature virion, and finally release out from the host

Binding and Entry of Viral Particles

Coronavirus life cycle initiates when the viral surface spike glycoprotein binds to the complementary host cell receptor. To date, two possible ways for entry of viral particles are studied viz., ACE2 and CD147 surface receptor-mediated access. In humans, SARS-CoV-2 infection induces when the virion anchors to ACE2 (angiotensin-converting enzyme 2) receptor. Post-attachment, a host cell protease cleaves and activates the receptor-bound spike protein. The host cell protease governs the viral entry by endocytosis or fusion through the host cell membrane [8, 26–28]. Detail study of CD147 surface receptor-mediated entry of viral particles is not yet done; hence, further research is needed to verify and strengthen its involvement [30, 31].

Translation of Viral RNA

Post host cell entry, the uncoated virus particles and their genome start functioning. The 5′ methylated cap and a 3′ poly (A) tail of the coronavirus RNA genome permit its attachment to the host ribosome for subsequent translation. The overlapping open reading frame of the viral genome after translation forms a long polyprotein. The polyprotein proteases further cleave the polyprotein into various nonstructural proteins (NSPs). Furthermore, genomic RNA is synthesized from (+) sense RNA template while the subgenomic RNA translates into other supplementary and structural proteins like spike (S), envelope (E), membrane (M), and nucleocapsid [32–34].

Assembly of Structural Proteins

Translated structural proteins S, E, and M are associated with the ER (endoplasmic reticulum) membrane. Next, these structural proteins along with nucleocapsid enter the endoplasmic reticulum-Golgi intermediate compartment (ERGIC) which is known as a gathering unit that assembles the structural proteins and nucleocapsid proteins in the cytoplasm [12, 35–37].

Maturation of Virion and Exocytosis

Maturation and exocytosis of virus particles are the last two steps in their life cycle. After proper assembly of viral components in ERGIC, the assembled units are transferred to the Golgi body for maturation and packaging. Finally, matured viral particles after exocytosis either spread to other cells of the host or transmitted to the other host. Although human-to-human physical contacts or airborne transmission are the two major routes of transmission, ocular and oro-fecal are two other possible additional paths of transmission, because the virus is documented in the digestive tract and ocular surface [34, 38–41]. There may be other possible routes of transmission, yet to be established.

Viral entry and replication are the prime steps to induce the immunological disbalances in the innate as well as humoral immunity. The immunological changes are crucial for the development of therapeutic as well as preventive strategies against this infection. Therefore, in the next section of this review authors are interested to describe the immunological responses of SARS-CoV-2 infection.

Pathogenesis and Clinical Manifestations

Understanding the clinical manifestations of this virus is crucial for effective management and control of the disease. SARS-CoV-2 exhibits a diverse array of symptoms, ranging from mild respiratory symptoms to severe pneumonia and multiorgan dysfunction. This section aims to provide a comprehensive overview of the clinical manifestations of SARS-CoV-2 infection.

Fever is one of the most common symptoms of COVID-19, reported in a majority of patients. Dry cough is another hallmark symptom, often persistent and irritating. Dyspnea or difficulty in breathing may range from mild to severe, often requiring hospitalization in critical cases. Some patients may experience a sore throat, which can vary in severity. Gastrointestinal Symptoms such as nausea and vomiting may occur, particularly in mild cases. Diarrhea has been reported in a subset of patients, sometimes preceding respiratory symptoms. Neurological Manifestations such as headache, anosmia (loss of smell) and ageusia (loss of taste), dizziness and confusion, and stroke may appear. Cardiovascular symptoms such as chest pain, palpitations, and myocarditis may be observed. COVID-19 pneumonia is characterized by bilateral infiltrates on chest imaging, often progressing rapidly in severe cases. Severe cases of COVID-19 can lead to ARDS, a life-threatening condition requiring mechanical ventilation. SARS-CoV-2 infection has been associated with an increased risk of pulmonary embolism, possibly due to hypercoagulability. Immunological manifestations such as cytokine Storm and multiorgan dysfunction syndrome, involving dysfunction of multiple organ systems, including the lungs, heart, kidneys, and liver. Long-term effects such as fatigue, cognitive impairment, and postural orthostatic tachycardia syndrome have been reported in a subset of COVID-19 survivors [20, 39, 40].

The clinical manifestations of SARS-CoV-2 infection are wide-ranging and can affect multiple organ systems. While respiratory symptoms predominate, COVID-19 can also manifest with gastrointestinal, neurological, cardiovascular, dermatological, and immunological manifestations. Moreover, the long-term effects of COVID-19, often referred to as long COVID, pose significant challenges for healthcare systems worldwide [3, 4]. A better understanding of these clinical manifestations is essential for timely diagnosis, appropriate management, and development of targeted therapeutic strategies to mitigate the impact of this global pandemic.

Conclusion

The pathogenesis and biology of SARS-CoV-2 reveal a complex interplay between viral factors and host responses, showcasing the virus's sophisticated mechanisms for infection and immune evasion. The virus's ability to efficiently enter host cells via the ACE2 receptor, coupled with its adept replication strategy and ability to suppress immune detection, underscores its high transmissibility and pathogenicity. This has been evident in the global spread and severe impact of COVID-19. Continued research into the molecular mechanisms of SARS-CoV-2 infection, immune evasion, and host-pathogen interactions is paramount. Such research not only enhances our understanding of the virus but is also critical for developing effective therapeutic and preventive measures. By unraveling these molecular intricacies, scientists can innovate targeted treatments, improve vaccine designs, and ultimately control and mitigate the effects of COVID-19, paving the way for better preparedness against future pandemics.

Author's Contributions SK and MB conceived the idea and review structure; SK and MB wrote the manuscript; SK and MB revised and edited the manuscript. Authors have read, agreed, and approved the final draft of the manuscript.

Data Availability The software Biorender (https://www.biorender.com/) is used to draw some elements of the few figures.

Declarations

Conflicts of Interests There is no competing and conflict of interests.

References

1. Ciotti M, Ciccozzi M, Terrinoni A, Jiang WC, Wang CB, Bernardini S. The COVID-19 pandemic. Crit Rev Clin Lab Sci. 2020;57(6):365–88. https://doi.org/10.1080/10408363.2020.1783198.
2. Pokhrel S, Chhetri R. A literature review on impact of COVID-19 pandemic on teaching and learning. High Educ Future. 2021;8(1):133–41. https://doi.org/10.1177/2347631120983481.
3. Ghosh MK, Kumar S, Ganguly KK, et al. COVID-19 and cancer: insights into their association and influence on genetic and epigenetic landscape. Epigenomics. 2023;15(4):227–48. https://doi.org/10.2217/epi-2023-0052.
4. Kumar S, Basu M, Ghosh P, Pal U, Ghosh MK. COVID-19 therapeutics: clinical application of repurposed drugs and futuristic strategies for target-based drug discovery. Genes Dis. 2023;10(4):1402–28. https://doi.org/10.1016/j.gendis.2022.12.019.
5. Angelini M, Teglia F, Astolfi L, Casolari G, Boffetta P. Decrease of cancer diagnosis during COVID-19 pandemic: a systematic review and meta-analysis. Eur J Epidemiol. 2023;38(1):31–8. https://doi.org/10.1007/s10654-022-00946-6.
6. Kumar S, Basu M, Ghosh MK. Chaperone-assisted E3 ligase CHIP: a double agent in cancer. Genes Dis. 2021;9(6). https://doi.org/10.1016/j.gendis.2021.08.003.
7. Kumar S, Chatterjee M, Ghosh P, Ganguly KK, Basu M, Ghosh MK. Targeting PD-1/PD-L1 in cancer immunotherapy: an effective strategy for treatment of triple-negative breast cancer (TNBC) patients. Genes Dis. 2023;10(4):1318–50.

8. Walls AC, Park YJ, Tortorici MA, Wall A, McGuire AT, Veesler D. Structure, function, and antigenicity of the SARS-CoV-2 spike glycoprotein. Cell. 2020;181(2):281–92.
9. Angeletti S, Benvenuto D, Bianchi M, Giovanetti M, Pascarella S, Ciccozzi M. COVID-2019: the role of the nsp2 and nsp3 in its pathogenesis. J Med Virol. 2020;92(6):584–8. https://doi.org/10.1002/jmv.25719.
10. Krishnan A, Mukherjee D, Hamilton JP, Alqahtani SA, Woreta TA. Coronavirus disease 2019: taxonomy, genomic structure, replication and pathogenesis. J Infect Dis Ther. 2021;3:2.
11. Naqvi AAT, Fatima K, Mohammad T, et al. Insights into SARS-CoV-2 genome, structure, evolution, pathogenesis and therapies: structural genomics approach. Biochimica et Biophysica Acta (BBA). 2020;1866(10):165878.
12. Mousavizadeh L, Ghasemi S. Genotype and phenotype of COVID-19: their roles in pathogenesis. J Microbiol Immunol Infect. 2021;54(2):159–63.
13. Raskin S. Genetics of COVID-19. J Pediatr. 2021;97:378–86.
14. Barrantes FJ. Structural biology of coronavirus ion channels. Acta Crystallogr Sect Struct Biol. 2021;77(4):391–402.
15. Aleebrahim-Dehkordi E, Ghoshouni H, Koochaki P, et al. Targeting the vital non-structural proteins (NSP12, NSP7, NSP8 and NSP3) from SARS-CoV-2 and inhibition of RNA polymerase by natural bioactive compound naringenin as a promising drug candidate against COVID-19. J Mol Struct. 2023;1287:135642.
16. Ni X, Han Y, Zhou R, Zhou Y, Lei J. Structural insights into ribonucleoprotein dissociation by nucleocapsid protein interacting with non-structural protein 3 in SARS-CoV-2. Commun Biol. 2023;6(1):193.
17. Züst R, Cervantes-Barragán L, Kuri T, et al. Coronavirus non-structural protein 1 is a major pathogenicity factor: implications for the rational design of coronavirus vaccines. PLoS Pathog. 2007;3(8):e109.
18. Shimizu Y. Understanding the immunopathogenesis of COVID-19: its implication for therapeutic strategy. World J Clin Cases. 2020;8(23):5835.
19. Acosta MAT, Singer BD. Pathogenesis of COVID-19-induced ARDS: implications for an ageing population. Eur Respir J. 2020;56(3). Accessed 28 May 2024. https://erj.ersjournals.com/content/56/3/2002049.short.
20. Li C, He Q, Qian H, Liu J. Overview of the pathogenesis of COVID-19 (Review). Exp Ther Med. 2021;22(3):1011. https://doi.org/10.3892/etm.2021.10444.
21. Nile SH, Nile A, Qiu J, Li L, Jia X, Kai G. COVID-19: pathogenesis, cytokine storm and therapeutic potential of interferons. Cytokine Growth Factor Rev. 2020;53:66–70.
22. Zhou S, Lv P, Li M, et al. SARS-CoV-2 E protein: pathogenesis and potential therapeutic development. Biomed Pharmacother. 2023;159:114242.
23. Canova V, Lederer Schläpfer H, Piso RJ, et al. Transmission risk of SARS-CoV-2 to healthcare workers–observational results of a primary care hospital contact tracing. Swiss Med Wkly. 2020;150(1718):w20257.
24. Harrison AG, Lin T, Wang P. Mechanisms of SARS-CoV-2 transmission and pathogenesis. Trends Immunol. 2020;41(12):1100–15. Published online 2020.
25. Tang S, Mao Y, Jones RM, et al. Aerosol transmission of SARS-CoV-2? Evidence, prevention and control. Environ Int. 2020;144:106039.
26. Hoffmann M, Kleine-Weber H, Schroeder S, et al. SARS-CoV-2 cell entry depends on ACE2 and TMPRSS2 and is blocked by a clinically proven protease inhibitor. Cell. 2020;181(2):271–80.
27. Letko M, Marzi A, Munster V. Functional assessment of cell entry and receptor usage for SARS-CoV-2 and other lineage B betacoronaviruses. Nat Microbiol. 2020;5(4):562–9.
28. Ou X, Liu Y, Lei X, et al. Characterization of spike glycoprotein of SARS-CoV-2 on virus entry and its immune cross-reactivity with SARS-CoV. Nat Commun. 2020;11(1):1–12.
29. Kumar S, Basu M, Ghosh P, Ansari A, Ghosh MK. COVID-19: clinical status of vaccine development to date. Br J Clin Pharmacol. 2023;89(1):114–49. https://doi.org/10.1111/bcp.15552.

30. Acosta Saltos F, Acosta Saltos AD. Entry of SARS-CoV2 through the basal surface of alveolar endothelial cells–a proposed mechanism mediated by CD147 in Covid-19. Published online 2020.
31. Qiao J, Li W, Bao J, et al. The expression of SARS-CoV-2 receptor ACE2 and CD147, and protease TMPRSS2 in human and mouse brain cells and mouse brain tissues. Biochem Biophys Res Commun. 2020;533(4):867–71.
32. Knoops K, Kikkert M, van den Worm SH, et al. SARS-coronavirus replication is supported by a reticulovesicular network of modified endoplasmic reticulum. PLoS Biol. 2008;6(9):e226.
33. Baranov PV, Henderson CM, Anderson CB, Gesteland RF, Atkins JF, Howard MT. Programmed ribosomal frameshifting in decoding the SARS-CoV genome. Virology. 2005;332(2):498–510.
34. Ogando NS, Dalebout TJ, Zevenhoven-Dobbe JC, et al. SARS-coronavirus-2 replication in Vero E6 cells: replication kinetics, rapid adaptation and cytopathology. J Gen Virol. 2020;101(9):925–40. https://doi.org/10.1099/jgv.0.001453.
35. Fischer F, Stegen CF, Masters PS, Samsonoff WA. Analysis of constructed E gene mutants of mouse hepatitis virus confirms a pivotal role for E protein in coronavirus assembly. J Virol. 1998;72(10):7885–94.
36. Lai MM, Cavanagh D. The molecular biology of coronaviruses. Adv Virus Res. 1997;48:1–100.
37. Masters PS. The molecular biology of coronaviruses. Adv Virus Res. 2006;66:193–292.
38. Ghinai I, McPherson TD, Hunter JC, et al. First known person-to-person transmission of severe acute respiratory syndrome coronavirus 2 (SARS-CoV-2) in the USA. Lancet. 2020;395(10230):1137–44.
39. Han Y, Yang H. The transmission and diagnosis of 2019 novel coronavirus infection disease (COVID-19): a Chinese perspective. J Med Virol. 2020;92(6):639–44.
40. Chan JFW, Zhang AJ, Yuan S, et al. Simulation of the clinical and pathological manifestations of coronavirus disease 2019 (COVID-19) in a golden Syrian hamster model: implications for disease pathogenesis and transmissibility. Clin Infect Dis. 2020;71(9):2428–46.
41. Yeo C, Kaushal S, Yeo D. Enteric involvement of coronaviruses: is faecal–oral transmission of SARS-CoV-2 possible? Lancet Gastroenterol Hepatol. 2020;5(4):335–7.

COVID-19: Modulation of Host's Genetic and Epigenetic Machineries

3

Kirat Kumar Ganguly

Abstract

The COVID-19 pandemic caused by the novel Severe Acute Respiratory Syndrome coronavirus 2 (SARS-CoV-2) has posed a unique threat to global clinical and research scenarios and economies. Beyond the direct effects of the virus, understanding how it interacts with the host's genetic and epigenetic machinery is crucial for developing effective treatments. Understanding the interplay between the virus and the host's genetic and epigenetic machinery is crucial for unraveling the mechanisms of viral pathogenesis and host response and shedding light on potential therapeutic avenues. This chapter explores the modulation of host genetic and epigenetic processes by SARS-CoV-2 infection, highlighting the intricate molecular interactions contributing to disease progression and variability in clinical outcomes. Drawing insights from recent research articles, this chapter delves into the impact of viral infection on gene expression, epigenetic modifications, and immune responses, shedding light on potential therapeutic targets for mitigating COVID-19 severity.

Keywords

COVID-19 · SARS-CoV-2 · Genetics · Epigenetics · Gene expression · Immune response · Pathogenesis · Therapeutic targets

K. K. Ganguly (✉)
Michael Madhusudan Memorial College,, Durgapur, West Bengal, India

M. K. Ghosh, M. Basu (eds.), *Pathogenesis, Management and Socio-Economic Impact of COVID-19 Pandemic*, Medical Virology: From Pathogenesis to Disease Control, https://doi.org/10.1007/978-981-95-1426-7_3

Introduction

The coronavirus disease-2019 (COVID-19) pandemic instigated by severe acute respiratory syndrome coronavirus-2 (SARS-CoV-2) has speedily blowout throughout the globe, resulting in significant morbidity and mortality. The novel Coronavirus (COVID-19) outbreak, which originated in Wuhan City, China, was announced as a "Public Health Emergency of International Concern" by the World Health Organization on January 30, 2020. While the clinical manifestations of COVID-19 contrast broadly, ranging from asymptomatic infection to severe respiratory distress and multiorgan failure, emerging evidence suggests that host genetic and epigenetic factors are crucial in determining disease susceptibility, severity, and outcomes. While genetic factors play a role in susceptibility and severity of infection, epigenetic alterations also significantly influence disease outcomes. Epigenetics involves the study of phenotypic changes caused without any alterations in the DNA nucleotide sequence. The essential mechanism of epigenetics lies in gene expression regulation through reversible modifications of several DNA-associated molecules like methylation, acetylation of histone, remodeling of chromatin, noncoding RNAs, and methylation of DNA/RNA. These modifications impact gene accessibility and transcriptional activity, ultimately shaping cellular responses.

This chapter delves into the modification of the host epigenomic landscape as a result of coronavirus infections and discusses emerging therapeutic strategies over genetics. This explores how existing drug candidates can serve as potent weapons against SARS-CoV-2 through molecular-dynamics simulation studies and molecular docking. The chapter also describes the intricate interplay between SARS-CoV-2 and the host's genetic and epigenetic machinery, shedding light on the molecular mechanisms underlying COVID-19 pathogenesis and host response.

Host Genetic Variability and COVID-19 Susceptibility

Genetic susceptibility to infectious diseases, including COVID-19, is governed by variations in host genes involved in viral entry, immune response, and inflammation. Genome-wide association studies have pointed to a number of genetic loci associated with COVID-19 susceptibility and severity (Fig. 3.1). For instance, polymorphisms in the gene encoding the receptor used by SARS-CoV-2 for cell entry, called angiotensin-converting enzyme 2 (ACE2), have been implicated in modulating viral infectivity and disease severity [9]. Additionally, gene variations related to the host immune response, such as those encoding cytokines, chemokines, and human leukocyte antigens (HLAs), influence individual susceptibility to severe COVID-19 outcomes [15]. Furthermore, host genetic factors may contribute to interindividual differences in vaccine efficacy and adverse reactions, highlighting the importance of personalized medicine approaches in combating the pandemic [38]. Interestingly, investigations show the putative relationship between 25-hydroxy Vitamin-D (25OHD) levels and COVID-19 outcomes using a Mendelian randomization (MR) approach. The study found no evidence to support the notion that

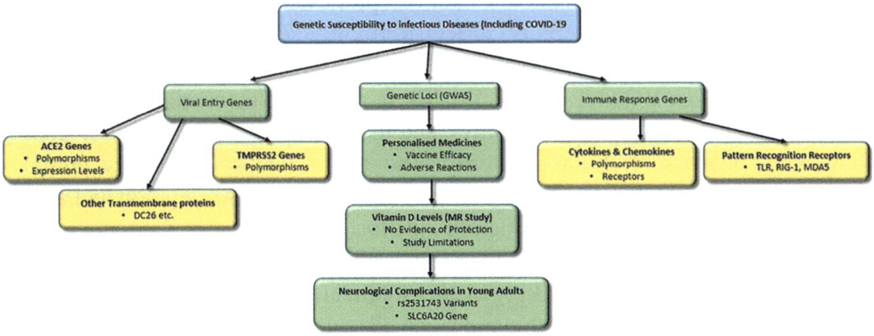

Fig. 3.1 Types of genetic susceptibilities in COVID-19. The diagram illustrates the genetic factors contributing to susceptibility to infectious diseases, including COVID-19. It categorizes these genetic factors into three main groups: Viral Entry Genes, Genetic Loci (GAS), and Immune Response Genes

increasing 25OHD levels would protect against COVID-19 susceptibility, hospitalization, or severity. The study acknowledges limitations such as excluding individuals with vitamin D deficiency and the potential for weak instrument bias in the analysis. Despite these limitations, the study leverages a large cohort of COVID-19 cases and controls and a comprehensive analysis of genetic determinants of 25OHD levels. The findings align with the results of the largest randomized trial on vitamin D supplementation to date, supporting the credibility of the MR results. Overall, the study contributes valuable insights to the ongoing discussion on the involvement of vitamin D in COVID-19 outcomes, also highlights the importance of rigorous research methods in evaluating potential interventions [8]. Another study investigates the association between host genetic variants and COVID-19 prognosis in young adults. A cohort among 336 university students of European ancestry was analyzed for COVID-19 infection history, clinical symptoms, and neurological complications. The rs2531743 variant near the SLC6A20 gene was identified as associated with COVID-19 susceptibility and post-recovery neurological complications. The study highlights the genetic basis of COVID-19 outcomes in young adults and underscores the importance of genetic factors in understanding disease susceptibility and neurological sequelae post-infection [26].

Host genetic variability significantly affects the susceptibility and severity of COVID-19. There are several key genetic factors influencing the disease including. Variations in the ACE2 gene affect susceptibility to infection. Some polymorphisms may increase the binding affinity of the virus to ACE2, facilitating easier entry and potentially leading to more severe disease.

Additionally, ACE2 expression levels vary among individuals and populations, contributing to differences in disease outcomes [23, 45]. Transmembrane protease serine 2 (TMPRSS2) is essential for priming the spike protein of SARS-CoV-2, essential for viral entry. Polymorphisms in the TMPRSS2 gene can influence the efficiency of this process, affecting disease severity [20, 65]. Other transmembrane proteins, such as DC26, also play roles in viral entry and pathogenesis. Variations in genes coding cytokines, chemokines, their receptors, and pattern recognition receptors (PRRs) like Toll-like receptors (TLRs), Retinoic acid-inducible Gene I (RIG-1), and melanoma differentiation-associated gene 5 (MDA5) can impact the immune trigger to SARS-CoV-2 [11, 35, 54, 59]. These polymorphisms may alter cytokine production, immune signaling pathways, and the overall immune response, leading to varying degrees of disease severity. The major histocompatibility complex (MHC), including MHCI and MHCII, is crucial for antigen presentation and activation of T cells. HLA polymorphisms can influence the efficiency of antigen presentation and the subsequent immune response, affecting the ability to clear the virus and the severity of COVID-19. Epigenetic modifications, such as alterations in histone modification and DNA methylation, can also influence the expression of genes involved in the immune response to SARS-CoV-2. For example, epigenetic regulation of ACE2 and cytokine genes like IL-6 can affect disease outcomes. Understanding these genetic and epigenetic factors is crucial for predicting disease susceptibility and severity, guiding personalized treatment strategies, and developing effective therapeutic interventions [57]. The effect of host genetic variability on COVID-19 infection is discussed in several ways in the provided articles. The gene of ACE2 receptor shows variations that can impact susceptibility to COVID-19. Certain ACE2 variants can affect the stability and binding affinity of the ACE2 protein to the viral spike protein, influencing the likelihood of infection and severity of the disease. For instance, specific ACE2 missense variants such as p.(Lys26Arg), p.(Asn720Asp), and p.(Gly211Arg) may interfere with protein structure and stabilization, while others like p.(Leu351Val) and p.(Pro389His) may interfere with the internalization process of the virus. Variations in other transmembrane proteins and enzymes, such as TMPRSS2 and CD26, also play a role in COVID-19 susceptibility. TMPRSS2, for instance, is crucial for priming the SARS-CoV-2 spike protein for entry into cells. Polymorphisms in TMPRSS2 and CD26 can increase susceptibility to the virus by affecting this entry process [44, 52]. Polymorphisms in cytokines, chemokines, their receptors, pattern recognition receptors (PRRs), and major histocompatibility complex (MHC) coding genes are associated with the immune response to SARS-CoV-2. For example, polymorphisms in Toll-like receptors (TLRs) and other PRRs can affect the recognition of viral RNA, impacting the innate immune response. Variations in cytokine genes can influence the severity of the inflammatory reaction to the infection [59, 60]. The human leukocyte antigen (HLA) machinery, which plays a crucial role in immune reaction, also shows polymorphisms that affect COVID-19 outcomes. Certain HLA alleles are associated with either increased susceptibility or resistance to the virus, influencing the severity and progression of the disease [15, 39]. Overall, these genetic variations can help explain the differences in COVID-19 susceptibility and severity observed among individuals and populations, highlighting the importance of considering genetic factors in understanding and managing the disorder [13, 42]. Figure 3.1 is a diagram illustrating the key points from the provided text about genetic susceptibility to COVID-19 and related findings.

Epigenetic Regulation of COVID-19

The COVID-19 pandemic has emerged as the maximum clinical crisis of global health in recent years, necessitating the development of effective treatments. The novel coronavirus SARS-CoV-2 causes this infectious disease. Autophagy, a vital and highly conserved catabolic process, is crucial for the growth and replication of coronaviruses. Consequently, understanding how autophagy is regulated is of great interest. The regulation of autophagy is intricate and influenced by multiple factors, including various epigenetic modifications such as histone alteration and DNA methylation, which are also associated with SARS-CoV-2 replication. Thus, an investigation of the molecular regulation of the epigenetic pathways regulating autophagy and COVID-19 could reveal new therapeutic interventions. This review examines the role of epigenetic regulation of autophagy in managing COVID-19 and emphasizes potential therapeutic applications [4]. Epigenetic alterations like histone modifications, DNA methylation, and noncoding RNA-induced gene regulation play a crucial role in modulating host-virus interactions and immune responses during SARS-CoV-2 infection. Studies suggests that SARS-CoV-2 infection induces widespread changes in host epigenetic profiles, which may contribute to disease pathogenesis and clinical outcomes [24]. For example, dysregulation of DNA methylation patterns in immune-related genes has been observed in COVID-19 patients, leading to aberrant cytokine production and inflammatory responses [64]. Moreover, histone modifications, such as acetylation and methylation, regulate the expression of genes involved in antiviral defense mechanisms and inflammatory pathways, thereby influencing the host's ability to control viral replication and dissemination [48]. Additionally, microRNAs (miRNAs) and long noncoding RNAs (lncRNAs) serve as key regulators of immune signaling pathways and may modulate the host response to SARS-CoV-2 infection through post-transcriptional gene regulation [63]. SARS-CoV-2 appeared in late 2019 and led to a global pandemic, partly due to its ability immune suppressive efficiency. In rare instances, viral proteins inhibit antiviral defenses by imitating essential areas of human histone proteins, especially those involved in transcriptional regulation through post-translational modifications. Recent studies have shown that SARS-CoV-2 significantly disrupts the host cell's epigenetic regulation. However, the precise mechanisms by which SARS-CoV-2 manipulates the host cell's epigenome, including the potential use of histone mimicry, remain unclear. Researchers have found that the ORF8 protein of SARS-CoV-2 mimics histones, specifically imitating the ARKS motifs (Arginine-Arginine-Lysine-Serine/Threonine) in histone H3, which disrupts the host's epigenetic regulation. ORF8 interacts with chromatin, interfering with the regulation of crucial histone post-translational modifications and promoting chromatin compaction. Eliminating either the ORF8 gene or its histone mimicry site reduces the virus's ability to modify host cell chromatin, impacts the transcriptional response to infection, and decreases the viral genome copy number. These discoveries highlight a novel function of ORF8 and a mechanism through which SARS-CoV-2 interferes with host cell epigenetic regulation, offering a molecular explanation for the observed reduction in COVID-19 severity in cases where SARS-CoV-2 lacks ORF8

[27]. Epigenetics is an expanding field of science, with research into the role of epigenetic regulation in diseases, including infectious diseases, becoming increasingly prominent. MicroRNAs (miRNAs), an epigenetic mechanism, interact with their target messenger RNAs (mRNAs) to affect their stability and/or translation. This study investigated noncoding RNAs (ncRNAs), especially miR-200c, in the context of COVID-19. It begins by defining epigenetics and miRNAs, then delves into the role of miRNAs in diseases, particularly lung diseases. It then delves into the specific role and significance of miR-200c in viral infections and SARS-CoV-2. A database search pipeline, including MEDLINE, PubMed, Web of Science, Embase, and the Cochrane Library, was conducted to review relevant papers published from 2000 to 2021. Ultimately, 128 articles were selected that best aligned with the study's objectives, with 5 specifically examining the impact of miR-200c. The study highlights the importance of epigenetic processes, particularly miRNAs, due to varying therapeutic outcomes across different races and populations. The results indicate that various miRNAs can affect lung diseases through different target pathways and genes. Increased levels of miR-200c may reduce the expression of ACE2, potentially heightening infection risk, inflammation, and COVID-19 complications. miR-200c and its association with ACE2 could potentially serve as early prognostic and diagnostic markers [51]. ACE2 acts as the receptor for three coronaviruses: HCoV-NL63, SARS-CoV, and SARS-CoV-2. It plays a critical role in regulating the renin-angiotensin system, blood pressure, and several signaling pathways, including integrin signaling. ACE2 expression is controlled both transcriptionally and post-transcriptionally, governed by two promoters, with tissue-specific promoter usage. ACE2 is most abundantly expressed in the small intestine, kidney, and heart, but is also present in various other tissues. This review examines the chemical and mechanical signal transduction pathways that regulate ACE2 gene expression, along with the epigenetic and chromatin features associated with its expression.

The COVID-19 pandemic, caused by SARS-CoV-2, has led to a global health crisis over the past 3 years, claiming nearly 6.3 million lives. Numerous studies have been conducted to understand the mechanisms SARS-CoV-2 uses to exacerbate the viral outbreak. The virus utilizes ACE2 receptors and TMPRSS2 to enter host cells. Once inside, it hijacks the host's machinery to replicate and disrupt normal cellular functions, leading to infection-related health problems. Additionally, several epigenetic factors, including DNA methylation, acetylation, histone modifications, microRNAs, and variables like age and sex, significantly influence viral entry, immune evasion, and cytokine responses, thereby impacting COVID-19 severity. Findings on the epigenetic regulation of viral pathogenicity suggest that epi-drugs could be a promising therapeutic strategy against COVID-19 [3].

The COVID-19 pandemic, driven by SARS-CoV-2, has significantly disrupted global social and economic stability. Researchers are employing omics approaches, particularly epitranscriptomics, to curb infections and develop effective treatments. Multiomics approaches reveal the dynamics of the host's response across different disease phases, uncovering molecular and cellular landscapes [14]. Epitranscriptomics specifically investigates gene transcription mechanisms in cells and tissues and the interaction between genetic material and epigenetic regulation.

A review highlights the importance of post-transcriptional regulation in SARS-CoV-2, influencing processes like viral infection, replication, immunogenicity, and pathogenicity [55].

Impact of SARS-CoV-2 on Host Gene Expression Profiles

SARS-CoV-2 infection induces significant changes in host gene expression, affecting both innate and adaptive immune responses as well as pathways related to viral replication, host cell survival, and tissue damage. The global pandemic instigated by SARS-CoV-2 has affected millions of people worldwide. Retrospective studies have shown a link between elevated inflammatory cytokines and disease severity. Through detailed epigenetic, transcriptional, in vitro, and in vivo analyses, it has been demonstrated that inhibiting topoisomerase 1 (TOP1) can effectively suppress the severe inflammation caused by SARS-CoV-2. In hamsters, treatment with the FDA-approved TOP1 inhibitor, topotecan (TPT), reduced inflammation induced by the infection. Administering TPT up to 4 days post-infection decreased morbidity and rescued mortality in a transgenic model of mouse. These propose that TOP1 inhibition could be a promising host-directed therapy against severe SARS-CoV-2 infection. TPT and its derivatives are widely available, cost-effective, and accessible in most countries, making them suitable candidates for clinical trials to assess their efficacy in treating severe COVID-19 in humans [5, 21, 41].

Recent research has highlighted the significance of disrupted cytokine and immune cell networks in worsening COVID-19 severity and causing organ damage. SARS-CoV-2 infection, like other viral infections, triggers cellular senescence, leading to a senescence-associated secretory phenotype (SASP). The release of proinflammatory cytokines, extracellular matrix-modifying factors, and procoagulatory agents marks this phenotype. COVID-19 patients exhibit signs of senescence in their airway mucosa along with increased serum levels of SASP factors [25]. Laboratory studies have shown that virus-induced senescence (VIS) leads to macrophage activation, complement-mediated cell lysis, and secondary senescence of endothelial cells, which aligns with critical aspects of COVID-19 pathology such as immune cell infiltration, endothelial damage, and extensive lung thrombosis. Supernatants from VIS cells, including those induced by SARS-CoV-2, stimulate the formation of neutrophil extracellular traps and activate platelets and the clotting cascade. Senolytic treatments, such as navitoclax and a dasatinib-quercetin combination, have been found to selectively target and eliminate VIS cells, reducing inflammation and alleviating COVID-19-like lung disease in animal models infected with SARS-CoV-2. These findings suggest that targeting VIS cells could be an effective therapeutic approach against SARS-CoV-2 and potentially other viral infections [28, 47].

Recent studies have also emphasized the role of dysregulated cytokine and immune cell networks in contributing to COVID-19 severity and organ damage. SARS-CoV-2 infection has been found to induce cellular senescence, similar to other viruses, leading to a senescence-associated secretory phenotype (SASP) characterized by the release of proinflammatory cytokines, extracellular matrix-active factors, and procoagulatory mediators. COVID-19 patients have shown markers of senescence in their airway mucosa and elevated serum levels of SASP factors. In vitro experiments have demonstrated that virus-induced senescence (VIS) results in macrophage activation, complement lysis, and secondary senescence of endothelial cells, aligning with key features of COVID-19 pathology such as macrophage and neutrophil infiltration, endothelial damage, and widespread thrombosis in the lungs. Supernatants from VIS cells, including those induced by SARS-CoV-2, promote neutrophil extracellular trap formation and activation of platelets and the clotting cascade. Targeting VIS cells with senolytics like navitoclax and a combination of dasatinib plus quercetin has been shown to selectively eliminate these cells, reducing inflammation and mitigating COVID-19-like lung disease in SARS-CoV-2-infected hamsters and mice. This underscores the potential of targeting VIS cells as a therapeutic strategy against SARS-CoV-2 and possibly other viral infections.

Disruption of genome architecture has been linked to persistent anosmia in COVID-19 patients [7]. The downregulation of key transcription factors Lhx2 and Ebf, essential for olfactory sensory neuron (OSN) physiology, leads to reduced expression of genes involved in odor perception. In hamsters, disruption of olfactory receptor (OR) compartments occurs before the downregulation of Lhx2/Ebf and persists even after their restoration. Similarly, in infected human olfactory epithelium, both OR transcription and compartmentalization are disrupted, while Lhx2 and Ebf levels remain relatively unchanged. The initial disruption of OR compartments, which form during differentiation and may be irreversible in mature OSNs, acts as a "nuclear memory," delaying the restoration of OR transcription. This disruption may necessitate the replacement of affected OSNs for the recovery of the sense of smell, a process that can take weeks to months. The downregulation of Adcy3, involved in OSN axon guidance and OR expression stabilization, may further exacerbate sensory confusion. Long-term deficits in nuclear architecture in OSNs may have broader implications for other neuronal populations. Other mechanisms such as sustained antiviral programs, tissue vascular damage, and hypoxia may also contribute to long-lasting neurological deficits, including anosmia. These findings highlight the fragility of genomic interactions in olfactory function and suggest that olfaction could serve as an indicator for various human conditions, ranging from viral infections to neurodegeneration [62].

In summary, SARS-CoV-2 infection profoundly impacts host gene expression and immune responses, contributing to the severity of COVID-19. The identification of mutations in the nucleocapsid protein associated with increased disease severity underscores the need for further research into the molecular mechanisms underlying these effects. Understanding these processes may lead to the development of more effective therapies and vaccines for COVID-19.

A study categorized nasopharyngeal swab samples from COVID-19 patients into two groups based on the presence or absence of KR mutations in the SARS-CoV-2 nucleocapsid protein. Researchers conducted a nasopharyngeal transcriptome analysis on age, gender, and ethnicity-matched COVID-19 patients, comparing those infected with the KR mutation (KR patients, $n = 39$) and those with the wild-type N protein (RG patients, $n = 39$) to healthy controls ($n = 34$). The study further explored the impact of the KR mutation on the immune response through transcriptomic and proteomic profiling of cells exposed to virus-like particles (VLPs). KR patients exhibited significantly higher expression levels of proinflammatory cytokines, chemokines, and interferon-stimulated genes (ISGs) compared to RG patients. The nasopharyngeal transcriptome data revealed that KR patients had markedly elevated neutrophil levels and a higher neutrophil-to-lymphocyte ratio (NLR) than RG patients. Additionally, transcriptomic and proteomic profiling of VLP-exposed cells demonstrated a similar hyper-inflammatory response driven by the KR variant [62].

These findings reveal a previously unrecognized link between KR mutations in the nucleocapsid protein and an enhanced inflammatory immune response in severe COVID-19 cases, offering valuable insights into how SARS-CoV-2 mutations affect host immune responses and disease progression, which could inform more effective therapeutic and vaccine strategies. Since the discovery of SARS-CoV-2 and its rapid global spread, new Variants of Concern (VOC) have emerged. Understanding the impact of these variants on host response and disease progression is critical for vaccine and treatment development. Syrian golden hamsters, a preferred small animal model for studying severe COVID-19, were inoculated intranasally with either the original virus (nCoV-WA1-2020) or VOCs first identified in the UK (B.1.1.7, Alpha) and South Africa (B.1.351, Beta). Researchers then analyzed viral loads and host responses [46].

All three variants caused similar pulmonary lesions, both gross and histopathologic. Although differences in viral genomic copy numbers were observed in the lungs and oral swabs, infectious titers in the lungs were comparable across variants. The effectiveness of antibodies in neutralizing the virus varied based on the initial challenge virus, affecting cross-variant protection. Transcriptional profiling of lung samples 4 days post-challenge (DPC) showed significant activation of antiviral pathways in response to all variants, with a stronger inflammatory response noted for B.1.1.7. No additional spike protein mutations were detected at 4 DPC. While there were no significant differences in disease severity and viral shedding, the VOCs induced distinct humoral responses and transcriptional profiles compared to the original virus. These results suggest that emerging VOCs may trigger different early immune responses or alter immune modulation [40].

Immune Evasion Strategies and Host Genetic Determinants of COVID-19

Various immune evasion strategies are observed to be utilized by SARS-CoV-2 to penetrate the host immune defense and initiate proliferation and spread. These strategies include modulation of host antiviral pathways, inhibition of interferon (IFN) responses, and suppression of antigen presentation and T cell activation. Host genetic determinants, such as polymorphisms in genes encoding viral identification receptors (e.g., Toll-like receptors), its downstream signal molecules (e.g., IRF3, NF-κB), and interferon-stimulated genes (e.g., IFITM3, OAS1), may influence susceptibility to severe COVID-19 outcomes by affecting the efficiency of viral clearance and the magnitude of inflammatory responses [59, 60]. Understanding the genetic basis of host susceptibility to SARS-CoV-2 infection and immune evasion mechanisms is key for developing targeted therapies and vaccines against COVID-19.

Human Coronaviruses (HCoV) are a recurring global concern, posing potential threats to human health, with notable examples including SARS-CoV-2, the causative virus for COVID-19. Efforts to combat the current SARS-CoV-2 outbreak have focused on targeting various viral proteins for therapeutic purposes. However, endeavors aimed at inhibiting these proteins through drug and vaccine development have faced significant challenges. Evidence suggests that coronaviruses are capable of rapid genetic recombination, leading to the emergence of new strains with altered virulence. Moreover, they can evade both the host's antiviral defenses and humoral immune responses, resulting in severe illness such as cytokine storms [12]. Understanding the phenotypic and genotypic characteristics, as well as the pathogenesis of SARS-CoV-2, is vital for the improvement of effective therapies. Given the limited clinical data on COVID-19 pathogenesis, comparing immunological responses associated with previous pandemic HCoVs can offer valuable insights. This review aims to outline the potential origins and transmission modes of coronaviruses, as well as the current understanding of viral genome integrity in relation to SARS-CoV-2. Additionally, it explores host immune responses and viral evasion strategies based on available clinical evidence, with the ultimate goal of informing the development of therapeutics capable of targeting a broad spectrum of coronaviruses [29]. A weakened interferon (IFN) response to SARS-CoV-2 infection has been recognized as a contributing factor to severe COVID-19 cases. To identify the specific molecular elements involved in IFN's regulation of SARS-CoV-2 infection, an in-depth study was undertaken. This analysis involved investigating how human IFN-stimulated genes (ISGs) influence viral replication. The results revealed that only a subset of ISGs effectively controlled viral infection. These included factors related to endosomes that hindered viral entry, RNA-binding proteins that suppressed viral RNA synthesis, and a cluster of ISGs located in the endoplasmic reticulum and Golgi apparatus, which inhibited viral assembly and release. Some ISGs displayed broad antiviral effects, while others specifically targeted SARS-CoV-2 and its related virus, SARS-CoV-1. Notably, one of these broad-acting ISGs, BST2/tetherin, impeded viral release but was countered by the SARS-CoV-2 Orf7a protein. These findings shed light on the key ISGs involved in the innate immune response against SARS-CoV-2/SARS-CoV-1 infection. Understanding these host factors can provide insights into disease severity and may lead to the development of potential therapeutic interventions for COVID-19 [16, 34]. Severe acute respiratory syndrome coronavirus 2 (SARS-CoV-2) exhibits distinct genetic variations across different regions, which can influence its transmissibility and the severity of coronavirus disease 2019 (COVID-19). These variations, along with genetic differences among human populations, can shape the evolution of the virus. Recent SARS-CoV-2 variants have mutations in the spike (S) glycoprotein's receptor binding domain, affecting its interaction with angiotensin 1-converting enzyme 2 (ACE2) and leading to immune evasion. Since ACE2 and transmembrane protease serine 2 (TMPRSS2) are crucial for viral entry, genetic differences in these proteins may drive positive selection in the SARS-CoV-2 S glycoprotein. Additionally, intercellular adhesion molecule-3-grabbing nonintegrin, found in dendritic and liver/lymph cells, plays a role in various pathogens, and genetic variations in these host proteins could influence susceptibility to SARS-CoV-2. This review consolidates recent findings on the impact of genetic variations in the viral S glycoprotein and key host proteins, aiming to enhance the understanding of transmission and pathogenesis. Furthermore, it seeks to inform the development of vaccines, antiviral drugs, and precision medicine strategies, particularly for populations at higher genetic risk [65]. This study utilized a comprehensive approach to analyze the humoral immune responses to SARS-CoV-2. By employing a high-density peptide array covering the entire proteomes of both SARS-CoV-2 and other human coronaviruses made it possible to pinpoint specific B cell epitopes and understand their evolutionary and structural characteristics. This enabled to identify regions where pre-existing immunity exists and to uncover cross-reactive epitopes that contribute to the overall immune response against SARS-CoV-2 [30]. By analyzing a large dataset containing over 38,000 viral genomes from the early stages of the pandemic, it was investigated how these epitopes evolve and vary across different proteins, virus waves, and SARS-CoV-2 variants. These findings revealed that mutations occurring in epitopes located in the spike and nucleocapsid proteins are subject to stronger selection pressure between hosts rather than within hosts. This suggests that the majority of evolutionary changes aimed at evading the immune system occur during transmission between individuals [37]. New genetic variants of SARS-CoV-2 continue to arise as the COVID-19 pandemic persists. The Omicron variant, designated as B.1.1.529, has gained notable attention due to its rapid spread and heightened infectivity, quickly becoming the predominant variant in recent times. The spike glycoprotein of SARS-CoV-2 plays a crucial role in attaching to and entering host cells through angiotensin-converting enzyme 2 receptors. The Omicron variant carries a cluster of mutations, with over 30 changes found in the spike protein alone. Some of these mutations, such as K417N, T478K, N501Y, and P681H, are shared with previous variants like Alpha, Beta, Gamma, or Delta, and have been linked to increased transmissibility, infectivity, and potential evasion of immune responses (Fig. 3.2). Studies have indicated that the Omicron variant displays partial resistance to neutralizing antibodies from therapies and convalescent sera, posing challenges for vaccine effectiveness and antibody treatments. This review offers a comprehensive analysis of Omicron's epidemiology and immune evasion mechanisms, along with suggesting potential therapeutic approaches against this variant. The aim is to provide valuable insights for ongoing research endeavors aimed at mitigating the impact of new variants during the pandemic [1]. Despite the progress made by global vaccination campaigns in curbing the spread of COVID-19, the emergence of new SARS-CoV-2 variants capable of bypassing existing immunity has posed challenges. The continuous mutation and evolution of the virus underscore the need for ongoing efforts in vaccine development. The Spike protein of SARS-CoV-2 has been a primary focus for COVID-19 vaccine development, yet it is also a frequent target for mutations directly impacting host susceptibility and

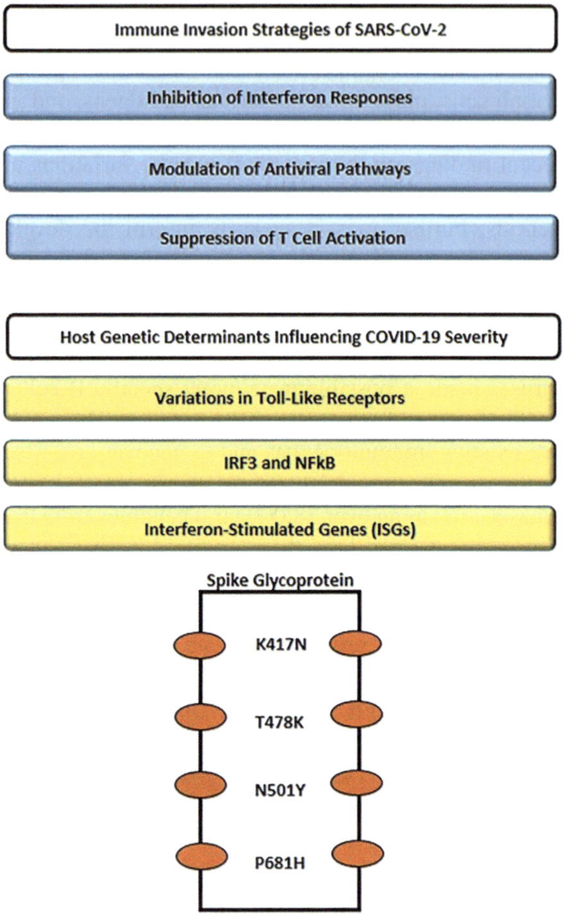

Fig. 3.2 The key points related to SARS-CoV-2 immune evasion strategies. Host genetic determinants influencing COVID-19 severity, and the impact of genetic variability in the spike glycoprotein on virus transmissibility and immune evasion

immune evasion. Predicting emerging variants and identifying conserved epitopes are crucial for developing broad-spectrum therapies or universal vaccines. In a study, general patterns of immune responses triggered by COVID-19 vaccines, with a focus on the Spike protein's immunological epitopes associated with generating protective immunity post-vaccination, were investigated. The study analyzed the structural and evolutionary features of the SARS-CoV-2 Spike protein that influence immune activation and function through toll-like receptors (TLRs), B cells, and T cells. Here, the aim was to offer a comprehensive analysis of Spike protein immune epitopes, thereby contributing to the exploration of novel strategies for broad neutralization or universal vaccination [33]. The gene Orf8 presents a unique distinction in genome organization between SARS-CoV-2 and its predecessor SARS-CoV-1, adding to the complexity of the coronavirus lineage.

Through extensive sequence comparisons, significant similarities have been identified between the Orf8 protein of SARS-CoV-2 and its counterparts in

SARS-CoV-1, as well as with X4-like genes present in other coronaviruses. This includes a notable paralog gene, Orf7a, which produces a potential immune antagonist with a well-characterized structure. The analysis reveals that, despite their distant sequence similarities, these genes show subtle conservation patterns within a newly identified superfamily, characterized by an immunoglobulin-like beta sandwich structure. Interestingly, while Orf7a and Orf8 in SARS-CoV-1 are more similar to sequences from bat viruses, their counterparts in SARS-CoV-2 are more closely related to pangolin virus homologs, indicating distinct conservation patterns in SARS-CoV-2 genomes. The divergence between Orf7a and Orf8 is particularly notable, with Orf7a appearing more constrained and Orf8 undergoing more frequent changes. This distinctive feature suggests possible unknown viral infection strategies associated with Orf8. Despite their common origins, Orf7a and Orf8 follow different evolutionary paths within the coronavirus lineage, likely due to their complex interactions with mammalian host cells. These interactions are reflected in the varied functional roles of Orf8 in SARS-CoV-2 compared to Orf7a [36].

Therapeutic Implications and Future Directions

Since its emergence in 2019, the COVID-19 pandemic has caused global disruption and is considered the most significant epidemic of the twenty-first century. Despite the introduction of Remdesivir as the first antiviral drug for COVID-19, effective treatments for SARS-CoV-2 infection remain limited. Recently, the FDA-approved drug Paxlovid, a combination of Nirmatrelvir and Ritonavir, has shown promise against the SARS-CoV-2 main protease (Mpro). Additionally, epigenetic-targeting drugs (epi-drugs) have demonstrated significant inhibitory potential against SARS-CoV-2 Mpro. These epi-drugs, which are traditionally used in combination with cancer treatments to regulate gene expression, were systematically screened for their ability to inhibit SARS-CoV-2 Mpro using advanced molecular docking programs and free energy calculations. Several epi-drugs exhibited binding affinities to SARS-CoV-2 Mpro comparable to Nirmatrelvir. Notably, the histone methyltransferase inhibitor EPZ005687 and the DNA methyltransferase inhibitor Guadecitabine emerged as the most promising candidates. Extensive molecular dynamics (MD) simulations and subsequent molecular mechanics/generalized born surface area (MM-GBSA) calculations confirmed the stability of the EPZ005687-Mpro complex, with a binding free energy (ΔGbind) of −48.2 kcal/mol for EPZ005687 compared to −44.7 kcal/mol for Nirmatrelvir. These findings highlight the untapped antiviral potential of epi-drugs (Table 3.1), which could be especially beneficial for patients with chronic immunosuppressive disorders, expanding their use beyond traditional cancer therapy [53].

The modulation of host genetic and epigenetic mechanisms by SARS-CoV-2 infection offers new insights into potential therapeutic targets for reducing COVID-19 severity and improving patient outcomes. Targeted strategies that restore immune balance, block viral entry and replication, and reduce hyperinflammatory responses hold promise for COVID-19 treatment. Moreover, personalized medicine

Table 3.1 Describing the drugs against SARS-CoV-2 modulated epigenetic changes

Sl. no.	Drug	Mechanism of action	Key findings	Citations
1	Decitabine	DNA methyltransferase inhibitor	Inhibits SARS-CoV-2 replication by demethylation of viral RNA	[31]
2	Azacitidine	DNA methyltransferase inhibitor	Reduces viral load and inflammatory response in cell models	[2]
3	Vorinostat (SAHA)	Histone deacetylase inhibitor	Suppresses SARS-CoV-2 replication by modulating host gene expression	[18]
4	Romidepsin	Histone deacetylase inhibitor	Inhibits viral replication and cytokine storm in preclinical studies	[56]
5	Belinostat	Histone deacetylase inhibitor	Demonstrates antiviral activity and immune modulation in vitro	[61]
6	Givinostat	Histone deacetylase inhibitor	Shows potential in reducing inflammation and viral replication in models	[50]
7	Entinostat	Class I histone deacetylase inhibitor	Reduces inflammatory cytokine production and viral replication in vitro	[6]
8	Mocetinostat	Class I/IV histone deacetylase inhibitor	Decreases expression of ACE2, the receptor for SARS-CoV-2 entry	[19]
9	Zebularine	DNA methylation inhibitor	Inhibits SARS-CoV-2 by targeting viral epigenetic machinery	[43]
10	Trichostatin A	Histone deacetylase inhibitor	Blocks SARS-CoV-2 replication and inflammatory responses	[49]

approaches, tailored to individual genetic and epigenetic profiles, could help identify high-risk populations and optimize treatment strategies. Future research should focus on understanding the long-term effects of SARS-CoV-2 on host epigenetic regulation, exploring the role of host genetics in vaccine-induced immunity, and developing precision medicine approaches to combat emerging viral variants [17].

COVID-19 continues to pose a significant global threat, particularly to the elderly and those with preexisting conditions. While vaccines are available, effective therapeutic drugs are crucial for controlling the outbreak and preventing severe health consequences. Epigenetic modifications, which regulate gene expression by altering chromatin structure, have been linked to viral pathogenesis. These modifications influence both the virus's life cycle and the host's immune response, making epigenetic drugs promising candidates for COVID-19 treatment. Vitamin D, a multifunctional secosteroid hormone, is another area of interest, as its deficiency is a global health concern. Vitamin D and its receptor play a role in regulating genes involved in immunity, apoptosis, proliferation, differentiation, and inflammation. Evidence suggests that vitamin D, modulated by epigenetic mechanisms, may reduce disease risk and serve as a therapeutic agent for COVID-19. This review underscores the epigenetic impacts of COVID-19 and explores the potential role of vitamin D in treating infections.

Previous research has identified angiotensin-converting enzyme 2 (ACE2) as the key cell-surface receptor for both SARS-CoV and SARS-CoV-2, playing a crucial role in viral entry and infection. An in silico investigation of ACE2's mechanism in the lungs identified potential new drugs for COVID-19 treatment. Gene expression profiles from public datasets, including Gene Expression Omnibus, The Cancer Genome Atlas, Genotype-Tissue Expression, and Gene Ontology, were analyzed through pathway enrichment to explore the primary functions of ACE2-related genes. This led to the development of a protein-protein interaction network including genes co-expressed with ACE2, which was then linked to established drugs to assess their potential for COVID-19 therapy. The analysis indicated that sterol biosynthesis, aryl-dialkyl-phosphatase activity, adenosylhomocysteinase activity, trialkylsulfonium hydrolase activity, and acetate-CoA and CoA ligase activity are predominantly associated with genes related to ACE2. A network comprising 193 genes, 222 interactions, and 36 potential drugs was identified, with Nimesulide, Fluticasone Propionate, Thiabendazole, Photofrin, Didanosine, and Flutamide emerging as promising candidates for COVID-19 treatment [10].

A drug repurposing strategy was employed using a library of antiviral drugs and FDA-approved drugs from Selleckchem Inc. and the DrugBank database, respectively. Computational techniques, including molecular docking with LeadIT FlexX software and 100 ns molecular dynamics simulations, were utilized to assess the conformational stability of the drug-Mpro complexes. Among the FDA-approved drugs, Mitoxantrone and Leucovorin, and from the antiviral drug library, Birinapant and Dynasore, demonstrated high interaction efficiency with SARS-CoV-2 Mpro due to enhanced steric and hydrophobic interactions within the binding cavity, forming stable complexes. The 100 ns molecular dynamics simulations indicated a mean RMSD value of 2.25 Å for all complex systems, suggesting that the lead compounds bind tightly within the Mpro cavity, maintaining conformational stability. Glutamic acid (Glu166) in Mpro plays a key role in stabilizing these complexes by forming hydrogen bonds and salt bridges with the lead compounds. These findings suggest that Mitoxantrone, Leucovorin, Birinapant, and Dynasore are potential inhibitors of SARS-CoV-2 Mpro [10, 32].

Conclusion

In conclusion, the modulation of host genetic and epigenetic machinery by SARS-CoV-2 infection plays a crucial role in determining COVID-19 susceptibility, severity, and outcomes. Understanding the molecular mechanisms underlying host-virus interactions is essential for developing effective therapeutic strategies and mitigating the impact of the COVID-19 pandemic. Future research efforts aimed at unravelling the complex interplay between viral and host factors hold promise for advancing the knowledge of COVID-19 pathogenesis and facilitating the development of targeted interventions to combat the disease. Targeted therapies may be developed that complement existing antiviral strategies by deciphering these intricate mechanisms. Epigenetic regulation holds the key to unraveling the mysteries of COVID-19 pathogenesis and improving patient outcomes.

Acknowledgments I would like to acknowledge Dr. G.M. Helaluddin, Honourable Principal Michael Madhusudan Memorial College; my colleagues in the dept. of Microbiology, Michael Madhusudan Memorial College, Durgapur, West Bengal, India, for continuous support and encouragement. Dr. Mrinal Kanti Ghosh, Chief Scientist, CSIR-IICB, for giving opportunity to write this chapter.

Author's Contributions Dr. Kirat Kumar Ganguly himself contributed the chapter for data collection, analysis, figure preparation, and writing this chapter.

Data Availability Data were mostly procured from open-accessible journals and internet-based databases.

Declarations

Conflicts of Interests There is no conflict of interest in this chapter.

References

1. Ao D, Lan T, He X, Liu J, Chen L, Baptista-Hon DT, Zhang K, Wei X. SARS-CoV-2 Omicron variant: immune escape and vaccine development. MedComm (2020). 2022;3(1):e126. https://doi.org/10.1002/mco2.126. PMID: 35317190; PMCID: PMC8925644.
2. Lin X, Ke X, Jian X, et al. Azacytidine targeting SARS-CoV-2 viral RNA as a potential treatment for COVID-19. Sci Bull (Beijing). 2022;67(10):1022–25. https://doi.org/10.1016/j.scib.2022.02.002.
3. Beacon TH, Delcuve GP, Davie JR. Epigenetic regulation of ACE2, the receptor of the SARS-CoV-2 virus[1]. Genome. 2021;64(4):386–99. https://doi.org/10.1139/gen-2020-0124. Epub 2020 Oct 21. PMID: 33086021.
4. Behrouj H, Vakili O, Sadeghdoust A, Aligolighasemabadi N, Khalili P, Zamani M, Mokarram P. Epigenetic regulation of autophagy in coronavirus disease 2019 (COVID-19). Biochem Biophys Rep. 2022;30:101264. https://doi.org/10.1016/j.bbrep.2022.101264. Epub 2022 Apr 21. PMID: 35469237; PMCID: PMC9021360.
5. Blanco-Melo D, et al. Imbalanced host response to SARS-CoV-2 drives development of COVID-19. Cell. 2020;181(5):1036–1045.e9. https://doi.org/10.1016/j.cell.2020.04.026.
6. Shah VK, Firmal P, Alam A, Ganguly D, Chattopadhyay S. Overview of Immune Response During SARS-CoV-2 Infection: Lessons From the Past. Front Immunol. 2020;11:1949. Published 2020 Aug 7. https://doi.org/10.3389/fimmu.2020.01949.
7. Brann DH, et al. Non-neuronal expression of SARS-CoV-2 entry genes in the olfactory system suggests mechanisms underlying COVID-19-associated anosmia. Sci Adv. 2020;6(31):eabc5801. https://doi.org/10.1126/sciadv.abc5801.
8. Butler-Laporte G, Nakanishi T, Mooser V, Morrison DR, Abdullah T, Adeleye O, Mamlouk N, Kimchi N, Afrasiabi Z, Rezk N, Giliberti A, Renieri A, Chen Y, Zhou S, Forgetta V, Richards JB. Vitamin D and COVID-19 susceptibility and severity in the COVID-19 Host Genetics Initiative: a Mendelian randomization study. PLoS Med. 2021;18(6):e1003605. https://doi.org/10.1371/journal.pmed.1003605. PMID: 34061844; PMCID: PMC8168855.

9. Cao W, Fang Z, Hou G, Han M, Xu X, Dong J, Zheng J. The psychological impact of the COVID-19 epidemic on college students in China. Psychiatry Res. 2020;287:112934. https://doi.org/10.1016/j.psychres.2020.112934. Epub 2020 Mar 20. PMID: 32229390; PMCID: PMC7102633.
10. Cava C, Bertoli G, Castiglioni I. In silico discovery of candidate drugs against covid-19. Viruses. 2020;12(4):404. https://doi.org/10.3390/v12040404. PMID: 32268515; PMCID: PMC7232366.
11. Channappanavar R, Perlman S. Pathogenic human coronavirus infections: causes and consequences of cytokine storm and immunopathology. Semin Immunopathol. 2017;39(5):529–39. https://doi.org/10.1007/s00281-017-0629-x.
12. Singh S, Singh KK. Valproic Acid in Prevention and Treatment of COVID-19. Int J Respir Pulm Med. 2020;7:138. https://doi.org/10.23937/2378-3516/1410138.
13. Choi B, et al. Persistence and evolution of SARS-CoV-2 in an immunocompromised host. N Engl J Med. 2020;383(23):2291–3. https://doi.org/10.1056/nejmc2031364.
14. Dey A, Vaishak K, Deka D, Radhakrishnan AK, Paul S, Shanmugam P, Daniel AP, Pathak S, Duttaroy AK, Banerjee A. Epigenetic perspectives associated with COVID-19 infection and related cytokine storm: an updated review. Infection. 2023;51(6):1603–18. https://doi.org/10.1007/s15010-023-02017-8. Epub 2023 Mar 12. PMID: 36906872; PMCID: PMC10008189.
15. Ellinghaus D, Degenhardt F, Bujanda L, Severe Covid-19 GWAS Group, et al. Genomewide association study of severe covid-19 with respiratory failure. N Engl J Med. 2020;383(16):1522–34. https://doi.org/10.1056/nejmoa2020283. Epub 2020 Jun 17. PMID: 32558485; PMCID: PMC7315890.
16. Kentaro Uemura, Haruaki Nobori, Akihiko Sato, Takao Sanaki, Shinsuke Toba, Michihito Sasaki, Akiho Murai, Noriko Saito-Tarashima, Noriaki Minakawa, Yasuko Orba, Hiroaki Kariwa, William W. Hall, Hirofumi Sawa, Akira Matsuda, Katsumi Maenaka, 5-Hydroxymethyltubercidin exhibits potent antiviral activity against flaviviruses and coronaviruses, including SARS-CoV-2, iScience. 2021;24(10):103120. ISSN 2589-0042. https://doi.org/10.1016/j.isci.2021.103120.
17. Foolchand A, Mazaleni S, Ghazi T, Chuturgoon AA. A review: highlighting the links between epigenetics, COVID-19 infection, and vitamin D. Int J Mol Sci. 2022;23(20):12292. https://doi.org/10.3390/ijms232012292. PMID: 36293144; PMCID: PMC9603374.
18. Hegde S, Tang Z, Zhao J, Wang J. Inhibition of SARS-CoV-2 by Targeting Conserved Viral RNA Structures and Sequences. Front Chem. 2021;9:802766. Published 2021 Dec 23. https://doi.org/10.3389/fchem.2021.802766.
19. Rosal RJZ, Paderes MC. Inhibiting SARS-CoV-2 viral entry by targeting spike:ACE2 interaction with O-modified quercetin derivatives. RSC Med Chem. Published online August 16, 2024. https://doi.org/10.1039/d4md00286e.
20. Hoffmann M, et al. SARS-CoV-2 cell entry depends on ACE2 and TMPRSS2 and is blocked by a clinically proven protease inhibitor. Cell. 2020;181(2):271–280.e8. https://doi.org/10.1016/j.cell.2020.02.052.
21. Huang C, et al. Clinical features of patients infected with 2019 novel coronavirus in Wuhan, China. Lancet. 2020;395(10223):497–506. https://doi.org/10.1016/s0140-6736(20)30183-5.
22. Huang SW, Wang SF. SARS-CoV-2 Entry Related Viral and Host Genetic Variations: Implications on COVID-19 Severity, Immune Escape, and Infectivity. Int J Mol Sci. 2021;22(6):3060. https://doi.org/10.3390/ijms22063060. Published 2021 Mar 17.
23. Hussain M, et al. Structural variations in human ACE2 may influence its binding with SARS-CoV-2 spike protein. J Med Virol. 2020;92(9):1580–6. https://doi.org/10.1002/jmv.25832.
24. Jiang F, Deng L, Zhang L, Cai Y, Cheung CW, Xia Z. Review of the clinical characteristics of coronavirus disease 2019 (COVID-19). J Gen Intern Med. 2020;35(5):1545–9. https://doi.org/10.1007/s11606-020-05762-w. Epub 2020 Mar 4. PMID: 32133578; PMCID: PMC7088708.

25. Justice JN, et al. Cellular senescence biomarker p16INK4a+ cell burden in thigh adipose is associated with poor physical function in older women. J Gerontol Ser A. 2019;74(3):385–92. https://doi.org/10.1093/gerona/gly220.
26. Kazantseva A, Enikeeva R, Takhirova Z, Davydova Y, Mustafin R, Malykh S, Karunas A, Kanapin A, Khusnutdinova E. Host genetic variants linked to COVID-19 neurological complications and susceptibility in young adults-a preliminary analysis. J Pers Med. 2023;13(1):123. https://doi.org/10.3390/jpm13010123. PMID: 36675784; PMCID: PMC9860613.
27. Kee J, Thudium S, Renner DM, Glastad K, Palozola K, Zhang Z, Li Y, Lan Y, Cesare J, Poleshko A, Kiseleva AA, Truitt R, Cardenas-Diaz FL, Zhang X, Xie X, Kotton DN, Alysandratos KD, Epstein JA, Shi PY, Yang W, Morrisey E, Garcia BA, Berger SL, Weiss SR, Korb E. SARS-CoV-2 disrupts host epigenetic regulation via histone mimicry. Nature. 2022;610(7931):381–8. https://doi.org/10.1038/s41586-022-05282-z. Epub 2022 Oct 5. Erratum in: Nature. 2023 Jan;613(7945):E5. Erratum in: Nature. 2023 Feb;614(7949):E44. PMID: 36198800; PMCID: PMC9533993.
28. Kirkland JL, Tchkonia T. Cellular senescence: a translational perspective. EBioMedicine. 2017;21:21–8. https://doi.org/10.1016/j.ebiom.2017.04.013.
29. Liu K, Zhu Y, Cao X, et al. Curcumin as an antiviral agent and immune-inflammatory modulator in COVID-19: A scientometric analysis. Heliyon. 2023;9(11):e21648. Published 2023 Nov 2. https://doi.org/10.1016/j.heliyon.2023.e21648.
30. Bhat S, Rishi P, Chadha VD. Understanding the epigenetic mechanisms in SARS CoV-2 infection and potential therapeutic approaches. Virus Res. 2022;318:198853. https://doi.org/10.1016/j.virusres.2022.198853.
31. Li, et al. Decitabine suppresses SARS-CoV-2 replication by epigenetic modulation. J Mol Med. 2021.
32. Lokhande KB, Doiphode S, Vyas R, Swamy KV. Molecular docking and simulation studies on SARS-CoV-2 Mpro reveals Mitoxantrone, Leucovorin, Birinapant, and Dynasore as potent drugs against COVID-19. J Biomol Struct Dyn. 2021;39(18):7294–305. https://doi.org/10.1080/07391102.2020.1805019. Epub 2020 Aug 20. PMID: 32815481; PMCID: PMC7484567.
33. Magazine N, Zhang T, Bungwon AD, McGee MC, Wu Y, Veggiani G, Huang W. Immune epitopes of SARS-CoV-2 spike protein and considerations for universal vaccine development. Immunohorizons. 2024;8(3):214–26. https://doi.org/10.4049/immunohorizons.2400003. PMID: 38427047; PMCID: PMC10985062.
34. Martin-Sancho L, Lewinski MK, Pache L, Stoneham CA, Yin X, Becker ME, Pratt D, Churas C, Rosenthal SB, Liu S, Weston S, De Jesus PD, O'Neill AM, Gounder AP, Nguyen C, Pu Y, Curry HM, Oom AL, Miorin L, Rodriguez-Frandsen A, Zheng F, Wu C, Xiong Y, Urbanowski M, Shaw ML, Chang MW, Benner C, Hope TJ, Frieman MB, García-Sastre A, Ideker T, Hultquist JF, Guatelli J, Chanda SK. Functional landscape of SARS-CoV-2 cellular restriction. Mol Cell. 2021;81(12):2656–2668.e8. https://doi.org/10.1016/j.molcel.2021.04.008. Epub 2021 Apr 13. PMID: 33930332; PMCID: PMC8043580.
35. Merad M, Martin JC. Pathological inflammation in patients with COVID-19: a key role for monocytes and macrophages. Nat Rev Immunol. 2020;20(6):355–62. https://doi.org/10.1038/s41577-020-0331-4.
36. Neches RY, Kyrpides NC, Ouzounis CA. Atypical divergence of SARS-CoV-2 Orf8 from Orf7a within the coronavirus lineage suggests potential stealthy viral strategies in immune evasion. MBio. 2021;12(1):e03014–20. https://doi.org/10.1128/mBio.03014-20. PMID: 33468697; PMCID: PMC7845636.
37. N'Guessan A, Kailasam S, Mostefai F, Poujol R, Grenier JC, Ismailova N, Contini P, De Palma R, Haber C, Stadler V, Bourque G, Hussin JG, Shapiro BJ, Fritz JH, Piccirillo CA. Selection for immune evasion in SARS-CoV-2 revealed by high-resolution epitope mapping and sequence analysis. iScience. 2023;26(8):107394. https://doi.org/10.1016/j.isci.2023.107394. PMID: 37599818; PMCID: PMC10433132.
38. Nguyen VT, Nguyen MQ, Le NT, Nguyen TNH, Huynh G. Predictors of intention to get a COVID-19 vaccine of health science students: a cross-sectional study. Risk Manag Healthc

Policy. 2021;14:4023–30. https://doi.org/10.2147/RMHP.S328665. PMID: 34594141; PMCID: PMC8478496.
39. Novelli A, et al. HLA allele frequencies and susceptibility to COVID-19 in a group of 99 Italian patients. HLA. 2020;96(5):610–4. https://doi.org/10.1111/tan.14047.
40. O'Donnell KL, Pinski AN, Clancy CS, Gourdine T, Shifflett K, Fletcher P, Messaoudi I, Marzi A. Pathogenic and transcriptomic differences of emerging SARS-CoV-2 variants in the Syrian golden hamster model. EBioMedicine. 2021;73:103675. https://doi.org/10.1016/j.ebiom.2021.103675. Epub 2021 Nov 7. PMID: 34758415; PMCID: PMC8572342.
41. Olagnier D, et al. Inhibition of SARS-CoV-2 infection by the antiviral topoisomerase I inhibitor topotecan. Int J Infect Dis. 2020;103:732–9. https://doi.org/10.1016/j.ijid.2020.12.057.
42. Kgatle MM, Lawal IO, Mashabela G, et al. COVID-19 Is a Multi-Organ Aggressor: Epigenetic and Clinical Marks. Front Immunol. 2021;12:752380. Published 2021 Oct 8. https://doi.org/10.3389/fimmu.2021.752380.
43. Patel, et al. Zebularine targets viral epigenetic machinery to inhibit SARS-CoV-2. J Virol. 2021.
44. Schön E, et al. CD26 in immunity and disease. Biomolecules. 2021;11(1):147. https://doi.org/10.3390/biom11010147.
45. Shang J, et al. Structural basis of receptor recognition by SARS-CoV-2. Nature. 2020;581(7807):221–4. https://doi.org/10.1038/s41586-020-2179-y.
46. Shuaib M, Adroub S, Mourier T, Mfarrej S, Zhang H, Esau L, Alsomali A, Alofi FS, Ahmad AN, Shamsan A, Khogeer A, Hashem AM, Almontashiri NAM, Hala S, Pain A. Impact of the SARS-CoV-2 nucleocapsid 203K/204R mutations on the inflammatory immune response in COVID-19 severity. Genome Med. 2023;15(1):54. https://doi.org/10.1186/s13073-023-01208-0. PMID: 37475040; PMCID: PMC10360309.
47. Siddiqi HK, Mehra MR. COVID-19 illness in native and immunosuppressed states: a clinical-therapeutic staging proposal. J Heart Lung Transplant. 2020;39(5):405–7. https://doi.org/10.1016/j.healun.2020.03.012.
48. Bouyahya A, El Omari N, Bakha M, Aanniz T, El Menyiy N, El Hachlafi N, El Baaboua A, El-Shazly M, Alshahrani MM, Al Awadh AA, Lee L.-H, Benali T, Mubarak MS. Pharmacological Properties of Trichostatin A, Focusing on the Anticancer Potential: A Comprehensive Review. Pharmaceuticals. 2022;15(10):1235. https://doi.org/10.3390/ph15101235.
49. Mousavi S, Zare S, Mirzaei M, Feizi A. Novel Drug Design for Treatment of COVID-19: A Systematic Review of Preclinical Studies. Can J Infect Dis Med Microbiol. 2022;2022:2044282. https://doi.org/10.1155/2022/2044282. Published 2022 Sep 25.
50. Sinha, et al. Givinostat shows promise in reducing COVID-19 inflammation and replication. Virol J. 2021.
51. Sodagar H, Alipour S, Hassani S, Aziz SG, Ansari MHK, Asghari R. The role of microRNAs in COVID-19 with a focus on miR-200c. J Circ Biomark. 2022;11:14–23. https://doi.org/10.33393/jcb.2022.2356. PMID: 35356072; PMCID: PMC8939267.
52. Stopsack KH, et al. TMPRSS2 and COVID-19: serendipity or opportunity for intervention? Cancer Discov. 2020;10(6):779–82. https://doi.org/10.1158/2159-8290.cd-20-0451.
53. Uzuner U, Akkus E, Kocak A, Çelik Uzuner S. Exploring epigenetic drugs as potential inhibitors of SARS-CoV-2 main protease: a docking and MD simulation study. J Biomol Struct Dyn. 2023:1–12. https://doi.org/10.1080/07391102.2023.2236714. Epub ahead of print. PMID: 37458994.
54. van der Made CI, et al. Presence of genetic variants among young men with severe COVID-19. JAMA. 2020;324(7):663–73. https://doi.org/10.1001/jama.2020.13719.
55. Liu K, Zou R, Cui W, et al. Clinical HDAC Inhibitors Are Effective Drugs to Prevent the Entry of SARS-CoV2. ACS Pharmacol Transl Sci. 2020;3(6):1361-1370. Published 2020 Nov 11. https://doi.org/10.1021/acsptsci.0c00163.
56. Zang, et al. Romidepsin's role in controlling SARS-CoV-2 replication and cytokine storm. J Clin Invest. 2021.
57. Zepeda-Cervantes J, Martínez-Flores D, Ramírez-Jarquín JO, Tecalco-Cruz ÁC, Alavez-Pérez NS, Vaca L, Sarmiento-Silva RE. Implications of the immune polymorphisms of the

host and the genetic variability of SARS-CoV-2 in the development of COVID-19. Viruses. 2022;14(1):94. https://doi.org/10.3390/v14010094. PMID: 35062298; PMCID: PMC8778858.
58. Zhang Y, Ma ZF. Impact of the COVID-19 pandemic on mental health and quality of life among local residents in Liaoning Province, China: a cross-sectional study. Int J Environ Res Public Health. 2020;17(7):2381. https://doi.org/10.3390/ijerph17072381. PMID: 32244498; PMCID: PMC7177660.
59. Zhang Q, et al. Inborn errors of type I IFN immunity in patients with life-threatening COVID-19. Science. 2020;370(6515):eabd4570. https://doi.org/10.1126/science.abd4570.
60. Gil C, Ginex T, Maestro I, Nozal V, Barrado-Gil L, Cuesta-Geijo MÁ, Urquiza J, Ramírez D, Alonso C, Campillo NE, Martinez A. COVID-19: drug targets and potential treatments. Journal of medicinal chemistry 2020:63(21):12359–86. https://doi.org/10.1021/acs.jmedchem.0c00606.
61. Zhang, et al. Belinostat as a potential antiviral agent against COVID-19. Antivir Res. 2021.
62. Zhang X, et al. KR mutations in the SARS-CoV-2 nucleocapsid protein associated with increased disease severity. J Virol. 2021;95(3):e01850–20. https://doi.org/10.1128/JVI.01850-20.
63. Zhao J, Yuan Q, Wang H, Liu W, Liao X, Su Y, Wang X, Yuan J, Li T, Li J, Qian S, Hong C, Wang F, Liu Y, Wang Z, He Q, Li Z, He B, Zhang T, Fu Y, Ge S, Liu L, Zhang J, Xia N, Zhang Z. Antibody responses to SARS-CoV-2 in patients with novel coronavirus disease 2019. Clin Infect Dis. 2020;71(16):2027–34. https://doi.org/10.1093/cid/ciaa344. PMID: 32221519; PMCID: PMC7184337.
64. Zhou B, Perel P, Mensah GA, Ezzati M. Global epidemiology, health burden and effective interventions for elevated blood pressure and hypertension. Nat Rev Cardiol. 2021;18(11):785–802. https://doi.org/10.1038/s41569-021-00559-8. Epub 2021 May 28. PMID: 34050340; PMCID: PMC8162166.
65. Ziegler CGK, et al. SARS-CoV-2 receptor ACE2 is an interferon-stimulated gene in human airway epithelial cells and is detected in specific cell subsets across tissues. Cell. 2020;181(5):1016–1035.e19. https://doi.org/10.1016/j.cell.2020.04.035.

SARS-CoV-2 Mutants and Variants of Interest

4

Troyee Das, Byapti Ghosh, and Zhumur Ghosh

Abstract

The emergence of SARS-CoV-2 mutants and variants of interest has significantly impacted the global COVID-19 pandemic landscape. Rapid viral evolution has led to diverse variants globally, challenging containment efforts due to heightened transmissibility and potential immune evasion. Effective surveillance and containment strategies are crucial given the variants' widespread distribution. These mutations can substantially impact virus transmission and virulence, posing obstacles to public health strategies. Concerns arise regarding vaccine effectiveness against some variants, necessitating ongoing research to assess efficacy and explore the need for booster doses or updated formulations. Bridging knowledge gaps in variant understanding is critical for guiding future research and public health responses. Overall, a comprehensive understanding of SARS-CoV-2 mutants and variants is essential for preparing future viral challenges.

Keywords

COVID-19 · SARS-CoV-2 mutants & variants · Transmission · Virulence

Troyee Das and Byapti Ghosh contributed equally with all other contributors.

T. Das · B. Ghosh · Z. Ghosh (✉)
Department of Biological Sciences, Unified Academic Campus, Bose Institute, Kolkata, WB, India
e-mail: zhumur@jcbose.ac.in

M. K. Ghosh, M. Basu (eds.), *Pathogenesis, Management and Socio-Economic Impact of COVID-19 Pandemic*, Medical Virology: From Pathogenesis to Disease Control, https://doi.org/10.1007/978-981-95-1426-7_4

Introduction

The novel coronavirus known as SARS-CoV-2 emerged in late 2019 and has captured the attention of the international community, causing an unprecedented health catastrophe and testing the resilience of healthcare systems across the globe. One of the defining features of this virus is its ability to evolve rapidly through genetic mutations, giving rise to new variants with diverse characteristics.

SARS-CoV-2 belongs to the family of coronaviruses, enveloped RNA viruses known for their propensity to mutate. Genetic mutations occur naturally during the replication process of viruses, allowing them to adapt with changing environments and host immune responses. Comprehending the importance of these changes is essential for grasping the dynamics of viral transmission, the intensity of the disease, and the efficacy of preventative measures like vaccinations and treatments.

The evolutionary history of coronaviruses is multidecadal and includes a wide range of viruses that can infect both people and animals. Coronaviruses belong to the Coronaviridae family and are distinguished by having large, enveloped, positive-sense RNA genomes [1]. These viruses exhibit a remarkable capacity for genetic diversity, facilitated by their unique replication machinery and high mutation rates.

Before the advent of SARS-CoV-2, coronaviruses were mainly linked to minor respiratory diseases in humans, such as the common cold. However, notable exceptions emerged with the outbreaks of severe acute respiratory syndrome (SARS) [2] in 2002–2003 and Middle East respiratory syndrome (MERS) [3] in 2012, both brought on by new coronaviruses that breached the species barrier from animals. These zoonotic transmissions underscored the potential for coronaviruses to spark devastating pandemics with significant public health and socioeconomic implications.

The evolutionary source of SARS-CoV-2 can be traced back to zoonotic transmission events involving coronaviruses circulating in wildlife reservoirs, likely bats [4]. According to genetic investigations, SARS-CoV-2 and bat coronaviruses, especially those belonging to the genus Betacoronavirus, exhibit a great deal of sequence similarity. However, the exact intermediate host responsible for transmitting the virus to humans remains elusive, with speculation focusing on potential involvement of pangolins or other mammalian species.

The adaptive landscape of coronaviruses is characterized by continual genetic diversification driven by selective pressures exerted by host immune responses, interspecies transmission events, and environmental factors. Mutations accumulated over time can confer phenotypic changes that enhance viral fitness, transmissibility, or pathogenicity, shaping the trajectory of viral evolution and pandemic potential.

In the context of infectious diseases, the terms "mutants" and "variants" are often used interchangeably, but they denote distinct concepts. Mutants refer to individual viruses harboring genetic changes, whereas variants denote a collection of viruses sharing specific mutations or genomic characteristics. Concerns have been raised by the appearance of SARS—variants of concern (VOCs) and CoV-2 variants of interest (VOIs) because of possible effects on immune evasion, pathogenicity, and transmissibility. Monitoring and understanding these variants are essential for guiding public health strategies and vaccine development efforts.

Key Mutations and Variants

Comprehending the key mutations and variants of SARS-CoV-2 is imperative in tracking the virus's progression and evaluating its influence on public health. This section looks at significant VOCs, VOIs, and their functional significance in relation to the SARS-CoV-2 genome alterations.

Prominent Mutations Observed in the SARS-CoV-2 Genome Affecting the Structural and Nonstructural (NSP) Proteins

Several structural, nonstructural (NSP), and accessory proteins are present in the genome of SARS-CoV-2, and each is essential for the pathogenesis, host contact, and viral replication. SARS-CoV-2 structural proteins, including spike (S), envelope (E), membrane (M), and nucleocapsid (N), contribute to viral structure and replication, while the replication and transcription complex is made up of nonstructural proteins (Nsp1–Nsp16) that are transcribed from Orf1a and Orf1b [5]. Notable changes found in the SARS-CoV-2 genome have garnered attention because of their possible effect on the virus's ability to spread, its pathogenicity, and the effectiveness of vaccinations against it. These mutations, which arise spontaneously during viral replication, can result in amino acid substitutions, deletions, or insertions within viral proteins, influencing various aspects of virus-host interactions.

The investigation of global mutational landscape using over ten million SARS-CoV-2 genomic sequences from the Global Initiative on Sharing Avian Influenza Data (GISAID) database [6], highlighted six mutations exhibiting frequencies exceeding 50%. Notably, these mutations included P323L in NSP12, D614G in the S protein, T492I in NSP4, R203M in the N protein, T60A in Orf9b, and P1228L in NSP3. Interestingly, certain regions of the genome, such as NSP7-NSP11 and NSP16, exhibited no mutations exceeding 90% of the strain. Conversely, N, NSP3, Spike, NSP4, NSP12, and M proteins displayed the highest mutation rates.

Mutations in Structural Proteins

2020 saw the emergence of D614G, a missense mutation in the S protein of SARS-CoV-2 that replaced aspartic acid (D with glycine (G) at position 614. This mutation, linked to increased viral transmission and infectivity soon spread around the world. Furthermore, reports of anosmia, or loss of smell, as a symptom of COVID-19 are correlated with its global incidence. This correlation may be caused by enhanced protein stability in the olfactory epithelium or higher binding of the spike protein to the ACE2 receptor, which would boost the tissue's infectivity [7].

Another noteworthy mutation is the E484K, which occurs at position 484 as a replacement in the receptor-binding domain (RBD) of the spike protein. Because this mutation has been linked to a decreased susceptibility to neutralizing antibodies, questions have been raised about how it may affect the effectiveness of vaccinations. Variants with E484K mutation have shown resistance to convalescent plasma and certain monoclonal antibodies, emphasizing the need for continued surveillance and the modification of vaccination approaches [8].

Mutations within the N-terminal domain (NTD) of the spike protein, such as the deletion at position 69–70 (Δ69–70), have been linked to immune escape and enhanced virus binding to host cell receptors [9]. This mutation may cause "spike gene target failure" and produce a PCR viral test result that is false negative.

The deletion mutation RSYLTPGD246-253N (Δ246-253N) in NTD replaces amino acids 246 to 252 with asparagine (N) and alters position 253 from aspartic acid (D) to asparagine (N). This particular mutation is associated with the variant's increased resistance to neutralizing antibodies [10].

When combined with the E484K mutation, K417N mutation significantly boosts virus-cell fusion. This is because the E484K mutation changes carboxylic acid (E) with the amine group (K) and improves virus's electrostatic complementarity with the negatively charged ACE2. The SARS-CoV-2 may become more transmissible due to this physicochemical effect, even while its infectivity stays the same [11]. Additionally, deletion Δ144 and K417N significantly impair the antigenic peptide's capacity to bind with the pertinent human leukocyte antigen (HLA-A class I), which eliminates host CD8+ T cell responses [12]. Leucine to arginine substitution L452R at position 452 results in a higher affinity of the spike protein for the ACE2 receptor and a reduced immune system recognition capacity [13]. The mutation F490S at amino acid position 490 and is linked to decreased susceptibility to antibodies produced by individuals infected with other strains [14]. The P681R mutation, residing at a furin cleavage site separating the spike S1 and S2 subunits, increases cleavage of the full-length spike protein, leading to greater cell surface access and infection, as demonstrated by significant replication reduction upon reversion to wild-type P681 [15].

The multifunctional N protein is engaged in many phases of the viral life cycle. It forms the ribonucleoprotein complex which is necessary for both transcription and viral replication—by interacting with viral RNA. The R203K/G204R mutation within the nucleocapsid protein enhances its agglutination capacity, while the R203M variant amplifies virus replication. Furthermore, the G215C mutation enhances dimer-dimer interactions and may promote the formation of disulfide bonds between protomers. This mutation is located in the region connecting the SR-rich and leucine-rich domains. The importance of these mutations in viral pathogenesis is shown by the critical functions played by these regions in interactions with viral RNA and proteins [16].

Mutations in Nonstructural Proteins (NSP)

NSP4, sometimes referred to as main protease (Mpro) or 3C-like protease (3CLpro), is an essential enzyme for viral maturation and replication. The T492I mutation, situated adjacent to the NSP4|5 cleavage site (also termed the N-terminal autoprocessing site of NSP5), emerged early in the pandemic and has exhibited a notable rise in prevalence since May 2021 across global regions, except Asia. This substitution is pivotal in facilitating the swift dissemination and bolstering the immune evasion prowess of SARS-CoV-2 [17].

NSP9 is a small, single-stranded RNA-binding protein that is essential for viral RNA replication. Its main function is to facilitate the replication of the viral RNA genome by interacting with other viral and host factors. The mutation G104E has the potential to disrupt its protein structure, consequently impairing its function and leading to the arrest of viral replication [18].

In SARS-CoV-2, NSP12 functions as an RNA-dependent RNA polymerase (RdRp), which is essential for promoting viral RNA replication, and also possesses proofreading capability. The P323L mutation within NSP12 is pivotal, linked to a heightened mutation rate and increased severity of COVID-19 [19].

Notable Variants of Concern and Any Emerging Variants Till Now

A great deal of attention has been paid to some SARS-CoV-2 strains because of their increased transmissibility, potential to evade immune responses, or correlation with severe illness. In response, WHO (World Health Organization) has devised categorization schemes to identify emerging variants, distinguishing them as either VOCs or VOIs.

Variants of Concern (VOCs)

These variations are seen to represent a serious risk to the public's health because of things like enhanced transmissibility, more severe illness outcomes, decreased efficacy of vaccinations or therapies, or the capacity to elude diagnostic identification. Noteworthy VOCs among them are the following:

Alpha Variant (B.1.1.7)

After a thorough genomic study of patient samples in September 2020, the Alpha variant of SARS-CoV-2 was discovered in the United Kingdom, making it one of the first altered strains from the original strain. Known as the B.1.1.7 variant, it harbors 17 mutations altering proteins within the viral genome. Notably, nine of these mutations (Δ69–70, Δ144, N501Y, A570D, D614G, P681H, T716I, S982A,

D1118H) affect the spike (S) protein. Among them, N501Y is particularly significant as it increases the spike protein's affinity to ACE2 receptors, thereby augmenting viral attachment and facilitating entry into host cells [20, 21].

Beta Variant (B.1.3.51)

The second wave of COVID-19 infections began in 2020 when the Beta variant was discovered for the first time in South Africa in May of the same year [22]. Notably, this variant of concern acquired key mutations, including nine substitutions and one deletion (L18F, D80A, D215G, Δ242–244, R246I, K417N, E484K, N501Y, D614G, and A701V) in the spike protein. Among these mutations, three (K417N, E484K, and N501Y) located in the receptor-binding domain (RBD) notably enhance the affinity for ACE receptors. Additionally, the variant harbour mutations K1655N (NSP3: K837N) and Δ3675–3677(NSP6: Δ106–108) in Orf1ab, P71L in the E protein, and T205I in the N protein. Numerous studies have indicated that the neutralizing antibody levels against the beta variant in vaccinated individuals' sera were lower compared to those against the original strain [23]. Patients infected with the Beta form frequently suffer from more severe respiratory symptoms, like excruciating chest discomfort and shortness of breath.

Gamma Variant (P.1)

The Gamma variant, initially detected in Brazil in December 2020, emerged with notable genetic changes [24]. It has been revealed via genetic sequencing of many virus samples from infected people that the Gamma variant contains around 22 mutations total, with over 12 of those mutations occurring on the spike protein. Among these mutations are L18F, T20N, P26S, D138Y, R190S, K417T, E484K, N501Y, D614G, H655Y, T1027I, and V1176. Notably, three mutations (K417N, E484K, and N501Y) are situated in the receptor-binding domain (RBD), mirroring similarities to the B.1.351 variant [25]. Patients infected with the Gamma variant commonly had neurological symptoms, such as confusion, seizures, or loss of consciousness.

Delta Variant (B.1.617.2)

The B.1.617.2 variant, commonly known as the Delta variant, emerged as the fourth variant of concern. It was first identified in October 2020 in India, and it became well-known because it contributed to the catastrophic second wave of COVID-19 infections that occurred there in April 2021 [26]. The Delta variant quickly garnered attention for its heightened transmissibility compared to previously identified variants. Among its numerous mutations, key changes include the L452R and T478K mutations in the spike RBD, along with the P681R mutation at the cleavage site.

Studies have indicated a higher prevalence of the Delta variant among younger age groups. Additionally, notable mutations within the spike protein include T19R, G142D, Δ156–157, R158G, D614G, and D950N. Symptoms associated with the Delta variant are often severe and encompass a range of manifestations, including intense fatigue, persistent coughing, and gastrointestinal issues such as nausea, vomiting, and diarrhoea.

Omicron (B.1.1.529 Lineage)

On November 23, 2021, during a spike in COVID-19 cases, the fifth variant of concern—known by the WHO as B.1.1.529 or the Omicron variant—was discovered for the first time in South Africa [27]. Numerous unique mutations, including 30 amino acid alterations, three small deletions, and one short insertion in the spike protein, define this variation. Notably, 15 mutations are situated in RBD indicating potential implications for viral infectivity and immune evasion. Additionally, various alterations and deletions have been observed in other genomic regions. The reported mutations include T9I in the E-protein, D3G, Q19E, A63T in the M protein, P13L, Δ31–33, R203K, G204R, S413R in the N protein, A67V, Δ69–70, T95I, G142D, Δ143–145, N211I, Δ212, 215EPEins in the NTD, G339D, S371L, S373P, S375F, K417N, N440K, G446S, S477N, T478K, E484A, Q493R, G496S, Q498R, N501Y, Y505H in the RBD, D796Y in the S1/S2 site as well as multiple other mutations in the S-protein and several others in the non-structural proteins [28]. Symptoms associated with the Omicron variant range from mild to severe, the most common of which are fever, coughing, and exhaustion. It is interesting to note that those infected with this variety had a lower incidence of taste loss.

Table 4.1 lists the VOC's characteristics, the nation where it first appeared on record, the date of the earliest reporting, and the mutations that affect structural, nonstructural, and accessory proteins.

The exclusive and shared mutations in structural, nonstructural, and accessory proteins among the VOCs are illustrated in Fig. 4.1.

Upon deeper investigation into Omicron's genetic composition, a variety of sublineages have emerged within the Omicron lineage. The primary sublineage, now recognized as BA.1 (or B.1.1.529.1), serves as the standard reference. Additionally, two other notable sublineages have been identified: BA.2 (or B.1.1.529.2) and BA.3 (or B.1.1.529.3). In mid-2022, the emergence of BA.4 (or B.1.1.529.4) and BA.5 (or B.1.1.529.5) was observed across several countries [29]. Interestingly, the WHO has classified every one of these sublineages, along with their descendants, as VOCs.

In response to the prevalence and evolution of the Omicron family, the WHO introduced a new category called VOC lineages under monitoring (VOC-LUMs) on May 25, 2022. This decision aimed to acknowledge the dominance of the Omicron lineage in sequenced samples and the substantial evolution within this family. The name, characteristics, and date of outbreak of the Omicron subvariants are tabulated in Table 4.2.

Table 4.1 Details of variants of concern (VOCs) of SARS-CoV-2, which include the name of the VOCs, country where detected first, date of earliest documentation and key mutations in the structural (S, E, M, N), non-structural (Orf1ab) and accessory proteins

		Mutations					
Variant of concern	Earliest documented	ORF1ab (NSP)	S	E	M	N	Accessory
Alpha (B.1.1.7)	UK September 2020	PLpro: T183I, A890D, I1412T Nsp6: Δ106–108 RdRP: P323L	Δ69–70, Δ144, N501Y, A570D, D614G, P681H, T716I, S982A, D1118H	–	–	D3L, S235F	ORF8: R52I, Q27, Y73C
Beta (B.1.351)	South Africa May 2020	Nsp2: T85I PLpro: K837N Nsp4: H26Y, S137L 3CL: K90R Nsp6: Δ106–108 RdRP: D135Y, P323L Nsp13: T588I	L18F, D80A, D215G, Δ242–244, R246I, K417N, E484K, N501Y, D614G, A701V	P71L	–	T205I	ORF3a: Q57H, S171L
Gamma (P.1)	Brazil December 2020	PLpro: S370L, K977Q Nsp6: Δ106–108 RdRP: P323L Nsp13: E341D	L18F, T20N, P26S, D138Y, K417T, E484K, N501Y, D614G, H655Y, T1027I, V1176F	–	–	P80R, R203K	ORF3a: S253P

Delta (B.1.617.2)	India October 2020	RdRP: P323L, G671S Nsp13:P77L	T19R, G142D, Δ156–157, R158G, L452R, T478K, D614G, P681R, D950N	–	I82T	R203M, D377Y	ORF3: S26L ORF7a: T120I, V82A ORF8: Δ119–120
Omicron (B.11.0.529)	South Africa November 2021	PLpro: K38R, S1265I, Δ1266, A1892T Nsp4: T492I 3CL: P132H Nsp6: L105F, Δ106–108, I189V RdRP: P323L Nsp14:I42V	A67V, Δ69–70, T95I, G142D, Δ143–145, N211I, Δ212, 215EPEins, G339D, S371L, S373P, S375F, K417N, N440K, G446S, S477 N, T478K, E484A, Q493R, G496S, Q498R, N501Y, Y505H, T547K, D614G, H655Y, N679K, P681H, N764K, D796Y, N856K, Q954H, N969K, L981F	T9I	D3G, Q19E, A63T	P13L, Δ31–33, R203K, G204R	

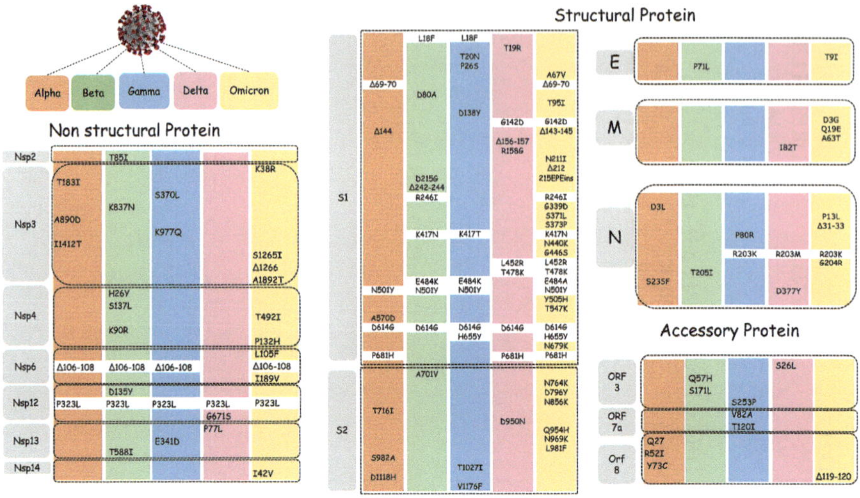

Fig. 4.1 Unveiling the genetic landscape of SARS-CoV-2 variants of concern (VOCs): in-depth exploration of mutational patterns in structural, nonstructural, and accessory proteins. The mutations shared across two or more variants are highlighted in white. Abbreviations used: Nsp: non-structural proteins, E: envelope protein, M: membrane protein, N: nucleocaspid, S1/S2: The two subunits of Spike protein

As of March 15, 2023, WHO has revised the classification of all these variants, downgrading their status. Presently, no variants of SARS-CoV-2 meet the criteria to be classified as VoCs.

Variants of Interest (VOIs)

These are variants that are under surveillance and investigation by public health authorities due to specific genetic changes that may affect virus characteristics, but they have not yet demonstrated the same level of impact as VOCs. VOIs may have potential implications for public health, but more data is needed to assess their significance fully.

WHO identified several variants of interest (VOIs) that were under surveillance and monitoring. The details of the VOIs, date of outbreak synonyms, along with their mutations in structural, non-structural and accessory proteins are tabulated in Table 4.3.

Table 4.2 The name, date of outbreak and characteristics of the different Omicron subvariants of SARS-CoV-2

Omicron sublineage	Earliest detected	Characteristics
BA.1	November 2021	Associated with increased transmissibility and potential immune evasion, rapidly spread globally
BA.2	November 2021	Initially believed to be more transmissible and cause severe disease, but later found to cause less severe illness than BA.1; resistant to certain monoclonal antibodies. Spread globally, dominating in various countries
XE	January 2022 (UK)	Recombinant of BA.1 and BA.2, potentially 10% more transmissible than BA.2, making it highly contagious
BA.2.12 and BA.2.12.1	May 2022 (USA)	Two subvariants detected in New York, with significant growth advantage over BA.2, contributing to rise in infections
BA.2.75 and BA.2.75.2	May 2022 (India)	Detected in India, classified as a variant under monitoring by WHO, potentially capable of escaping neutralizing antibodies
XBB and XBB.1	August 2022	Descendants of BA.2.10.1.1 and BA.2.75.3.1.1.1 sublineages, with XBB.1.5 potentially becoming dominant in Europe
EG.5 and EG.5.1	February 2023	Descendants of XBB.1.9.2, EG.5 identified as a variant of interest by WHO, potentially more transmissible, but no impact on disease severity or transmissibility
HV.1	July 2023	Sublineage of XBB.1.9.2, dominant in U.S., no evidence of increased severity or transmissibility
BA.2.86	August 2023	A subvariant of BA.2, with over 30 mutations, reported cases in multiple countries, potentially spreading internationally
JN.1	August 2023	A subvariant of BA.2.86, emerged in in Luxembourg. By December 2023, detected in 12 countries, including the UK and the USA, with low overall public health risk
BA.3	Unknown	Rare sublineage of Omicron with S-gene target failure, similar to BA.1
BA.4 and BA.5	April 2022	Significant growth advantage over BA.2, became dominant globally, particularly in Europe and the United States
BQ.1 and BQ.1.1	October 2022	Subvariants of BA.5, prevalent in France and the United States, associated with increased vaccine escape compared to previous variants

Table 4.3 Details of variants of interest (VOIs) of SARS-CoV-2, which include the name of the VOIs, country, date of earliest documentation, and key mutations in the structural (S, E, M, N), non-structural (Orf1ab) and accessory proteins. The WHO lists XBB.1.5, XBB.1.16, EG.5, BA.2.86, and JN.1 as circulating VOIs as of December 20, 2023

		Mutations					
Variants of Interest	Earliest documented	ORF1ab	Spike	E	M	N	Accessory
Epsilon (B.1.427) [30]	USA July 2020	Nsp2:T85I Nsp9: I64V RdRp: P323L Nsp13:D260Y	S13I, W152C, L452R, D614G	–	–	T205I	ORF3a: Q57H
Zeta (P.2) [31]	Brazil April 2020	3CL: L205V Nsp7: L71F RdRp: P323L	E484K, D614G, V1176F	–	–	A119S	
Eta (B.1.525) [32]	UK Nigeria December 2020	PLpro: T1189I RdRp: P323F	A67V, Δ69–70, Δ144, E484K, D614G, Q677H, F888L	L21F	I82T	Δ3, A12G, T205I	ORF6:Δ2
Iota (B.1.526) [33]	USA November 2020	Nsp6: Δ106–108	L5F, T95I, D253G, E484K, D614G, A701V	–	–	–	
Theta (P.3) [34]	Philippines February 2021	PLpro: D736G Nsp4: L438P Nsp6: D112E Nsp7: L71F RdRp: P323L Nsp13:A368V	E484K, N501Y, D614G, P681H, E1092K, H1101Y, V1176F	–	–	R203K, G204R	ORF8: K2Q

Kappa (B.1.617.1) [35]	India October 2020	PLPro: T749I Nsp6: T77A RdRp: P323L Nsp13: M429I Nsp15: K259R	T95I, G142D, E154K, L452R, E484Q, D614G, P681R, Q1071H	–	–	R203M, D377Y	ORF3: S26L ORF7a: V82A
Lambda (C.37) [10]	Peru December 2020	PLpro: T428I, P1469S, F1569V Nsp4: L438P, T492I 3CL: G15S Nsp6: S106K, Δ107–109 RdRP: P323L	G75V, T76I, Δ246–253 N, L452Q, F490S, D614G, T859N	–	–	P13L, R203K, G204R, G214C	
Mu (B.1.621) [36]	Columbia January 2021	PLpro: T237A, T720I Nsp4: T492I Nsp6: Q160R RdRP: P323L Nsp13: P419S	T95I, Y144S, Y145N, R346K, E484K, N501Y, D614G, P681H, D950N			T205I	ORF3a: Q57H, Δ256–257 ORF8: T11K, P38S, S67F

Impact of Mutations on Transmission and Virulence: Public Health Implications

Genetic alterations within SARS-CoV-2 have significantly affected its behavior during the COVID-19 pandemic. These mutations have diverse impacts on transmission, with certain variants demonstrating heightened transmissibility compared to the original strain. In contrast to ancestral strains, variations in the spike protein, as seen in Alpha, Beta, Gamma, and Delta, have exhibited evidence of enhanced virulence, transmissibility, or immunity-evasion, leading to higher viral loads and more effective person-to-person transmission [37, 38]. Additionally, mutations in other parts of the viral genome can influence replication rates and the duration of viral shedding, further shaping transmission dynamics. While some mutations may not directly influence disease severity, changes in transmission dynamics can indirectly impact public health outcomes by straining healthcare systems and contributing to higher rates of illness and mortality.

The Impact of Specific Mutations and Variants on SARS-CoV-2 Transmissibility Compared to the Original Strain

Mutations within the viral genome occur ubiquitously, giving rise to variants with diverse characteristics. While some mutations are silent, others confer distinct advantages, such as heightened infectivity, transmissibility, and evasion of immune responses. Prominent alterations such as R203K/G204R in the N protein have been linked to heightened viral infectivity and replication. A deletion in NSP1 (Δ500–532) correlates with reduced plasma IFN-β levels and viral load. Mutations in ORF8 boost viral transmission and immune evasion. ORF8 gene products regulate host cell major histocompatibility complex class I, and a common premature stop codon (Q27stop) contributes to the increased transmission of the B.1.1.7 variation. Furthermore, the early pandemic detection of the widely distributed D614G mutation in the S gene improves viral properties by raising infectivity, ACE-2 receptor binding affinity, and viral replication rates. Increased protease-induced cleavage of the S protein, which facilitates viral entrance into host cells, is associated with this mutation. Variants like the B.1.1.7 lineage carry the S RBD N501Y mutation, which has been linked to higher transmissibility and greater binding affinity to the ACE-2 receptor. Originating in South Africa, the B.1.351 lineage expanded over the world and is known to carry significant mutations in the RBD of the S protein, such as K417N, E484K, and N501Y, which improve the fitness and adaptability of the virus. The B.1.617 lineage, encompassing the delta variant, demonstrates elevated transmissibility due to mutations like L452R and T478K which stabilizes the RBD-ACE-2 complex. Despite accumulating mutations, the Omicron variant, with mutations like Q498R and S477N, may not necessarily lead to increased disease severity or fatality rates. These insights provide light on the intricate relationship between viral mutations and the course of disease, guiding public health initiatives to lessen the effects of newly developing variants [39].

The Challenges Posed by SARS-CoV-2 Variants to Public Health Strategies, Including Testing, Contact Tracing, Containment Measures, and Societal Responses

Despite extensive efforts and resources dedicated to mitigating SARS-CoV-2 transmission, containing the COVID-19 pandemic remained a significant challenge. The emergence of new virus variants has consistently led to fresh waves of infections and the reinstatement of containment measures. Between April 2020 and May 2022, 131 outbreaks, each comprising at least 50 cases and lasting a minimum of 7 days, were reported across 90 prefecture-level Chinese cities. These comprised 27 outbreaks of the Delta variation, 92 outbreaks of the Omicron lineages, and 12 outbreaks attributable to pre-Delta variants (e.g., lineages B.1.1, B.2, and B.1.1.7). Notably, social distancing measures exhibited significant effectiveness in curbing transmission prior to the onset of Delta-variant outbreaks, with reductions exceeding 50% [40].

Amid the presence of COVID-19 variants, an important consideration regarding the contact tracing strategy was the timing of easing restrictions for economic benefits. This demanded improved strategies for controlling COVID-19 transmission without resorting to additional lockdowns. While lockdowns and testing strategies played pivotal roles, Automated Contact Tracing emerged as a valuable tool that, as several nations have shown, expedites and improves the efficacy of containment measures. Training staff, increasing capacities for effective data analytics, and improving diagnostic testing skills in underdeveloped countries all required addressing major logistical and capacity challenges [41].

Findings from Epidemiological Studies on the Spread and Impact of Different Variants

Amidst the global health crisis, five significant variants—Alpha, Beta, Gamma, Delta, and Omicron—have been identified, with Delta becoming the predominant strain worldwide in June 2021 and Omicron BA.2 becoming the dominant global strain. According to WHO's February 2022 report, Omicron was identified in November 2021, with approximately 130 million people diagnosed within 2.5 months. China witnessed over 500,000 local infections of the Omicron variant from March 1 to April 22, 2022. The emergence of these variants has led to numerous epidemics caused by different virus strains in various countries and regions.

Vaccine Evasion and Immune Escape

Presently, over 200 vaccines are undergoing trials at various clinical stages, according to the COVID-19 Treatment and Vaccine Tracker website (https://covid-19tracker.milkeninstitute.org/). Many hazardous variations with a variety of epidemiological characteristics and immune-evading capacities have been discovered in different countries. Nonetheless, majority of the vaccines created up to this point of time have been created using the wild-type DNA that became available on January 5, 2020. As a result, new SARS-CoV-2 variations are found to avoid immunity brought on by earlier infections or vaccinations against previous versions, which presents difficulties for vaccine treatments [42].

Evidence Regarding the Ability of SARS-CoV-2 Variants to Evade Immunity Acquired by Vaccination or Natural Infection

Limited antigenic change was observed in the Alpha variant, while the Gamma, Beta, and Delta variants showed significant evasion of vaccine-induced antibodies and convalescent serum. However, epidemiological investigations have indicated that the effectiveness of the vaccine against Beta and Delta variants remained largely intact. A globally used vaccine Sanofi-GSK is based on the beta variant. Consequently, first-generation SARS-CoV-2 vaccinations provide considerable protection from severe illness, allowing most of the global population to resume regular activities, even though they were created using an early spike sequence. The Omicron "complex" with sublineages BA.1, BA.2, BA.3, BA.4, and BA.5 posed a challenge to current control strategies by infecting vaccinated and previously infected individuals. With over 15 spike RBD mutations, BA.1, BA.2, BA.4, and BA.5 are hard for first-generation vaccines and pre-Omicron antibodies to neutralize. Vaccines approved in multiple countries, such as mRNA-based vaccines viz. Moderna and Pfizer-BioNTech, are now produced using both the original strain and the Omicron variant. Moreover, most current therapeutic monoclonal antibodies are ineffective against these variants, except for bebtelovimab, which continues to be effective against all SARS-CoV-2 strains and targets the spike protein's RBD. Researchers have suggested that Omicron lineages should be regarded as a distinct serotype or strain from pre-Omicron lineages due to this notable antigenic "shift" [37].

Effectiveness of Current Vaccines Against Emerging Variants and the Need for Booster Doses or Updated Vaccines

WHO panel supports the development of monovalent and polyvalent COVID-19 vaccines targeting single and multiple variants. For instance, Pfizer-BioNTech is creating multivalent vaccinations (such as bivalent Alpha/Omicron/Delta and Alpha) in addition to variant-specific vaccines (e.g., Alpha, Beta, and Delta). With the exception of the beta-specific vaccine, booster doses of these vaccinations have demonstrated noticeably higher neutralizing effectiveness against the Omicron variant than the initial shot. Research on animals has demonstrated that bivalent COVID-19 vaccinations with booster doses that contain mRNAs expressing Omicron BA.1 or BA.4/5 SARS-CoV-2 spike proteins can improve immunogenicity and protect mice against the BA.5 variant. Bivalent booster doses after 2, 3, or 4 monovalent shots provide further protection against Omicron sub-lineages like BA.5, XBB, and XBB.1.5. Therefore, it is recommended that everyone who qualifies should maintain up-to-date COVID-19 vaccination records. Omicron-specific formulations of the Novavax and Sputnik V vaccines are also being developed, as well as modified forms of the Oxford-AstraZeneca vaccine that target the Beta and Omicron types. To evaluate the current vaccination and booster implementation campaigns, substantial data on the practical effectiveness of these variant-specific vaccinations must be continuously gathered [43].

The FDA has put forth a proposal for updating COVID-19 vaccines annually, mirroring the same methodology as taken with influenza vaccines. On important issues including the optimal vaccination schedule, frequency, and composition, there are differing opinions among scientists. Currently undergoing preclinical and/or clinical assessment, universal or pan-SARS-CoV-2 vaccines may provide the best long-term defence against future variations. Therefore, gaining a better grasp of the virus's evolutionary trends is crucial to creating more efficient programs that adequately anticipate the potential emergence of highly infectious and dangerous strains. A comparative overview of mutation rates, rates of transmission, and vaccine efficacy of different VOCs of SARS-CoV-2 is depicted in Fig. 4.2.

The above figure illustrates the comprehensive approach taken in the chapter to analyze SARS-CoV-2 variants. The key variants of concerns, including Alpha, Beta, Gamma, Delta, and Omicron, are highlighted. The gradient representation indicates the mutation rate, transmission rate, and vaccine efficacy for each variant. This visual aids in understanding the relative differences among the variants in terms of their evolutionary adaptations and impact on public health measures.

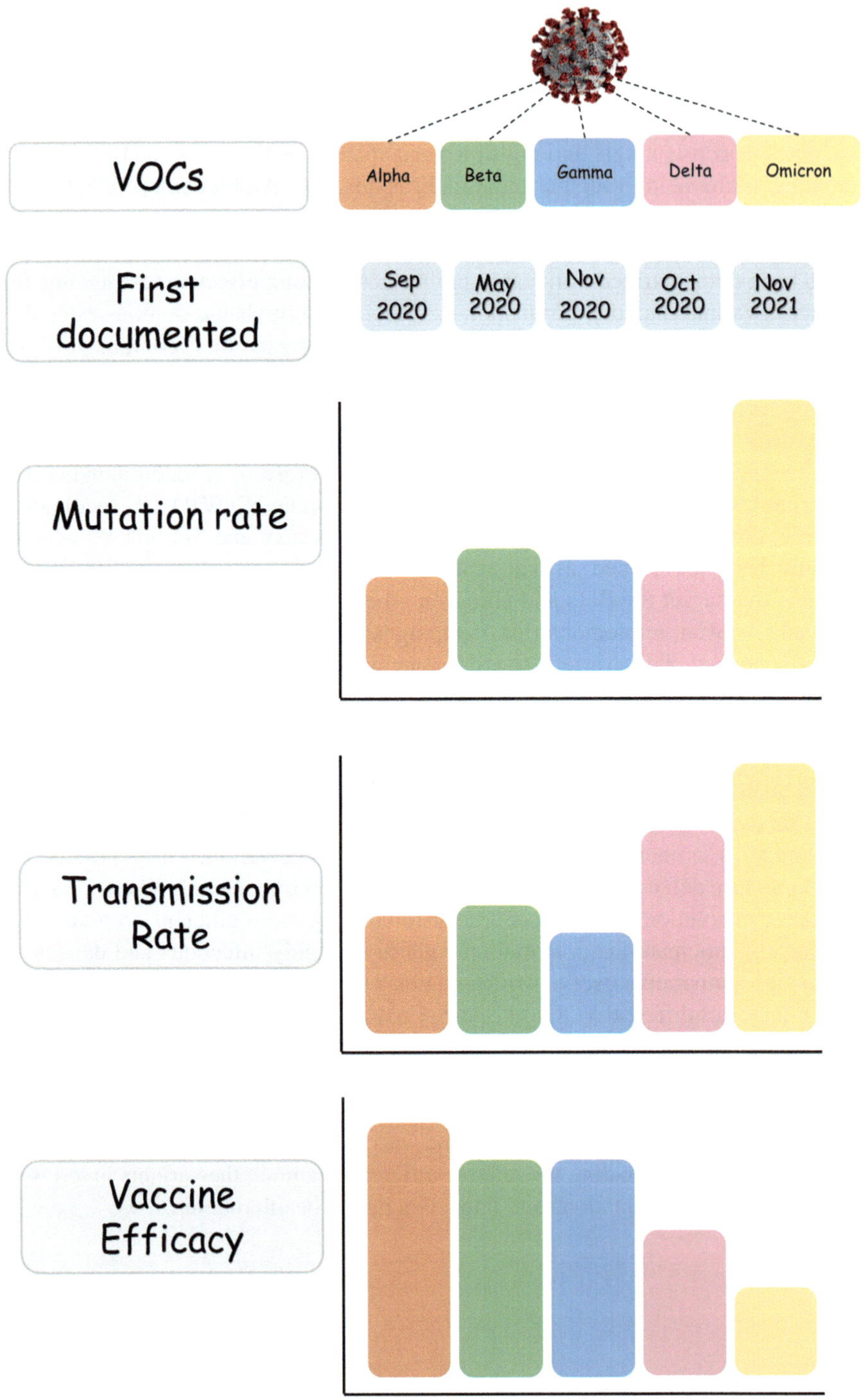

Fig. 4.2 Comprehensive overview of SARS-CoV-2 variants of concerns: a comparative perspective on mutation rates, transmission rates, and vaccine efficacy gradients

Future Directions and Research Needs

Gaps in Current Knowledge Regarding SARS-CoV-2 Mutants and Variants

While significant strides have been made in understanding SARS-CoV-2 mutants and resulting variants, there remain notable gaps in our current knowledge. The long-term effects of new variations on the effectiveness of currently available vaccinations and treatments represent a significant area of uncertainty. Additionally, the precise mechanisms underlying the increased transmissibility and virulence of certain variants, such as Delta and Omicron, require further investigation. Furthermore, there is limited perception of the potential impact of viral mutations on diagnostic testing accuracy and effectiveness. Moreover, the role of host factors, environmental conditions, and viral evolution dynamics in shaping the emergence and spread of new variants necessitates a comprehensive investigation. Addressing these knowledge gaps is important for guiding effective public health interventions and refining strategies to mitigate the ongoing COVID-19 pandemic.

Understanding the Biological Mechanisms Driving Variant Emergence and Exploring Novel Therapeutic Strategies

Investigating the biological mechanisms underlying variant emergence is crucial for devising effective therapeutic strategies against evolving strains of SARS-CoV-2. While the virus continues to mutate, understanding the genetic and molecular determinants driving these changes is paramount. Research efforts are focused on deciphering how mutations in key viral proteins, such as the spike protein, impact viral infectivity, transmissibility, and immune evasion. Furthermore, exploring the host-virus interactions and host factors influencing viral evolution provides valuable insights into the emergence and spread of variants. Novel therapeutic approaches, including antiviral drugs, monoclonal antibodies, and other immunomodulatory agents, are being explored to target specific viral vulnerabilities and host pathways associated with viral replication and pathogenesis. By elucidating the biological mechanisms driving variant emergence and developing tailored therapeutic interventions, we can enhance our ability to combat the evolving threat posed by SARS-CoV-2 variants.

The Importance of Genomic Surveillance in Tracking the Evolution and Spread of SARS-CoV-2 Variants

Genomic surveillance of SARS-CoV-2 involves sequencing and analyzing virus genomes from COVID-19 patients to track mutations and variations, revealing the virus's evolution. Genomic monitoring assists in tracking the genesis and distribution of SARS-CoV-2 variants by comparing sequences from various samples over time and across geographical locations. This approach enables the detection of VOCs and VOIs, which may exhibit altered transmissibility, virulence, or resistance to vaccines and therapeutics. Genomic surveillance also helps identify potential clusters of transmission, track the introduction of variants from international travel, and assess the effectiveness of control measures. Particularly since the availability of SARS-CoV-2 vaccines, genomic surveillance has been essential for the early recognition of mutations, tracking the viral evolution, and determining the degree of similarity between vaccine strains and circulating variants. As new variants continue to emerge, sustained genomic surveillance efforts are essential for informing public health policies, guiding vaccination strategies, and lessen the impact of the evolving pandemic. Various platforms play a crucial role in facilitating genomic surveillance for SARS-CoV-2, aiding researchers in sharing and analyzing sequencing data. Global Initiative on Sharing Avian Influenza Data (GISAID) [44], a global database for sharing influenza and coronavirus genome sequences, provides a platform for researchers to upload, access, and analyze viral genome sequences, enabling real-time monitoring of viral evolution and the detection of emerging variants. Another valuable resource is Nextstrain [45], an open-source project that visualizes the genetic diversity and transmission patterns of SARS-CoV-2 through phylogenetic analysis. Nextstrain integrates genomic data from multiple sources, including GISAID, to track the global spread of the virus and identify hotspots of transmission. Additionally, the COVID-19 Genomics UK (COG-UK) consortium has played a pivotal role in genomic surveillance within the UK, sequencing a large proportion of positive COVID-19 cases and sharing data with public health agencies and researchers worldwide. Outbreak.info tracks 40 million combinations among Pango lineages and mutations across 7000 sites. It helps researchers, public health professionals, and the public with genomic surveillance and hypothesis generation. Further platforms that work well for tracking SARS-CoV-2 variation in genome are Coronapp [46], CoV-Spectrum [47], and CoVigator [48].

Genomic surveillance effectiveness depends on sampling methodology, case identification accuracy, diagnostic test use, viral load quantification, and case-to-sequence interval. COVID-19 highlighted these challenges and emphasized genomics' vital role in real-time monitoring, policy development, and disease prevention. Technological and infrastructural advancements during the pandemic must be sustained. Funding, integration of SARS-CoV-2 sequencing networks, and bioinformatics are crucial for public health, research, and surveillance frameworks.

Developing Computational Pipelines for the Rapid Detection and Characterization of Novel Variants from Large-Scale Genomic Datasets

Sequencing the genomes of SARS-CoV samples offers a plethora of information about the virus's genetic composition. Bioinformatics can be utilized for the analysis of these sequences to detect mutations and variations in the viral genome. AI algorithms, like deep learning models, are able to handle large datasets more effectively than conventional approaches. They have the ability to identify slight patterns and connections in the genetic sequences, aiding in forecasting future mutation patterns. Researchers can make conclusions about evolutionary trends and predict future mutations by studying how mutations occur and spread through time and across various geographical areas.

Machine Learning Models

Predictive models can be created using machine learning techniques with viral genomic data and metadata. This data could contain details about aspects like host species, geographical location, methods of transmission, and results of treatment. Through the use of historical data, researchers can train machine learning algorithms to create models that can predict the potential evolution of the virus in different scenarios. These models have the ability to include factors like selective pressures, population dynamics, and environmental factors in order to produce predictions that are more precise.

Structural Bioinformatics

Bioinformatics software has the capability to anticipate the 3D arrangement of viral proteins from their amino acid sequences. AI algorithms can analyze these structures in order to forecast the functional impacts of mutations. Molecular dynamics simulations and computational modeling can be used to determine how mutations impact protein stability, binding affinity, and interactions with host cells. Identifying shared regions of vulnerability within various strains of the virus can assist in designing antiviral medications and vaccinations.

Population Genomics

Combining genomic data with epidemiological data enables scientists to develop complete models of viral transmission and evolution. AI-powered epidemiological models have the ability to replicate the transmission of various viral strains among populations. Taking into account the factors like population demographics, mobility patterns, social behavior, and public health interventions, these models are able to forecast the emergence and spread of various variants.

Monitoring in Real Time

Continuous real-time data integration allows researchers to update and improve models, offering guidance for public health policies and interventions. AI-driven surveillance systems are capable of monitoring genomic databases and spotting emerging mutations or new variants of concern instantly. Natural language processing (NLP) algorithms have the ability to extract important data from scientific literature, social media, and other sources in order to give advance notice of possible dangers. By identifying abnormal sets of mutations, these systems can signal researchers and public health authorities to look into the issue more closely and respond accordingly. In short, combining AI and bioinformatics allows researchers to analyze large genomic datasets to comprehend the evolutionary trends of SARS-CoV and forecast forthcoming mutation trends. These forecasts can advise public health tactics, steer the creation of treatments and vaccines, and ultimately aid in reducing the repercussions of pandemics on global health.

Conclusion

The continuous evolution of SARS-CoV-2 presents ongoing challenges to global health, driven by mutations that affect the virus's transmissibility, virulence, and immune evasion. This has necessitated updated vaccines, booster doses, and new therapeutic strategies. While genomic surveillance has been crucial in tracking and responding to these variants, predicting the long-term efficacy of current interventions remains uncertain. Advanced computational tools, like AI and machine learning, have enhanced our ability to detect and characterize new variants quickly, informing public health policies. Moving forward, sustained investment in research, genomic surveillance, and technology will be essential to anticipate and combat future variants effectively.

Declarations

Author's Contributions ZG conceived the idea and review structure; TD, BG, and ZG wrote the manuscript; TD, BG, and ZG revised and edited the manuscript. Author has read, agreed, and approved the final draft of the manuscript.

Funding None.

Data Availability Not applicable.

Conflicts of Interests There is no competing and conflict of interest.

References

1. V'kovski P, Kratzel A, Steiner S, Stalder H, Thiel V. Coronavirus biology and replication: implications for SARS-CoV-2. Nat Rev Microbiol. 2021;19(3):155–70. https://doi.org/10.1038/s41579-020-00468-6.
2. Yang Y, Peng F, Wang R, et al. The deadly coronaviruses: the 2003 SARS pandemic and the 2020 novel coronavirus epidemic in China. J Autoimmun. 2020;109:102434. https://doi.org/10.1016/j.jaut.2020.102434.

3. Omrani AS, Al-Tawfiq JA, Memish ZA. Middle East respiratory syndrome coronavirus (MERS-CoV): animal to human interaction. Pathogens Global Health. 2015;109(8):354–62. https://doi.org/10.1080/20477724.2015.1122852.
4. Lau SK, Li KS, Huang Y, et al. Ecoepidemiology and complete genome comparison of different strains of severe acute respiratory syndrome-related Rhinolophus bat coronavirus in China reveal bats as a reservoir for acute, self-limiting infection that allows recombination events. J Virol. 2010;84(6):2808–19. https://doi.org/10.1128/jvi.02219-09.
5. Jahirul Islam M, Nawal Islam N, Siddik Alom M, Kabir M, Halim MA. A review on structural, non-structural, and accessory proteins of SARS-CoV-2: highlighting drug target sites. Immunobiology. 2023;228(1):152302. https://doi.org/10.1016/j.imbio.2022.152302.
6. Abbasian MH, Mahmanzar M, Rahimian K, et al. Global landscape of SARS-CoV-2 mutations and conserved regions. J Transl Med. 2023;21(1):152. https://doi.org/10.1186/s12967-023-03996-w.
7. Bhattacharya M, Chatterjee S, Sharma AR, Agoramoorthy G, Chakraborty C. D614G mutation and SARS-CoV-2: impact on S-protein structure, function, infectivity, and immunity. Appl Microbiol Biotechnol. 2021;105(24):9035–45. https://doi.org/10.1007/s00253-021-11676-2.
8. Yang WT, Huang WH, Liao TL, Hsiao TH, Chuang HN, Liu PY. SARS-CoV-2 E484K mutation narrative review: epidemiology, immune escape, clinical implications, and future considerations. Infect Drug Resist. 2022;15:373–85. https://doi.org/10.2147/idr.s344099.
9. Meng B, Kemp SA, Papa G, et al. Recurrent emergence of SARS-CoV-2 spike deletion H69/V70 and its role in the Alpha variant B.1.1.7. Cell Rep. 2021;35(13):109292. https://doi.org/10.1016/j.celrep.2021.109292.
10. Kimura I, Kosugi Y, Wu J, et al. The SARS-CoV-2 Lambda variant exhibits enhanced infectivity and immune resistance. Cell Rep. 2022;38(2):110218. https://doi.org/10.1016/j.celrep.2021.110218.
11. Pondé RAA. Physicochemical effect of the N501Y, E484K/Q, K417N/T, L452R and T478K mutations on the SARS-CoV-2 spike protein RBD and its influence on agent fitness and on attributes developed by emerging variants of concern. Virology. 2022;572:44–54. https://doi.org/10.1016/j.virol.2022.05.003.
12. Zhang H, Deng S, Ren L, et al. Profiling CD8(+) T cell epitopes of COVID-19 convalescents reveals reduced cellular immune responses to SARS-CoV-2 variants. Cell Rep. 2021;36(11):109708. https://doi.org/10.1016/j.celrep.2021.109708.
13. Khan A, Khan SA, Zia K, Altowyan MS, Barakat A, Ul-Haq Z. Deciphering the impact of mutations on the binding efficacy of SARS-CoV-2 omicron and Delta variants with human ACE2 receptor. Front Chem. 2022;10:892093. https://doi.org/10.3389/fchem.2022.892093.
14. Guo H, Fan Q, Song S, et al. Increased resistance of SARS-CoV-2 lambda variant to antibody neutralization. J Clin Virol. 2022;150-151:105162. https://doi.org/10.1016/j.jcv.2022.105162.
15. Khatri R, Siddqui G, Sadhu S, et al. Intrinsic D614G and P681R/H mutations in SARS-CoV-2 VoCs Alpha, Delta, Omicron and viruses with D614G plus key signature mutations in spike protein alters fusogenicity and infectivity. Med Microbiol Immunol. 2023;212(1):103–22. https://doi.org/10.1007/s00430-022-00760-7.
16. Wu W, Cheng Y, Zhou H, Sun C, Zhang S. The SARS-CoV-2 nucleocapsid protein: its role in the viral life cycle, structure and functions, and use as a potential target in the development of vaccines and diagnostics. Virol J. 2023;20(1):6. https://doi.org/10.1186/s12985-023-01968-6.
17. Lin X, Sha Z, Trimpert J, et al. The NSP4 T492I mutation increases SARS-CoV-2 infectivity by altering non-structural protein cleavage. Cell Host Microbe. 2020;31(7):1170–1184.e7. https://doi.org/10.1016/j.chom.2023.06.002.
18. Miknis ZJ, Donaldson EF, Umland TC, Rimmer RA, Baric RS, Schultz LW. Severe acute respiratory syndrome coronavirus nsp9 dimerization is essential for efficient viral growth. J Virol. 2009;83(7):3007–18. https://doi.org/10.1128/jvi.01505-08.
19. Biswas SK, Mudi SR. Spike protein D614G and RdRp P323L: the SARS-CoV-2 mutations associated with severity of COVID-19. Genom Informat. 2020;18(4):e44. https://doi.org/10.5808/GI.2020.18.4.e44.

20. Volz E, Mishra S, Chand M, et al. Assessing transmissibility of SARS-CoV-2 lineage B.1.1.7 in England. Nature. 2021;593(7858):266–9. https://doi.org/10.1038/s41586-021-03470-x.
21. Tao K, Tzou PL, Nouhin J, et al. The biological and clinical significance of emerging SARS-CoV-2 variants. Nat Rev Genet. 2021;22(12):757–73. https://doi.org/10.1038/s41576-021-00408-x.
22. Tegally H, Wilkinson E, Giovanetti M, et al. Detection of a SARS-CoV-2 variant of concern in South Africa. Nature. 2021;592(7854):438–43. https://doi.org/10.1038/s41586-021-03402-9.
23. Choi JY, Smith DM. SARS-CoV-2 variants of concern. Yonsei Med J. 2021;62(11):961–8. https://doi.org/10.3349/ymj.2021.62.11.961.
24. Faria NR, Mellan TA, Whittaker C, et al. Genomics and epidemiology of the P.1 SARS-CoV-2 lineage in Manaus, Brazil. Science. 2021;372(6544):815–21. https://doi.org/10.1126/science.abh2644.
25. Farhud DD, Mojahed N. SARS-COV-2 notable mutations and variants: a review article. Iran J Public Health. 2022;51(7):1494–501. https://doi.org/10.18502/ijph.v51i7.10083.
26. Mlcochova P, Kemp SA, Dhar MS, et al. SARS-CoV-2 B.1.617.2 Delta variant replication and immune evasion. Nature. 2021;599(7883):114–9. https://doi.org/10.1038/s41586-021-03944-y.
27. Vaughan A. Omicron emerges. New Scientist. 2021;252(3363):7. https://doi.org/10.1016/s0262-4079(21)02140-0.
28. Callaway E. Heavily mutated omicron variant puts scientists on alert. Nature. 2021;600(7887):21. https://doi.org/10.1038/d41586-021-03552-w.
29. Rhoden J, Hoffmann AT, Stein JF, et al. Diversity of omicron sublineages and clinical characteristics in hospitalized patients in the southernmost state of Brazil. BMC Infect Dis. 2024;24(1):193. https://doi.org/10.1186/s12879-024-09089-3.
30. Tchesnokova V, Kulasekara H, Larson L, et al. Acquisition of the L452R mutation in the ACE2-binding Interface of spike protein triggers recent massive expansion of SARS-CoV-2 variants. J Clin Microbiol. 2021;59(11):e0092121. https://doi.org/10.1128/jcm.00921-21.
31. Fumagalli MJ, Castro-Jorge LA, Fraga-Silva TFC, et al. Protective immunity against gamma and zeta variants after inactivated SARS-CoV-2 virus immunization. Viruses. 2021;13(12) https://doi.org/10.3390/v13122440.
32. Yadav PD, Nyayanit DA, Gupta N, et al. Detection and isolation of SARS-CoV-2 eta variant from the international travelers and local residents of India. J Med Virol. 2022;94(7):3404–9. https://doi.org/10.1002/jmv.27676.
33. West AP, Wertheim JO, Wang JC, et al. Detection and characterization of the SARS-CoV-2 lineage B.1.526 in New York. Nat Commun. 2021;12(1):4886. https://doi.org/10.1038/s41467-021-25168-4.
34. Rose R, Nolan DJ, LaFleur TM, Lamers SL. Outbreak of P.3 (Theta) SARS-CoV-2 emerging variant of concern among service workers in Louisiana. J Infect Public Health. 2022;15(1):7–9. https://doi.org/10.1016/j.jiph.2021.11.011.
35. Ren W, Ju X, Gong M, et al. Characterization of SARS-CoV-2 variants B.1.617.1 (Kappa), B.1.617.2 (Delta), and B.1.618 by cell entry and immune evasion. MBio. 2022;13(2):e0009922. https://doi.org/10.1128/mbio.00099-22.
36. Halfmann PJ, Kuroda M, Armbrust T, et al. Characterization of the SARS-CoV-2 B.1.621 (Mu) variant. Sci Transl Med. 2022;14(657):eabm4908. https://doi.org/10.1126/scitranslmed.abm4908.
37. Carabelli AM, Peacock TP, Thorne LG, et al. SARS-CoV-2 variant biology: immune escape, transmission and fitness. Nat Rev Microbiol. 2023;21(3):162–77. https://doi.org/10.1038/s41579-022-00841-7.
38. Ramesh S, Govindarajulu M, Parise RS, et al. Emerging SARS-CoV-2 variants: a review of its mutations, its implications and vaccine efficacy. Vaccine. 2021;9(10) https://doi.org/10.3390/vaccines9101195.
39. Hirabara SM, Serdan TDA, Gorjao R, et al. SARS-COV-2 variants: differences and potential of immune evasion. Review. Front Cell Infect Microbiol. 2022;11 https://doi.org/10.3389/fcimb.2021.781429.

40. Ge Y, Wu X, Zhang W, et al. Effects of public-health measures for zeroing out different SARS-CoV-2 variants. Nat Commun. 2023;14(1):5270. https://doi.org/10.1038/s41467-023-40940-4.
41. Ojokoh BA, Aribisala B, Sarumi OA, et al. Contact tracing strategies for COVID-19 prevention and containment: a scoping review. Big Data Cognit Comput. 2022;6(4) https://doi.org/10.3390/bdcc6040111.
42. Zhan C, Zheng Y, Shao L, Chen G, Zhang H. Modeling the spread dynamics of multiple-variant coronavirus disease under public health interventions: a general framework. Inform Sci. 2023;628:469–87. https://doi.org/10.1016/j.ins.2023.02.001.
43. Zabidi NZ, Liew HL, Farouk IA, et al. Evolution of SARS-CoV-2 variants: implications on immune escape, vaccination, therapeutic and diagnostic strategies. Viruses. 2023;15(4) https://doi.org/10.3390/v15040944.
44. Shu Y, McCauley J. GISAID: Global initiative on sharing all influenza data - from vision to reality. Euro Surveill. 2017;22(13) https://doi.org/10.2807/1560-7917.es.2017.22.13.30494.
45. Hadfield J, Megill C, Bell SM, et al. Nextstrain: real-time tracking of pathogen evolution. Bioinformatics. 2018;34(23):4121–3. https://doi.org/10.1093/bioinformatics/bty407.
46. Mercatelli D, Triboli L, Fornasari E, Ray F, Giorgi FM. Coronapp: a web application to annotate and monitor SARS-CoV-2 mutations. J Med Virol. 2021;93(5):3238–45. https://doi.org/10.1002/jmv.26678.
47. Chen C, Nadeau S, Yared M, et al. CoV-Spectrum: analysis of globally shared SARS-CoV-2 data to identify and characterize new variants. Bioinformatics. 2021;38(6):1735–7. https://doi.org/10.1093/bioinformatics/btab856.
48. Bukur T, Riesgo-Ferreiro P, Sorn P, et al. CoVigator-a knowledge base for navigating SARS-CoV-2 genomic variants. Viruses. 2023;15(6) https://doi.org/10.3390/v15061391.

Part II

COVID-19: Diagnosis and Development of Diagnostics

Imaging-Based COVID-19 Diagnosis

5

Abdolreza Esmaeilzadeh, Reza Elahi, Vahid Mohammadi, Siavash Taremi, Armin Jahani Maleki, and Parsa Karami

Abstract

As of December 2019, coronavirus disease (COVID-19) has become a serious worldwide health dilemma. The main organ affected by COVID-19 is the lungs, but other organs, e.g., cardiac, brain, gastrointestinal system, etc., can also be affected. COVID-19 symptoms may vary from upper respiratory infection to severe pulmonary infection and even death in some cases. The gold standard test for diagnosing COVID-19 is reverse transcriptase–polymerase chain reaction; however, imaging techniques also have a crucial role in its diagnosis. Chest computed tomography (chest CT) and chest X-ray (CXR) are the primary imaging techniques used to screen, diagnose, and manage patients. More recently, the chest CT has only been limited to specific indications, e.g., severe infection and COVID-19 complications. Various reporting protocols and classification recommendations have been defined for the standard-of-care diagnosis and screening of COVID-19 based on CXR and chest CT scans. Moreover, ultrasound, magnetic resonance imaging, and nuclear medicine can also be used for the diagnosis and evaluation of COVID-19 complications and extrapulmonary manifestations. Moreover, with the advancement of artificial intelligence (AI) in healthcare, sev-

A. Esmaeilzadeh (✉)
Department of Immunology, Zanjan University of Medical Sciences, Zanjan, Iran

Cancer Gene Therapy Research Center, Zanjan University of Medical Sciences, Zanjan, Iran
e-mail: a46reza@zums.ac.ir

R. Elahi
Department of Radiology, Zanjan University of Medical Sciences, Zanjan, Iran

School of Medicine, Zanjan University of Medical Sciences, Zanjan, Iran

V. Mohammadi · S. Taremi · A. J. Maleki · P. Karami
School of Medicine, Zanjan University of Medical Sciences, Zanjan, Iran

M. K. Ghosh, M. Basu (eds.), *Pathogenesis, Management and Socio-Economic Impact of COVID-19 Pandemic*, Medical Virology: From Pathogenesis to Disease Control, https://doi.org/10.1007/978-981-95-1426-7_5

eral algorithms have been developed for computer-aided diagnosis of COVID-19. In this chapter, we aim to overview the current knowledge of imaging patterns and standard recommendations for imaging-based diagnosis of COVID-19. We provide case illustrations of different COVID-19 patterns in each modality to enhance the understanding of the imaging patterns. We also briefly discuss the expanding role of AI in imaging-based diagnosis and prediction of prognosis in COVID-19.

Keywords

Coronavirus disease 2019 · COVID-19 · Diagnosis · Imaging · Artificial intelligence (AI)

Introduction

Since late December 2019, coronavirus disease 2019 (COVID-19), caused by severe acute respiratory syndrome coronavirus 2 (SARS-CoV-2), has become a critical health dilemma. Since then, this highly contagious virus has affected over 650 million people and has resulted in the death of over seven million people since the onset of the outbreak (https://www.who.int/emergencies/diseases/novel-coronavirus-2019/situation-reports) (Fig. 5.1) [1]. The spectrum of clinical presentations of COVID-19 can range from asymptomatic cases to pneumonia-associated severe lung injury, and even death in critically ill patients. However, the most commonly reported symptoms are fever, nausea/vomiting, shortness of breath, diarrhea, and gradual worsening dry coughs. In addition to pulmonary manifestations, systematic inflammation, hypoxia, and hypercoagulative state can contribute to extrapulmonary complications, such as cerebrovascular, cardiovascular, and gastrointestinal diseases [2].

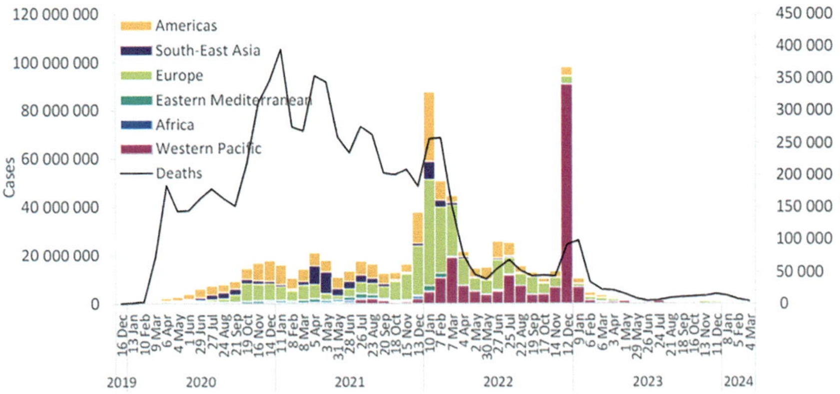

Fig. 5.1 Current WHO report on COVID-19 cases and deaths worldwide, as of March 2024 (https://www.who.int/emergencies/diseases/novel-coronavirus-2019/situation-reports)

Early diagnosis of COVID-19 remains challenging owing to nonspecific symptoms and rapid progression of the disease. The current gold standard test for COVID-19 detection is the reverse transcription–polymerase chain reaction (RT-PCR) carried out on the swabs from the nasopharynx. Despite its high specificity, the sensitivity of the test is reportedly low which may be caused by failure of the extraction process during sampling or inadequate viral load [3]. In this regard, imaging techniques are considered an important tool in early diagnosis and disease risk classification. At the beginning of the epidemic, imaging modalities, specifically chest X-ray (CXR) and chest computed tomography (chest-CT) scans, served a crucial role in the early diagnosis of pulmonary manifestations and screening of the patients [4]. Since then, radiographic patterns have been extensively studied, and standard imaging reporting systems have been defined, revealing a diverse spectrum of findings in both acute and chronic stages of the disease. Several imaging techniques, such as magnetic resonance imaging (MRI), positron emission tomography (PET), and ultrasound can be applied for diagnosis, severity stratification, and identifying the prognosis of COVID-19. In addition, multimodality imaging protocols have shed light on cardiovascular, cerebrovascular, and abdominopelvic complications following COVID-19 infection [5]. This chapter describes the key imaging patterns, currently established recommendations and reporting systems, indications of different imaging techniques, and the pros and cons of each technique for diagnosing COVID-19. We will also briefly review the current progress and challenges of using artificial intelligence (AI)-based networks for the diagnosis of COVID-19.

Chest X-Ray (CXR)

Since the beginning of the pandemic, CXR has been used as a beneficial modality for the screening of COVID-19. Although CXR is not the gold-standard imaging tool for diagnosing COVID-19, it is a cost-effective and highly accessible tool worldwide [6]. According to the limitations of CT in preventing COVID-19 infections and cross-infections, the American College of Radiology (ACR) has recommended that the portable CXR benefits obviate the cross-infection problems. CXR has an essential role in detecting clinically suspected patients at the initial phases of the disease, especially when there are limitations in accessibility to RT-PCR. CXR cannot precisely rule out an infection, and the significant characteristics observed in COVID-19 can be similar to other respiratory diseases; however, it can be helpful with clinical decision-making [7].

The sensitivity of CXR in COVID-19 detection improves during the disease process. Its sensitivity varies from 55% within 2 days of the disease up to 79% after 11 days, but the specificity has a reduction during the disease (83–70%). The false negative rate is increased in African-American ethnicity and young-aged patients. It correlates with the normal and mild severity of CXR findings, including 40% false-negative for normal patients and 87% for combined mild/normal [8]. Even with the accessibility and global use of CXR, it has to be mentioned that CXR has limitations in detecting early stages of infection since some cases with mild clinical symptoms might have normal imaging patterns [9].

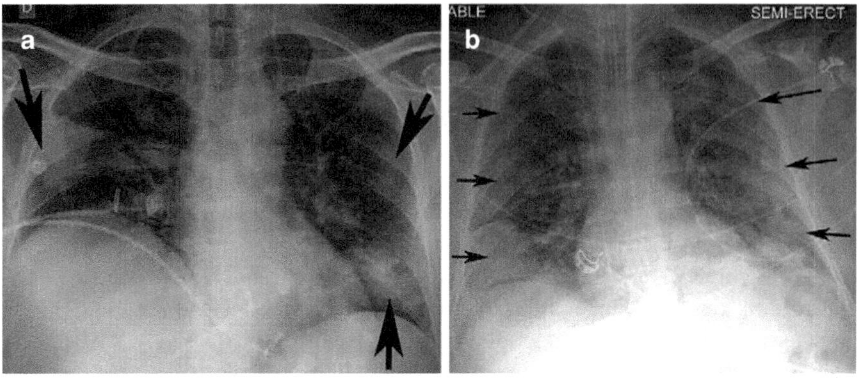

Fig. 5.2 CXR examples of patients with COVID-19 obtained from the RICORD dataset [11]. (**a**) Peripheral reticulonodular airspace GGOs and consolidations (arrows). (**b**) Multiple peripheral consolidations and GGO patterns in the lung periphery, with nodular opacifications (arrows)

Compared to chest CT, studies have reported lower sensitivity for CXR in detecting COVID-19 lung injury in the early phases of the infection [10]. The first common COVID-19 manifestations reported in CXR are peripheral and bilateral ground glass opacities (GGOs) that manifest as hazy areas, indicating inflammation and fluid accumulation in the lungs (Fig. 5.2) (Table 5.1).

Nevertheless, there is a huge limitation for CXR in detecting some of the opacities that could be detected in chest CT because of the correlation of GGOs. Bilateral consolidations in the lower lobar zones of the lung are the second important manifestation of COVID-19 in CXR. COVID-19 infiltrations can be differentiated from bacterial pneumonia, which has single lobar and unilateral consolidations [10]. The third common pattern is the peripheral air space opacities [12]. In progressive stages of COVID-19, necessitating intensive care unit (ICU) admission and intubation, diffused GGOs indicating acute respiratory distress syndrome (ARDS) may also be observed [13]. Furthermore, some uncommon presentations could be observed in CXR, including pleural effusion, which has been reported in rare cases at the late stages of the disease [14], and tension pneumothorax and lung cavitation, which could be found in rare cases [14]. Several reporting systems have been defined for COVID-19, of which the most important ones are the COVID-19 Structured Reporting System, COVID-19 Imaging Reporting and Data System (COVID-RADS), British Society of Thoracic Imaging System, Radiological Society of North America Expert Consensus Reporting System for COVID-19, and COVID-19 Reporting and Data System (CO-RADS) (for more information, refer to [4]).

Table 5.1 Key pulmonary imaging findings of different modalities in COVID-19

Imaging modality	Key pulmonary findings	Strengths	Limitations
CXR	GGO: hazy areas and fluid accumulation. Bilateral and unilateral consolidations in lower lobar zones of the lung. Peripheral air space opacities. Pleural effusion(rare).	Low-dose radiation. Could be available at bedside. Appropriate tool for screening.	Low specificity. Overlap of CXR findings with other infectious disease.
CT-scan	Bilateral or unilateral GGOs. Crazy-paving patterns distributed in the peripheral, sub-pleural, and posterior regions. Consolidations. Reverse halo sign.	Provides more sensitivity in the detection of lung abnormalities. Concise structural information. Provides additional benefits for diagnosing COVID-19 complications.	Increases the risk of infectious transmission to healthcare staff and other patients. Many of the CT findings of COVID-19 overlap with those of other viral infections. High radiation dosage.
MRI	Similar to CT.	High specificity and sensitivity. High spatial resolution.	Increases the risk of infectious transmission to healthcare staff and other patients. It is expensive and may not be available at many centers.
Ultrasound	Subpleural and trans-lobar consolidations. Pleural thickening. Air bronchograms. B-line artifacts.	Bedside imaging using POCUS limits disease transfection. Do not use ionized radiation. Highly sensitive results comparable to CT-scan. Affordable. It can be used for risk stratification and prediction of ICU admission.	Requires an experienced operator.

CXR chest X-ray, *CT* computed tomography, *MRI* magnetic resonance imaging, *GGO* ground glass opacity, *POCUS* point-of-care ultrasound, *COVID-19* coronavirus disease 2019, *ICU* intensive care unit

Computed Tomography (CT) Scan

Chest CT Scan

Compared to CXR, chest CT provides more sensitivity in detecting COVID-19. Studies have shown that chest CT has a specificity of 46% and sensitivity of 87%, with a negative predictive value (NPV) of 89% and positive predictive value (PPV) of 69% for COVID-19 diagnosis [15]. Nevertheless, while using CT can be benevolent in areas with a high COVID-19 prevalence, in regions with a low prevalence, CT-based diagnosis will lead to an increased number of false positive cases [15].

Typical imaging patterns of COVID-19 on chest CT include peripheral GGOs, areas of consolidation in the middle and lower lobes, nodular involvement, and reticular and crazy-paving patterns mainly distributed bilaterally in the peripheral, subpleural, and posterior regions (Fig. 5.3). In some cases, the consolidation, manifested by high-density opacities, can involve both lungs, the "white lung" [17]. Another notable pattern resembling COVID-19 pneumonia includes a reversed halo sign, characterized by a focal and central GGO surrounded by a peripheral ring of dense consolidation [18]. COVID-19 CT findings are dynamic and can change within the course of the disease as the symptoms aggravate. In earlier stages, GGO is the indeterminate finding appearing on the CT images. As the disease progresses, crazy-paving patterns become present, and GGOs increase in size and number and transform into multifocal areas of consolidation. In the healing phase, consolidations and crazy-paving patterns gradually resolve [17]. However, lung abnormalities on chest CT can persist beyond the resolution of symptoms, with a study reporting residual findings on 94% of final CT scans obtained from discharged patients [19]. In the recovery phase of severe COVID-19 infection, lung fibrosis might be predominantly located in the lower lobes, peripheral and subpleural regions [20].

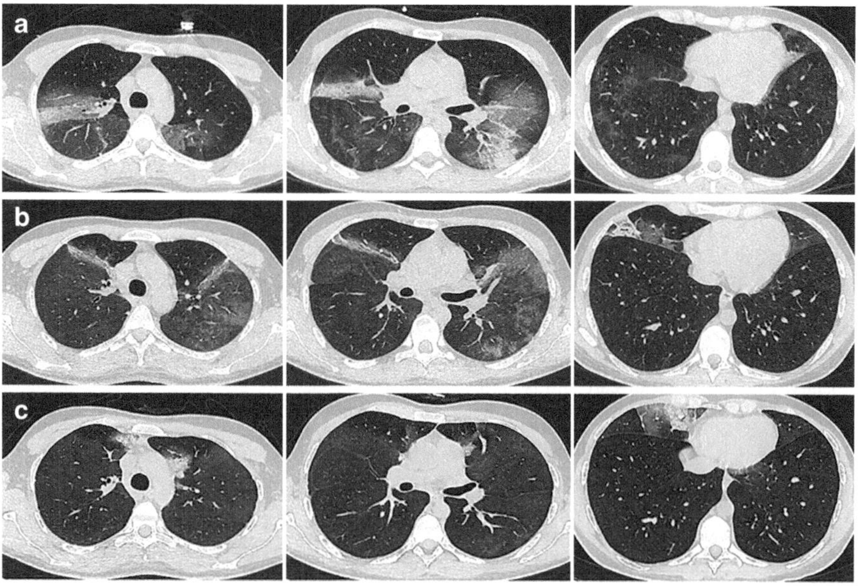

Fig. 5.3 Axial chest-CT images in a COVID-19 patient demonstrating transient and migratory ground-glass opacities with small areas of consolidation in both lungs with gradual resolution of findings from (**a**): day 1 to (**b**): day 15 and (**c**): day 28 post-infection [16]

The ACR recommends against using CT as the first-line modality for the diagnosis of COVID-19. Experts believe frequent CT utilization will accelerate infection transmission to other patients and healthcare providers. Besides, many of the COVID-19 CT findings overlap with other viral infections, e.g., H1N1 influenza. Moreover, 15–50% of RT-PCR affirmed COVID-19 patients tend to have normal CT findings [21]. Therefore, chest CT is not suggested for routine screening for COVID-19. Nevertheless, chest CT may help diagnose highly suspected patients with moderate-to-severe pulmonary symptoms found to be negative on RT-PCR tests. Furthermore, planning chest CT might be needed in cases where symptoms are suspicious of another concurrent infectious, inflammatory disorder, or neoplasm. For instance, in areas that are endemic for tuberculosis, it is important to rule out active/latent pulmonary TB before ordering immune modulators or corticosteroids. Hence, in such cases, CT is detrimental in designing the therapeutic plan. Chest CT can also be indicated in triaging the need for hospital administration of elderly (age > 65) patients with comorbidities (such as cardiovascular diseases, diabetes, respiratory diseases, obesity, and hypertension) who are at an elevated risk of COVID-related mortality and complications [22].

Cardiovascular CT Scan

Cardiovascular adverse events related to COVID-19, including acute heart failure, myocarditis, myocardial infarction, and thromboembolism, have been an important concern due to the increasing rate of mortality and morbidity. Cardiac CT with contrast favors detecting COVID-19-associated myocarditis, especially in patients undergoing lung CT. Typical CT findings in myocarditis include myocardial hypertrophy in combination with subendocardial perfusion defects [23]. CT angiography (CTA) is a noninvasive and accurate imaging modality that combines a CT scan with contrast injection to visualize arteries and veins. Several hospitals have replaced coronary angiography with CTA due to lower costs and adverse effects. In hospitalized patients, CTA has been rapidly used to detect pulmonary and aortic thromboembolism (Fig. 5.4). According to European Society of Cardiology guidelines, in stable COVID-19 subjects with elevated troponin biomarkers, coronary CTA may pave the way for identifying patients at risk and avoiding invasive approaches. However, regarding the high presence of tachycardia among COVID-19 patients, the feasibility of CTA might be hampered [25]. Additionally, pulmonary CTA might be conducted to rule out pulmonary thromboembolism when unenhanced CT scans cannot adequately demonstrate the severity of respiratory failure [26].

Brain CT Scan

Research has reported various neurological manifestations during and/or after COVID-19 infection. Although uncommon, the most reported neurological manifestations are encephalitis, encephalopathy, and stroke [27]. While neurological

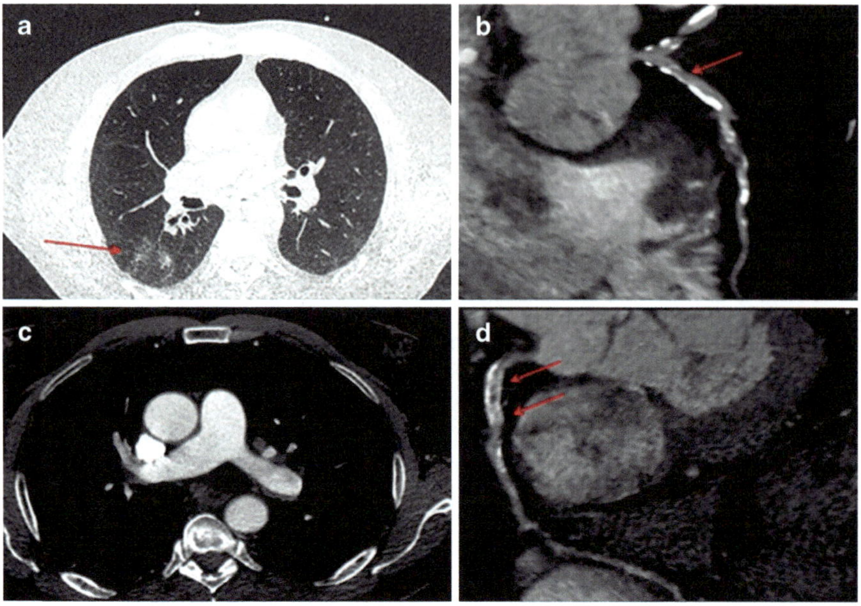

Fig. 5.4 Chest-CT and cardiac CT-scan of a patient with rapid onset dyspnea, positive D-dimer test, and SARS-CoV-2 PCR. (**a**): "ground glass" opacities on the right posterior basal segments (arrow); (**b**): mixed (arrow) and calcified atherosclerotic plaques with severe stenosis on the left anterior descending artery (LAD); (**c**): permeable pulmonary arteries; (**d**): stent on the proximal right coronary artery with moderate in-stent restenosis (arrow) [24]

symptoms are well-recognized, neuroimaging findings help us to contemplate COVID-19 infection as a differential diagnosis in patients with atypical symptoms or who have not yet been tested for COVID-19. Several brain CT findings have been reported in COVID-19 individuals, e.g., hypodense areas indicative of brain edema, subarachnoid hemorrhage, ventricular hemorrhage, and vascular lesions representative of stroke. Moreover, focal hypoattenuation in thalami, indicating acute necrotizing encephalopathy, and hyperdensity of venous structures, indicating thrombosis, have also been observed [28]. Brain edema is another possible complication of COVID-19 that can be observed with shrinkage of the cerebral ventricles and blurring of the corticosubcortical junction in severe cases and is thought to be provoked by the cytokine release syndrome (Fig. 5.5) [2].

Abdominal CT Scan

Abdominal CT has been commonly requested for unexplained, otherwise prominent abdominal pain for detecting complications of COVID-19, e.g., pancreatitis, bowel ischemia, or mesenteric thrombosis. Findings observed on abdominal CT of COVID-19 patients vary from fluid-filled colon in early phases of the disease to severe enterocolitis with wall thickening, segmental ischemia, necrosis, and

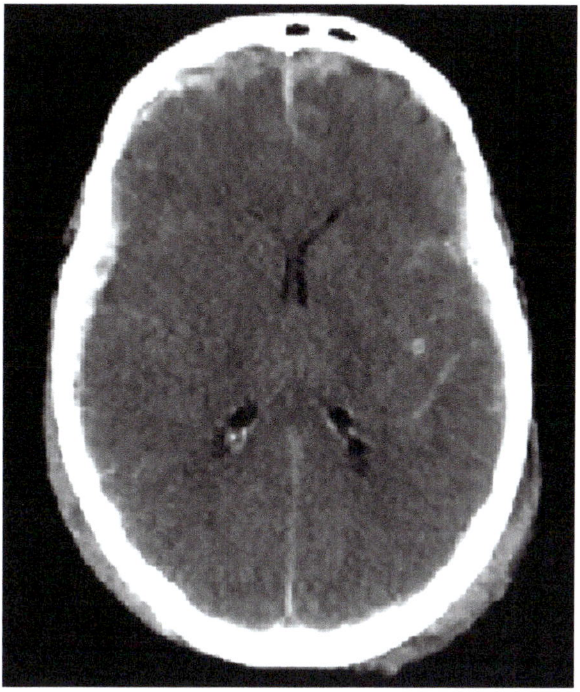

Fig. 5.5 Generalized benign cerebral edema in a 26-year-old patient with nausea, vertigo, and double vision. Shrinkage of the cerebral ventricles and blurring of the corticosubcortical junction can be observed on axial brain CT [2]

pneumatosis intestinalis. Also, pneumoperitoneum, ascites, and intussusception of the intestine have been reported [29]. Besides, regarding the inflammation in the bowel wall, a hyperdense area surrounding the intestinal wall resembling a COVID-19-specific pattern can be seen on CT images (Fig. 5.6a, b) [31]. Acute pancreatitis in COVID-19 patients can be observed as pancreatic edema, pancreatic

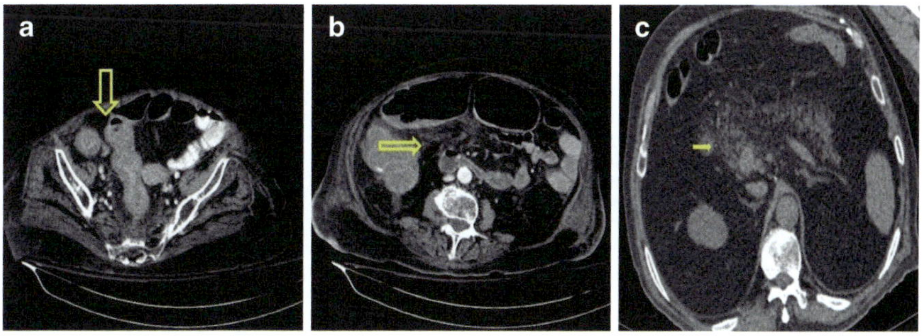

Fig. 5.6 Examples of abdominal CT scan in COVID-19 patients. (**a**) Abdominal contrast-enhanced CT (CECT) in a COVID-19+ 75-year-old male patient with abdominal pain and diarrhea diagnosed with inflammatory colitis. The arrow depicts the thickening of the bowel wall. (**b**) Arrow showing mesenteric fat stranding in the same patient. (**c**) Acute pancreatitis in an 81-year-old patient with COVID-19 pneumonia developed epigastric pain on the second day of hospitalization [30]

duct dilation, fat stranding, and pancreatic necrosis on CT images (Fig. 5.6c) [32]. However, because of the lack of etiological workups, it is still being investigated whether there are any relations between COVID-19 infection and acute pancreatitis.

Ultrasound

Ultrasound is an accessible and portable imaging modality that enables clinicians to carry out bedside examinations without transferring patients to other hospital sections. Ultrasound has a higher accuracy for the detection of lung pathologies than CXR [33]. The PPV and NPV of lung ultrasound in COVID-19 patients are 100% and 85.71%, respectively [34]. Nevertheless, Lu et al. reported that the sensitivity and specificity of lung ultrasound (LUS) were found to be different in various phases of COVID-19. They demonstrated that ultrasound had the highest sensitivity (100%) and specificity (85.7%) for screening severe lung lesions. In comparison, the diagnostic efficacy is relatively low in mild (with a specificity of 85% and sensitivity of 68.8%) and moderate (with a specificity of 76.2% and sensitivity of 77.8%) lung lesions [35]. Despite the promising features described above, performing and interpreting ultrasound requires an adequately expertized operator, which is its most important limitation.

Common lung ultrasound findings of COVID-19 patients include subpleural and trans-lobar and peripheral consolidations, irregular pleural thickening, air bronchograms, B-line artifacts, and the emergence of A-lines in the recovery time (Fig. 5.7)

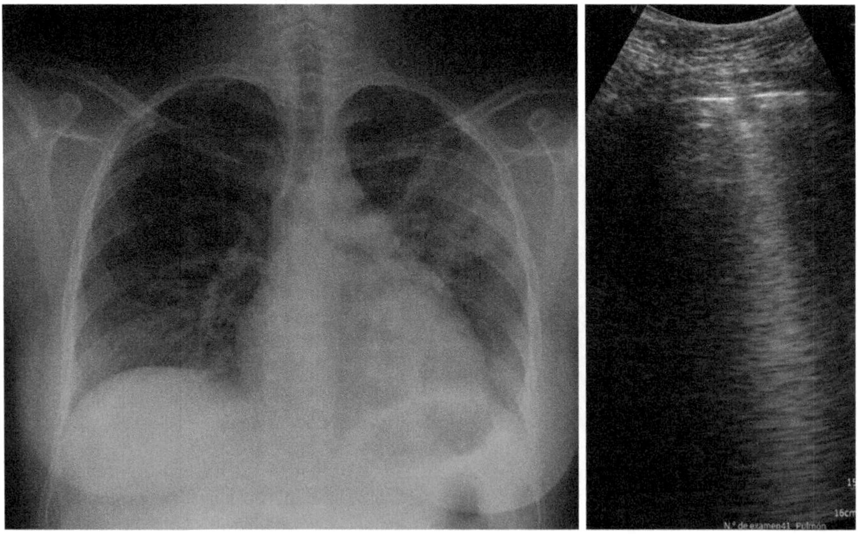

Fig. 5.7 CXR and lung ultrasonography findings in the same subject with bilateral infiltrates and small consolidation areas [36]

[37]. B-lines are vertical artifacts that appear as hyperechoic lines extending from areas of consolidation or pleura into the depth of the screen without losing intensity. B lines indicate fluid in the alveoli or interstitial space of the lungs. While pleural thickening and bilateral or unilateral B-lines are more common in earlier stages, air bronchograms and subpleural thickening dominate in later stages. In the remission phase, A-lines emerge, and prior findings tend to regress [17].

Point-of-Care Ultrasound (POCUS)

Currently, no guidelines have been defined for the use of ultrasound in COVID-19 patients. Nevertheless, one of the promising tools for both diagnosis and management of COVID-19 is the POCUS [38]. POCUS has gained attraction in several medical communities as it is affordable, does not use ionized radiation, and is also bedside available. POCUS can not only be used for COVID-19 diagnosis but also to monitor disease progression and assess related complications. These include assessing cardiogenic pleural edema, pleural and pericardial effusions, ARDS, pneumothorax, and evaluation of pulmonary embolism (PE) and deep vein thrombosis. Besides, POCUS can be used for risk stratification and prediction of ICU admission in emergency departments based on the LUS score [38].

Since the COVID-19 outbreak, several POCUS scanning protocols have been introduced. The most frequent protocols include 6-, 8-, 12-, and 14-zone on the patient's chest, with the 12-zone being commonly used [38]. This includes scanning six sites on each hemithorax: (1) lower posterior, (2) upper posterior, (3) lower lateral, (4) upper lateral, (5) lower anterior, and (6) upper anterior [39]. Scanning results will ultimately be calculated as LUS scores. LUS score is an efficient tool for assessing the aeration of the lungs. Based on ultrasound findings, each zone scanned will have a score from 0 to 3, with a total score tabulated from 12 zones. The highest score from each zone is summed up to give a total score, which ranges from 0 (all zones clean and aerated) to 36 (consolidation of all zones). Moreover, while it has been demonstrated that an LUS score of less than 5 has an NPV of 100% for the requirement of ICU admission, an elevated LUS score has been correlated with an increased mortality rate in COVID-19 [40]. Ciurba et al. showed that a LUS score of 29 was proposed as a cut-off for prediction of ICU admission in COVID-19, with a sensitivity and specificity of 80% and 97.7%, respectively [41].

Echocardiography

Echocardiography can be used in evaluating COVID-19-related cardiovascular adverse events such as thromboembolism, myocardial infarction, myocarditis, and right ventricular dysfunction [17]. During the pandemic, point-of-care echocardiography has been frequently used at the bedside to hamper the transmission risk. Hyperdynamic cardiac functions have been reported in the echocardiography of COVID-19 patients, possibly due to the inflammatory status [42]. Other echocardiographic features that have been reported include I) stress-induced cardiomyopathy (takotsubo) defined as hyperkinetic left ventricular (LV) basal segment with LV apical ballooning; II) pulmonary hypertension and right ventricle (RV) enlargement seen as intraventricular septum extended to the LV (D-sign) which is followed by LV systolic and diastolic dysfunction; III) decreased myocardial function of the whole heart in later stages [43]. Although ultrasound does not have a high sensitivity for the detection of PE in its early phases, signs of RV dysfunction, such as paradoxical movement of the intraventricular septum and severe RV-free wall akinesis with sparing of the apex (McConnell sign), have been shown to be a prognostic marker for PE detection [44].

Magnetic Resonance Imaging (MRI)

MRI has been used in different investigations to evaluate COVID-19 severity and complications, including COVID-19-related neurological and cardiac complications, rather than chest MRI, which does not provide additional information compared to chest CT.

Lung MRI

Although chest CT is broadly applied for COVID-19 imaging, advantages of MRI has made it an alternative diagnostic tool. For instance, in a study, 32 COVID-19 patients underwent chest MRI and CT (Fig. 5.8). A total of 31 patients had signs of lung involvement in both MRI and CT. GGOs were the most common CT finding, seen in 90.6% of cases. The second common pattern was consolidation observed in 43.75%. In all these cases, MRIs also showed GGOs and consolidation. 34.4% and 37.5% of patients had nodules seen on MRIs and CT scans, respectively. It was

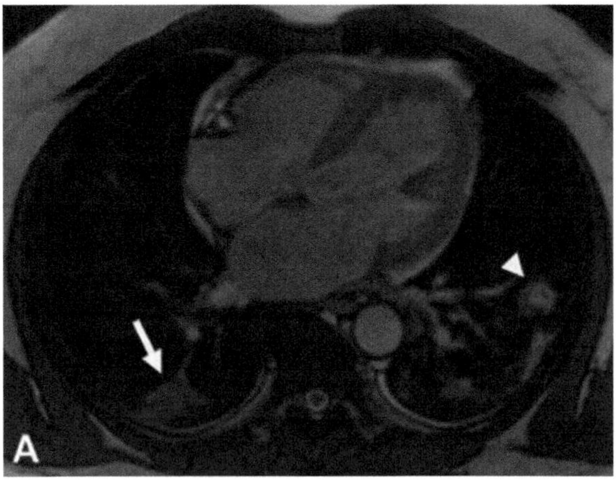

Fig. 5.8 Lung MRI examples in a patient with pulmonary COVID-19. Left arrow shows consolidation and right arrow shows nodular involvement [45]

determined that MRI has a specificity of 100% and a sensitivity of 91.67% for nodule identification. Moreover, it was demonstrated that the MRI and CT agreed nearly perfectly for consolidation and changes from previous CT, significantly for prominent pattern and interlobular septal thickening, and poorly for GGOs, fibrosis, pleural indentation, and vessel enlargement. The MRI's specificity was generally high, but its sensitivity was high for consolidation, interlobular septal thickening, and GGOs, and was low for vascular enlargement, pleural indentation, and fibrotic changes [46].

Brain MRI

Over one-third of COVID-19 patients have neurological findings, which indicates the involvement of the nervous system. Nonetheless, the majority of COVID-19 patients who have neuroimaging do not exhibit specific imaging findings [5]. As an example, in a study, 185 COVID-19 patients underwent neuroimaging, and the most frequent finding among individuals who had brain MRIs was intra-axial susceptibility abnormalities [5]. In a meta-analysis and systematic review study, researchers obtained MRIs from 3118 COVID-19 patients with neurological involvement. The most frequent abnormal MRI findings in individuals were grey matter abnormalities, subacute/acute infarcts, cerebral microbleeds (Fig. 5.9b), periventricular white matter abnormalities (Fig. 5.9c), and olfactory bulb (OB) abnormalities, with an incidence of 12%, 22%, 17%, 20%, and 22%, respectively. Other MRI findings that were observed at a lower incidence included cranial neuropathies affecting the optic, oculomotor, facial, and vestibulocochlear nerves, leptomeningeal enhancement, acute disseminated encephalomyelitis (ADEM) or ADEM-like lesions, non-traumatic intracranial hemorrhage (ICH), abnormalities in the basal ganglia and thalamus (Fig. 5.9a), posterior reversible encephalopathy syndrome (PRES),

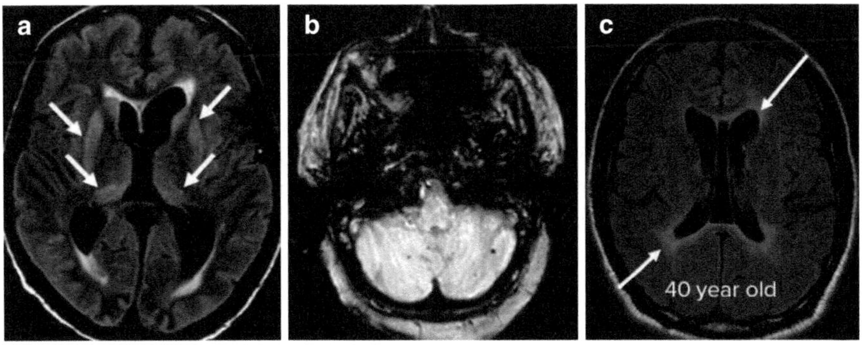

Fig. 5.9 Brain MRI findings in COVID-19 patients with central nervous system involvement. (**a**) Axial fluid-attenuated inversion recovery (FLAIR) image showing bilateral basal ganglia and thalami in a COVID-19 patient, (**b**) Isolated cerebellar microhemorrhages on susceptibility-weighted imaging (SWI) in a patient with COVID-19. (**c**) Axial FLAIR image demonstrating periventricular FLAIR hyperintensities in a 40-year-old patient [47]

cytotoxic lesions of the corpus callosum, venous thrombosis, and hypoxic-ischemic changes [48].

Another meta-analysis study evaluated the brain MRI obtained from 2125 COVID-19 patients. Encephalopathy/encephalitis, acute spontaneous ICH, cerebral microhemorrhages, and acute/subacute infarcts were reported in 3.3%, 6.9%, 5.4%, and 24.0% of patients, respectively. Furthermore, the incidence of encephalitis/encephalopathy decreased considerably in individuals with a mean or median age beyond 65 [49].

A retrospective analysis assessed the brain MRI of severe COVID-19 patients. The most common findings on MRI were medial temporal lobe signal abnormalities (43% of patients), nonconfluent multifocal white matter hyperintense lesions with variable enhancement seen with diffusion-weighted imaging and fluid-attenuated inversion recovery (FLAIR), extensive and isolated white matter microhemorrhages (24% of patients), and associated hemorrhagic lesions (30% of patients), and. The majority of patients exhibited intracerebral hemorrhage lesions, which were linked to a higher risk of ICU admission, ARDS development, and severity of clinical presentation [50]. Various case studies have depicted brain MRI findings in COVID-19 patients with simultaneous PRES. In a study that evaluated 30 COVID-19 patients with PRES, the average age was 57, and 80% needed mechanical ventilation. On brain MRI findings, restricted diffusion and superimposed foci of hemorrhage were observed in 23% and 50% of patients, respectively [51].

Cardiac MRI

COVID-19-associated cardiac injury is a common consequence observed by several studies. It is crucial to determine if cardiac damage seen during the acute stage of COVID-19 will result in cardiac morbidity and dysfunction. Therefore, as a noninvasive diagnostic technique, MRI has been used in recent investigations to evaluate whether cardiac involvement has occurred in COVID-19-infected individuals. A comprehensive assessment of 34 papers, including 199 COVID-19 patients, found that myocarditis was the most frequent finding, with an incidence of 40.2%, and 21% of the patients had normal cardiac MRIs. The most frequent imaging patterns included T1 and T2 anomalies, late gadolinium enhancement (LGE) pericardial effusion, edema on T2/short Tau inversion recovery (STIR), perfusion deficits, and extracellular volume mapping abnormalities. In inferior segments, LGE was most frequently observed in the basal-mid part of the left ventricle and subepicardial area. Ventricular function was normal in most patients. Acute myocarditis developed during a week of the initiation of symptoms. Diffuse myocardial edema without LGE was seen on native T1 mapping and T2/STIR sequences in cardiac MRI, indicating myocardial inflammation without necrosis and fibrosis [52]. Significant anomalies in wall motion were also seen in 38% of COVID-19 individuals, which were in line with the patterns of stress-induced cardiomyopathy (Fig. 5.10). Furthermore, minor pericardial effusions and pericardial enhancement were seen in 38% of subjects with COVID-19 [54].

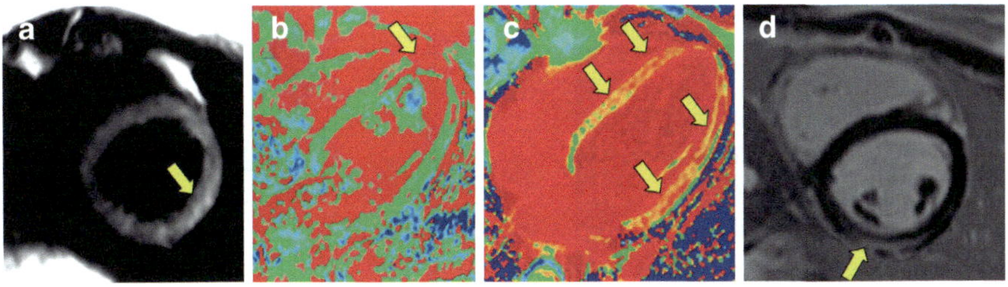

Fig. 5.10 Example of cardiac MRI in a 16-year-old patient with COVID-19-associated myocarditis. (**a**) Inferior apical wall T2 hypersignal on a short-axis plane. (**b**) High native T2 on a four-chamber T2 mapping sequence. (**c**) High native T1 on a four-chamber T1 mapping sequence. (**d**) Inferior wall late enhancement on the intramyocardial and epicardial layers of the inferior wall on a PSIR short-axis plane [53]

Nuclear Imaging

Nuclear medicine imaging is a technique for detecting radiation from different body parts after the radioactive tracer is given to the patients. PET/CT single-photon emission computed tomography (SPECT) or SPECT/CT, and planar scintigraphy are the main imaging nuclear medicine techniques. Nuclear medicine modalities help physicians assess the onset, progression, and treatment of patients who are COVID-19^{+}. They can also evaluate inflammation, fibrosis, and changes in metabolic rate in the affected organs (Fig. 5.11). Moreover, it can be effective in the early disease phase when the clinical symptoms are absent [56].

Fludeoxyglucose F18 (18-FDG) is the most common radiotracer in PET/CT imaging, showing inflammatory cell metabolic activity. The findings of FDG-PET/CT range from the early to late stages of the disease, from the low FDG uptake in GGO to higher uptake in pulmonary consolidation. Consistently, high persistent FDG uptake may be associated with delayed recovery. Also, compared to patients who received high-dose corticosteroids and asymptomatic patients, higher FDG uptake was seen in untreated patients [57]. However, FDG-PET/CT has low specificity for pulmonary involvement of COVID-19. Moreover, the maximum standardized uptake value (SUV max) and degree of FDG uptake had no established correlation with the presence, absence, severity, and duration of the disease [58]. Nuclear medicine physicians need to notice other findings in FDG-PET/CT of patients with cancer who did this imaging to follow up on their disease progression. Furthermore, early detection of the infection in these immunodeficient patients can play an important role in fighting effectively against the virus.

The extrapulmonary involvement of the COVID-19 disease, including myocarditis, can be diagnosed with FDG-PET/CT, which is noninvasive and more applicable than endomyocardial biopsy as a gold standard of myocarditis diagnosis. Few reports on the FDG-PET/CT findings indicated the elevated focal and diffuse uptake of the FDG as a sign of myocardial involvement [56]. FDG-PET/CT can also diagnose neurologic involvement, while other modalities, such as MRI, are negative. In patients with COVID-19-related neuropathy, despite the normal MRI and

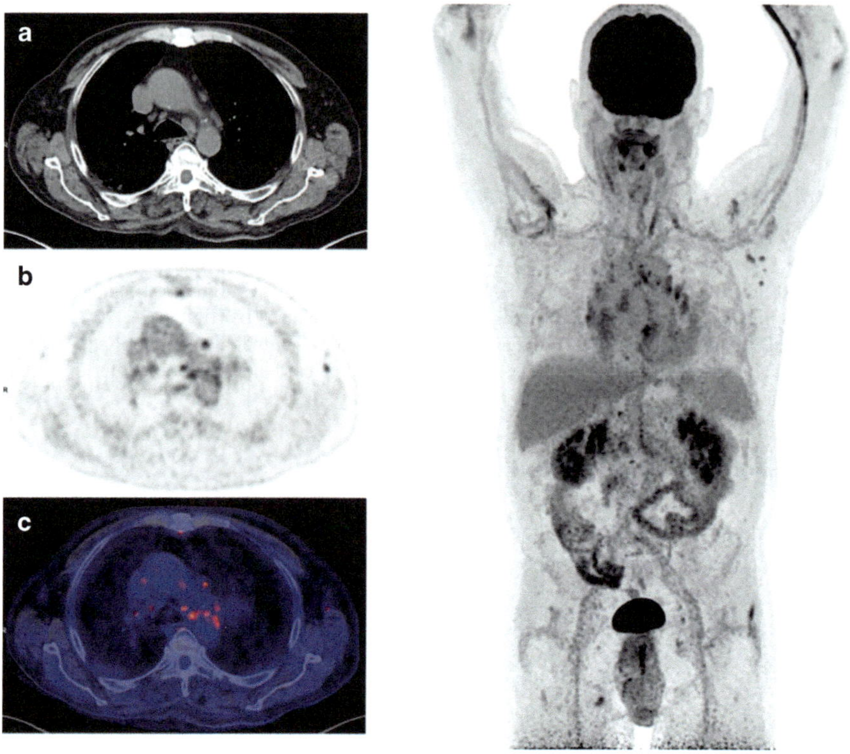

Fig. 5.11 A 64-year male with multi-lobar COVID-19 pneumonia. (**a**) Chest-CT, (**b**) 18-FDG-PET, (**c**) fused 18-FDG-PET and CT images. Right: whole-body, maximal intensity projection image, displaying mediastinal lymph nodes with 18-FDG uptake. The 18-FDG-PET/CT 3 months after symptom onset shows increased 18-FDG uptake in residual pulmonary lesions and mediastinum lymph node. Pulmonary function tests evidenced severe pulmonary diffusion impairment, with a diffusing capacity of the lungs for carbon monoxide at 41% of the predicted value [55]

cerebrospinal fluid analysis, FDG hypermetabolism in the cerebellar vermis and FDG hypometabolism in the prefrontal or orbitofrontal cortices were reported [59]. Diffuse hypermetabolism of the cerebellum and focal FDG uptake in the putamen were reported in COVID-19 patients with myoclonus and subacute cerebellar involvement [60]. However, the FDG-PET/CT limitation is that it could not be first-line imaging in an emergency setting; it could be used in complicated cases of COVID-19 with multi-organ involvement as a complementary diagnostic tool [61]. Another modality of nuclear-based imaging, SPECT, using radiolabeled technetium 99m (Tc99m), is critical in diagnosing PE and COVID-19-related cardiac involvement. SPECT is also effective for following up on patients with persistent pulmonary abnormalities due to microthrombosis and small airway disease [62].

Comparing Different Modalities for the Diagnosis of COVID-19 and Its Complications

Each imaging modality has its pros and cons in diagnosing COVID-19. Its bedside availability and acceptable sensitivity make CXR an appropriate tool for screening COVID-19 in mild-to-moderate cases. Nevertheless, CXR has low sensitivity in the early phases (<2 to 4 days), as in some cases, CXR may be normal in mild COVID-19, while chest CT in the same patient may show COVID-19-specific findings (Fig. 5.12) [63]. Chest CT has the highest sensitivity for COVID-19 diagnosis, has high accuracy, and provides high resolution. Therefore, it can be used to evaluate disease severity, early detection of lung lesions, and assessment of treatment response in hospitalized patients. Nevertheless, due to radiation exposure, the risk of hospital staff contamination, and low specificity in differentiating COVID-19 from non-COVID pneumonia, it is recommended to be used in severe cases, when COVID-19 complications are strongly suspected and in elderly patients with other comorbidities [17]. Lung MRI, on the other hand, does not provide additional benefits than a chest CT scan; however, it could be helpful in severe COVID-19 patients with CT contraindications. Chest CT is also helpful in detecting cardiac complications of COVID-19, including myocarditis [17].

On the other hand, CTA is also considered the choice method for diagnosing PTE in COVID-19 [26]. Brain CT can also help detect COVID-19-associated neurologic complications, i.e., hemorrhage and stroke [28], while abdominopelvic CT-scan can be used for assessing patients with GI COVID-19 complications, e.g., pancreatitis and colitis [32]. Ultrasonography, on the other side, is more accessible at the bedside and has high specificity and sensitivity for diagnosing severe COVID-19 lung infections but has lower accuracy for detecting mild-to-severe cases (Fig. 5.7) [35]. Nevertheless, LUS scoring based on POCUS provides more focused ICU admission prediction and accurate risk stratification in COVID-19 patients. After all, LUS is proposed as an alternative to CXR and chest CT for assessing pulmonary findings in COVID-19 patients. Nevertheless, LUS has some

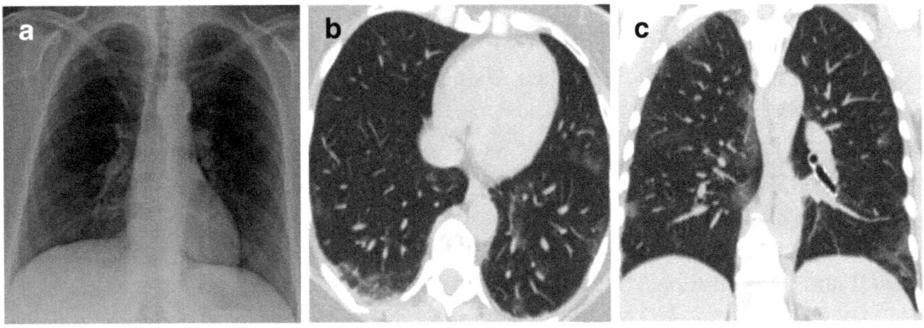

Fig. 5.12 Comparison of CXR and chest-CT findings in a 62-year-old woman with confirmed COVID-10. (**a**) No COVID-19-specific findings are observed on anteroposterior (AP) CXR, while on (**b**) axial chest-CT and (**c**) coronal chest-CT, multiple bilateral slight GGOs can be observed, especially in the subpleural areas [63]

limitations, e.g., it has limited accuracy and resolution in the aerated lungs, which suspends the evaluation of deep lung lesions [38].

Nuclear imaging techniques, specifically FDG-PET, are indicated for early-stage infections (with increased focal uptake) and long-term lung injury mediated by COVID-19. Nevertheless, it is not recommended as a standard imaging tool in an emergency during COVID-19. However, it could provide a complementary method for diagnosing different organ involvement, assessing the severity of infection, and predicting treatment response [58].

The Expanding Role of AI for Computer-Aided Diagnosis (CAD) of COVID-19

Due to radiologists' limitations in distinguishing healthy, bacterial, or other viral lung pneumonia, many studies have established AI and deep learning (DL) networks, which could improve radiologist decision-making, even providing superior performance. Studies have approved the efficacy and accuracy of DL networks, which can accurately differentiate COVID-19 in CXR images with up to 99% accuracy. For example, Mousavi et al. evaluated a convolutional neural network (CNN) model and achieved promising results in separating COVID-19 against viral infections with 90% and 99% accuracy in healthy against COVID-19 scenarios [64]. In other similar experiences, Loannis et al. and Narin et al. have reached 98% accuracy in differentiating CXR of healthy and pneumonia cases by utilizing transfer learning networks [65, 66]. Sethy et al. used a support vector machine and ResNet50 to classify COVID+ from COVID- patients and reached sensitivity, accuracy, and FPR of 95.33%, 95.33%, and 2.33%, respectively [67]. For more information on the current advances of DL in COVID-19 diagnosis, refer to: https://doi.org/10.1007/s42979-022-01326-3.

In 2023, Foysal et al. [68] used a deformable CNN-based ResNet-50 model to diagnose COVID-19 on 2481 chest CT scans. The proposed model achieved 97.6% accuracy, with 98.5% specificity and 96.5% sensitivity. More information on the current advances of DL models for COVID-19 diagnosis based on chest CT images can be found here: https://doi.org/10.1007/s12559-022-10076-6. AI has been applied less in MRI in a few studies. Using AI-based MRI volumetry, the potential effect of COVID-19 on brain volumes was evaluated in patients who recovered from mild and severe COVID-19/asymptomatic versus healthy controls. The percentiles of different brain regions and brain volumes varied statistically significantly between groups. In COVID-19 patients, there were notable reductions in brain volume, which increased in more severe cases and primarily impacted the right thalamus, frontal and parietal lobes, and supratentorial grey matter [69].

Conclusion

Imaging techniques have a crucial role in diagnosing COVID-19. CXR is a reliable tool for screening patients, specifically when RT-PCR is unavailable. Chest-CT scan has high sensitivity and should be applied in severe COVID-19 patients and when other comorbidities are suspected. LUS is not routinely recommended but can be used as an alternative to CXR and CT in emergencies. In addition to the pulmonary findings of COVID-19, nonpulmonary imaging also provides important clues for possible complications and comorbidities. Cardiovascular and cerebrovascular CT scans, brain MRI, FDG-PET, PET-CT, and cardiac MCI are indicated in case of suspected COVID-19 complications. Imaging patterns of COVID-19 are essential for radiologists since non-COVID patients, e.g., cancer, might have COVID-19-specific findings as well. To conclude, imaging modalities have a critical role in diagnosing COVID-19 infection and related complications and assessing treatment response.

Acknowledgments The authors would like to acknowledge Dr. Yas Shahbakhsh, MD, Shahid Beheshti University of Medical Sciences, and Dr. Zahra Ghahremani, MD, Zanjan University of Medical Sciences for their support in reporting the imaging findings of the patients.

Conflicts of Interests The authors declare that they have no commercial or scientific issues considered conflicts of interest.

References

1. Elahi R, Karami P, Bazargan M, Ahmadi S, Azhideh A, Esmaeilzadeh A. Analysis of the current, past, and future evolution of COVID-19. In: Fractal signatures in the dynamics of an epidemiology. CRC Press; 2023. p. 1–18.
2. Heidary A, Elahi R, Nazari M. Benign cerebral edema and increased intracranial pressure (ICP) as manifestations of COVID-19 reinfection; a case report. J Clin Lab Med. 2021;6(1):138.
3. Ahmadi S, Bazargan M, Elahi R, Esmaeilzadeh A. Immune evasion of severe acute respiratory syndrome coronavirus-2 (SARS-CoV-2); molecular approaches. Mol Immunol. 2023;156:10–9.
4. Kanne JP, Bai H, Bernheim A, Chung M, Haramati LB, Kallmes DF, et al. COVID-19 imaging: what we know now and what remains unknown. Radiology. 2021;299(3):E262–e79.
5. Klironomos S, Tzortzakakis A, Kits A, Öhberg C, Kollia E, Ahoromazdae A, et al. Nervous system involvement in coronavirus disease 2019: results from a retrospective consecutive neuroimaging cohort. Radiology. 2020;297(3):E324–e34.
6. Costa YM, Silva JSA, Teixeira LO, Pereira RM, Bertolini D, Britto JAS, et al. Covid-19 detection on chest x-ray and ct scan: a review of the top-100 most cited papers. Sensors. 2022;22(19):7303.
7. Jacobi A, Chung M, Bernheim A, Eber C. Portable chest X-ray in coronavirus disease-19 (COVID-19): a pictorial review. Clin Imaging. 2020;64:35–42.
8. Stephanie S, Shum T, Cleveland H, Challa SR, Herring A, Jacobson FL, et al. Determinants of chest radiography sensitivity for COVID-19: a multi-institutional study in the United States. Radiol Cardiothorac Imaging. 2020;2(5):e200337.

9. López-Cabrera JD, Orozco-Morales R, Portal-Diaz JA, Lovelle-Enríquez O, Pérez-Díaz M. Current limitations to identify COVID-19 using artificial intelligence with chest X-ray imaging. Health Technol. 2021;11(2):411–24.
10. Wong HYF, Lam HYS, Fong AH-T, Leung ST, Chin TW-Y, Lo CSY, et al. Frequency and distribution of chest radiographic findings in patients positive for COVID-19. Radiology. 2020;296(2):E72–E8.
11. Tsai EB, Simpson S, Lungren MP, Hershman M, Roshkovan L, Colak E, et al. The RSNA international COVID-19 open radiology database (RICORD). Radiology. 2021;299(1):E204–e13.
12. Ng M-Y, Lee EY, Yang J, Yang F, Li X, Wang H, et al. Imaging profile of the COVID-19 infection: radiologic findings and literature review. Radiol Cardiothorac Imaging. 2020;2(1):e200034.
13. Bernheim A, Mei X, Huang M, Yang Y, Fayad ZA, Zhang N, et al. Chest CT findings in coronavirus disease-19 (COVID-19): relationship to duration of infection. Radiology. 2020;295(3):685–91.
14. Salehi S, Abedi A, Balakrishnan S, Gholamrezanezhad A. Coronavirus disease 2019 (COVID-19): a systematic review of imaging findings in 919 patients. Am J Roentgenol. 2020;215(1):87–93.
15. Khatami F, Saatchi M, Zadeh SST, Aghamir ZS, Shabestari AN, Reis LO, Aghamir SMK. A meta-analysis of accuracy and sensitivity of chest CT and RT-PCR in COVID-19 diagnosis. Sci Rep. 2020;10(1):22402.
16. Degli Antoni M, Crosato V, Pennati F, Borghesi A, Cristini G, Allegri R, et al. COVID-19 pneumonia with migratory pattern in agammaglobulinemic patients: a report of two cases and review of literature. Tomography. 2023;9(3):894–900.
17. Varadarajan V, Shabani M, Ambale Venkatesh B, Lima JAC. Role of imaging in diagnosis and management of COVID-19: a multiorgan multimodality imaging review. Front Med. 2021:8.
18. McLaren TA, Gruden JF, Green DB. The Bullseye sign: a variant of the reverse halo sign in COVID-19 pneumonia. Clin Imaging. 2020;68:191–6.
19. Wang Y, Dong C, Hu Y, Li C, Ren Q, Zhang X, et al. Temporal changes of CT findings in 90 patients with COVID-19 pneumonia: a longitudinal study. Radiology. 2020;296(2):E55–e64.
20. Pan F, Ye T, Sun P, Gui S, Liang B, Li L, et al. Time course of lung changes at chest CT during recovery from coronavirus disease 2019 (COVID-19). Radiology. 2020;295(3):715–21.
21. Bernheim A, Mei X, Huang M, Yang Y, Fayad ZA, Zhang N, et al. Chest CT findings in coronavirus disease-19 (COVID-19): relationship to duration of infection. Radiology. 2020;295(3):200463.
22. Garg M, Prabhakar N, Bhalla AS, Irodi A, Sehgal I, Debi U, et al. Computed tomography chest in COVID-19: when & why? Indian J Med Res. 2021;153(1 & 2):86–92.
23. Kim IC, Kim JY, Kim HA, Han S. COVID-19-related myocarditis in a 21-year-old female patient. Eur Heart J. 2020;41(19):1859.
24. Mester A, Benedek I, Rat N, Tolescu C, Polexa SA, Benedek T. Imaging cardiovascular inflammation in the COVID-19 era. Diagnostics. 2021;11(6):1114.
25. Oudkerk M, Büller HR, Kuijpers D, van Es N, Oudkerk SF, McLoud T, et al. Diagnosis, prevention, and treatment of thromboembolic complications in COVID-19: report of the National Institute for Public Health of The Netherlands. Radiology. 2020;297(1):E216–e22.
26. European Society of Cardiology guidance for the diagnosis and management of cardiovascular disease during the COVID-19 pandemic: part 1-epidemiology, pathophysiology, and diagnosis. Eur Heart J. 2022;43(11):1033–58.
27. Kremer S, Lersy F, Anheim M, Merdji H, Schenck M, Oesterlé H, et al. Neurologic and neuroimaging findings in patients with COVID-19: a retrospective multicenter study. Neurology. 2020;95(13):e1868–e82.
28. Mahammedi A, Saba L, Vagal A, Leali M, Rossi A, Gaskill M, et al. Imaging of neurologic disease in hospitalized patients with COVID-19: an Italian multicenter retrospective observational study. Radiology 2020;297(2):E270–E273.
29. Lui K, Wilson MP, Low G. Abdominal imaging findings in patients with SARS-CoV-2 infection: a scoping review. Abdom Radiol (NY). 2021;46(3):1249–55.

30. Mavropoulou X, Psoma E, Papachristodoulou A, Pyrrou N, Spanou E, Alexandratou M, et al. Gastrointestinal imaging findings in the era of COVID-19: a pictorial review. Medicina (B Aires). 2023;59(7):1332.
31. Takifuji K, Terasawa H, Oka M, Sahara M, Hara T, Itoh H. Computerized tomography scan findings of a patient with severe enterocolitis associated with the coronavirus disease 2019: a case report; 2020.
32. Hadi A, Werge M, Kristiansen KT, Pedersen UG, Karstensen JG, Novovic S, Gluud LL. Coronavirus Disease-19 (COVID-19) associated with severe acute pancreatitis: case report on three family members. Pancreatology. 2020;20(4):665–7.
33. Mojoli F, Bouhemad B, Mongodi S, Lichtenstein D. Lung ultrasound for critically ill patients. Am J Respir Crit Care Med. 2019;199(6):701–14.
34. Lin MD, Zhang MD, Kou MD, Zhao MD, Li MD, Wu MD, et al. Application value of lung ultrasound in asymptomatic patients with confirmed COVID-19. Adv Ultras Diagn Ther. 2020;4:67.
35. Lu W, Zhang S, Chen B, Chen J, Xian J, Lin Y, et al. A clinical study of noninvasive assessment of lung lesions in patients with coronavirus disease-19 (COVID-19) by bedside ultrasound. Ultraschall Med. 2020;41(3):300–7.
36. Caroselli C, Blaivas M, Tung Chen Y, Marcosignori M, Cherubini A, Longo D. A modified corona score using lung ultrasound to identify COVID-19 patients. Diagnostics. 2024;14(1):51.
37. Xing C, Li Q, Du H, Kang W, Lian J, Yuan L. Lung ultrasound findings in patients with COVID-19 pneumonia. Crit Care. 2020;24(1):174.
38. Chua MT, Boon Y, Yeoh CK, Li Z, Goh CJM, Kuan WS. Point-of-care ultrasound use in COVID-19: a narrative review. Ann Transl Med. 2024;12(1):13.
39. Kok B, Schuit F, Lieveld A, Azijli K, Nanayakkara PW, Bosch F. Comparing lung ultrasound: extensive versus short in COVID-19 (CLUES): a multicentre, observational study at the emergency department. BMJ Open. 2021;11(9):e048795.
40. Bonadia N, Carnicelli A, Piano A, Buonsenso D, Gilardi E, Kadhim C, et al. Lung ultrasound findings are associated with mortality and need for intensive care admission in COVID-19 patients evaluated in the emergency department. Ultrasound Med Biol. 2020;46(11):2927–37.
41. Ciurba BE, Sárközi HK, Szabó IA, Ianoşi ES, Grigorescu BL, Csipor-Fodor A, et al. Applicability of lung ultrasound in the assessment of COVID-19 pneumonia: diagnostic accuracy and clinical correlations. Respir Investig. 2022;60(6):762–71.
42. Mahmoud-Elsayed HM, Moody WE, Bradlow WM, Khan-Kheil AM, Senior J, Hudsmith LE, Steeds RP. Echocardiographic findings in patients with COVID-19 pneumonia. Can J Cardiol. 2020;36(8):1203–7.
43. Churchill TW, Bertrand PB, Bernard S, Namasivayam M, Churchill J, Crousillat D, et al. Echocardiographic features of COVID-19 illness and association with cardiac biomarkers. J Am Soc Echocardiogr. 2020;33(8):1053–4.
44. Manna S, Wruble J, Maron SZ, Toussie D, Voutsinas N, Finkelstein M, et al. COVID-19: a multimodality review of radiologic techniques, clinical utility, and imaging features. Radiol Cardiothorac Imaging. 2020;2(3):e200210.
45. Camastra G, Arcari L, Ciolina F, Danti M, Ansalone G, Cacciotti L, Sbarbati S. Characterization of COVID-19-related lung involvement in patients undergoing magnetic resonance T1 and T2 mapping imaging: a pilot study. J Imaging. 2022;8(12):314.
46. Ates OF, Taydas O, Dheir H. Thorax magnetic resonance imaging findings in patients with coronavirus disease (COVID-19). Acad Radiol. 2020;27(10):1373–8.
47. Tanwar M, Singhal A, Alizadeh M, Sotoudeh H. Magnetic resonance imaging (MRI) findings in COVID-19 associated encephalitis. Neurol Int. 2023;15(1):55–68.
48. Afsahi AM, Norbash AM, Syed SF, Sedaghat M, Afsahi G, Shahidi R, et al. Brain MRI findings in neurologically symptomatic COVID-19 patients: a systematic review and meta-analysis. J Neurol. 2023;270(11):5131–54.
49. Choi Y, Lee MK. Neuroimaging findings of brain MRI and CT in patients with COVID-19: a systematic review and meta-analysis. Eur J Radiol. 2020;133:109393.

50. Kremer S, Lersy F, de Sèze J, Ferré JC, Maamar A, Carsin-Nicol B, et al. Brain MRI findings in severe COVID-19: a retrospective observational study. Radiology. 2020;297(2):E242–e51.
51. Yeahia R, Schefflein J, Chiarolanzio P, Rozenstein A, Gomes W, Ali S, et al. Brain MRI findings in COVID-19 patients with PRES: a systematic review. Clin Imaging. 2022;81:107–13.
52. Ojha V, Verma M, Pandey NN, Mani A, Malhi AS, Kumar S, et al. Cardiac magnetic resonance imaging in coronavirus disease 2019 (COVID-19): a systematic review of cardiac magnetic resonance imaging findings in 199 patients. J Thorac Imaging. 2021;36(2):73–83.
53. Militaru S, Mihu A, Genunche-Dumitrescu AV, Neagoe CD, Avramescu TE, Istratoaie O, et al. Multimodality cardiac imaging in COVID-19 infection. Medicina (Kaunas). 2023;59(7):1223.
54. Luetkens JA, Isaak A, Öztürk C, Mesropyan N, Monin M, Schlabe S, et al. Cardiac MRI in suspected acute COVID-19 myocarditis. Radiol Cardiothorac Imaging. 2021;3(2):e200628.
55. Triviño-Ibáñez EM, Jiménez-Rodríguez BM, Rudolphi-Solero T, García-Rivero EY, Rodríguez-Fernández A, Llamas-Elvira JM, et al. [18F]FDG PET/CT in short-term complications of COVID-19: metabolic markers of persistent inflammation and impaired respiratory function. Diagnostics. 2022;12(4):835.
56. Halsey R, Priftakis D, Mackenzie S, Wan S, Davis LM, Lilburn D, et al. COVID-19 in the act: incidental 18F-FDG PET/CT findings in asymptomatic patients and those with symptoms not primarily correlated with COVID-19 during the United Kingdom coronavirus lockdown. Eur J Nucl Med Mol Imaging. 2021;48:269–81.
57. Thornton A, Fraioli F, Wan S, Garthwaite HS, Ganeshan B, Shortman RI, et al. Evolution of 18F-FDG PET/CT findings in patients after COVID-19: an initial investigation. J Nucl Med 2022;63(2):270–273.
58. Wakfie-Corieh CG, Blanes Garcia AM, Ferrando-Castagnetto F, Valhondo-Rama R, Ortega Candil A, Rodriguez Rey C, et al. Assessment of extra-parenchymal lung involvement in asymptomatic cancer patients with COVID-19 pneumonia detected on 18 F-FDG PET-CT studies. Eur J Nucl Med Mol Imaging. 2021;48:768–76.
59. Delorme C, Paccoud O, Kas A, Hesters A, Bombois S, Shambrook P, et al. COVID-19-related encephalopathy: a case series with brain FDG-positron-emission tomography/computed tomography findings. Eur J Neurol. 2020;27(12):2651–7.
60. Grimaldi S, Lagarde S, Harlé J-R, Boucraut J, Guedj E. Autoimmune encephalitis concomitant with SARS-CoV-2 infection: insight from 18F-FDG PET imaging and neuronal autoantibodies. J Nucl Med. 2020;61(12):1726–9.
61. Elsakka A, Yeh R, Das J, editors. The clinical utility of molecular imaging in COVID-19: an update. Seminars in Nuclear Medicine. Elsevier; 2023.
62. Evbuomwan O, Endres W, Tebeila T, Engelbrecht G. Identification and follow-up of COVID-19 related matching ventilation and perfusion defects on functional imaging using VQ SPECT/CT. Nucl Med Mol Imaging. 2023;57(1):9–15.
63. Maino C, Franco PN, Talei Franzesi C, Giandola T, Ragusi M, Corso R, Ippolito D. Role of imaging in the management of patients with SARS-CoV-2 lung involvement admitted to the emergency department: a systematic review. Diagnostics. 2023;13(11):1856.
64. Mousavi Z, Shahini N, Sheykhivand S, Mojtahedi S, Arshadi A. COVID-19 detection using chest X-ray images based on a developed deep neural network. SLAS Technol. 2022;27(1):63–75.
65. Narin A, Kaya C, Pamuk Z. Automatic detection of coronavirus disease (covid-19) using x-ray images and deep convolutional neural networks. Pattern Anal Applic. 2021;24:1207–20.
66. Apostolopoulos ID, Mpesiana TA. Covid-19: automatic detection from x-ray images utilizing transfer learning with convolutional neural networks. Phys Eng Sci Med. 2020;43:635–40.
67. Sethy PK, Behera SK. Detection of coronavirus disease (covid-19) based on deep features. 2020.

68. Foysal M, Hossain A, Yassine A, Hossain MS. Detection of COVID-19 case from chest CT images using deformable deep convolutional neural network. J Healthc Eng. 2023;2023:4301745.
69. Bendella Z, Widmann CN, Layer JP, Layer YL, Haase R, Sauer M, et al. Brain volume changes after COVID-19 compared to healthy controls by artificial intelligence-based MRI volumetry. Diagnostics (Basel). 2023;13(10):1716.

Nucleic Acid Amplification Testing (NAAT)-Based COVID-19 Kits

6

Bhumi M. Shah and Palmi Modi

Abstract

SARS-CoV-2 is a novel coronavirus which is responsible for the infectious disease COVID-19. Infection is transmitted when infected individual's respiratory droplets meet healthy person via sneezing, coughing, or speaking, affect respiratory symptoms, and cause pneumonia and in some cases respiratory failure. Early and accurate diagnosis is the main requirement to reduce its spread, mitigation of potential consequences as well as initiation of timely treatment due to its high level of contagiousness. Initially, reverse transcriptase polymerase chain reaction was employed for the detection of COVID-19. Nowadays, other alternative techniques have also gained popularity for the detection, i.e., colloidal gold immunochromatographic assay, loop-mediated isothermal amplification, clustered regularly interspaced short palindromic repeats, electrochemical sensors, etc. However, they vary in their efficacy, specificity, sensitivity, accuracy, and cost. This summarization of literature on COVID-19 diagnosis tools showed the advantages and disadvantages of these methods since accurate diagnosis plays a pivotal role in interrupting the chains of transmission and improving patients' outcomes. These compilations can also help to enhance detection efficiency as well as quality in greater detail.

Keywords

SARS-CoV-2 · Molecular assay · RT-PCR · RT-LAMP · TMA · CRISPR · Nucleic Acid Hybridization · Amplicon-Based Metagenomic Sequencing

B. M. Shah (✉) · P. Modi
Department of Pharmaceutical Chemistry, L. J. Institute of Pharmacy, LJ University, Ahmedabad, Gujarat, India

M. K. Ghosh, M. Basu (eds.), *Pathogenesis, Management and Socio-Economic Impact of COVID-19 Pandemic*, Medical Virology: From Pathogenesis to Disease Control, https://doi.org/10.1007/978-981-95-1426-7_6

Introduction

COVID-19, caused by the SARS-CoV-2 virus, first appeared in China in December 2019. The disease rapidly spread across the globe, leading the World Health Organization (WHO) to declare it a global pandemic in March 2020 [1]. The primary indications of COVID-19 infection are fever, cough, and other respiratory symptoms. In some cases, these symptoms can progress into severe pneumonia, and unfortunately, fatalities have been observed [35]. As these are the foremost indications of the disease, some patients do not report any related symptoms and become potential candidates to transmit the virus to other healthy individuals. SARS-CoV-2 primarily spreads between humans through respiratory droplets. This transmission occurs when an infected person sneezes, coughs, or speaks, releasing droplets that can reach a healthy person within a distance of up to 2 meters. According to the World Health Organization (WHO), by April 28, 2024, there were 775,379,864 confirmed cases of COVID-19 and 7,047,396 related deaths globally [7]. Because of many undetected asymptomatic cases, the original number of infections could be higher than reported. Asymptomatic individuals are just as contagious as those with symptoms, and it is estimated that each symptomatic patient can infect up to three other people. Consequently, developing rapid and accurate detection methods for SARS-CoV-2 is essential to prevent further spread of the virus.

Structural proteins that form SARS-CoV-2 are surface glycoprotein (S), envelop protein (E), membrane protein (M), and nucleocapsid protein (N). S protein is made up of S1 and S2 subunits. Among these, the S1 subunit is the major surface antigen involved in interaction with the host cell receptor, while the S2 subunit facilitates membrane fusion, allowing the entry of SARS-CoV-2 RNA genomes into host cells. The smallest structural protein is the E protein (8.4 to 12 kDa), which plays a role in the meeting and release of viral particles in host cells [36].

The M protein is the most abundant in viral particles and is essential for viral assembly and internal stability. The N protein is a key structural component essential for the replication and transcription of viral RNA. Additionally, the N protein contributes to the construction and maintenance of ribonucleoprotein complexes, wrapping the enveloped genome into the virus, and regulating the cycle of the host cell, all of which encourage the proliferation as well as transmission of the virus [11].

Currently, COVID-19 diagnostic tests are categorized into two main types. The first category includes molecular assays, which detect viral RNA through polymerase chain reaction (PCR)-based methods or nucleic acid hybridization techniques. The second category encompasses serological and immunological assays that identify antibodies produced in response to the virus. Both types of tests are essential for managing the COVID-19 pandemic. PCR tests typically target three conserved regions of the SARS-CoV-2 genome: E, N, and ORF1ab, while serological assays focus on detecting antibodies against the S and N proteins [46]. In addition to RT-PCR and immunoassays, other detection methods such as electrochemical sensors and point-of-care testing have also been developed for identifying SARS-CoV-2 infections.

In this summary, we reviewed the advantages and limitations of these methods, including their cost, the requirement for trained personnel, time, efficiency, quality, and sensitivity. Since accurate diagnosis is pivotal for interrupting transmission chains and improving patient outcomes, this compilation can help enhance detection efficiency and quality in greater detail.

Molecular Assay for Detection of Nucleic Acid

Molecular-level diagnostics, such as nucleic acid testing, are among the most significant areas of modern medicine. Nucleic acid testing is especially well-suited for accurate COVID-19 diagnosis due to its ability to specifically identify pathogens. These tests include reverse transcriptase polymerase chain reaction (RT-PCR), gene sequencing, loop-mediated isothermal amplification (LAMP), cartridge-based nucleic acid amplification tests (CBNAAT), and clustered regularly interspaced short palindromic repeats (CRISPR) [14]. Among these, PCR is highly sensitive and specific for virus detection, requiring thermal cycling. In contrast, isothermal nucleic acid amplification operates at a persistent temperature, offering rapid detection without requiring thermal cycling. Phylogenetic analysis and complete genome sequencing indicate that SARS-CoV-2 is more closely correlated to bat-SARS-CoV than to human coronaviruses (HCoVs), though the symptoms are like those caused by HCoVs. Most molecular diagnostic methods rely on the viral genome, its protein compositions, and structural changes before and after infection. Combining and optimizing different methods can enhance clinical detection rates. Table 6.1 depicts the molecular diagnostic tests for SARS-CoV-2 authorized by emergency use authorizations (EUAs) over the past two years [24].

Reverse Transcription Polymerase Chain Reaction (RT-PCR)

RT-PCR, a cutting-edge nuclear-derived technique, amplifies even small amounts of viral genetic material from a sample and is widely regarded as the gold standard for detecting the SARS-CoV-2 virus. RT-PCR tests typically use samples from the upper respiratory tract, although some studies have also utilized saliva, serum, stool, or ocular secretions [43, 46]. To detect the COVID-19 virus, the process begins with reverse transcriptase, an RNA-dependent DNA polymerase, converting the viral RNA into complementary DNA (cDNA). In this reaction, specially designed DNA primers recognize the viral RNA genome sequences and initiate the synthesis of a short cDNA strand using reverse transcriptase. The cDNA is then amplified using sequence-specific DNA probes or fluorescent dyes. These samples are placed in a PCR thermocycler, where temperature changes trigger the chemical reactions necessary to replicate the genetic material. The amplification process is automatically repeated for approximately 40 cycles, or until the viral cDNA is detectable by a fluorescent or electrical signal [41].

Table 6.1 Authorized SARS-CoV-2 molecular diagnostic tests approved by emergency use authorizations (EUAs)

No	Entity	Diagnostic	Attributes	Sample source	Date of EUA issue
1	Roche Molecular Systems, Inc.	cobas liat SARS-CoV-2, Influenza A/B & RSV nucleic acid test	RT-PCR	Anterior nasal and Nasopharyngeal swab	07-06-2024
2	LMSI, LLC d/b/a Lighthouse Lab Services	SalivaNow SARS-CoV-2 Assay	Real-time RT-PCR	Saliva	01-04-2024
3	Tetracore, Inc.	EZ-SARS-CoV-2 Real-Time RT-PCR	Real-time RT-PCR	Mid-turbinate nasal swab	19-03-2024
4	RCA Laboratory Services LLC dba GENETWORx	Gx HTIQ SARS-CoV-2 Test	Real-time RT-PCR	Mid-turbinate and anterior nasal swab	08-12-2023
5	Roche Molecular Systems	cobas SARS-CoV-2 & Influenza A/B v2	Real-time RT-PCR	Nasal or nasopharyngeal swab	16-11-2023
6	MAWD Laboratories	MAWD Laboratories SARS-CoV-2 Dual Target by RT-PCR	Real-time RT-PCR	Nasopharyngeal swab	13-10-2023
7	3EO Health, Inc.	3EO Health COVID-19 Test	RT-LAMP	Anterior nasal swab	19-09-2023
8	Alphadera Labs, LLC	ALPHADx SARS-COV-2 RT-PCR Test	Real-time RT-PCR	Nasopharyngeal and Mid-turbinate swab	20-07-2023
9	Discover Labs	Discover Labs COVID-19 Assay	Real-time RT-PCR	nasal, nasopharyngeal, and oropharyngeal swab	30-06-2023
10	Michigan State University laboratories	In-Dx SARS-CoV-2 RT-LAMP Assay	RT-LAMP	Nasopharyngeal swab	08-06-2023
11	Acutis Diagnostics	SARS-CoV-2 Acutis Multiplex Assay	Real-time RT-PCR	Nasopharyngeal swab	05-06-2023
12	Access Medical Laboratories, Inc.	Global Direct RT-PCR Test	Real-time RT-PCR	Anterior nasal and Nasopharyngeal swab	03-05-2023

(continued)

Table 6.1 (continued)

No	Entity	Diagnostic	Attributes	Sample source	Date of EUA issue
13	University of Massachusetts	ICTC SARS-CoV-2 RT-PCR Assay	Real-time RT-PCR	Anterior nasal swab	26-04-2023
14	MedArbor, LLC dba MedArbor Diagnostics	MedArbor Diagnostics SARS-CoV-2 Assay	Real-time RT-PCR	nasopharyngeal, oropharyngeal, anterior nasal, mid-turbinate nasal swab, nasopharyngeal wash/aspirate, and nasal aspirate	19-04-2023
15	Quest Diagnostics Nichols Institute	Quest COVID-19 Nucleic Acid Test Collection Kit	Home Collection Kit, Screening	Anterior nasal swab	10-03-2023
16	Pfizer Inc.	Lucira by Pfizer COVID-19 & Flu Home Test	RT-LAMP	Nasal swab	24-02-2023
17	ADL Diagnostics, Inc.	The AscencioDx COVID-19 Test and The AscencioDx Molecular Detector	RT-LAMP	Nasal swab	08-02-2023
18	LumiraDx UK Ltd.	LumiraDx SARS-CoV-2 & Flu A/B RNA STAR Complete Assay	RT, Non-isothermal Nucleic Acid Amplification qSTAR	Anterior nasal and Nasopharyngeal swab	03-02-2023

Traditionally, RT-PCR has been performed using either a single-step or double-step process. All necessary primers and reagents are combined in a single tube to carry out the whole RT-PCR reaction in single-step real-time RT-PCR. In contrast, double-step real-time RT-PCR involves using separate tubes for the reverse transcription and amplification reactions. Double-step RT-PCR offers higher flexibility and sensitivity compared to the single-step method. It needs minimal starting material and allows for the storage of cDNA, which can later be used to quantify multiple targets [41]. However, single-step RT-PCR is often preferred due to its limited sample handling, quick setup, reduced bench time, and decreased risk in the error of pipetting as well as cross-contamination during the reverse transcription and real-time PCR steps. However, the single-step method requires a higher RT reaction temperature than the double-step approach.

Protein E, protein N, RNA-dependent RNA polymerase (RdRp), and ORF1ab are key viral targets of SARS-CoV-2 for real-time RT-PCR testing. The detection accuracy is influenced by the design of the primer probes and the selection and combination of amplification sites in primer design. The WHO recommends RT-PCR testing targeting the E protein, validated using the RdRp gene [13]. Research by Chan et al. found that methods based on ORF1B-NSP14 provide better detection than those using RdRp [8]. Additionally, some investigators have advanced with multiplex PCR methods that amplify two gene regions within the same PCR reaction. These methods have demonstrated results consistent with those obtained from detection systems approved by the Centers for Disease Control and Prevention (CDC) [39].

RT-PCR results can be affected by mutations in the target gene region, potentially leading to false-negative results, particularly when the viral load falls below the detection threshold. Large-scale testing typically shows a 30% positive detection rate. The type of sample and timing of collection significantly impact the viral load. Swab specimens from the nasopharynx and oropharynx are typically collected after symptoms appear. Over time, even as patients continue to shed the SARS-CoV-2 virus, the positive detection rate decreases [31].

In conclusion, RT-PCR remains the "gold standard" for the diagnosis of SARS-CoV-2 because of its excellent specificity and sensitivity. However, it has several drawbacks, including vulnerability to RNA mutations, stringent requirements for machinery, reagents, sample cleanliness, and specialized expertise. Additionally, there are potential issues with false negatives and false positives. Despite these challenges, Chavda et al. suggest that RT-PCR-based detection methods still warrant further research and development [10].

Isothermal Nucleic Acid Amplification

General RT-PCR techniques require temperature changes during the amplification cycle, necessitating a thermal cycler. In contrast, isothermal nucleic acid amplification techniques enable amplification continuously at a similar temperature, eliminating the need for a thermal cycler. Several methods have been developed based on this principle.

Reverse Transcription Loop-Mediated Isothermal Amplification (RT-LAMP)

RT-LAMP is a fast as well as cost-friendly substitute for SARS-CoV-2 diagnosis. This technique uses four primers targeting six distinct genes or regions and operates at a constant temperature of 65 °C to improve sensitivity. By combining loop-mediated isothermal amplification (LAMP) with reverse transcription, RT-LAMP enables the detection of RNA which has shown in Fig. 6.1 [25, 26]. These amplification methods produce various detection outcomes, such as measuring the turbidity resulting from magnesium pyrophosphate, a byproduct of DNA amplification. Detection can then be determined by measuring turbidity, fluorescence, or color

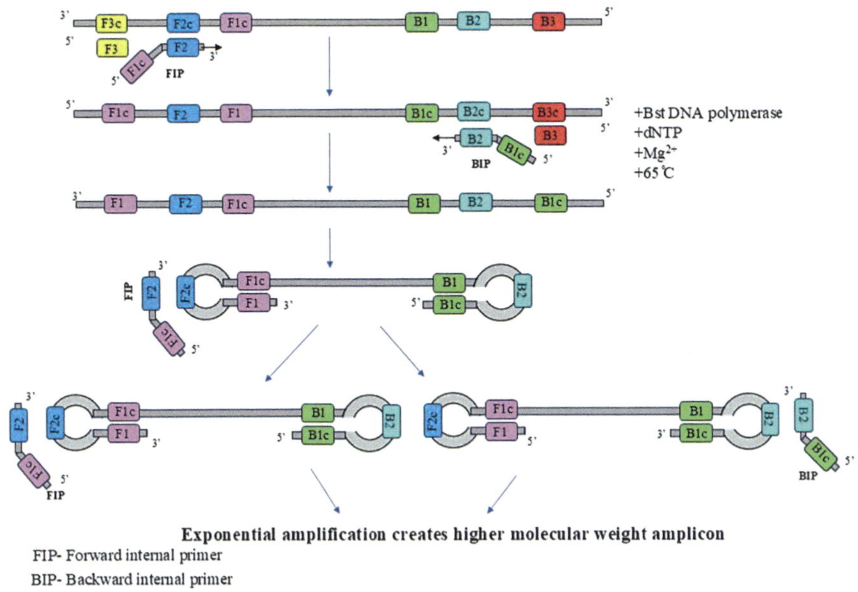

Fig. 6.1 Reverse transcription loop-mediated isothermal amplification (RT-LAMP)

changes using intercalating dyes for analysis [12]. The simplicity and sensitivity of real-time RT-LAMP make it a promising method for viral detection, as it requires only heating and visual inspection.

RT-LAMP involves reverse transcription of RNA into cDNA, followed by isothermal nucleic acid amplification using the LAMP technique. The results of this testing can be obtained within one hour, with a minimum virus detection limit of 20 copies. In a comparison study by Jiang et al., involving 213 negative and 47 positive clinical samples, RT-LAMP outperformed RT-PCR in both sensitivity and specificity, delivering results in half the time [27]. Yu et al. developed an RT-LAMP method that directly indicates results through a color change in the solution [47]. Additionally, Zhang et al. developed a method combining RT-LAMP with an Au-nanoparticle-based colorimetric approach and Cas12a, known as the RT-LAMP CLAP test [48]. Sherrill-Mix et al. enhanced RT-LAMP with molecular beacons, reducing costs and increasing screening capacity by implementing a simple purification step and customized polymerase [38].

In conclusion, the diagnostic tool LAMP is highly trustful and affordable which directly uses patient specimens [9]. Its shorter detection time compared to RT-PCR makes it especially suitable for diagnostic settings. To enhance accessibility to COVID-19 diagnostic tools, Yang et al. developed a mobile app. LAMP is particularly beneficial for developing and low-income nations due to its faster testing times and increased effectiveness [44]. However, the primary challenge for LAMP is the design and screening of primers, as there is a risk of aerosol contamination during experiments. Despite the swift progress in RT-LAMP-based assays, problems with cross-reactivity have impeded the commercialization of several methods.

Transcription-Mediated Amplification (TMA)

TMA (transcription-mediated amplification) is a patented as an isothermal amplification technique with a single tube which amplifies specific regions of DNA or RNA more effectively than RT-PCR. Modeled after retroviral replication, TMA employs T7 RNA polymerase and retroviral reverse transcriptase. This technique has been applied to detect nucleic acids from various infections. RT-PCR and TMA can be performed using Hologic's Panther Fusion platform [19]. This platform allows for the simultaneous screening of other respiratory viruses with symptoms similar to COVID-19, utilizing the same patient samples and collection vials. Its ability to process up to 1000 tests in 24 hours makes it highly efficient.

TMA involves multiple transcript creation using RNA polymerase and isothermal rRNA amplification via reverse transcription. For detection, with the help of complementary oligonucleotide probe, RNA copies are amplified and hybridized. Further, it is labeled with a chemiluminescent tag. Initially, the viral RNA target is hybridized with a specified capture probe and another oligonucleotide that contains a T7 promoter primer. Further, with the help of magnetic microparticles, these complexes are captured using a magnetic field. The captured RNA target is reverse transcribed to complementary cDNA, and hybridized to the T7 promoter primer. The target RNA strand from the hybrid RNA-cDNA duplex is cleaved using the RNase H activity of the reverse transcriptase, resulting in single-stranded cDNA containing the T7 promoter. A second primer is used to create double-stranded DNA, which is then transcribed by T7 RNA polymerase into RNA amplicons. These new RNA amplicons re-enter the TMA process, enabling exponential amplification and generating billions of RNA amplicons in less than an hour. For detection, single-stranded nucleic acid "lamps" are used, which precisely and rapidly hybridize to the RNA amplicon. Each lamp has a conjugate quencher and fluorophore. Upon hybridization to the RNA amplicon, the fluorophore is excited and generates a detectable signal. An automated detection technique, the Procleix SARS-CoV-2 assay based on TMA, has been developed and evaluated by Sauleda et al. using inactivated and clinical samples. Their study indicated that the Procleix SARS-CoV-2 assay is a highly sensitive and specific tool, effective at detecting the virus even in asymptomatic individual [37].

Day et al. introduced a novel transcription-mediated isothermal RNA amplification (TMIRA) combined with gold nanoparticles as a biosensor approach for visual, colorimetric lateral flow tests. This method utilizes anti-RNA: DNA hybrid (anti-RDH) antibodies to detect SARS-CoV-2 RNA with high specificity and sensitivity. Initially, biotinylated oligonucleotides (DNAB) specific to SARS-CoV-2 RNA (vRNA) form a vRNA-DNAB hybrid. Streptavidin coupled with gold nanoparticles then binds to the biotin-tagged vRNA-DNAB hybrids [17]. The hybrid complex is captured by an anti-RDH antibody immobilized on a nitrocellulose membrane, producing a pink color signal that becomes visible to the naked eye within 1 minute. The coupling of RDH-LFA with TMIRA demonstrated approximately 96% sensitivity and 100% specificity compared to RT-qPCR.

TMA has been identified by Gorzalski et al. as a highly sensitive and efficient method for detecting SARS-CoV-2 in nasopharyngeal specimens compared to RT-PCR [22]. Furthermore, Vahidnia et al. explored saliva as a less invasive alternative sampling method and found comparable sensitivity to nasopharyngeal and oropharyngeal samples [40]. In summary, TMA's high analytical sensitivity enables the detection of the SARS-CoV-2 genome in human specimens where RT-PCR may fall short, highlighting its potential for diagnostic settings. However, TMA's drawbacks include its expense and its provision of qualitative rather than quantitative results.

Clustered Regularly Interspaced Short Palindromic Repeats (CRISPR)-Based Assays

Prokaryotic organisms like bacteria possess a nucleic acid sequence-based immune defense mechanism called CRISPR. This system includes CRISPR-associated enzymes (Cas9, Cas12, and Cas13), which can identify and cleave specific nucleic acid sequences. The CRISPR system functions with a Cas protein and guide RNA (gRNA). Additionally, the Cas protein, recognized by the gRNA, binds to the target RNA or DNA sequence, leading to the degradation of foreign nucleic acids [4]. This molecular tool, often referred to as molecular scissors, shows promise as a diagnostic tool by targeting and cleaving viral RNA sequences for detecting SARS-CoV-2.

For the recognition and degradation within the CRISPR-Cas system, Cas12a and Cas13a of CRISPR RNA enzymes are used. According to Kellner et al., when the guide RNA (gRNA) binds to the target RNA or DNA sequence, Cas12a or Cas13a specifically cleaves the single-stranded target nucleic acid sequence [29]. Subsequently, these enzymes activate their non-specific endonuclease activity, cleaving nearby nontarget nucleic acid sequences and generating detectable signals. These signals are utilized to target and identify specific nucleic acids. Various detection techniques have emerged for detecting SARS-CoV-2 by integrating the CRISPR-Cas system with other molecular assays, as illustrated in Fig. 6.2.

CRISPR-based SHERLOCK detection method was developed by Sherlock Biosciences [28]. The CRISPR-Cas13a system consists of two main components: nucleic acid identification and isothermal target nucleic acid amplification. Cas13a can distinguish SARS-CoV-2 from other viruses with high specificity (in the range of 10–100 samples per microliter). The dipstick test has not yet been tried on patients, even though it simplifies the process and lowers testing expenses. Similarly, the CRISPR-based DETECTR detection method was developed by Mammoth Biosciences but the only difference is that they used CRISPR-Cas12a-based technology with RT-LAMP with 100% specificity and 95% positive predictive power [6]. Also, Ali et al. developed the iSCAN test which is equally efficient and accurate for early diagnosis [3]. Furthermore, Ding et al. developed a CRISPR-Cas12a (AIOD-CRISPR) assay which selectively targeted the cDNA of the SARS-CoV-2 N gene [18]. The Indian Council of Medical Research (ICMR) has authorized the CRISPR-based FnCas9 Editor Linked Uniform Detection Assay (FELUDA)

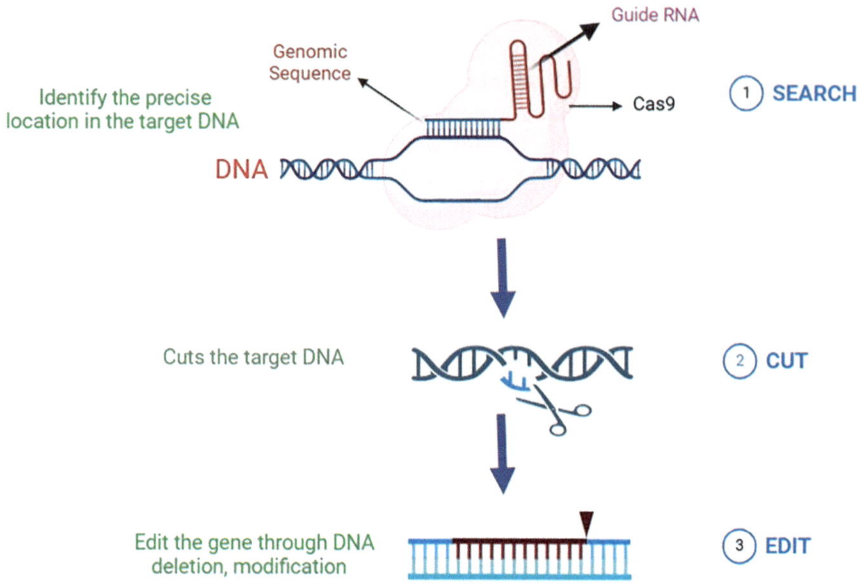

Fig. 6.2 Clustered regularly interspaced short palindromic repeats (CRISPR) method

detection method. This methodology utilizes Cas9 from Francisella novicida (FNCas9), known for its high mismatch sensitivity, making it effective in distinguishing between viruses with closely related genomes such as SARS-CoV and SARS-CoV-2 [5]. Compared to methods like SHERLOCK, DETECTR, and qPCR assays, FELUDA is fast, user-friendly, and inexpensive. Its SARS-CoV-2 detection test demonstrates high sensitivity (100%) and specificity (97%).

Furthermore, following standard PCR procedures, researchers have employed fluorescence probes in conjunction with CRISPR-Cas12a/RNA complexes to detect target amplification sequences. Huang et al. demonstrated that this method offers greater sensitivity and accuracy compared to CDC-approved quantitative RT-PCR results, achieving a detection limit of 2 copies [25, 26]. Recently, Yoshimi et al. developed in vitro nucleic acid assay Cas3-operated Nucleic Acid detection CONAN by using of Cas3a, which demonstrates accurate and precise recognition of single base pairs [45].

In conclusion, CRISPR-based detection methods provide an alternative to established techniques, offering advantages in terms of efficiency, affordability, and user-friendliness. However, additional validation is needed to confirm their accuracy, specificity, and reliability, as well as to address potential issues such as false positives and false negatives.

Rolling Circle Amplification

Rolling circle amplification (RCA) is utilized to amplify circular DNAs, effectively amplifying circular DNA fragments, plasmids, and circular virus genomes. This isothermal enzymatic method involves creating a long single-stranded DNA or

RNA from short RNA or DNA primers using a circular DNA template and specific RNA or DNA polymerases [2]. RCA depends on the activity of a DNA polymerase that can perform strand displacement. Initiated by a short DNA primer binding to a specific location on the circular DNA, the process begins with the DNA polymerase enzyme adding complementary nucleotides unidirectionally along the circular DNA, causing the original strand to displace. This results in the formation of a long single-stranded DNA molecule that contains multiple copies of the original circular template sequence. Subsequent cycles of RCA amplify the dsDNA, leading to substantial amplification [15]. RCA is particularly effective for amplifying very small amounts of circular DNA, making it ideal for samples with limited starting material. Consequently, this technology shows promise for improving sensitive diagnostic procedures targeting a range of entities, including DNA, RNA, and proteins.

Khan et al. introduced a rapid, efficient, and highly specific method known as three-way junction-induced exponential rolling circle amplification, integrated with matrix-assisted laser desorption ionization–time of flight (MALDI-TOF) MS for targeting SARS-CoV-2 RNA [30]. This approach targets the nucleocapsid (N) and ORF1ab genes of the virus, enabling detection of multiple SARS-CoV-2-associated genes in less than 30 minutes at 55 °C. This distinguishes it from other coronaviruses such as SARS-CoV, MERS-CoV, and bat SARS-like coronavirus (bat-SL-CoVZC45). The method involves forming three-way junctions that include a complementary sequence to the COVID-19 viral target sequence at the 3′ end and a rolling circle primer sequence at the 5′ end, with the active site of a nicking endonuclease positioned in between. The precise formation of stable junctions with SARS-CoV-2 target RNA ensures accurate detection. Additionally, Deng et al. developed a dual-gene-controlling RCA technique employing an electrochemical method for the simultaneous detection of N and RdRp genes [16]. This innovative approach confirms the accuracy and specificity of SARS-CoV-2 viral studies by detecting both genes concurrently.

Kumari et al. developed a one-step detection method for SARS-CoV-2 based on RCA, known as the RCA-based multipadlock technique (MP-RCA) [32]. This innovative approach utilizes 40 padlock-based probes targeting various sequences across the viral genome, ensuring accurate detection even of mutated strains. Their optimized isothermal reaction protocol, coupled with fluorescent output detection using a plate reader, demonstrated linear sensitivity proportional to the number of padlock probes. MP-RCA requires minimal equipment and enhances sensitivity, particularly in identifying SARS-CoV-2 variants with S-gene target failure, making it a valuable alternative to qPCR for detecting rapidly mutating viruses. Furthermore, Zhang et al. developed an analytical method for detecting the SARS-CoV-2 receptor-binding domain (RBD) using RCA with fluorometric and colorimetric measurements [49]. This method utilizes streptavidin-coated magnetic beads modified with an RBD aptamer, which binds a biotin-tagged complementary DNA strand (biotin-cDNA). Upon RBD binding, the aptamer dissociates from the biotin-cDNA, allowing cDNA to initiate RCA. The detection method employs two signal outputs: fluorescence, where the RCA product hybridizes with a dsDNA probe labeled with a fluorophore emitting at $\lambda em = 520$ nm, and

UV–vis absorbance at λabs = 450 nm, generated by extending RCA to form a DNA flower encapsulating horseradish peroxidase, catalyzing the reaction between H_2O_2 and 3,3′,5,5′-tetramethylbenzidine TMB to produce signals. Park et al. developed an oil- and bead-free RCA method on a substrate, akin to ELISA, enabling localized signal amplification at the antigen capture point for visualizing individual antigen molecules [34]. This method detects SARS-CoV-2 N protein at concentrations below 1 pg/mL, demonstrating remarkable sensitivity.

Recently, Wang et al. introduced an ultraspecific point-of-care technique called microchannel cantilever array detection (mCAD) for detecting SARS-CoV-2 RNA [42]. By integrating multiplex RCA and toehold-mediated strand displacement reactions, mCAD achieves RNA amplification while maintaining high sensitivity. This method employs a layered specificity approach using sequence-specific hybridization, effectively eliminating false positive results. Similarly, Zhu et al. introduced a one-pot diagnostic technique named OPERATOR, which combines enzyme-catalyzed RCA with a CRISPR-FnCas12a assay [50]. OPERATOR utilizes a single-strand padlock DNA containing a protospacer adjacent motif site and a complementary sequence to the target RNA is used to convert and amplify genomic RNA into DNA through RNA-templated DNA ligation and multiply-primed rolling circle amplification (MRCA). The single-stranded DNA amplicon produced by MRCA is then cleaved by the FnCas12a/crRNA complex and detected using a fluorescence reader or lateral flow strip.

In conclusion, RCA has gained significant attention due to its minimal reagent requirements and reduced likelihood of false positives. However, it has not yet been extensively applied for detecting SARS-CoV-2.

Nucleic Acid Hybridization Using Microarray

Microarray assays have been employed for the rapid detection of SARS-CoV nucleic acids. Initially, viral RNA undergoes reverse transcription to produce cDNA, which is subsequently labeled with specific probes. This labeled cDNA is applied to the surfaces of microarray trays where solid-phase oligonucleotides are immobilized. Upon hybridization, the bound cDNA remains attached after washing away unbound DNA, resulting in detectable signals indicating the presence of virus-specific nucleic acids, as illustrated in Fig. 6.3. The microarray assay has demonstrated capability in detecting 24 single-nucleotide polymorphisms associated with mutations in the spike (S) gene of SARS-CoV with 100% accuracy. As COVID-19 evolves, it undergoes various mutational changes that can be effectively detected using microarray assays. A nonfluorescent, cost-effective, low-density oligonucleotide array test has been developed for identifying mutated strains of coronavirus. This test demonstrates sensitivity comparable to individual real-time RT-PCR, although the high cost associated with microarray testing remains a limitation. Moreover, microarray chips have been utilized to detect specific nucleic acids of the MERS coronavirus, as well as respiratory syncytial and influenza viruses [23].

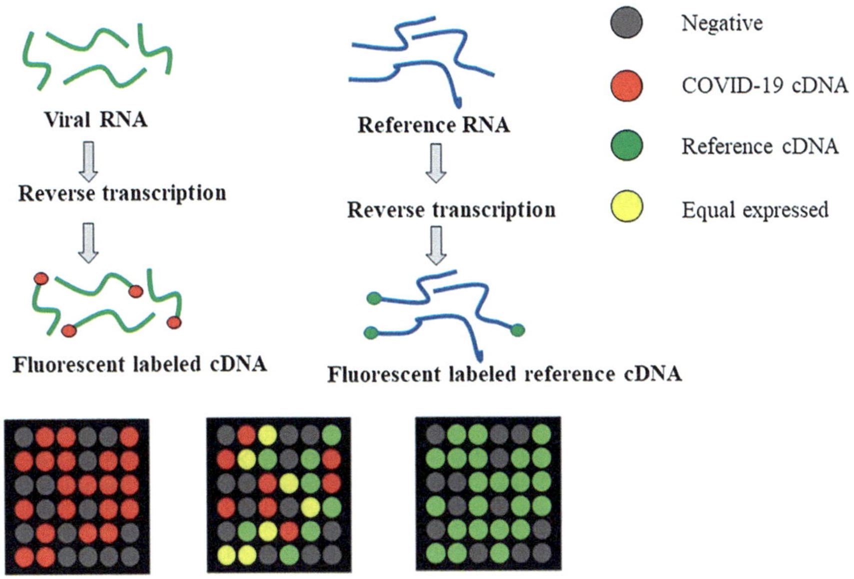

Fig. 6.3 Nucleic acid hybridization using microarray

Amplicon-Based Metagenomic Sequencing

Often, for diagnosing SARS-CoV-2, amplicon-based sequencing is combined with metagenomic sequencing. Amplicon-based sequencing is utilized to study viral evolution, molecular epidemiology, and contact tracing of SARS-CoV-2. Meanwhile, metagenomic sequencing helps identify the background microbiota associated with the disease, aiding in identifying pathogens responsible for secondary infections. Among metagenomic sequencing methods, sequence-independent single primer amplification provides additional checks on sequence divergence. This combined approach is particularly relevant for SARS-CoV-2 as it enables the evaluation of virus mutations and potential recombination with other human coronaviruses, crucial for vaccine development and assessing medication efficacy.

Building on this approach, Moore et al., through the international severe acute respiratory and emerging infection consortium (ISARIC) 4C collaboration, developed a MinION-based sequencing technique capable of rapidly (in less than 8 hours) identifying the sequence of the SARS-CoV-2 genome and other microbiota in nasopharyngeal swabs from COVID-19 patients [33]. They selected 16 primer binding sites across conserved areas of the SARS-CoV-2 genome for amplicon-based sequencing, enabling successive amplification of approximately 1000 bp segments with 200 bp overlap regions. The resulting cDNA was amplified using these primers, generating 30 amplicons that were sequenced using MinION. Additionally, Illumina has developed next-generation shotgun metagenomic sequencing technology, which not only detects various coronavirus strains but also investigates numerous pathogenic organisms present in a sample. Illumina's metagenomics systems require specific components for operation, such as the TruSeq Ribo-Zero Gold rRNA depletion kit for sample preparation, TruSeq stranded total RNA for library preparation, an Illumina benchtop sequencing system for sequencing, and tools like LRM Resequencing module or IDby DNA Explify Platform for final data analysis.

However, amplicon techniques have drawbacks, including incomplete coverage of the viral genome ends due to primer limitations, as well as cost and time consumption associated with library preparation. To address these challenges, Gohl et al. introduced a tailed amplicon methodology capable of achieving balanced amplicon coverage, coverage metrics, and variant calls. They propose this approach as highly scalable and cost-effective for sequencing SARS-CoV-2 [20].

Cartridge-Based Nucleic Acid Amplification Test (CBNAAT)

The World Health Organization (WHO) has approved an automated molecular diagnostic technique known as CBNAAT for diagnosing tuberculosis in children. In response to the increasing number of SARS-CoV-2 infections, the Indian Council of Medical Research (ICMR) has adopted and approved the use of the CBNAAT assay for detecting SAR-CoV-2 [21]. This technique integrates SAR-CoV-2 sample preparation, RNA extraction, amplification, and target sequence detection through real-time qPCR, typically completed in about 45–60 minutes. Samples collected from the nasal cavity in viral transport media are first prepared by thoroughly mixing in an inverted tube. They are then transferred to the laboratory under cold chain conditions and loaded into the sample chamber of an Xpert Xpress SAR-CoV-2 cartridge. The GeneXpert instrument system is employed for real-time qPCR analysis. In summary, this method is straightforward, rapid, and minimizes the risk of cross-contamination. However, it necessitates a consistent power supply for operation and may yield false negative results if the selected S and N_2 target genes are mutated.

Future Perspective

Various diagnostic techniques like RT-PCR, RT-LAMP, TMA, CRISPR-based assays, RCA, microarray assays, and CBNAAT play an important role in COVID-19 detection and tracking. Future innovation in diagnostic techniques has the potential to completely transform the management of infectious disease, even better than what they have been today. Innovation in case of scalability, cost, and accessibility is the topmost priority. By combining real-time data analysis with artificial intelligence, it may be possible to identify newer variant quickly. Even future advancements can be aimed for multipathogen detection, integrated sample preparation, and enhanced compatibility with mobile health technologies, ensuring widespread accessibility and usability in community settings. These developments promise to transform diagnostic capabilities, playing a pivotal role in global health preparedness and response strategies against infectious diseases.

Conclusion

Over the years, a wide range of molecular assays has been developed for detecting viral nucleic acids. However, most tests used to identify the SARS-CoV-2 virus rely on RT-PCR and isothermal nucleic acid amplification technologies. Currently, there are 274 molecular diagnostic techniques approved under emergency use authorization. Among these, approximately 80% are based on PCR or RT-PCR technology, 8% utilize isothermal amplification and home collection methods, while the remainder employ techniques such as TMA, nonisothermal amplification, RT-qSTAR, and sequencing.

Declaration

Author's Contribution BMS conceived the idea and review structure; BMS and PM wrote the manuscript and revised and edited the manuscript. Author has read, agreed, and approved the final draft of the manuscript.

Funding None.

Data Availability All the data are included within the manuscript.

Conflicts of Interests There is no conflict of interest.

References

1. Acter T, Uddin N, Das J, Akhter A, Choudhury TR, Kim S. Evolution of severe acute respiratory syndrome coronavirus 2 (SARS-CoV-2) as coronavirus disease 2019 (COVID-19) pandemic: a global health emergency. Sci Total Environ. 2020;730:138996. https://doi.org/10.1016/j.scitotenv.2020.138996.
2. Ali MM, Li F, Zhang Z, Zhang K, Kang DK, Ankrum JA, Le XC, Zhao W. Rolling circle amplification: a versatile tool for chemical biology, materials science and medicine. Chem Soc Rev. 2014;43:3324–41. https://doi.org/10.1039/C3CS60439J.
3. Ali Z, Aman R, Mahas A, Rao GS, Tehseen M, Marsic T, Salunke R, Subudhi AK, Hala SM, Hamdan SM, Pain A. iSCAN: an RT-LAMP-coupled CRISPR-Cas12 module for rapid, sensitive detection of SARS-CoV-2. Virus Res. 2020;288:198129. https://doi.org/10.1016/j.virusres.2020.198129.
4. Antropov DN, Stepanov GA. Molecular mechanisms underlying CRISPR/Cas-based assays for nucleic acid detection. Curr Issues Mol Biol. 2023;45:649–62. https://doi.org/10.3390/cimb45010043.
5. Bhardwaj P, Kant R, Behera SP, Dwivedi GR, Singh R. Next-generation diagnostic with CRISPR/Cas: beyond nucleic acid detection. Int J Mol Sci. 2022;23:6052. https://doi.org/10.3390/ijms23116052.
6. Broughton JP, Deng X, Yu G, Fasching CL, Servellita V, Singh J, Miao X, Streithorst JA, Granados A, Sotomayor-Gonzalez A, Zorn K. CRISPR–Cas12-based detection of SARS-CoV-2. Nat Biotechnol. 2020;38:870–4. https://doi.org/10.1038/s41587-020-0513-4.
7. Burukanli M, Yumuşak N. TfrAdmCov: a robust transformer encoder based model with Adam optimizer algorithm for COVID-19 mutation prediction. Connect Sci. 2024;36:2365334. https://doi.org/10.1080/09540091.2024.2365334.

8. Chan JFW, Yip CCY, To KKW, Tang THC, Wong SCY, Leung KH, Fung AYF, Ng ACK, Zou Z, Tsoi HW, Choi GKY. Improved molecular diagnosis of COVID-19 by the novel, highly sensitive and specific COVID-19-RdRp/Hel real-time reverse transcription-PCR assay validated in vitro and with clinical specimens. J Clin Microbiol. 2020;58:10–1128. https://doi.org/10.1128/jcm.00310-20.
9. Chaouch M. Loop-mediated isothermal amplification (LAMP): an effective molecular point-of-care technique for the rapid diagnosis of coronavirus SARS-CoV-2. J Med Virol. 2021;31:e2215. https://doi.org/10.1002/rmv.2215.
10. Chavda VP, Valu DD, Parikh PK, Tiwari N, Chhipa AS, Shukla S, Patel SS, Balar PC, Paiva-Santos AC, Patravale V. Conventional and novel diagnostic tools for the diagnosis of emerging SARS-CoV-2 variants. Vaccine. 2023;11:374. https://doi.org/10.3390/vaccines11020374.
11. Cheng L, Lan L, Ramalingam M, He J, Yang Y, Gao M, Shi Z. A review of current effective COVID-19 testing methods and quality control. Arch Microbiol. 2023;205:239. https://doi.org/10.1007/s00203-023-03579-9.
12. Choi MH, Lee J, Seo YJ. Dual-site ligation-assisted loop-mediated isothermal amplification (dLig-LAMP) for colorimetric and point-of-care determination of real SARS-CoV-2. Microchim Acta. 2022;189:176. https://doi.org/10.1007/s00604-022-05293-7.
13. Corman VM, Landt O, Kaiser M, Molenkamp R, Meijer A, Chu DK, Bleicker T, Brünink S, Schneider J, Schmidt ML, Mulders DG. Detection of 2019 novel coronavirus (2019-nCoV) by real-time RT-PCR. Euro Surveill. 2020;25:2000045.
14. Das P, Mondal S, Pal S, Roy S, Vidyadharan A, Dadwal R, Bhattacharya S, Mishra DK, Chandy M. COVID diagnostics by molecular methods: a systematic review of nucleic acid based testing systems. Indian J Med Microbiol. 2021;39:271–8. https://doi.org/10.1016/j.ijmmb.2021.05.012.
15. Demidov VV. Rolling-circle amplification in DNA diagnostics: the power of simplicity. Expert Rev Mol Diagn. 2002;2:542–8. https://doi.org/10.1586/14737159.2.6.542.
16. Deng Y, Zhou T, Peng Y, Wang M, Xiang L, Zhang Y, Li J, Yang J, Li G. Dual-gene-controlled rolling circle amplification strategy for SARS-CoV-2 analysis. Anal Chem. 2023;95:3358–62. https://doi.org/10.1021/acs.analchem.2c04572.
17. Dey A, Prakash J, Das R, Shelar S, Saini A, Cherian S, Patel SC, Hassan PA, Khandekar A, Dasgupta K, Misra HS. Development of a rapid and ultra-sensitive RNA: DNA hybrid immunocapture based biosensor for visual detection of SARS-CoV-2 RNA. PNAS Nexus. 2023;2:031. https://doi.org/10.1093/pnasnexus/pgad031.
18. Ding X, Yin K, Li Z, Lalla RV, Ballesteros E, Sfeir MM, Liu C. Ultrasensitive and visual detection of SARS-CoV-2 using all-in-one dual CRISPR-Cas12a assay. Nat Commun. 2020;11:4711. https://doi.org/10.1038/s41467-020-18575-6.
19. Fernández-Huerta M, Salmerón P, Hernández-Hermida Y, Andrés C, Niubó J, Calatayud L, Domínguez MÁ, Pumarola T, Ardanuy C, Antón A, Càmara J. Multicenter clinical evaluation of a novel transcription-mediated amplification assay for SARS-CoV-2 molecular testing. Enfermedades infecciosas y microbiologia clinica (English ed.). 2023;41:462–7. https://doi.org/10.1016/j.eimce.2022.01.007.
20. Gohl DM, Garbe J, Grady P, Daniel J, Watson RH, Auch B, Nelson A, Yohe S, Beckman KB. A rapid, cost-effective tailed amplicon method for sequencing SARS-CoV-2. BMC Genomics. 2020;21:1–10. https://doi.org/10.1186/s12864-020-07283-6.
21. Gogoi S, Bora I, Debnath E, Sarkar S, Jais MB, Sharma A. Perplexity vs clarity in choosing the right molecular diagnostic techniques for SARS-COV2 detection in Indian setup. J Family Med Prim Care. 2021;10:615–24. https://doi.org/10.4103/jfmpc.jfmpc_1793_20.
22. Gorzalski AJ, Tian H, Laverdure C, Morzunov S, Verma SC, VanHooser S, Pandori MW. High-throughput transcription-mediated amplification on the Hologic panther is a highly sensitive method of detection for SARS-CoV-2. J Clin Virol. 2020;129:104501. https://doi.org/10.1016/j.jcv.2020.104501.
23. Hardick J, Metzgar D, Risen L, Myers C, Balansay M, Malcom T, Rothman R, Gaydos C. Initial performance evaluation of a spotted array Mobile Analysis Platform (MAP) for

the detection of influenza A/B, RSV, and MERS coronavirus. Diagn Microbiol Infect Dis. 2018;91:245–7. https://doi.org/10.1016/j.diagmicrobio.2018.02.011.
24. https://www.fda.gov/medical-devices/covid-19-emergency-use-authorizations-medical-devices/in-vitro-diagnostics-euas-molecular-diagnostic-tests-sars-cov-2. Accessed on 10 June 2024.
25. Huang WE, Lim B, Hsu CC, Xiong D, Wu W, Yu Y, Jia H, Wang Y, Zeng Y, Ji M, Chang H. RT-LAMP for rapid diagnosis of coronavirus SARS-CoV-2. Microb Biotechnol. 2020a;13:950–61. https://doi.org/10.1111/1751-7915.13586.
26. Huang Z, Tian D, Liu Y, Lin Z, Lyon CJ, Lai W, Fusco D, Drouin A, Yin X, Hu T, Ning B. Ultra-sensitive and high-throughput CRISPR-p owered COVID-19 diagnosis. Biosens Bioelectron. 2020b;164:112316. https://doi.org/10.1016/j.bios.2020.112316.
27. Jiang M, Pan W, Arasthfer A, Fang W, Ling L, Fang H, Daneshnia F, Yu J, Liao W, Pei H, Li X. Development and validation of a rapid, single-step reverse transcriptase loop-mediated isothermal amplification (RT-LAMP) system potentially to be used for reliable and high-throughput screening of COVID-19. Front Cell Infect Microbiol. 2020;10:331. https://doi.org/10.3389/fcimb.2020.00331.
28. Joung J, Ladha A, Saito M, Kim NG, Woolley AE, Segel M, Barretto RP, Ranu A, Macrae RK, Faure G, Ioannidi EI. Detection of SARS-CoV-2 with SHERLOCK one-pot testing. N Engl J Med. 2020;383:1492–4. https://doi.org/10.1056/NEJMc202617.
29. Kellner MJ, Koob JG, Gootenberg JS, Abudayyeh OO, Zhang F. SHERLOCK: nucleic acid detection with CRISPR nucleases. Nat Protoc. 2019;14:2986–3012.
30. Khan SF, Rathod P, Gupta VK, Khedekar PB, Chikhale RV. Evolution and impact of nucleic acid amplification test (NAAT) for diagnosis of coronavirus disease. Anal Chem. 2024; https://doi.org/10.1021/acs.analchem.3c05225.
31. Kim KW, Wang X, Adhikari A, Yeang M, Jenkins F, Naing Z, Walker GJ, Foster CS, Stelzer-Braid S, Deveson I, Craig ME. Persistent high-level shedding of cultivable SARS-CoV-2 Delta virus 33 days after onset of COVID-19 in a hospitalized patient with pneumonia. J Med Virol. 2022;94:4043–6. https://doi.org/10.1002/jmv.27832.
32. Kumari R, Lim JW, Sullivan MR, Malampy R, Baush C, Smolina I, Robin H, Demidov VV, Ugolini GS, Auclair JR, Konry T. A novel rolling circle amplification-based detection of SARS-CoV-2 with multi-region padlock hybridization. Diagnostics. 2022;12:2252. https://doi.org/10.3390/diagnostics12092252.
33. Moore SC, Penrice-Randal R, Alruwaili M, Dong X, Pullan ST, Carter DP, Bewley K, Zhao Q, Sun Y, Hartley C, Zhou EM. Amplicon based MinION sequencing of SARS-CoV-2 and metagenomic characterisation of nasopharyngeal swabs from patients with COVID-19. MedRxiv. 2020;2020-03 https://doi.org/10.1101/2020.03.05.20032011.
34. Park J, Park M, Kim J, Heo Y, Han BH, Choi N, Park C, Lee R, Lee DG, Chung S, Kang JY. Beads-and oil-free single molecule assay with immuno-rolling circle amplification for detection of SARS-CoV-2 from saliva. Biosens Bioelectron. 2023;232:115316. https://doi.org/10.1016/j.bios.2023.115316.
35. Park SE. Epidemiology, virology, and clinical features of severe acute respiratory syndrome-coronavirus-2 (SARS-CoV-2; Coronavirus Disease-19). Clin Exp Pediatr. 2020;63:119. https://doi.org/10.3345/cep.2020.00493.
36. Satarker S, Nampoothiri M. Structural proteins in severe acute respiratory syndrome coronavirus-2. Arch Med Res. 2020;51:482–91. https://doi.org/10.1016/j.arcmed.2020.05.012.
37. Sauleda S, Palacios L, Bres V, Pinana M, Alonso-Hernandez L, Bes M, Piron M, Entrena E, Minguez-Micolau AM, Marimon JM, Gurrola A. Clinical evaluation of the Procleix SARS-CoV-2 assay, a sensitive, high-throughput test that runs on an automated system. Diagn Microbiol Infect Dis. 2022;102:115560. https://doi.org/10.1016/j.diagmicrobio.2021.115560.
38. Sherrill-Mix S, Hwang Y, Roche AM, Glascock A, Weiss SR, Li Y, Haddad L, Deraska P, Monahan C, Kromer A, Graham-Wooten J. Detection of SARS-CoV-2 RNA using RT-LAMP and molecular beacons. Genome Biol. 2021;22:169. https://doi.org/10.1186/s13059-021-02387-y.

39. Tombuloglu H, Sabit H, Al-Suhaimi E, Al Jindan R, Alkharsah KR. Development of multiplex real-time RT-PCR assay for the detection of SARS-CoV-2. PLoS One. 2021;16:0250942. https://doi.org/10.1371/journal.pone.0250942.
40. Vahidnia A, Souverein D, Euser SM, Haverkort M, Noordhuis E, van Zijl TZ, Kalpoe J, Sinnige JC, Herpers BL. Similar sensitivity of SARS-CoV-2 detection in oropharyngeal/nasopharyngeal and saliva samples on the hologic panther platform. Diagnostics. 2023;13:347. https://doi.org/10.3390/diagnostics13030347.
41. VanGuilder HD, Vrana KE, Freeman WM. Twenty-five years of quantitative PCR for gene expression analysis. BioTechniques. 2008;44:619–26. https://doi.org/10.2144/000112776.
42. Wang F, Ma X, Ye J, Shi C, Chen Y, Yu Z, Li T, Yang D, Li M, Wang P. Precise detection of viral RNA by programming multiplex rolling circle amplification and Strand displacement. Anal Chem. 2023;95:17699–707. https://doi.org/10.1021/acs.analchem.3c03548.
43. Xia J, Tong J, Liu M, Shen Y, Guo D. Evaluation of coronavirus in tears and conjunctival secretions of patients with SARS-CoV-2 infection. J Med Virol. 2020;92:589–94. https://doi.org/10.1002/jmv.25725.
44. Yang T, Wang YC, Shen CF, Cheng CM. Point-of-care RNA-based diagnostic device for COVID-19. Diagnostics. 2020;10:165. https://doi.org/10.3390/diagnostics10030165.
45. Yoshimi K, Takeshita K, Yamayoshi S, Shibumura S, Yamauchi Y, Yamamoto M, Yotsuyanagi H, Kawaoka Y, Mashimo T. CRISPR-Cas3-based diagnostics for SARS-CoV-2 and influenza virus. Iscience. 2022;25 https://doi.org/10.1016/j.isci.2022.103830.
46. Younes N, Al-Sadeq DW, Al-Jighefee H, Younes S, Al-Jamal O, Daas HI, Yassine HM, Nasrallah GK. Challenges in laboratory diagnosis of the novel coronavirus SARS-CoV-2. Viruses. 2020;12:582. https://doi.org/10.3390/v12060582.
47. Yu L, Wu S, Hao X, Dong X, Mao L, Pelechano V, Chen WH, Yin X. Rapid detection of COVID-19 coronavirus using a reverse transcriptional loop-mediated isothermal amplification (RT-LAMP) diagnostic platform. Clin Chem. 2020;66:975–7. https://doi.org/10.1093/clinchem/hvaa102.
48. Zhang Y, Chen M, Liu C, Chen J, Luo X, Xue Y, Liang Q, Zhou L, Tao Y, Li M, Wang D. Sensitive and rapid on-site detection of SARS-CoV-2 using a gold nanoparticle-based high-throughput platform coupled with CRISPR/Cas12-assisted RT-LAMP. Sens Actuators B Chem. 2021;345:130411. https://doi.org/10.1016/j.snb.2021.130411.
49. Zhang M, Ye L. Detection of SARS-CoV-2 receptor binding domain using fluorescence probe and DNA flowers enabled by rolling circle amplification. Microchim Acta. 2023;190:163. https://doi.org/10.1007/s00604-023-05747-6.
50. Zhu Z, Guo Y, Wang C, Yang Z, Li R, Zeng Z, Li H, Zhang D, Yang L. An ultra-sensitive one-pot RNA-templated DNA ligation rolling circle amplification-assisted CRISPR/Cas12a detector assay for rapid detection of SARS-CoV-2. Biosens Bioelectron. 2023;228:115179. https://doi.org/10.1016/j.bios.2023.115179.

Immunological Assay-Based COVID-19 Detection

7

Bhumi M. Shah and Palmi Modi

Abstract

SARS-CoV-2 is the novel coronavirus responsible for COVID-19, a highly infectious disease. The virus spreads through respiratory droplets from an infected person when they sneeze, cough, or speak, leading to symptoms like pneumonia and, in severe cases, respiratory failure. Due to its high contagion, early and accurate diagnosis is vital for controlling the spread, mitigating consequences, and initiating timely treatment. Initially, COVID-19 detection relied on reverse transcriptase polymerase chain reaction. Now, various serological and immunological assays are used for early detection, including the enzyme-linked immunosorbent assay, the lateral flow immunoassay, the neutralization assay, the luminescent immunoassay, the biosensor test, and the rapid antigen test. These methods vary in efficacy, specificity, sensitivity, accuracy, and cost. This review of the literature on serological and immunological assays outlines their advantages and disadvantages, emphasizing the importance of precise diagnosis in breaking transmission chains and improving patient outcomes. This compilation aims to enhance detection efficiency and quality in greater detail.

Keywords

Serological and immunological assays · ELISA · Lateral flow immunoassay · Neutralization assay · Luminescent immunoassay · Biosensor test · Rapid antigen test

B. M. Shah (✉) · P. Modi
Department of Pharmaceutical Chemistry, L. J. Institute of Pharmacy, LJ University, Ahmedabad, Gujarat, India
e-mail: bhumi.shah_ljip@ljinstitutes.edu.in

M. K. Ghosh, M. Basu (eds.), *Pathogenesis, Management and Socio-Economic Impact of COVID-19 Pandemic*, Medical Virology: From Pathogenesis to Disease Control, https://doi.org/10.1007/978-981-95-1426-7_7

Introduction on Serological and Immunological Assays

A serology test analyzes a blood or plasma sample to detect antibodies or other substances. The immune system produces antibodies in response to foreign substances like viruses [26]. To check whether a person has been exposed to or infected by a virus or other infectious agent, these tests are performed to detect antibodies. Polymerase chain reaction (PCR)-based assays can be used early to detect viral RNA in a patient's sample and identify carriers, whether they are symptomatic or asymptomatic [45]. In addition to PCR tests, immune-based assays determine immunity and exposure to SARS-CoV-2, aiding decision-making for reintegrating low-risk individuals into normal activities. Reverse transcriptase (RT)-PCR can detect viral RNA for diagnosis but cannot track the disease's course, past infections, or immunity [17]. Therefore, serological tests have gained importance due to their quicker detection time, ease of use, and cost-effectiveness [96]. Various serological tests are available for detecting SARS-CoV-2, such as enzyme-linked immunosorbent assay (ELISA), chemiluminescence immunoassay (CLIA), lateral flow immunoassay (LFIA), neutralization assay, and rapid antigen test [28]. This study discusses the functioning, potential issues, benefits, and future directions for these tests.

The Principle of Serological Test

When a person meets a virus and releases virus antigen in bloodstream, the human immune system starts producing antibodies (IgA, IgM, IgG) which last for long time and are more concentrated than antigens [10]. For this reason, researchers preferred to use serological antibodies rather than viral antigen for the development of safe, quick, and sensitive serological diagnostic tests. The immunology principle states that circulating virus-specific antibodies are more easily detected than total immunoglobulin. Therefore, various researchers have chosen IgM/IgA/IgG antibodies specific to N and S proteins as diagnostic targets for SARS-CoV-2 [33]. Further, to increase specificity, some researchers chose the receptor binding domain (RBD) on the S1 subunit as an antigen [47].

Amongst the different antibodies, IgA (5 days after infection) and IgM (10–30 days after infection) are produced during the early stage of infection while IgG (20–90 days after infection) is produced at the later stage of infection [49]. So, the serological test's efficacy for SARS-CoV-2 infection at the early stage of infection is not satisfactory. According to a study by Wolfol et al., only 50% of SARS-CoV-2 patients showed positive levels of IgM and IgG in their blood serum a week since the onset of symptoms [54]. However, by this time, serum viral load decreases, while the number of antibodies increases significantly, indicating better effectiveness of the serological test than the RT-PCR test in the middle or late stages of infections [55]. Furthermore, serological tests can predict the severity of patients' symptoms which RT-PCR cannot do. Major serological tests used IgM and IgG as detection targets. However, some tests used IgA as the detection target due to the higher amount of IgA in the early stage of infection compared to IgM and IgG [59]. Moreover, there is significant interest in developing diagnostic techniques that combine antibody detection to improve accuracy. In April 2020, the FDA gave emergency use authorization (EUA) certification to the first serological test for SARS-CoV-2, developed by Cellex Inc., called the qSARS-CoV-2 IgG/IgM Rapid Test [38]. Since then, many clinical laboratories and commercial manufacturers have developed additional serological tests as listed in Table 7.1 (last 2 years).

Enzyme-Linked Immunosorbent Assay (ELISA)

The popular microwell plate-based ELISA method is used to identify and detect substances such proteins, peptides, hormones, and antibodies. It provides both qualitative and quantitative analysis and results are prepared in 1–5 hours. Depending on the target proteins and detection techniques, varieties of ELISA have evolved such as sandwich, indirect, double antibody, competitive, blocking, and others. Among all, the indirect ELISA approach stands out as being very effective, demonstrating its versatility and accuracy in diagnostics over the years. For SARS-CoV-2, viral proteins (N, S), a subunit of S protein (S1), or RBD are coated on the plate wells. This protein is bound to the antiviral antibodies in the sample, which then combine to create a complex that can be detected using another tracer antibody, resulting in a readout that is colorimetric or fluorescent.

Many researchers tried different diagnostic methods based on ELISA for the detection of SARS-CoV-2 using different samples. Methods that detect IgM, IgG, and IgA antibodies targeting RBD of SARS-CoV-2 using serum samples have been developed [51]. The study demonstrated that the detection of only IgG levels can suffice for the diagnosis [22]. To identify antibodies in blood against the S protein of SARS-CoV-2 and the S1/S2 subunit, Waleed developed an ELISA test and evaluated its effectiveness. They observed that S1 had the highest specificity and the full-length S protein had the strongest response to IgG [32]. Luo et al. developed an indirect ELISA technique using nucleocapsid (N) protein [30]. Also, Isho et al. detected SARS-CoV-2 in serum samples using S Protein and RBD-specific IgG/IgA/IgM antibodies [23]. Further, Bhuiyan et al. developed an AI-controlled microfluidic ELISA device using oropharyngeal saliva samples. They proposed an Internet of Disease (IOD) platform associated with the lab on a chip device for the diagnosis and the resulting diagnostic data is uploaded on the cloud-based system which is connected to the Centers for Disease Control and Prevention which monitors territories in providing public health guidance. Using the IOD platform, the infection rate of different provinces can be easily forecasted [4]. Humbert et al. developed the nucleocapsid ELISA technique using nose, throat swab, and sputum samples [21].

Capillary blood is a safer and more cost-effective method for the detection of antibodies than venous blood testing due to its dependability. Baggio et al. examined the viability of capillary blood self-testing in individuals under 65 vs those over 65. They showed that an effective and dependable substitute for venous blood

Table 7.1 Authorized SARS-CoV-2 serological tests approved by emergency use authorizations (EUAs)

No	Entity	Diagnostic	Attributes	Date of EUA issue
1	Healgen Scientific, LLC	Healgen COVID-19/Flu A&B Ag Combo Rapid Test Cassette	Lateral flow immunoassay	10-06-2024
2	iHealth Labs, Inc.	iHealth COVID-19/Flu A&B Rapid Test Pro	Lateral flow immunoassay	31-05-2024
3	Watmind USA	Speedy Swab Rapid COVID-19 + Flu A&B Antigen Self-Test	Lateral flow immunoassay	24-05-2024
4	Watmind USA	Speedy Swab Rapid COVID-19 + Flu A&B Antigen Test	Lateral flow immunoassay	08-05-2024
5	iHealth Labs, Inc.	iHealth COVID-19/Flu A&B Rapid Test	Lateral flow immunoassay	07-05-2024
6	Wondfo USA Co., Ltd	WELLlife COVID-19 / Influenza A&B Home Test	Lateral flow immunoassay	30-04-2024
7	Wondfo USA Co., Ltd.	WELLlife COVID-19 / Influenza A&B Test	Lateral flow immunoassay	19-04-2024
8	CorDx, Inc.	CorDx TyFast Flu A/B & COVID-19 At Home Multiplex Rapid Test	Lateral flow immunoassay	05-04-2024
9	OSANG LLC	QuickFinder COVID-19/Flu Antigen Self Test	Lateral flow immunoassay	03-04-2024
10	CorDx, Inc.	CorDx Tyfast Flu A/B & COVID-19 Multiplex Rapid Test 0	Lateral flow immunoassay	21-03-2024
11	OSANG LLC	OHC COVID-19/Flu Antigen Test Pro	Lateral flow immunoassay	21-03-2024
12	SEKISUI Diagnostics, LLC	OSOM Flu SARS-CoV-2 Combo Test	Lateral flow immunoassay	29-02-2024
13	SEKISUI Diagnostics, LLC	OSOM Flu SARS-CoV-2 Combo Home Test	Lateral flow immunoassay	29-02-2024
14	SD Biosensor, Inc.	STANDARD Q COVID-19 Ag Test 2.0	Lateral flow immunoassay	29-09-2023
15	Princeton BioMeditech Corp.	ViraDx SARS-CoV-2/Flu A+B Rapid Antigen Test	Lateral flow immunoassay	08-09-2023
16	Immunostics Inc.	Swab-N-Go Home Test COVID-19 Ag	Lateral flow immunoassay	17-07-2023
17	BioTeke USA, LLC	Bio-Self COVID-19 Antigen Home Test	Lateral flow immunoassay	22-05-2023
18	Nano-Ditech Corporation	Nano-Check COVID-19 Antigen At-Home Test	Lateral flow immunoassay	12-05-2023
19	Princeton BioMeditech Corp.	Status COVID-19 Antigen Rapid Test for Home Use	Lateral flow immunoassay	24-04-2023
20	Azure Biotech, Inc.	Azure FaStep® COVID-19 Antigen Pen Home Test	Lateral flow immunoassay	14-04-2023
21	Biosynchronicity Corporation	C-Sync COVID-19 Antigen Test	Lateral flow immunoassay	24-03-2023

(continued)

Table 7.1 (continued)

No	Entity	Diagnostic	Attributes	Date of EUA issue
22	Mologic, Inc.	COVI-Go SARS-CoV-2 Ag Self-Test	Lateral flow immunoassay	22-02-2023
23	GenBody Inc.	GenBody COVID-19 Ag Home Test	Lateral flow immunoassay	17-02-2023

testing has emerged as capillary blood self-testing. However, while utilizing self-sampling equipment, one should consider the age differences [3]. Based on this principle, a capillary-driven microfluidic sequential flow device as a point-of-care was prepared [7]. The device is made up of microfluidic channels that are composed of transparent film, double-sided adhesive, and a paper pump that drives flow. With just two user steps, automated sequential washing and reagent addition are made possible by the channel and storage pad design. An enzyme label and colorimetric substrate provide an enhanced, visible signal for higher sensitivity while integrated washing steps improves repeatability and decrease false positives. The human eye can interpret qualitative data, while a smartphone camera is used for quantitative analysis. The capillary-driven immunoassay method showed a noteworthy progression in equipment-free point-of-care technology when it comes to SARS-CoV-2 antibody detection. Further, to test the presence of the SARS-CoV-2 antibodies in animals, Yuan et al. have created a monoclonal antibody-based blocking ELISA test using nucleocapsid protein. Similarly, Mariotti et al. developed a monoclonal antibody specific for epitopes of spike protein. A newly developed method, dIgA-ELISA, and dIgA-multiplex bead assay (MBA) can differentiate between monomeric IgA and dimeric Ig A (dIgA) using a chimeric secretory component [53]. According to the study, after 20 days of the infection, 70% of ELISA can be found dIgA, while 50% can be found by dIgA-MBA which indicates that, compared to moderate infections, severe infections are associated with persistent plasma dIgA levels.

In conclusion, ELISA showed high sensitivity and high specificity due to antibody-antigen reaction. However, it cannot be able to detect infections in the early stages. Initially, it was labor driven but nowadays various ELISA-based point of care detection techniques are available which makes it more popular.

Lateral Flow Immunoassay (LFIA)

LFIA is a simple, paper-based detection test that requires a small sample load and 3–30 min to complete the entire process [6]. In 1988, the first commercial home-based kits were developed by Unilever subsidiary. This test can be performed using a variety of biological samples like blood, serum, plasma, tears, saliva, urine, etc. [40]. Five elements are required to form LFIA equipment, such as membrane with immobilized antibodies, different pads like a sample pad, a conjugate release pad, an adsorbent pad, and an adhesive pad [34].

The LFIA process is unquestionably straightforward for the diagnosis of SARS-CoV-2. First, a sample drop is placed into the sample pad and allowed to travel through the conjugate release pad, which has control and S protein-specific IgG that has been previously linked with fluorescent or colored particles. These particle-conjugate complexes are then attached to certain antibodies which have been immobilized in the detecting area accompanied by the flow of the solution [16]. According to Anfossi et al., the antibodies, which are previously immobilized on the dipstick, interact with their target molecule just by moving a liquid sample [1]. The SARS-CoV-2 antibody may capture the viral antigen from the sample after finding the matching antibody [31]. As a result, rapid and extensive serological COVID-19 diagnosis can be possible using LFIA. However, due to their instability and susceptibility to photobleaching, fluorescence-based LFIA methods need to find alternatives. Chen et al. developed an LFIA using lanthanide-doped polystyrene nanoparticles to measure anti-SARS-CoV-2 IgG in human blood. Subsequently, many LFAI tests using colloidal gold labelled with SARS-CoV-2 antigen have been developed for the diagnosis of COVID-19. However, the use of gold nanoparticle-based LFIA for COVID-19 diagnosis has become controversial due to its low sensitivity and high rates of false negatives [60]. Thus, for the improvement of sensitivity and specificity, Hu and his colleagues formed a chelate of EDTA-K2 with colloidal gold [20]. 100% specificity and sensitivity were observed when the LFAI device was used with a fluorescent nanoparticle as a signal for qualitatively detecting IgM, IgA, and IgG antibodies against SARS-CoV-2. This study's findings required further evaluation; however, this approach will have more commercial application if more researchers can improve the LFIA's detection accuracy.

According to a thorough review by Ye et al., the methods include surface-enhanced Raman scattering (SERS), photothermal detection (PTD), and photoacoustic (PA) have been developed to improve the sensitivity of colloidal gold nanoparticles-based LFIA. Amongst them, PTD has demonstrated a tenfold improvement in detection ability [58]. Therefore, Azuma et al. developed a thermometric lateral flow immunoassay using non-fluorescent red and black colored latex beads (CLBs) as reporters deposited on a nitrocellulose membrane. They discovered that red and black latex beads increase light absorption by 1.3 and 3.2 times, respectively, at 520 nm, based on measuring the dependence of temperature rise on the number density of membrane-bound CLBs [2]. Further, a real-time two-channel SERS-based LFIA method was developed by Liu et al. for the detection of anti-SARS-CoV-2 IgM/IgG which showed 800 times better sensitivity compared to gold-based LFIA [29]. Also, quantum nanobeads-based LFIA was developed by Zhou and coworkers.

In general, LFIA is user-friendly, cost-effective, quick, and can screen large numbers of samples in a short period. However, there are certain limitations that we need to consider while selecting this method. Results from LFIA are more often falsely positive and falsely negative. LFIA sensitivity was reduced in a patient with mild illness, and a false negative result may appear early in the course of the disease [43]. Despite this, LFIA is effective in screening asymptomatic persons. To overcome these limitations, further work should be done to improve LFIA's sensitivity and specificity.

Neutralization Assay

After SARS-CoV-2 infection or vaccination weather adaptive immunity is produced or not can be detected by measuring antibodies against S protein, N protein or RBD in recovering patients [56]. Various antibodies are generated once the person meets viral epitopes, including S protein. As viral epitope is not the only entrance mechanism, each antibody cannot effectively block the virus's entry in the host cell by binding to an epitope. But neutralizing antibodies can easily prevent the entry of virus through binding with the S Protein RBD to the ACE2 receptor [11]. Neutralization test measures the ability of an antibody to prevent viral infection as well as cytopathic effects due to viral infection. Patient's samples like blood, serum, plasma, etc., can be used to measure neutralizing antibody levels by simply adding them to cell cultures.

There are different types of neutralization assays to identify viruses: surrogate virus neutralization test (sVNTs), pseudovirus neutralization test (pVNTs), and plaque reduction neutralization test (PRNT) [15]. PRNT is the gold standard method; however, it requires trained persons and BSL3 facilities as deals with live viruses rather pVNTs reduce that requirement to BL2 labs with high throughput [36]. Even that requirement is reduced using automated chemiluminescence techniques called sVNTs which inhibit interactions between RBD and ACE2 receptors [13]. In conclusion, neutralizing antibody detection is important for the long-term development of vaccines as well as for the immediate therapeutic use of recovered plasma.

Chemiluminescence Immunoassay (CLIA)

Luminescent immunoassay techniques lower the detection limits of antibody-based reagents which mainly include fluorescence and chemiluminescence [52]. Chemiluminescence immunoassay (CLIA) combines two susceptible methods. One of them is the chemiluminescent reaction which forms an excited state after undergoing a chemical reaction that returns to the ground state and releases photons. This generated luminous intensity is then measured using an instrument.

CLIA is such an immunolabeling technique, which makes use of either an enzyme or inorganic catalyst as the detection limits of antibody-based reagents, making it possible to detect with very high sensitivity and specificity a target antigen or antibody through a chemiluminescent reaction on a previously prepared substrate. To achieve quantitative analyte detection, the immune response bridge links the analyte's concentration with its chemiluminescence intensity [9]. Sekirov et al. developed six different CLIA for detecting COVID-19 in symptomatically infected patients after 14 days of onset of symptoms [42]. They found that CLIA showed more sensitivity and specificity when compared with PCR results but their sensitivity was reduced after 14 days.

Liu et al. evaluated a chemiluminescence microparticle immunoassay-based SARS-CoV-2 antibodies detection method [27]. They conducted a study for the detection of IgM and total antibodies using serum samples from 260 infected patients and 270 healthy individuals. The results suggested that CMIA was more sensitive for the detection of total Ab than IgM and could reduce the rate of misinterpretation [14]. Also, peptide-based magnetic CLIA for the diagnosis of COVID-19 was developed by Cai et al. Even, two fully automated serological tests for SAR-CoV-2 which run on the automated Diazyme DZ-lite 3000 and chemiluminescence analyzer were developed by Diazyme Laboratories [8]. Further, the electrochemiluminescence-based detection method was developed using Pt-coated Au core and Au nanoparticle ECL acceptor which showed high specificity for N protein and high detection limit up to pg/mL [50].

In conclusion, CLIA utilizes the same principle as ELISA, i.e., binding affinity between the viral antigen and the host antibody. However, CLIA uses a chemical reaction to produce a glowing chemical probe which is then utilized for the final detection. The results obtained using CLIA take half an hour to two hours with improved accuracy and lower signal-to-noise ratio [41].

Biosensor Test

One crucial analytical device is the biosensor, used for detecting biological molecules such as nucleic acids, enzymes, and antibodies. It has with transducer and detector which is used to categorize analytes and provide a digital output [35]. These sensors are applied in medical diagnostics, food processing, agricultural products, environmental monitoring, and water quality assessment. Viral biosensors present a promising alternative to traditional diagnostic assays, delivering quick and highly accurate results compared to conventional laboratory methods. This technique is also cost-effective. Biosensors function by converting biomolecular interactions into a definitive signal through various methods, including optical, electrical, enzymatic, electrical, optical, or other methods [37].

For the past many years, the development and performance growth of biosensor research have been remarkable. This progress is driven by advancements in transduction systems, genetic engineering, and nanotechnology, which have introduced several approaches to enhance the recognition capabilities of biosensors [18]. Consequently, there are several types of biosensors, categorized based on their innovations and specific applications for the recognition viral elements, as illustrated in Table 7.2.

Surface plasmon resonance-based methods are believed to be the most accurate biosensors known for identifying biomolecular measures related to food, milk adulteration, and blood groups. They have been in much more use within the last 2 to 3 years for the rapid detection of SARS-CoV-2. Utilizing coronaviral surface antigen, the SPR-based biosensing chip has revealed huge potential and efficiency in

Table 7.2 Types of biosensors for the detection of SARS-CoV-2 virus

Types of biosensors	Recognition element	Other viruses detected
Piezoelectric immunosensor	Spike protein S1	Influenza virus; adenovirus; RSV; MERS
Thermal biosensor	RNA-dependent RNA polymerase (*RdRp*) gene	MERS; SARS-CoV-2
Surface plasmonic resonance (SPR) biosensor	Nucleocapsid antibody	It discriminated against all other strains of H5N1 and subtypes of influenza virus
Field effect transistor (FET) based biosensor	Spike antigen proteins	Ebola virus antigen
Electrochemical Biosensors	the spike and nucleocapsid genes of the virus	HIV-1, HBV, Zika, Influenza and Dengue virus
Surface enhances Raman scattering	can identify SARS-CoV-2 in human nasopharyngeal aspirates	common human coronaviruses, influenza viruses

the rapid and highly accurate detection of the SARS-CoV-2 virus. The detection limit of SPR chips is 200 ng/mL for anti-SCVme antibodies, providing results within a couple of minutes [39].

Maciej Trzaskowski and co-workers used a portable SPR device to detect COVID-19 infection. They compared SARS-CoV-2-positive and -negative blood plasma samples with the ELISA test. The SPR device, tested on human plasma samples, effectively detected anti-SARS-CoV-2 antibodies with 40 ng/mL detection limit. The obtained results were reliable with those obtained using the reference diagnostic method [46]. Abdelhadi Djaileb and co-workers cross-validated ELISA with a portable-SPR instrument for IgG antibody serology in SARS-CoV-2 positive entities.

An SPR sensor coated with a peptide monolayer and various SARS-CoV-2 recombinant proteins detected human anti-SARS-CoV-2 IgG antibodies in clinical samples. The usage of a CHO cell line to direct spike ectodomain significantly enhanced assay performance. The portable SPR instrument measured four biological samples within 30 minutes, with the chip being reusable at least nine times [12].

Rapid Antigen Test

The SARS-CoV-2 rapid antigen test (RAT) is a qualitative rapid chromatographic immunoassay, which is used for the identification of SARS-CoV-2 antigens in the human nasopharynx. While RT-PCR is the best for the detection of SARS-CoV-2 because of its accurate precision in measuring the virus's genomic material [48], its limitations in terms of time, cost, and operational requirements make it less suitable for emergency and remote settings [19]. RAT has emerged as a crucial alternative, addressing these limitations by offering a low-cost, straightforward, and quick diagnostic tool that enables point-of-care testing, immediate isolation, and timely infection control measures essential for managing the pandemic. Currently, two major types of diagnostic tests are available for COVID-19: serological tests for anti-SARS-CoV-2 antibodies and direct methods for the detection of antigens, nucleic acids, or viral particles. While RAT generally shows excellent specificity, their sensitivity varies widely due to differences in study design, patient selection, specimen type, and disease stage at sample collection [25].

To measure the diagnostic accuracy of the RAT for SARS-CoV-2, Li-Rong Lin and colleagues conducted a systematic review and meta-analysis including with COVID-19 166,943 suspected patients. The major study aimed to compare RAT results with RT-PCR findings. RAT kits demonstrate high specificity and sensitivity at the very early stages of infection, particularly at the elevated viral load. This makes them an effective tool for hospital management, helping in the quick identification of positive patients during initial triage. Because of that, transmission can be prevented. However, the epidemiological benefits of RAT must be supplemented by cautious and accountable handling of negative test results. Sensitivity is markedly higher in samples with lower Ct values, with pooled sensitivity reaching 1.00 (95% CI: 0.70–1.00) among samples with Ct < 20. Testing within 10 days of symptom onset yields high sensitivity, achieving 0.91 (95% CI: 0.83–0.96) within ≤3 days, 0.89 (95% CI: 0.84–0.93) within ≤7 days, and 0.88 (95% CI: 0.83–0.92) within ≤10 days [57]. Katherine A. Bond and her team explored the effectiveness of SARS-CoV-2 RAT for triaging patients in emergency departments. Their research underscores the critical need for rapid and early detection of SARS-CoV-2 in healthcare settings to ensure effective infection control and promptly guide patients to the appropriate care pathways. While RAT is more sensitive than RT-PCR testing and its quick results can still enhance patient care [5].

Muhammad Fazli Khalid and colleagues conducted a systematic review and meta-analysis to evaluate the performance of rapid antigen tests (RATs) for diagnosing COVID-19. The analysis revealed that RATs have a pooled sensitivity of 70% and a specificity of 98%. Subgroup analyses showed varying sensitivities across different sample types: nasal swabs had the uppermost sensitivity at 83%, followed by nasopharyngeal swabs at 71%, throat swabs at 69%, and saliva at 68%. Sensitivity was notably higher in symptomatic patients (82%) compared to asymptomatic individuals (68%), and RATs achieved 96% sensitivity when the cycle threshold value was ≤25. Despite the ongoing need for improved sensitivity, RATs remain a valuable option, especially in settings lacking laboratory infrastructure, for early detection of COVID-19. Some RATs, such as the CoviNAg ELISA Kit and Sienna-Clarity COVID-19 Ag Rapid Test Cassette (RTC), demonstrated superior sensitivity, whereas MEDsan SARS-CoV-2 Ag showed lower performance [24]. Further research and technological advancements are crucial to enhance RAT sensitivity and reliability, particularly in diverse global regions where access to laboratory testing is limited.

Apurv Soni and colleagues conducted an observational study to evaluate the performance of rapid antigen tests (Ag-RDTs) in detecting SARS-CoV-2 infections among both symptomatic and asymptomatic individuals. The findings underscore the importance of repeated testing with Ag-RDTs to enhance diagnostic accuracy. Specifically, conducting two tests for symptomatic individuals, spaced 48 hours apart, significantly improved test performance. Similarly, performing consecutive tests at 48-hour intervals enhanced sensitivity in asymptomatic individuals. These results highlight the critical role of serial testing protocols, recommended in the study with a minimum 48-hour interval between tests, for optimal performance assessment. Moreover, the study revealed that more than one in ten new SARS-CoV-2 infections were detected only after repeated testing, underscoring the limitations of relying solely on a single-test approach with Ag-RDTs. It was noted that infections yielding a single positive result often had high cycle threshold (Ct) values above 30, indicating lower viral loads that could affect test sensitivity. Despite these insights, the study acknowledges several limitations, including its conduct during the circulation of the Delta and Omicron variants, which could potentially influence test outcomes [44].

Regarding the major benefits of RAT, it offers rapid results within 15–30 minutes, eliminating the need for follow-up appointments to discuss outcomes. RATs are designed to be user-friendly and require no specific training, making them accessible to a wide range of users. Moreover, RATs do not rely on instruments, enabling decentralized testing and improving access in areas without laboratory facilities.

Future Perspective

Serological and immunological assays encompass significant advancements driven by technological innovations, increasing demand for rapid and accurate diagnostic methods, and the growing importance of personalized medicine. The development of fully automated ELISA methods has increased throughput, reduced human error, and allowed for large-scale screening. Similarly, Lateral flow immunoassays have been enhanced with increased sensitivity and specificity. The integration of novel nanomaterials and advanced detection technologies, such as quantum dots and gold nanoparticles, will further improve the performance of these assays. Automation and high-throughput screening technologies in Neutralization assays will streamline the process of measuring neutralizing antibodies, which is particularly important for vaccine development and evaluation. Advances in luminescent materials and detection systems will increase the sensitivity and dynamic range of luminescent immunoassays. The advancement of wearable biosensors for continuous biomarker monitoring is poised to transform personalized healthcare and the management of chronic diseases. Continuous advancements in antigen detection technology will improve the accuracy and reduce the time to result of rapid antigen tests. These future perspectives highlight the transformative potential of serological and immunological assays in improving diagnostic capabilities, personalized medicine, and global health outcomes.

Conclusion

The future of serological and immunological assays is promising, with significant advancements across various methods. Automated ELISA systems will enhance throughput and accuracy, while lateral flow immunoassays will benefit from increased sensitivity and specificity through novel nanomaterials. Neutralization assays will see streamlined processes, crucial for vaccine development. Luminescent immunoassays will achieve greater sensitivity and dynamic range. Wearable biosensors will revolutionize personalized healthcare with continuous biomarker monitoring. Rapid antigen tests will become more accurate and faster. Collectively, these innovations will transform diagnostics, personalized medicine, and global health outcomes.

Declaration

Author's Contribution Conceived the idea and review structure: BMS; Wrote the manuscript: BMS & PM; Revised and edited the manuscript: BMS & PM. Author has read, agreed, and approved the final draft of the manuscript.

Funding None.

Data Availability All the data are included within the manuscript.

Conflicts of Interests There is no conflict of interest.

References

1. Anfossi L, Di Nardo F, Cavalera S, Giovannoli C, Baggiani C. Multiplex lateral flow immunoassay: an overview of strategies towards high-throughput point-of-need testing. Biosensors. 2018;9(1):2. https://doi.org/10.3390/bios9010002.
2. Azuma T, Hui YY, Chen OY, Wang YL, Chang HC. Thermometric lateral flow immunoassay with colored latex beads as reporters for COVID-19 testing. Sci Rep. 2022;12(1):3905. https://doi.org/10.1038/s41598-022-07963-1.
3. Baggio S, Togni G, Eckerle I, Vuillemier N, Kaiser L, Gétaz L. Feasibility of home-based ELISA capillary blood self-testing for anti-SARS-CoV-2 antibodies. Pract Lab Med. 2022;31:e00290. https://doi.org/10.1016/j.plabm.2022.e00290.
4. Bhuiyan NH, Uddin MJ, Lee J, Hong JH, Shim JS. An Internet-of-Disease System for COVID-19 Testing Using Saliva by an AI-Controlled Microfluidic ELISA Device. Adv Mater Technol. 2022;7(9):2101690. https://doi.org/10.1002/admt.202101690.
5. Bond KA, Smith B, Gardiner E, Liew KC, Williams E, Walsham N, Putland M, Williamson DA. Utility of SARS-CoV-2 rapid antigen testing for patient triage in the emergency department: a clinical implementation study in Melbourne, Australia. Lancet Reg Health West Pac. 2022;25 https://doi.org/10.1016/j.lanwpc.2022.100486.
6. Carrio A, Sampedro C, Sanchez-Lopez JL, Pimienta M, Campoy P. Automated low-cost smartphone-based lateral flow saliva test reader for drugs-of-abuse detection. Sensors. 2015;15(11):29569. https://doi.org/10.3390/s151129569.
7. Carrell C, Jang I, Link J, Terry JS, Call Z, Panraksa Y, Chailapakul O, Dandy DS, Geiss BJ, Henry CS. Capillary driven microfluidic sequential flow device for point-of-need ELISA:

COVID-19 serology testing. Anal Methods. 2023;15(22):2721–8. https://doi.org/10.1039/D3AY00225J.
8. Cai XF, Chen J, li Hu J, Long QX, Deng HJ, Liu P, Fan K, Liao P, Liu BZ, Wu GC, Chen YK. A peptide-based magnetic chemiluminescence enzyme immunoassay for serological diagnosis of coronavirus disease 2019. J Infect Dis. 2020;222(2):189. https://doi.org/10.1093/infdis/jiaa243.
9. Cinquanta L, Fontana DE, Bizzaro N. Chemiluminescent immunoassay technology: what does it change in autoantibody detection? Autoimmun Highlights. 2017;8:1. https://doi.org/10.1007/s13317-017-0097-2.
10. Combadiere B. Adaptive immunity against SARS-CoV-2. M S Med Sci. 2020;36(10):908. https://doi.org/10.1051/medsci/2020168.
11. Chen P, Nirula A, Heller B, Gottlieb RL, Boscia J, Morris J, Huhn G, Cardona J, Mocherla B, Stosor V, Shawa I. SARS-CoV-2 neutralizing antibody LY-CoV555 in outpatients with Covid-19. New Engl J Med. 2021;384(3):229. https://doi.org/10.1056/NEJMoa2029849.
12. Djaileb A, Jodaylami MH, Coutu J, Ricard P, Lamarre M, Rochet L, Cellier-Goetghebeur S, Macaulay D, Charron B, Lavallée É, Thibault V. Cross-validation of ELISA and a portable surface plasmon resonance instrument for IgG antibody serology with SARS-CoV-2 positive individuals. Analyst. 2021;146(15):4905. https://doi.org/10.1039/D1AN00893E.
13. Douxfils J, Gillot C, De Gottal É, Vandervinne S, Bayart JL, Dogné JM, Favresse J. Efficient maternal to neonate transfer of neutralizing antibodies after SARS-CoV-2 vaccination with BNT162b2: a case-report and discussion of the literature. Vaccine. 2021;9(8):907. https://doi.org/10.3390/vaccines9080907.
14. Grenache DG, Ye C, Bradfute SB. Correlation of SARS-CoV-2 neutralizing antibodies to an automated chemiluminescent serological immunoassay. J Appl Lab Med. 2021;6(2):491. https://doi.org/10.1093/jalm/jfaa195.
15. Gillot C, Favresse J, David C, Maloteau V, Dogne JM, Douxfils J. An evaluation of a SARS-CoV-2 Pseudovirus neutralization test and a comparison to a SARS-CoV-2 surrogate virus neutralization test in a COVID-19 long-term follow-up cohort. Microbiol Res. 2024;15(1):422. https://doi.org/10.3390/microbiolres15010028.
16. Ghaffari A, Meurant R, Ardakani A. COVID-19 serological tests: how well do they actually perform? Diagnostics. 2020;10(7):453. https://doi.org/10.3390/diagnostics10070453.
17. Gong F, Wei HX, Li Q, Liu L, Li B. Evaluation and comparison of serological methods for COVID-19 diagnosis. Front Mol Biosci. 2021;8:682405. https://doi.org/10.3389/fmolb.2021.682405.
18. Haleem A, Javaid M, Singh RP, Suman R, Rab S. Biosensors applications in medical field: a brief review. Sensors Int. 2021;2:100100. https://doi.org/10.1016/j.sintl.2021.100100.
19. Hameed K, Khan SB, Ahmed ST. Researches and applications of artificial intelligence to mitigate pandemics: history, diagnostic tools, epidemiology, healthcare, and technology. Academic Press; 2021.
20. Hu X, Wu C, Situ B, Tian P, An T, Li Q, Pan W, Zhang R, Yang B, Sun D, Hu Y. EDTA-K2 improves the detection sensitivity of SARS-CoV-2 IgM and IgG antibodies by chelating colloidal gold in the immunochromatographic assay. Int J Nanomedicine. 2021:715. https://doi.org/10.2147/IJN.S281594.
21. Humbert MV, Opurum PC, Brendish NJ, Poole S, He P, Katis I, Quaye J, Bediako Y, Duriez PJ, Eason RW, Sones C. A SARS-CoV-2 nucleocapsid ELISA represents a low-cost alternative to lateral flow testing for community screening in LMI countries. J Infect. 2022;84(1):48–55. https://doi.org/10.1016/j.jinf.2021.08.049.
22. Indenbaum V, Koren R, Katz-Likvornik S, Yitzchaki M, Halpern O, Regev-Yochay G, Cohen C, Biber A, Feferman T, Cohen Saban N, Dhan R. Testing IgG antibodies against the RBD of SARS-CoV-2 is sufficient and necessary for COVID-19 diagnosis. PLoS One. 2020;15(11):e0241164. https://doi.org/10.1371/journal.pone.0241164.
23. Isho B, Abe KT, Zuo M, Jamal AJ, Rathod B, Wang JH, Li Z, Chao G, Rojas OL, Bang YM, Pu A. Persistence of serum and saliva antibody responses to SARS-CoV-2 spike antigens in

COVID-19 patients. Sci Immunol. 2020;5(52):eabe5511. https://doi.org/10.1126/sciimmunol.abe5511.
24. Khalid MF, Selvam K, Jeffry AJN, Salmi MF, Najib MA, Norhayati MN, Aziah I. Performance of rapid antigen tests for COVID-19 diagnosis: a systematic review and meta-analysis. Diagnostics. 2022;12(1):110. https://doi.org/10.3390/diagnostics12010110.
25. Kubina R, Dziedzic A. Molecular and serological tests for COVID-19. A comparative review of SARS-CoV-2 coronavirus laboratory and point-of-care diagnostics. Diagnostics. 2020;10(6):434. https://doi.org/10.3390/diagnostics10060434.
26. Landsteiner K. The specificity of serological reactions. Courier Corporation; 2013.
27. Liu W, Kou G, Dong Y, Zheng Y, Ding Y, Ni W, Wu W, Tang S, Xiong Z, Zhang Y, Liu L. Clinical application of chemiluminescence microparticle immunoassay for SARS-CoV-2 infection diagnosis. J Clin Virol. 2020;130:104576. https://doi.org/10.1016/j.jcv.2020.104576.
28. Liu B, Su X, Yu G, Yang S, Wang F, Huang T, Zhou L, Hui Z, Liao Y, Qiu Y, Huang J. An automated chemiluminescent immunoassay (CLIA) detects SARS-CoV-2 neutralizing antibody levels in COVID-19 patients and vaccinees. Int J Infect Dis. 2022;115:116. https://doi.org/10.1016/j.ijid.2021.12.316.
29. Liu H, Dai E, Xiao R, Zhou Z, Zhang M, Bai Z, Shao Y, Qi K, Tu J, Wang C, Wang S. Development of a SERS-based lateral flow immunoassay for rapid and ultra-sensitive detection of anti-SARS-CoV-2 IgM/IgG in clinical samples. Sens Actuators B. 2021;329:129196. https://doi.org/10.1016/j.snb.2020.129196.
30. Luo J, Brakel A, Krizsan A, Ludwig T, Mötzing M, Volke D, Lakowa N, Gruenewald T, Lehmann C, Wolf J, Borte S. Sensitive and specific serological ELISA for the detection of SARS-CoV-2 infections. Virol J. 2022;19(1):50. https://doi.org/10.1186/s12985-022-01768-4.
31. Machado BAS, Hodel KVS, Barbosa-Júnior VC, Soares MBP, Badaró R. The main molecular and serological methods for diagnosing COVID-19: an overview based on the literature. Viruses. 2020;13(1):40. https://doi.org/10.3390/v13010040.
32. Mahallawi WH. Humoral immune responses in hospitalized COVID-19 patients. Saudi J Biol Sci. 2021;28(7):4055–61. https://doi.org/10.1016/j.sjbs.2021.04.032.
33. Mekonnen D, Mengist HM, Derbie A, Nibret E, Munshea A, He H, Li B, Jin T. Diagnostic accuracy of serological tests and kinetics of severe acute respiratory syndrome coronavirus 2 antibody: a systematic review and meta-analysis. Rev Med Virol. 2021;31(3):2181. https://doi.org/10.1002/rmv.2181.
34. Montesinos I, Gruson D, Kabamba B, Dahma H, Van den Wijngaert S, Reza S, Carbone V, Vandenberg O, Gulbis B, Wolff F, Rodriguez-Villalobos H. Evaluation of two automated and three rapid lateral flow immunoassays for the detection of anti-SARS-CoV-2 antibodies. J Clin Virol. 2020;128:104413. https://doi.org/10.1016/j.jcv.2020.104413.
35. Naresh V, Lee N. A review on biosensors and recent development of nanostructured materials-enabled biosensors. Sensors. 2021;21(4):1109. https://doi.org/10.3390/s21041109.
36. Nie J, Li Q, Wu J, Zhao C, Hao H, Liu H, Zhang L, Nie L, Qin H, Wang M, Lu Q. Establishment and validation of a pseudovirus neutralization assay for SARS-CoV-2. Emerg Microbes Infect. 2020;9(1):680. https://doi.org/10.1080/22221751.2020.1743767.
37. Nnachi RC, Sui N, Ke B, Luo Z, Bhalla N, He D, Yang Z. Biosensors for rapid detection of bacterial pathogens in water, food and environment. Environ Int. 2022;166:107357. https://doi.org/10.1016/j.envint.2022.107357.
38. Nehring M, Pugh S, Dihle T, Gallichotte E, Nett T, Weber E, Mayo C, Lynn L, Ebel G, Fosdick BK, VandeWoude S. Laboratory-based SARS-CoV-2 receptor binding domain serologic assays perform with equivalent sensitivity and specificity to commercial FDA-EUA approved tests. Viruses. 2022;15(1):106. https://doi.org/10.3390/v15010106.
39. Park TJ, Hyun MS, Lee HJ, Lee SY, Ko S. A self-assembled fusion protein-based surface plasmon resonance biosensor for rapid diagnosis of severe acute respiratory syndrome. Talanta. 2009;79(2):295. https://doi.org/10.1016/j.talanta.2009.03.051.
40. Patil AA, Kaushik P, Jain RD, Dandekar PP. Assessment of urinary biomarkers for infectious diseases using lateral flow assays: a comprehensive overview. ACS Infect Dis. 2022;9(1):9. https://doi.org/10.1021/acsinfecdis.2c00449.

41. Padoan A, Cosma C, Sciacovelli L, Faggian D, Plebani M. Analytical performances of a chemiluminescence immunoassay for SARS-CoV-2 IgM/IgG and antibody kinetics. Clin Chem Lab Med. 2020;58(7):1081. https://doi.org/10.1515/cclm-2020-0443.
42. Sekirov I, Barakauskas VE, Simons J, Cook D, Bates B, Burns L, Masud S, Charles M, McLennan M, Mak A, Chahil N. SARS-CoV-2 serology: validation of high-throughput chemiluminescent immunoassay (CLIA) platforms and a field study in British Columbia. J Clin Virol. 2021;142:104914. https://doi.org/10.1016/j.jcv.2021.104914.
43. Shafie MH, Antony Dass M, Ahmad Shaberi HS, Zafarina Z. Screening and confirmation tests for SARS-CoV-2: benefits and drawbacks. Beni-Suef Univ J Basic Appl Sci. 2023;12(1):6. https://doi.org/10.1186/s43088-023-00342-3.
44. Soni A, Herbert C, Lin H, Yan Y, Pretz C, Stamegna P, Wang B, Orwig T, Wright C, Tarrant S, Behar S. Performance of rapid antigen tests to detect symptomatic and asymptomatic SARS-CoV-2 infection: a prospective cohort study. Ann Intern Med. 2023;176(7):975. https://doi.org/10.7326/M23-0385.
45. Syal K. Guidelines on newly identified limitations of diagnostic tools for COVID-19 and consequences. J Med Virol. 2021;93(4):1837. https://doi.org/10.1002/jmv.26673.
46. Trzaskowski M, Mazurkiewicz-Pisarek A, Trzciński JW, Drozd M, Podgórski R, Zabost A, Augustynowicz-Kopeć E. Portable surface plasmon resonance detector for COVID-19 infection. Sensors. 2023;23(8):3946. https://doi.org/10.3390/s23083946.
47. Tian Y, Lian C, Chen Y, Wei D, Zhang X, Ling Y, Wang Y, Yeap LS. Sensitivity and specificity of SARS-CoV-2 S1 subunit in COVID-19 serology assays. Cell Discov. 2020;6(1):75. https://doi.org/10.1038/s41421-020-00224-3.
48. Tsang NNY, So HC, Ng KY, Cowling BJ, Leung GM, Ip DKM. Diagnostic performance of different sampling approaches for SARS-CoV-2 RT-PCR testing: a systematic review and meta-analysis. Lancet Infect Dis. 2021;21(9):1233. https://doi.org/10.1016/S1473-3099(21)00146-8.
49. Vashist SK. In vitro diagnostic assays for COVID-19: recent advances and emerging trends. Diagnostics. 2020;10(4):202. https://doi.org/10.3390/diagnostics10040202.
50. Villa-Manso AM, Guerrero-Esteban T, Pariente F, Toyos-Rodríguez C, de la Escosura-Muñiz A, Revenga-Parra M, Gutiérrez-Sánchez C, Lorenzo E. Bifunctional Au@ Pt/Au nanoparticles as electrochemiluminescence signaling probes for SARS-CoV-2 detection. Talanta. 2023;260:124614. https://doi.org/10.1016/j.talanta.2023.124614.
51. Villafañe L, Vaulet LG, Viere FM, Klepp LI, Forrellad MA, Bigi MM, Romano MI, Magistrelli G, Fermepin MR, Bigi F. Development and evaluation of a low cost IgG ELISA test based in RBD protein for COVID-19. J Immunol Methods. 2022;500:113182. https://doi.org/10.1016/j.jim.2021.113182.
52. Wang YF, Kobayashi M. Antibody detection: principles and applications. Adv Tech Diag Microb. 2012;53–73 https://doi.org/10.1007/978-1-4614-3970-7_4.
53. Wei Z, Angrisano F, Eriksson EM, Mazhari R, Van H, Zheng S, Center RJ, Boo I, McMahon J, Lau J, Kiernan- Walker N. Serological assays to measure dimeric IgA antibodies in SARS-CoV-2 infections. Immunol Cell Biol. 2023;101(9):857–66. https://doi.org/10.1111/imcb.12682.
54. Wölfel R, Corman VM, Guggemos W, Seilmaier M, Zange S, Müller MA, Niemeyer D, Jones TC, Vollmar P, Rothe C, Hoelscher M. Virological assessment of hospitalized patients with COVID-2019. Nature. 2020;581(7809):465. https://doi.org/10.1038/s41586-020-2196-x.
55. Xiang F, Wang X, He X, Peng Z, Yang B, Zhang J, Zhou Q, Ye H, Ma Y, Li H, Wei X. Antibody detection and dynamic characteristics in patients with coronavirus disease 2019. Clin Infect Dis. 2020;71(8):1930. https://doi.org/10.1093/cid/ciaa461.
56. Xie J, Ding C, Li J, Wang Y, Guo H, Lu Z, Wang J, Zheng C, Jin T, Gao Y, He H. Characteristics of patients with coronavirus disease (COVID-19) confirmed using an IgM-IgG antibody test. J Med Virol. 2020;92(10):2004. https://doi.org/10.1002/jmv.25930.
57. Xie JW, He Y, Zheng YW, Wang M, Lin Y, Lin LR. Diagnostic accuracy of rapid antigen test for SARS-CoV-2: a systematic review and meta-analysis of 166,943 suspected COVID-19 patients. Microbiol Res. 2022;265:127185. https://doi.org/10.1016/j.micres.2022.127185.

58. Ye H, Liu Y, Zhan L, Liu Y, Qin Z. Signal amplification and quantification on lateral flow assays by laser excitation of plasmonic nanomaterials. Theranostics. 2020;10(10):4359. https://doi.org/10.7150/thno.44298.
59. Yu HQ, Sun BQ, Fang ZF, Zhao JC, Liu XY, Li YM, Sun XZ, Liang HF, Zhong B, Huang ZF, Zheng PY. Distinct features of SARS-CoV-2-specific IgA response in COVID-19 patients. Eur Respir J. 2020;56(2) https://doi.org/10.1183/13993003.01526-2020.
60. Zhou Y, Chen Y, Liu W, Fang H, Li X, Hou L, Liu Y, Lai W, Huang X, Xiong Y. Development of a rapid and sensitive quantum dot nanobead-based double-antigen sandwich lateral flow immunoassay and its clinical performance for the detection of SARS-CoV-2 total antibodies. Sens Actuators B. 2021;343:130139. https://doi.org/10.1016/j.snb.2021.130139.

Variant-Specific Diagnosis of COVID-19

8

Meghna Bhattacharjee, Sunny Kumar, and Mrinal K. Ghosh

Abstract

The advent of the rapidly evolving severe acute respiratory syndrome coronavirus 2 variants has created havoc worldwide and brought the whole healthcare system to a standstill. It has fostered the need for more variant-specific, highly sensitive, user-friendly, and easily accessible diagnostic tools. For effective and prompt public health responses, accurate and routine detection of the virus is of utmost importance. With the emergence of the variants, identifying them with high accuracy has become imperative to mitigate the spread of the disease. This chapter delves into various diagnostic technologies, including reverse-transcriptase polymerase chain reaction, reverse transcription loop-mediated isothermal amplification, Clustered Regularly Interspaced Short Palindromic Repeats, lateral flow assay, enzyme-linked immunosorbent assay, and genome sequencing, and discusses the variants of concern and interest. It also explores the challenges these technologies face due to the rapidly mutating virus. Furthermore, it discusses the importance of combining genomic surveillance with diagnostic tools, as well as the role of databases like Global Initiative on Sharing All Influenza Data in tracking these variants through the rapid sharing of genomic data.

M. Bhattacharjee
Department of Life Science and Biotechnology, Jadavpur University, Kolkata, India

S. Kumar · M. K. Ghosh (✉)
Cancer Biology and Inflammatory Disorder Division, Council of Scientific and Industrial Research-Indian Institute of Chemical Biology (CSIR-IICB), TRUE Campus, CN-6, Sector–V, Salt Lake, Kolkata, India
e-mail: mrinalghosh@iicb.res.in

M. K. Ghosh, M. Basu (eds.), *Pathogenesis, Management and Socio-Economic Impact of COVID-19 Pandemic*, Medical Virology: From Pathogenesis to Disease Control, https://doi.org/10.1007/978-981-95-1426-7_8

Keywords

Variants of concern (VoC) · Molecular diagnostics · Immunoassay

Introduction

The emergence of a new pneumonia-causing virus in late 2019 marked a turning point in modern medical history [1]. The International Committee on Taxonomy of Viruses identified this novel virus, initially detected in Wuhan, China, as severe acute respiratory syndrome coronavirus 2 (SARS-CoV-2), and the World Health Organization (WHO) formally named the viral illness as coronavirus disease 2019 (COVID-19) [2]. It had an impact not only on public health but also on the global economy and daily life worldwide. COVID-19 affected millions of people across the globe in a short span of time, owing to its high transmissibility [3]. The symptoms of COVID-19 include respiratory distress, fevers, coughs, pneumonia, fatigue, hemoptysis, dyspnea, and muscle pain [2]. SARS-CoV-2 belongs to the coronavirus (CoV) family, which in turn belongs to the family Coronaviridae, genus Betacoronavirus, and order Nidovirales. This enveloped pathogen contains a positive-sense single-stranded RNA and shares structural homology with the other members of the CoV family. It is characterized by the presence of 16 nonstructural proteins and 4 structural proteins: (1) membrane (M), (2) spike (S), (3) envelope (E), and (4) nucleocapsid (N) [5]. The S protein is responsible for facilitating viral entry into the host cell. It consists of two subunits, S1 and S2. The digestive system mucosa, respiratory tract mucosa, and myocardial surface express the angiotensin-converting enzyme 2 (ACE2) receptor, to which the S1 subunit binds. Following this binding, the host protease furin or furin-like proteases cleave at the S1/S2 site. A host factor known as transmembrane serine protease 2 catalyzes this cleavage, exposing a fusion peptide that further facilitates the fusion with the host cell. The E, M, and N proteins are essential for viral replication and pathogenesis [5, 6].

According to the WHO's recommendations, rules like wearing masks, maintaining social distance, avoiding gatherings, and periodically washing hands were imposed to curb the virus's spread, as droplets in the air from coughing or sneezing serve as the primary transmission route [7]. The virus can spread not only through symptomatic individuals, but also silently through asymptomatic and presymptomatic carriers, making it all the more difficult to detect and control its spread [8]. Advancements in diagnostic testing have played an important role in detecting and isolating individuals. Additionally, the development, approval, and implementation of vaccines have significantly reduced mortality rates, thereby improving the situation [3].

Despite these successes, the advent of SARS-CoV-2 variants due to viral mutations has posed new challenges to vaccine efficacy as well as diagnosis, both of which, if compromised, can result in an uncontrolled spread of the virus. These variants exhibit changes in transmissibility and disease severity and are capable of evading immune responses, thereby impacting the effectiveness of existing diagnostic techniques and vaccination strategies. Thus, the development of new diagnostic tools that are capable of detecting the emerging variants is a must in order to keep the disease under control and ensure effective public health measures to mitigate the effect of the emerging SARS-CoV-2 variants [3].

Overview of SARS-CoV-2 Variants

SARS-CoV-2, like other viruses, is prone to mutations that can ultimately alter its structure and function, potentially affecting its virulence and the severity of the disease it causes [5]. It is a natural phenomenon to accumulate mutations resulting from continuous viral replication. The SARS-CoV-2 virus evolves at an approximate rate of 1.1×10^{-3} substitutions per site per year, which translates to nearly 1 substitution every 11 days. While most of the mutations do not have a significant impact on the virus, a few of them have led to the emergence of high-risk variants of the virus. With the advent of new variants, nomenclature and classification have become a big challenge for the WHO. However, in late 2020, the WHO came up with a classification system wherein the virus was classified into two groups, namely, variant of concern (VOC) and variant of interest (VOI) [2, 5]. VOCs include variants that have been demonstrated to exhibit greater disease severity, high transmissibility, and reduced neutralization by the antibodies produced, all of which greatly decrease the efficacy of vaccines and treatments. The Alpha (B.1.1.7), Beta (B.1.351), Gamma (P.1), Delta (B.1.617.2), and Omicron (B.1.1.529) variants fall under this category. On the other hand, VOIs include variants with mutations similar to those of VOCs, but do not exhibit such high pervasive transmission [9]. These variants arise from mutations that may alter key characteristics, such as receptor binding, efficacy of treatments, neutralization by antibodies, and a potential increase in the severity and transmissibility of the disease [2]. The Epsilon (B.1.427 and B.1.429), Mu (B.1.621), Kappa (B.1.617.1), Lamda (C.37), Eta (B.1.525), Zeta (P.2), Iota (B.1.526), and Theta (P.3) variants belong to this group [5].

The WHO has recommended the use of Greek alphabets for naming each novel variant in order to streamline and synchronize the tracking and identification of these emerging variants. This approach helps in avoiding the stigma associated with naming variants as per the geographical locations where they were first detected. This standardized nomenclature allows researchers and healthcare organizations to respond to the evolving landscape of SARS-CoV-2 variants more effectively, have a better understanding of their impacts, and develop effective public health strategies to control their spread [2, 10].

Alpha Variant

In September 2020, a new variant emerged in the United Kingdom, which was termed the Alpha variant by the WHO. This variant spread much more rapidly, indicating a fitness advantage over the original strain. Among the 17 mutations that are found in the viral genome of the Alpha variant, 8 mutations are in the S protein, namely S982A, N501Y, A570D, Δ69–70 deletion, Δ144 deletion, P681H, T716I, and D1118H [11]. Of these, N501Y, found within the receptor binding domain (RBD), is responsible for increasing the binding affinity for the ACE2 receptor. Additionally, P681H enhances the transmissibility of the virus, and ΔH69/ΔV70 in the S protein is associated with immune evasion and, therefore, decreased efficacy of diagnostic tools [12]. Studies have reported that the Alpha variant has a 55% (95% CI [39%–72%]) higher risk of death than the original variant, indicating that this variant not only exhibits higher transmissibility but also higher virulence [13].

Beta Variant

This variant first emerged in South Africa and was designated the Beta variant by the WHO. Among the 9 mutations (N501Y, D215G, E484K, L18F, D80A, D614G, R246I, K417N, and A701V) that are found in the S protein, 3 mutations (N501Y, E484K, and K417N) are present in the RBD. The N501Y mutation, as seen in the Alpha variant, increases the affinity of the virus towards the ACE2 receptor, and the presence of E484K and K417N mutations further enhances this binding affinity [11].

Studies have suggested that the E484K mutation is responsible for reduced neutralization by antibodies. Results have shown that low-to-moderate antibody titers have little-to-no neutralization capabilities, and thus, to improve the efficacy of treatments towards variants with this mutation, vaccine-induced antibody titers must be high [11, 14]. Regarding transmissibility, a study estimated that the variant is 50% (95% CI [20–113]) more infectious than previous variants in South Africa. Studies have also demonstrated that this variant can either completely or partially evade immune responses. When it comes to disease severity, a study from the European Centre for Disease Prevention and Control showed that the hospitalization rate of cases with the Beta variant is significantly higher than cases with non-variants. However, there is not much difference in the death risk associated with this variant versus non-variants [11].

Gamma Variant

The Gamma variant, also known as the P.1 lineage, first emerged in Brazil in January 2020. This variant shares several mutations with the Beta variant and harbors 17 mutations, of which 3 (N501Y, K417T, and E484K) are found in the RBD of the S protein. These 3 mutations, like the first 2 variants, are responsible for increasing the binding affinity for the ACE2 receptor in the host cell as well as enhancing transmissibility [13]. As mentioned earlier, SARS-CoV-2 with the E484K mutation has the capability to evade neutralization by antibodies and, thus, the immune response.

Delta Variant

In March 2021, a variant carrying the mutations E484Q and L452R in the S protein first emerged in India. This variant quickly spread to other countries, earning it the name B.1.617. It comprises three sublineages, one of which, B.1.617.2, was termed the Delta variant by the WHO [11]. Its transmissibility can be compared to that of the Alpha variant [15]. This variant has a mutation P681R in the furin cleavage site that results in increased pathogenicity and disease severity [16]. The Delta variant, when compared to the original variant, exhibits 656 unique mutations. The most prevalent mutation, D164G, confers a fitness advantage by enhancing the stability of the virions formed. Both the deletions of E484Q and D614G resulted in an increase in infectivity and a reduction in the neutralization capability of antibodies. In addition, this dominant variant has also been shown to evade immune responses. The Delta variant shows much higher disease severity than other variants, owing to the presence of a number of advantageous mutations in them [11, 13].

Omicron Variant

This variant was first detected in South Africa and Botswana in November 2021, and it quickly spread to other parts of the world [13]. The S protein accumulates over half of the 60 mutations found in this variant [17]. This variant shares multiple mutations (T95I, Δ69–70 deletion, G142D/Δ143–145 deletion, T478K, N501Y, N655Y, K417N, N679K, and P681H) with the other VOCs. The S protein harbors 15 mutations in the RBD, which are responsible for higher transmissibility and viral infectivity and enhance the ability to escape the host immune response [13]. A few of the key mutations include N501Y, which enhances the binding affinity towards the ACE2 receptor and thereby increases transmissibility; E484A, which, like E484K, reduces the effectiveness of immune responses due to their increased capability of evading neutralizing antibodies; K417N, which alters the structure of the S protein, making it difficult for antibodies to recognize and neutralize the virus; and T478K, which also contributes to immune evasion and increased infectivity [18]. Previous studies suggested that this variant spreads more rapidly than the other variants and is highly infectious, affecting even vaccinated individuals. However, recent preliminary data indicate that the Omicron variant causes less severe disease compared to the Delta variant [13].

Other Variants of Interest (VoI)

The Lamda (C.37) variant was first reported in South America in August 2020. This variant is highly infectious owing to the T76I and L452Q mutations. In August 2021, researchers detected another VOI, Mu (B.1.621), which has since spread to more than 20 countries. The Iota variant first emerged in New York in November 2020. It consists of several mutations in the S protein, one of which, S477N, may be associated with immune evasion. The Epsilon (B.1.427 and B.1.429) variant was first found in the US. It is 20% more infectious than the other variants. However, its pathogenicity remains unclear [19].

Diagnostic Methods for Variant Identification

With the advent of novel SARS-CoV-2 variants, developing newer variant-specific diagnostic tools for better and more rapid detection, as well as improved efficacy, is an absolute necessity. At the same time, creating more rapid and accurate technologies has become all the more difficult due to the emerging variants of SARS-CoV-2. The detection technologies primarily target viral proteins, viral nucleic acid, and host anti-SARS-CoV-2 antibodies and are classified based on these factors (Fig. 8.1) [2].

Genomic Sequencing

Whole-genome sequencing (WGS) of SARS-CoV-2 has played a pivotal role since the onset of the pandemic [20]. It has not only facilitated the identification of the variants but also contributed to the development of reverse-transcriptase polymerase chain reactions (RT-PCRs) and vaccine candidates, among others. WGS allows for deciphering new mutations, tracking variants, exploring lineage transmission, understanding how the virus evolves, and determining the effectiveness of various preventative interventions. WGS also assists in understanding the variants' transmissibility and virulence. It is used at an unprecedented rate worldwide through next-generation sequencing (NGS) methods [20]. The ThermoFisher Ion AmpliSeq SARS-CoV-2 Research Panel is an example that targets the whole genome with high accuracy and sensitivity, inclusive of numerous variants. This assay takes 24–48 h to sequence the whole genome. However, it is expensive and requires trained personnel [21]. Another assay that utilizes the WGS method is the Swift Biosciences SARS-CoV-2 Swift Normalase Amplicon Panel. This panel is known for its high sensitivity and a rapid turnaround time of 10 h, with an LOD of 40 copies per reaction. However, it requires expertise for the interpretation of data [22]. Although extremely useful, WGS has a few limitations. It requires specialized instruments and trained personnel. Additionally, it is time-consuming and not very cost-effective [20].

To address these issues, targeted NGS comes into play. By focusing on specific regions of the genome, targeted NGS reduces the cost. Sequencing can also be completed faster than WGS, rendering it suitable for high-throughput testing. Further, it can focus on specific genomic areas, such as the S gene, which is known to harbor mutations that are responsible for higher infectivity, immune evasion, and transmissibility. An example of a targeted NGS assay is the Illumina COVIDSeq Assay. This tool is capable of detecting the Omicron variant. It targets the ORF1ab gene and the N gene with high sensitivity. It sequences data from previous Omicron-infected

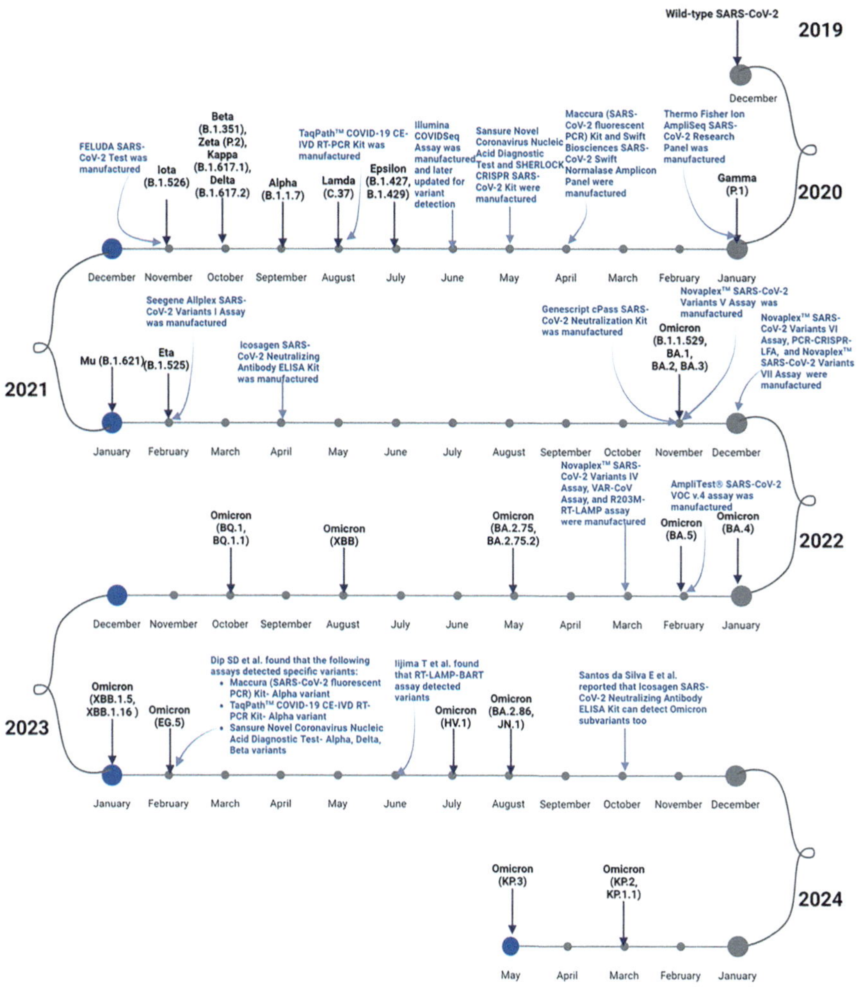

Fig. 8.1 Roadmap for variant-specific diagnostics development

cases and exhibits high performance, along with an easy-to-use quality control approach. This assay takes approximately 2 days for sequencing and analysis [23]. Targeted NGS, however, has its drawbacks. It does not provide information on the whole genome, thereby missing out on mutations present elsewhere, and it also requires NGS expertise [24] see Table 8.1.

Table 8.1 Methods and specifications in VoC diagnosis

Diagnostic technique and kits	Variants detected	Mutation detected or biomarkers targeted	Detection time	Sensitivity or Limit of detection (LOD)	Advantages	Disadvantages	References
RT-PCR							
Maccura (SARS-CoV-2 fluorescent PCR) Kit	Alpha (B.1.1.7)	E484K, N501Y, D614G; ORF1ab, N, E	>1 h	LOD: Low Sensitivity: 82.5%	High accuracy	Chances of getting false-positive results	[25]
TaqPath™ COVID-19 CE-IVD RT-PCR Kit	Alpha (B.1.1.7)	S:del69–70, N501Y; ORF1ab gene, S gene, N gene	>1 h	LOD: Low Sensitivity: 70.2%	High specificity, 81.1% accuracy and 0% false-positive rate	Low sensitivity	[25]
Seegene Allplex SARS-CoV-2 Variants I Assay	Alpha (B.1.1.7), Kappa (B.1.627.1)	HV69/70 del, N501Y, E484K; S gene, RdRp gene	Several hours	LOD: 698 cp/mL Sensitivity: high	Comprehensive mutation detection	Requires specialized equipment, longer turnaround time	[26, 27]
Novaplex™ SARS-CoV-2 Variants IV Assay	Delta (B.1.617.2)	L452R, P681R, K417N; S gene	Within a day	LOD: Low Sensitivity: high	High accuracy	Requires RNA extraction reagents for highest sensitivity	[27]
Novaplex™ SARS-CoV-2 Variants V Assay	Delta (B.1.617.2), Lamda (C.37)	F490S, P681R, L452Q, L452R; S gene	1–2 h	LOD: Approx 250 copies/mL Sensitivity: high	Simultaneous detection of multiple variants	Requires specialised equipment and trained personnel	[27]
Novaplex™ SARS-CoV-2 Variants VI Assay	Lamda (C.37), Mu (B.1.621)	F490S, L452Q, R346K, D950N; S gene	1–2 h	LOD: Approx 250 copies/mL Sensitivity: high	Rapid detection, high specificity	Less sensitivity with RNA extraction-free method, high cost of equipment	[27]

Novaplex™ SARS-CoV-2 Variants VII Assay	Omicron (B.1.1.529)	E484A, HV69/70, N501Y; S gene, RdRp gene	1–2 h	LOD: Approx 250 copies/mL Sensitivity: high	Quick detection of new variants	Requires constant updates, less sensitivity with RNA extraction-free method	[27]
Sansure Novel Coronavirus Nucleic Acid Diagnostic Test	Alpha (B.1.1.7), Delta (B.1.617.2), Beta (B.1.351)	P4715L, R203K, N501Y; ORF1ab gene, N gene	>1 h	LOD: Low Sensitivity: 91.2%	Low false-negative results	Low accuracy: 77.7%	[25]
AmpliTest® SARS-CoV-2 VOC v.4 assay	Delta (B.1.617.2), Omicron (B.1.1.529)	P681H, N679K, del69–70, P681R, L452R; S gene, N gene	Several hours	LOD: 1×10^3 copies/mL Sensitivity: high	Comprehensive mutation detection, high sensitivity	Results may not be reproducible if sample has low viral load	[26]
RT-LAMP							
RT-LAMP-BART assay	Alpha (B.1.1.7), Gamma (P.1), Omicron (BA.1)	N501Y, L452R, Q493R; S gene	30 min	LOD: 10 copies/reaction Sensitivity: high	Better sensitivity and specificity than RT-PCR	Not yet an established assay in real-world settings	[28]
VarDetect™	VOCs	Mutations in the S, N, E, ORF1ab genes	<60 min	LOD: 10 copies/reaction Sensitivity: >95%	High specificity and easy-to-use	May require specialized equipment	[29]

(continued)

Table 8.1 (continued)

Diagnostic technique and kits	Variants detected	Mutation detected or biomarkers targeted	Detection time	Sensitivity or Limit of detection (LOD)	Advantages	Disadvantages	References
R203M-RT-LAMP assay	Delta (B.1.617.2)	R203M; N gene	Approx 1 h	LOD: 481 copies/reaction Sensitivity: 100%	High accuracy and specificity	Not yet available in real-world setting	[30]
CovidDetect™	VOCs	Mutations in the S, N, ORF1ab genes	30–45 min	LOD: Low Sensitivity: moderate	Rapid detection, high specificity	May require specialized equipment	[31]
CRISPR							
SHERLOCK CRISPR SARS-CoV-2 Kit	Alpha (B.1.1.7) among others	N501Y among other mutations; ORF1ab, N gene, S gene,	60 min	LOD: 50 copies/μL Sensitivity: high	Rapid results, high specificity	Requires RNA extraction	[32, 33]
DETECTR CRISPR SARS-CoV-2 Kit	Alpha (B.1.1.7), Delta (B.1.617.2) among others	L452R, N501Y, T478K among other mutations; N gene, E gene	30 min	LOD: 10 copies/μL Sensitivity: high	Rapid and accurate detection	Requires specialized reagents	[34, 35]
FELUDA SARS-CoV-2 Test	Beta (B.1.351)	E484K, K417N; S gene, RdRp gene	1–2 h	High	Simple and inexpensive	Moderate LOD	[33, 36]
ELISA							
Icosagen SARS-CoV-2 Neutralizing Antibody ELISA Kit	Delta (B.1.617.2), Omicron (BA.1, BA.2, BA.5)	S protein	Rapid	Sensitivity: 87.76%	High specificity	Limited to detecting specific variants	[37]

Genescript cPass SARS-CoV-2 Neutralization Kit	Delta (B.1.617.2), Omicron (BA.1, BA.2)	S protein	Rapid	Sensitivity: high	FDA- approved, high specificity	May not detect all neutralizing antibodies, limited to known mutations	[37]
LFA							
VAR-CoV Assay	Alpha (B.1.1.7), Gamma (P.1), Delta (B.1.617.2)	N501Y, E484K, L452R, E484Q/A; S protein	Rapid	Sensitivity: high	High specificity and sensitivity	Requires laboratory setup and trained personnel	[38]
Bio-SCAN	Alpha (B.1.1.7), Beta (B.1.351), Delta (B.1.617.2)	SARS-CoV-2 genome	Within 1 h	LOD: 4 copies/μL Sensitivity: high	Rapid detection, no need for sophisticated equipment	Limited variant specificity	[39]
PCR-CRISPR-LFA	Alpha (B.1.1.7), Omicron (BA.1)	HV69–70 deletion; S protein	Rapid	LOD: 1 copy/μL Sensitivity: high	High sensitivity, no cross-reactivity with other pathogens	Requires PCR step and specific crRNA design	[40]
NGS							
Illumina COVIDSeq Assay [23]	Omicron (B.1.1.529)	Viral genome; ORF1ab, N gene	48 h	Sensitivity: high	High performance, sensitivity; easy-to-use quality control approaches	There may be chances of failure	

(continued)

Table 8.1 (continued)

Diagnostic technique and kits	Variants detected	Mutation detected or biomarkers targeted	Detection time	Sensitivity or Limit of detection (LOD)	Advantages	Disadvantages	References
Thermo Fisher Ion AmpliSeq SARS-CoV-2 Research Panel [21]	Multiple variants	Whole genome	24–48 h	Sensitivity: high	High accuracy, inclusive of numerous variants	Expensive, needs trained personnel	
Swift Biosciences SARS-CoV-2 Swift Normalase Amplicon Panel [22]	Multiple variants	Whole genome	10 h	LOD: 40 copies/reaction; Sensitivity: high	High sensitivity	Requires NGS expertise for interpretation	

Nucleic Acid-Based Methods

RT-PCR is considered a gold standard all over the world. It can be modified to detect variants by targeting mutations specific to the respective variants. These tests are highly specific and can provide results rapidly.

They are also less complex and much more cost-effective than WGS. Most RT-PCR kits target two genes of SARS-CoV-2 (ORF1ab and N), which often yield false-negative results and exhibit low sensitivity. To address this issue, researchers have developed assays targeting three genes, like the Maccura (SARS-CoV-2 fluorescent PCR) Kit. This assay is capable of detecting the Alpha variant with an accuracy of 82.2% and a sensitivity of 82.5%. It targets the ORF1ab gene, N gene, and E gene to enhance its detection capability [25]. Several other assays have been developed for variant-specific detection. One such example is the Seegene Allplex SARS-CoV-2 Variants I Assay, which is capable of detecting the Alpha and Kappa variants by targeting HV69/70 deletion, N501Y, and E484K mutations in the S gene and RNA-dependent RNA polymerase (RdRp) gene. It provides comprehensive mutation detection but needs specialized equipment and has a longer turnaround time [26, 27]. Another kit, the AmpliTest® SARS-CoV-2 VOC v.4 assay, can detect the Delta and Omicron variants with high sensitivity and an LOD of 1×10^3 copies/mL. It targets the S gene and the N gene, carrying the mutations P681H, N679K, del69–70, P681R, and L452R. It takes several hours to detect, but it provides comprehensive mutation detection. The results, however, may not be reproducible if the sample has a low viral load [41]. The Sansure Novel Coronavirus Nucleic Acid Diagnostic Test is yet another assay that is capable of detecting Alpha, Delta, and Beta variants by targeting the mutations P4715L, R203K, and N501Y in the ORF1ab gene and the N gene within an hour. This assay has a sensitivity of 91.2% and provides low false-negative results, but with an accuracy of 77.7%. For the detection of the Alpha variant, there is also another kit, the TaqPath™ COVID-19 CE-IVD RT-PCR Kit. Although the sensitivity is low, it can target the ORF1ab gene, S gene, and N gene with high specificity and 81% accuracy within an hour [25]. Additionally, several others kits have been manufactured by Seegene that are variant specific, such as Novaplex™ SARS-CoV-2 Variants IV Assay, which is capable of detecting the Delta variant by accurately targeting L452R, P681R, and K417N mutations in the S gene; Novaplex™ SARS-CoV-2 Variants V Assay, which is capable of rapidly detecting the Delta and Lamda variants by targeting F490S, P681R, L452Q, and L452R mutations in the S gene; Novaplex™ SARS-CoV-2 Variants VI Assay, which is capable of rapidly and specifically detecting Lamda and Mu variants by targeting F490S, L452Q, R346K, and D950N mutations in the S gene; and Novaplex™ SARS-CoV-2 Variants VII Assay that targets the E484A, HV69/70, and N501Y mutations in the S gene and RdRp gene and therefore, can detect the Omicron variant quickly with high specificity. However, this assay requires constant updates and has less sensitivity with RNA extraction-free method [27].

However, although useful, RT-PCR assays can only detect known mutations and are not useful tools for identifying emerging variants. Thus, the phenomenon of S gene target failure (SGTF) comes to the rescue. Several test kits targeting the S gene, among others, have failed to detect the S gene target sequence in samples positive for SARS-CoV-2 RNA. The TapPath COVID-19 combo test, for instance, is designed to detect the S, N, and ORF genes. The test produces a positive result if it detects two of the three genes. In 2020, the United Kingdom reported an increase in the detection of samples positive for the ORF gene and the N gene, but not for the S gene. This is referred to as SGTF, and it acts as a marker for the presence of emerging variants. Upon WGS of these samples carrying a novel variant, it was found that several mutations were present in the S gene, including the deletion of nucleotides leading to the deletion of amino acids at the 69th and 70th positions in the S protein. These deletions contributed to the SGTF observed on the TapPath kit. This variant was labeled the Alpha variant. As the prevalence of this variant increased, SGTF detection was used as a marker for its presence. On the basis of the presence or absence of the SGTF signature, the Alpha variant can be differentiated from the Delta variant, which is devoid of the Δ69–70 deletion. Similar to the Alpha variant, the Omicron variant, carrying the Δ69–70 deletion in the S protein, also exhibits this signature. This technique, therefore, is a valuable indicator of the presence of novel variants [42].

To avoid the high temperatures and sophisticated instrumentation required for RT-PCR, another method, reverse transcription loop-mediated isothermal amplification (RT-LAMP), was developed. This technique involves just a single step-reverse transcription, and the outcome can be visualized by a change in color. Additionally, it is highly specific, as it utilizes 6–8 specific primers for the identification of eight distinct sites of the target [2]. The RT-LAMP-BART assay is an RT-LAMP tool capable of detecting the Alpha, Gamma, and Omicron variants. This kit targets the N501Y, L452R, and Q493R mutations in the S gene. It has a 30-min turnaround time and an LOD of 10 copies per reaction. Despite its high sensitivity and specificity, it is not yet an established assay in real-world settings [28]. The VarDetect™ is another example that can detect VOCs with high sensitivity and specificity. It takes less than 60 min and has an LOD of 10 copies/reaction. It detects key mutations in ORF1ab, N, E, and S genes [29]. Furthermore, Yang J et al. designed an RT-LAMP assay that can detect the Delta variant by targeting the R203M mutation in the N gene. It takes approximately 1 h to detect and is not only highly accurate, but also highly specific and sensitive. However, this assay is not yet available in a real-world setting [30].

In recent years, Clustered Regularly Interspaced Short Palindromic Repeats (CRISPR), a highly sensitive and accurate technology, has also been used as a diagnostic tool for identifying variants [2]. The CRISPR-Cas system is relatively simple and requires only a molecule of Cas enzyme (usually Cas9, Cas12, or Cas13) and 1 crRNA. Cas 12 and Cas 13 are capable of cleaving both target as well as non-target sequences. Thus, in the presence of the target sequence, they get activated and start cleaving. The non-target sequences tagged with a reporter, upon cleavage, produce a signal indicating the presence of the nucleic acid of interest [43]. The SHERLOCK CRISPR SARS-CoV-2 Kit, which involves the isothermal amplification of the ssRNA along with the CRISPR-Cas13 system [3], for example, detects the Alpha variant among other variants by targeting the ORF1ab, N, and S genes. It provides rapid results with high specificity, but it necessitates RNA extraction. It has an LOD of 50 copies/μL and a turnaround time of 1 h [32, 33]. Another CRISPR assay, the FnCas9 Editor Linked Uniform Detection Assay (FELUDA) SARS-CoV-2 test, can detect the Beta variant by targeting the S and RdRp genes [33, 36]. This technique involves the reverse transcription of the ssRNA and subsequent amplification of the cDNA using biotin-labeled primers. Binding of the fluorescein amidite (FAM)-labeled Cas9 complex to the target sequences leads to its activation, which then cleaves the FAM-labeled tracrRNA. Finally, the addition of gold nanoparticles to the anti-FAM antibody results in antibody-specific binding, which makes the entire assay detectable [43]. It is simple and inexpensive but has a moderate LOD. It can detect within 1–2 h with high sensitivity [33, 36]. Yet another CRISPR-based assay, (DNA Endonuclease-Targeted CRISPR Trans Reporter) DETECTR, is able to detect the virus within 30 min. The basic principle of this assay involves the use of Cas12, which can recognize cDNA instead of ssRNA. Herein, the desired sequences are reverse transcribed and amplified, following which they are added to the CRISPR-Cas12 complex. Upon binding of Cas12 with the target sequence, it is activated and cleaves both the target as well as reported-labeled nontarget sequences, leading to their detection [43]. This kit can detect the Delta variant among other variants by targeting the N and E genes with high specificity and an LOD of 10 copies/μL. It is quick and accurate, but it requires specialized reagents [34, 35]. Nucleic acids in this technique can be detected either by lateral flow strips, fluorescence visualization, or by using agarose gel [3].

Rapid Diagnostic Tests (RDTs)

Enzyme-linked immunosorbent assay (ELISA) is a highly sensitive technique wherein multiwall plates are coated with antibodies against SARS-CoV-2 antigens. Upon the introduction of patient samples into these plates, enzyme-conjugated secondary antibodies are added, and a signal is generated [2]. The Icosagen SARS-CoV-2 neutralizing antibody ELISA Kit can detect the Delta and Omicron variants. It targets the S protein with high specificity and a sensitivity of 87.76%. However, despite its rapidity, this tool can only detect specific variants. Another kit capable of detecting the Omicron and Delta variants is the Genescript cPass SARS-CoV-2 Neutralization Kit. It targets the S protein with high specificity and is rapid. However, it is limited to known mutations [37].

Lateral flow assay (LFA), a convenient alternative to ELISA, offers rapid on-site testing and relies on antigen-antibody interactions. This is a user-friendly technique, with no sophisticated equipment required [44]. In this method, a sample containing SARS-CoV-2 antigens binds to antibodies with gold nanoparticles at the conjugate pad. Then, at the test line, secondary antibodies attached to the strip bind to the antigen-antibody complexes with the gold nanoparticles. Finally, at the control line, antibodies that can bind to the gold nanoparticle-conjugated antibodies capture the remaining antibodies, forming two lines: the test line and the control line [3]. An example of LFA is the VAR-CoV Assay, which rapidly detects the Alpha, Gamma, and Delta variants by targeting mutations in the S protein. It is highly specific and sensitive but requires a laboratory setup and trained personnel [38]. Bio-SCAN, another LFA kit, is capable of detecting the Alpha, Beta, and Delta variants rapidly within an hour with an LOD of 4 copies/μL. This assay, however, has limited variant specificity [39]. Despite the benefits of RDTs, there are some limitations, such as lower sensitivity than RT-PCR, particularly in asymptomatic cases or those with low viral loads.

To allow real-time tracking of variants and understand the epidemiological trends of SARS-CoV-2, the integration of RDTs with genomic surveillance is necessary. When an RDT detects a positive case, it can be further sent for genomic sequencing to identify the variant. The SGTF, for example, on certain RDTs acts as a marker for emerging variants like the Alpha variant and the Omicron variant. By flagging this, genomic surveillance can be focused more efficiently. Furthermore, integrating these two approaches has helped in monitoring the evolution and transmission of the virus, as well as the rapid identification of genetic variants that could ultimately lead to prompt public health measures, such as updating diagnostic tests and vaccination strategies on a large scale [45].

Bioinformatic Tools and Databases

As soon as SARS-CoV-2 was detected in 2019, researchers at the Centers for Disease Control and Prevention in China looked up to databases like the Global Initiative on Sharing All Influenza Data (GISAID) for the rapid sharing of the first whole genome sequence of the then unknown pathogen. Thanks to GISAID, the global research community was able to access the entire genome in about 48 h. GISAID, a high-quality platform, has since become a cornerstone in the efforts to track and identify the variants. The data shared on this platform helps governments trace the outbreak, initiate public health measures, and plan treatment strategies [46].

GISAID is a transparent platform, allowing researchers all over the world to share genomic data as well as access data uploaded from any part of the globe. This transparency is pivotal in tracking and identifying novel variants. This platform also assists in spreading the awareness necessary to mitigate the impact of SARS-CoV-2. Further, GISAID incorporates computational tools for sequence analysis, helping researchers analyze the mutations. The CoVsurver tool aligns sequences and annotates them to emphasize intriguing amino acid changes [46]. Tools like CoV-GLUE and CovMT focus on amino acid changes and mutations in the RBD, respectively. By utilizing SARS-CoV-2 protein sequences from GISAID, SARS2Mutant analyzes variant amino acids. Additionally, ViruSurf helps in searching for nucleotide and amino acid mutations. For phylogenetic tree construction, tools like PAUP have been employed. For large phylogenies, FastTree and IQTREE are efficient. BEAST2 and PhyDyn simultaneously estimate epidemiological parameters as well as pathogen phylogenies [47]. Pangolin assigns query SARS-CoV-2 genomes to global lineages on the basis of their genetic sequences. This helps in understanding the evolution of specific lineages and is useful for surveillance purposes, identifying which particular lineages are found in an area. Yet another tool, PoSeiDon, detects positive selection and recombination events, which helps in identifying protein changes and thus helps in determining the target sites for various drugs. There are several other tools that, along with databases like GISAID, can help the scientific community monitor, analyze, and respond to the ongoing viral evolution promptly, thereby reducing the impact of the pathogen, improving healthcare interventions, and preparing the community for future pandemics [47].

Future Directions and Challenges

Advancements in COVID-19 diagnostic techniques have paved the way for a transformative approach to managing infectious diseases. The improvement and integration of cutting-edge technologies determine the future of diagnostics for SARS-CoV-2 as well as other pathogens. Various techniques have been developed during the pandemic to enable efficient detection of the pathogen and its variants. Techniques like RT-PCR, RT-LAMP, and CRISPR, as well as ELISA, LFA, and biosensors, among others, have helped healthcare organizations diagnose and treat the disease in a more organized manner [3]. Artificial intelligence and machine learning have also been incorporated for the quick analysis of large datasets and the prediction of outbreaks to help manage the disease more promptly and efficiently [48]. Additionally, NGS and computational tools are being refined to improve specificity and sensitivity, allowing the detection of even low viral loads.

However, many limitations exist that must be addressed to deal with COVID-19 and future pandemics in a better way, such as the fluctuating accuracies and sensitivities across various kits, false positive and false negative results provided by the diagnostic tools, long turnaround time, complex instrumentation, high cost, the requirement of trained personnel in certain cases, the need for better diagnostics to identify and distinguish between variants, and the need to use multiple detection tools in order to identify the variants accurately. As disparities in genomic surveillance can hinder the rapid identification of emerging variants, there is also a need for global cooperation and transparency in data sharing [3, 49].

These challenges can be overcome by the introduction of efficient, portable, cost-effective, and ready-to-use devices, along with improved and flexible diagnostic platforms that will not only allow for widespread, accessible testing but also help in managing the scenario much more effectively. Ensuring the reliability of these test kits amidst a rapidly changing virus with continuously emerging variants is a persistent challenge. Furthermore, the emphasis on vaccination, public awareness, the readiness to impose and follow containment measures whenever necessary, and a global effort are of utmost importance to prepare the mass both at the community and individual levels to avoid yet another wave of future variants and therefore stay ahead of emerging threats [3].

Conclusion

The COVID-19 pandemic outbreak and the rapidly evolving SARS-CoV-2 variants have unquestionably presented major challenges in terms of diagnosis, treatment, and management of the disease. Not only has it sparked a global health emergency, but it has also revealed the weaknesses and vulnerabilities within the healthcare system. However, the pandemic has seen a leap in the advancement of better detection and therapeutic remedies, indicating that scientific research and technology are evolving at the same pace as the pandemic. This chapter discusses the emerging variants of SARS-CoV-2, a few of the variant-specific diagnostic techniques, their limitations, and ways to overcome these challenges.

Although the ability to detect variants with precision and high sensitivity is critical for disease diagnosis and further treatment, problems exist that result in a delay in implementing effective treatment measures. To overcome these obstacles and set the stage for future advancements, consistent research is necessary in order to keep abreast of any new variants and adapt diagnostic strategies to deal with the ever-evolving landscape of variants. The introduction of easily accessible and user-friendly kits is also important for the rapid detection of the pathogen. Furthermore, investing in advanced genomic sequencing techniques to enable better and faster variant identification, as well as setting up wider global surveillance networks to track the evolution and spread of the variants, will not only strengthen the healthcare framework but also prepare it for future pandemics. By embracing these challenges and implementing better treatment measures along with a collaborative and forward-thinking strategy, we can protect global health in the face of novel emerging threats.

Author's Contributions Conceived the idea and review structure: MB, SK, MKG; Wrote the manuscript: MB, SK; Revised and edited the manuscript: SK, MKG. Author has read, agreed, and approved the final draft of the manuscript.

Funding This work is jointly supported by the Department of Science and Technology (NanoMission: DST/NM/NT/2018/105(G); SERB: EMR/2017/000992) and Focused Basic Research (FBR), HCT, CSIR, Govt. of India.

Data Availability The software Biorender (https://www.biorender.com/) is used draw some elements of the few figures.

Declarations

Conflicts of Interests There is no competing and conflict of interest.

References

1. Zhou Y, Zhang L, Xie YH, Wu J. Advancements in detection of SARS-CoV-2 infection for confronting COVID-19 pandemics. Lab Investig. 2022;102(1):4–13. https://doi.org/10.1038/s41374-021-00663-w.
2. Fernandes Q, Inchakalody VP, Merhi M, et al. Emerging COVID-19 variants and their impact on SARS-CoV-2 diagnosis, therapeutics and vaccines. Ann Med. 2022;54(1):524–40. https://doi.org/10.1080/07853890.2022.2031274.
3. Mostafa M, Barhoum A, Sehit E, et al. Current trends in COVID-19 diagnosis and its new variants in physiological fluids: surface antigens, antibodies, nucleic acids, and RNA sequencing. TrAC Trends Anal Chem. Elsevier. 2022;157:116750. https://doi.org/10.1016/j.trac.2022.116750.
4. Ghosh MK, Tabassum S, Basu M. COVID-19 and cancer: dichotomy of the menacing dilemma. MedComm – Oncol. 2023;2:e58. https://doi.org/10.1002/mog2.58.
5. Gusev E, Sarapultsev A, Solomatina L, Chereshnev V. SARS-CoV-2-specific immune response and the pathogenesis of COVID-19. Int J Mol Sci. 2022;23(3):1716. https://doi.org/10.3390/ijms23031716. Published 2022 Feb 2.
6. Cheshmehzangi A, Su Z, Jin R, et al. Space and social distancing in managing and preventing COVID-19 community spread: an overview. Heliyon. 2023;9(3):e13879. https://doi.org/10.1016/j.heliyon.2023.e13879.
7. Oran DP, Topol EJ. Prevalence of asymptomatic SARS-CoV-2 infection: a narrative review. Ann Intern Med. 2020;173(5):362–7. https://doi.org/10.7326/M20-3012.
8. Chen KK, Tsung-Ning Huang D, Huang LM. SARS-CoV-2 variants – evolution, spike protein, and vaccines. Biom J. 2022;45(4):573–9. https://doi.org/10.1016/j.bj.2022.04.006.
9. Akhter N, Sana Sadia, Adnan Ahsan M, et al. Advances in diagnosis and treatment for SARS-CoV-2 variants. Infect Dis. IntechOpen. Available at: https://doi.org/10.5772/intechopen.107846.
10. Choi JY, Smith DM. SARS-CoV-2 variants of concern. Yonsei Med J. 2021;62(11):961–8. https://doi.org/10.3349/ymj.2021.62.11.961.
11. Davies NG, Abbott S, Barnard RC, et al. Estimated transmissibility and impact of SARS-CoV-2 lineage B.1.1.7 in England. Science. 2021;372(6538):eabg3055. https://doi.org/10.1126/science.abg3055.
12. Andre M, Lau LS, Pokharel MD, et al. From alpha to omicron: how different variants of concern of the SARS-Coronavirus-2 impacted the world. Biology (Basel). 2023;12(9):1267. https://doi.org/10.3390/biology12091267. Published 2023 Sep 21.
13. Jangra S, Ye C, Rathnasinghe R, et al. SARS-CoV-2 spike E484K mutation reduces antibody neutralisation. Lancet Microbe. 2021;2(7):e283–4. https://doi.org/10.1016/S2666-5247(21)00068-9.
14. Public Health England. SARS-CoV-2 variants of concern and variants under investigation in England: technical briefing. Available at https://khub.net/documents/135939561/405676950/Increased+Household+Transmission+of+COVID-19+Cases+-+national+case+study.pdf/7f7764fb-ecb0-da31-77b3-b1a8ef7be9aa.
15. Saito A, Irie T, Suzuki R, et al. Enhanced fusogenicity and pathogenicity of SARS-CoV-2 Delta P681R mutation. Nature. 2022;602(7896):300–6. https://doi.org/10.1038/s41586-021-04266-9.
16. Wang L, Cheng G. Sequence analysis of the emerging SARS-CoV-2 variant Omicron in South Africa. J Med Virol. 2022;94(4):1728–33. https://doi.org/10.1002/jmv.27516.
17. Guo Y, Han J, Zhang Y, et al. SARS-CoV-2 omicron variant: epidemiological features, biological characteristics, and clinical significance. Front Immunol. 2022;13:877101. https://doi.org/10.3389/fimmu.2022.877101. Published 2022 Apr 29.
18. Zhang Y, Zhang H, Zhang W. SARS-CoV-2 variants, immune escape, and countermeasures. Front Med. 2022;16(2):196–207. https://doi.org/10.1007/s11684-021-0906-x.

19. Zhu N, Zhang D, Wang W, et al. A novel coronavirus from patients with pneumonia in China, 2019. N Engl J Med. 2020;382(8):727–33. https://doi.org/10.1056/NEJMoa2001017.
20. Charre C, Ginevra C, Sabatier M, et al. Evaluation of NGS-based approaches for SARS-CoV-2 whole genome characterisation. Virus Evol. 2020;6(2):veaa075. https://doi.org/10.1093/ve/veaa075. Published 2020 Oct 5.
21. Plitnick J, Griesemer S, Lasek-Nesselquist E, Singh N, Lamson DM, St George K. Whole-genome sequencing of SARS-CoV-2: assessment of the Ion Torrent AmpliSeq Panel and comparison with the Illumina MiSeq ARTIC Protocol. J Clin Microbiol. 2021;59(12):e0064921. https://doi.org/10.1128/JCM.00649-21.
22. Shrestha L, Lin MJ, Xie H, et al. Clinical performance characteristics of the swift normalase amplicon panel for sensitive recovery of severe acute respiratory syndrome coronavirus 2 genomes. J Mol Diagn. 2022;24(9):963–76. https://doi.org/10.1016/j.jmoldx.2022.05.007.
23. Lowry K, Bauer MJ, Buckley C, et al. Evaluation of Illumina® COVIDSeq™ as a tool for Omicron SARS-CoV-2 characterisation. J Virol Methods. 2023;322:114827. https://doi.org/10.1016/j.jviromet.2023.114827.
24. Dip SD, Sarkar SL, Setu MAA, et al. Evaluation of RT-PCR assays for detection of SARS-CoV-2 variants of concern. Sci Rep. 2023;13(1):2342. https://doi.org/10.1038/s41598-023-28275-y. Published 2023 Feb 9.
25. Caza M, Hogan CA, Jassem A, Prystajecky N, Hadzic A, Wilmer A. Evaluation of the clinical and analytical performance of the Seegene allplex™ SARS-CoV-2 variants I assay for the detection of variants of concern (VOC) and variants of interests (VOI). J Clin Virol. 2021;144:104996. https://doi.org/10.1016/j.jcv.2021.104996.
26. Seegene Inc. n.d. https://www.seegene.com/advantages/complete_solution_for_the_covid_19_response.
27. Iijima T, Sakai J, Kanamori D, et al. A new method to detect variants of SARS-CoV-2 using reverse transcription loop-mediated isothermal amplification combined with a bioluminescent assay in real time (RT-LAMP-BART). Int J Mol Sci. 2023;24(13):10698. https://doi.org/10.3390/ijms241310698. Published 2023 Jun 27.
28. Prime's products for laboratories: COVID-19, variants (Omicron, Delta), Flu A/B, STIs. n.d. https://primediscoveries.com/products.html#vardetect.
29. Yang J, Hu X, Wang W, et al. RT-LAMP assay for rapid detection of the R203M mutation in SARS-CoV-2 Delta variant. Emerg Microbes Infect. 2022;11(1):978–87. https://doi.org/10.1080/22221751.2022.2054368.
30. Prime's products for laboratories: COVID-19, Variants (Omicron, Delta), Flu A/B, STIs. n.d. https://primediscoveries.com/products.html#coviddetect.
31. Kellner MJ, Koob JG, Gootenberg JS, Abudayyeh OO, Zhang F. SHERLOCK: nucleic acid detection with CRISPR nucleases [published correction appears in Nat Protoc. 2020 Mar;15(3):1311. doi: 10.1038/s41596-020-0302-z]. Nat Protoc. 2019;14(10):2986–3012. https://doi.org/10.1038/s41596-019-0210-2.
32. Fernandes RS, de Oliveira Silva J, Gomes KB, et al. Recent advances in point of care testing for COVID-19 detection. Biomed Pharmacother. 2022;153:113538. https://doi.org/10.1016/j.biopha.2022.113538.
33. Broughton JP, Deng X, Yu G, et al. CRISPR-Cas12-based detection of SARS-CoV-2. Nat Biotechnol. 2020;38(7):870–4. https://doi.org/10.1038/s41587-020-0513-4.
34. Fasching CL, Servellita V, McKay B, et al. COVID-19 variant detection with a high-fidelity CRISPR-Cas12 enzyme. J Clin Microbiol. 2022;60(7):e0026122. https://doi.org/10.1128/jcm.00261-22.
35. Azhar M, Phutela R, Kumar M, et al. Rapid and accurate nucleobase detection using FnCas9 and its application in COVID-19 diagnosis. Biosens Bioelectron. 2021;183:113207. https://doi.org/10.1016/j.bios.2021.113207.
36. Santos da Silva E, Servais JY, Kohnen M, et al. Validation of a SARS-CoV-2 surrogate neutralization test detecting neutralizing antibodies against the major variants of concern. Int J Mol Sci. 2023;24(19):14965. https://doi.org/10.3390/ijms241914965. Published 2023 Oct 6.

37. Gomez-Martinez J, Henry S, Tuaillon E, Van De Perre P, Fournier-Wirth C, Foulongne V, Brès JC. Novel lateral flow-based assay for simple and visual detection of SARS-CoV-2 mutations. Front Cell Infect Microbiol. 2022;12. https://doi.org/10.3389/fcimb.2022.902914.
38. Ali Z, Sánchez E, Tehseen M, et al. Bio-SCAN: a CRISPR/dCas9-based lateral flow assay for rapid, specific, and sensitive detection of SARS-CoV-2. ACS Synth Biol. 2022;11(1):406–19. https://doi.org/10.1021/acssynbio.1c00499.
39. Niu M, Han Y, Dong X, Yang L, Li F, Zhang Y, Hu Q, Xia X, Li H, Sun Y. Highly sensitive detection method for HV69-70del in SARS-CoV-2 alpha and omicron variants based on CRISPR/Cas13a. Front Bioeng Biotechnol. 2022;10. https://doi.org/10.3389/fbioe.2022.831332.
40. Shipulin GA, Savochkina Y, Shuryaeva AK, et al. Development and application of an RT–PCR assay for the identification of the delta and omicron variants of SARS-COV-2. Heliyon. 2023;9(6):e16917. https://doi.org/10.1016/j.heliyon.2023.e16917.
41. McMillen T, Jani K, Robilotti EV, Kamboj M, Babady NE. The spike gene target failure (SGTF) genomic signature is highly accurate for the identification of Alpha and Omicron SARS-CoV-2 variants. Sci Rep. 2022;12(1):18968. https://doi.org/10.1038/s41598-022-21564-y. Published 2022 Nov 8.
42. Ebrahimi S, Khanbabaei H, Abbasi S, et al. CRISPR-Cas system: a promising diagnostic tool for Covid-19. Avicenna J Med Biotechnol. 2022;14(1):3–9. https://doi.org/10.18502/ajmb.v14i1.8165.
43. Majdinasab M, Badea M, Marty JL. Aptamer-based lateral flow assays: current trends in clinical diagnostic rapid tests. Pharmaceuticals (Basel). 2022;15(1):90. https://doi.org/10.3390/ph15010090. Published 2022 Jan 13.
44. Tosta S, Moreno K, Schuab G, et al. Global SARS-CoV-2 genomic surveillance: what we have learned (so far). Infect Genet Evol. 2023;108:105405. https://doi.org/10.1016/j.meegid.2023.105405.
45. Elbe S, Buckland-Merrett G. Data, disease and diplomacy: GISAID's innovative contribution to global health. Glob Chall. 2017;1(1):33–46. https://doi.org/10.1002/gch2.1018. Published 2017 Jan 10.
46. Tan M, Xia J, Luo H, Meng G, Zhu Z. Applying the digital data and the bioinformatics tools in SARS-CoV-2 research. Comput Struct Biotechnol J. 2023;21:4697–705. https://doi.org/10.1016/j.csbj.2023.09.044. Published 2023 Oct 1.
47. Dogan O, Tiwari S, Jabbar MA, Guggari S. A systematic review on AI/ML approaches against COVID-19 outbreak. Complex Intell Syst. 2021;7(5):2655–78. https://doi.org/10.1007/s40747-021-00424-8.
48. Teymouri M, Mollazadeh S, Mortazavi H, et al. Recent advances and challenges of RT-PCR tests for the diagnosis of COVID-19. Pathol Res Pract. 2021;221:153443. https://doi.org/10.1016/j.prp.2021.153443.

Part III

COVID-19: Immunological Perspectives

Cytokine Storm, Inflammation and Immunity in COVID-19

9

Chandreyee Datta and Ashish Bhattacharjee

Abstract

COVID-19 began as a pneumonia outbreak in Wuhan, China, in 2019 and then it spread globally. SARS-CoV-2 virus possesses a receptor binding region which can bind to angiotensin converting enzyme 2 receptors of the host cells, allowing the virus to infect host cells. COVID-19 illness can range from normal cough and cold to life-threatening acute respiratory distress syndrome and multiorgan failure. Immune dysfunction and cytokine dysregulation, known as "cytokine storm syndrome," play a crucial role in progressing from mild to severe disease. In COVID-19 patients, elevation of cytokines and interferons is prevalent in SARS-CoV2-induced pneumonia, resulting in a cytokine storm syndrome. Identifying authorized medicines with appropriate safety profiles is crucial for treating hyperinflammation. Although some aspects of this topic have been explored previously, a thorough understanding of COVID cytokine storm (COVID-CS) is extremely necessary. In this chapter, we describe an updated and comprehensive picture of COVID-CS. Here, we reviewed the immunopathological aspects of COVID-CS, including its mechanism of action, the role of innate and adaptive immune responses, the individual cytokines and their associated signaling pathways involved in COVID-CS.

Keywords

SARS-CoV2 · Cytokine storm · Innate immunity · Adaptive immunity · Hyperinflammation · ARDS · PANoptosis

C. Datta · A. Bhattacharjee (✉)
Department of Biotechnology, National Institute of Technology Durgapur, Durgapur, West Bengal, India
e-mail: abhattacharjee.bt@nitdgp.ac.in

M. K. Ghosh, M. Basu (eds.), *Pathogenesis, Management and Socio-Economic Impact of COVID-19 Pandemic*, Medical Virology: From Pathogenesis to Disease Control, https://doi.org/10.1007/978-981-95-1426-7_9

Introduction

COVID-19 began as a pneumonia outbreak in Wuhan, China in December 2019 and then it spread globally. Next, a new virus, 2019-nCoV, was identified as the cause of the outbreak [1]. SARS-CoV-2 infects the host through angiotensin-converting enzyme 2 (ACE2) receptors found in a variety of organs and immune cells. The virus can produce a variety of clinical symptoms, ranging from mild forms such as fever, cough, and myalgia to moderate forms needing hospitalization (pneumonia and localized inflammation) to acute forms that are fatal. Pneumonia, disseminated intravascular coagulation (DIC), acute respiratory distress syndrome (ARDS), low blood pressure, and multiorgan failure these are the major manifestations of acute COVID infection [2]. COVID-19 severity is linked to levels of inflammatory cytokines such as IL-1β, TNF-α, IL-2, sIL-2RA, IL-6, IL-7, IL-17, IL-18, granulocyte colony stimulating factor, IP10, macrophage colony stimulating factor (M-CSF), MIP-1α/CCL3, MCP-3, and anti-inflammatory cytokines like IL-10 in patients' plasma/serum [3, 4]. Cytokine induction is primarily governed by different signaling pathways including (i) the angiotensin II/AT1R [5, 6]; (ii) ACE2 [7, 8]; and (iii) the innate immune signaling pathways like pattern recognition receptor (PRRs) such as TLRs, RIG-1, and MDA5 [9] and nucleotide-binding oligomerization domain, leucine-rich repeat domain, and pyrin domain containing protein 3 (Nlrp3) inflammasomes [10, 11]. Mounting evidences from several studies have suggested that in COVID-19, high levels of proinflammatory cytokines are produced which is known as Cytokine Storm (CS) [12]. So, in brief, hypercytokinemia or cytokine storm is defined mainly by the following three steps: persistent activation of lymphocytes and macrophages leading to immune dysregulation, excessive cytokine secretion, and severe systemic inflammation causing organ failure and high mortality. CS often may become life-threatening due to unfavorable clinical conditions caused by excessive activation of immune cells and overproduction of inflammatory cytokines. So, the timing of diagnosis and correct line of treatment plays a very crucial role in COVID-19-related disease conditions. Nevertheless, the profile and causal effect of CS in various circumstances considerably differ and the precise guideline for the treatment of CS is not available till now. So, comprehensive approaches are required to identify the immunomodulatory roles of the proinflammatory cytokines associated in COVID-CS.

Although it was partially discussed earlier, a thorough understanding of COVID-CS is needed for the appropriate diagnosis and therapy. In this chapter, we present a complete update of COVID-CS, beginning with discussion about the signaling molecules involved in CS, their mechanism of action and progressing to clinical diagnosis, therapy, and trials. In this chapter, we have discussed about the immunopathological aspects of COVID-CS; its mechanism of action; involvement of innate and adaptive immune responses, and the specific cytokines involved in COVID-CS, including their pathological significance, and downstream signaling. Furthermore, the challenges and opportunities for detecting and treating COVID-CS are explored and discussed in the conclusion and future perspective section.

Role of Immune Response During COVID-Cytokine Storm

SARS-CoV-2 infection leads to activation of innate immune cells to generate several inflammatory cytokines such IL-1, IL-6, IL-8, IL-12, and TNF-α. Next, inflammatory cytokines and chemokines recruit innate immune cells and subsequently activate adaptive immune cells from peripheral tissues. These cytokines produce sustained inflammatory cytokines such as IL-2, IFN-γ, and TNF-α, which can cause lung and epithelial damage [13].

Innate Immune Response and CS

Neutrophils, macrophages, and natural killer (NK) cells are key innate cells involved in cytokine storm pathogenesis.

Role of Neutrophils

In case of severe SARS-CoV-2 infection, neutrophils immediately respond to infections by phagocytosing pathogens and performing respiratory bursts [14]. These cells also kill infected cells by producing neutrophil extracellular trap or NETs, which help to generate thrombus and increase cytokine output during a cytokine storm. Based on clinical data, severe SARS-CoV-2 infection has been linked to NET caused by lung epithelial destruction and disease [15, 16]. Previous research has linked viral lung injury, higher serum NET levels in COVID-19 patients, and secreted cytokine profiles to severe pneumonia caused by SARS-CoV-2 infection. Studies found that COVID-19 patients had a higher rate of NETs compared to healthy control groups. This mechanism could be influenced by a variety of circumstances. Due to the elevation of viral load in lung tissue, viral proteins can activate NETosis. Damaged tissues and neutrophils produce cytokines, which boost neutrophil activity. During cytokine storm, the elevated levels of cytokines recruit neutrophil, which leads to increased production of ROS, and cause NETosis [15, 16].

Role of Monocyte/Macrophages

Macrophages, produced from circulating monocytes, are tissue-resident cells that perform several roles such as engulfing senescent cells, tissue healing, immunoregulation, and antigen presentation. During cytokine storms, macrophages release excessive cytokines, leading to tissue damage and organ failure. Single-cell RNA Seq of peripheral monocytes of severe COVID-19 patients admitted in ICU showed increased level of IL-6 because of a sharp increase in the peripheral $CD14^+$ $CD16^+$ monocytes [17]. In case of acute COVID-19 patients, monocytes and monocyte-derived macrophages in lung tissue contribute to disease development by causing a cytokine storm and destroying peripheral tissues. In these patients, bronchoalveolar fluid shows elevated levels of CCL2 and CCL7, leading to increased recruitment of $CCR2^+$ monocytes [14]. SARS-CoV-2 activated macrophages release inflammatory factors after phagocytizing dead tissue and cell debris, contributing to the development of fibrosis. These reactions are linked to the interaction of damage-associated molecular patterns (DAMPs) with PRR receptors. SARS-CoV-2 activates TLR4, TLR2, and TLR3, which releases inflammatory cytokines including IL-1β [16].

Role of Dendritic Cells

Apart from the involvement of neutrophils and monocytes, it was found that individuals with severe COVID-19, having CS, have nonclassical monocytes and $CD1C^{+}$, conventional dendritic cells that migrate from the circulation to the lungs. $CD141^{+}$ conventional and $CD123^{+}$ plasmacytoid DCs are not found in the peripheral blood or lung [17]. Myeloid cells, such as DCs, release IL-6 during innate immune responses to tissue infection or injury after sensing antigens using Toll-like receptor (TLRs). It has also been observed that in case of acute COVID-19 patient, having cytokine storm syndromes, functional impairment of dendritic cells is very common and have a higher ratio of conventional dendritic cells to plasmacytoid dendritic cells [14].

Role of Natural Killer Cells

In general, natural killer cells play a crucial role in the innate immune system, responding quickly and effectively to infections. So, these cells are altered for adaptive transmission and medication targeting *in vivo*. Patients diagnosed with severe COVID-19 and COVID-19 CS, exhibit reduced levels of CD107a, IFN-γ, and TNF-α in $TCD8^{+}$ and NK cells. Increased expression of Natural Killer Cell Group 2 D (NKG2D) in T and NK cells suggests cellular exhaustion in individuals. HLA-E on lung epithelial cells interacts with NKG2D, leading to NK cell exhaustion. *In vitro*, co-culture of peripheral blood NKs with lung epithelial cells transfected with SARS-CoV-2 spike protein gene resulted in enhanced production of NKG2D and decreased degranulation [17, 18].

Role of Mast Cells

Evidences from recent researches have demonstrated that acute COVID infection stimulate innate immune cells, tissue-resident endothelium and epithelial cells, and cause cytokine storms. COVID virus also stimulates respiratory mast cells during early infection.

During acute COVID-19 infection, mast cells produce several pro-inflammatory molecules like IL-1β, CCL2, IL-6, TNF-α, and bronchoconstrictor molecules like histamine, prostaglandin-D2, and leukotriene-C4, resulting in deadly inflammatory responses and pulmonary problems. Mast cells detect viral DAMPs using TLR receptors or IgE-FcεRI crosslinking, leading to pneumonia and fibrosis [19]. SARS-CoV-2 infection can produce abnormal inflammatory mediator production, leading to respiratory system inflammation and pulmonary problems.

Predictive Factors for COVID-CS Diagnosis

Uncontrolled production of cytokines, such as IL-2, IFN-γ, GM-CSF, and TNF-α, can cause macrophage activation syndrome (MAS) and erythro-phagocytosis (Haemophagocytic lymphohistiocytosis, HLH), leading to anemia [20, 21]. It can also disrupt coagulation and vascular hemostasis, leading to capillary leak syndrome, thrombosis [22], and DIC. These episodes collectively cause ARDS, multiorgan failure, and death [23]. Treg cells [24, 25] can generate cytokines such as IL-10 and tumor growth factor-β to suppress overactive immune responses [26, 27]. However, aggressive inflammatory disorders like CS cannot be controlled by the regulatory system's natural capabilities. COVID-CS is a complex inflammatory process triggered by cytokines that lead to immune cell hyperactivation, organ malfunction, and further complications. So, timely and precise therapeutic interventions are very crucial for effective management of COVID-CS. The therapy technique aims to decrease inflammation by targeting cytokines or signaling pathways and restoring immune regulation. Despite breakthroughs in immunology and pathology, precise approaches for developing diagnostic and therapeutic guidelines remain inadequate. Developing scoring systems such as the HScore, MScore, HLH-2004, Penn grading scale, and the Common Criteria for Adverse Events may be useful in predicting COVID-CS or linked outcomes. Caricchio et al. [28] developed predictive factors for COVID-CS diagnosis. These requirements include three clusters: the following criteria should be met: [a] albumin <2.87 mg/mL, lymphocytes <10.2%, neutrophil absolute count >11.4 × 10^3/mL, [b] alanine aminotransferase >60 IU/L, aspartate aminotransferase >87 IU/L, D-dimer >4930 ng/mL, lactate dehydrogenase >416 U/L, troponin I >1.09 ng/mL, [c] anion gap <6.8 mmol/L, chloride >106 mmol/L, potassium >4.9 mmol/L, and blood urea nitrogen. Ferritin >250 ng/mL and C-reactive protein (CRP) >4.6 mg/dL are also used to confirm continuous systemic inflammation. Another study suggests that a diagnostic criterion for COVID-CS should include peripheral blood oxygen saturation (SpO_2/FiO_2), CRP, ferritin, cytokines/chemokines, and neutrophil/lymphocyte ratio [29]. According to Mehta et al. [30], laboratory assays and the HScore should be conducted to screen hyperinflammation in all acute COVID-19 patients to detect COVID-CS. Although further validation is needed, these criteria offer useful proposals for creating formal COVID-CS guidelines.

Adaptive Immune Response During COVID-CS

The adaptive immune system consists of B cells and T lymphocytes. T cells are divided into subsets with specific effector roles that may contribute to cytokine storms. Th1 cells and CTLs play a key role in the immune system's defense against viruses. Th1 cells recruit macrophages, while Th2 cells recruit eosinophils and basophils, Th9 cells recruit mast cells, and Th17 cells recruit neutrophils [31]. During a cytokine storm, there may be an increased Th1-type inflammatory response. Th1 cells produce high levels of IFN-γ, trigger delayed hypersensitivity reactions, activate macrophages, and aid in defense against intracellular infections [32].

$CD4^+$ T cell subsets can affect the course of COVID-19, either positively or negatively. Individuals with severe COVID-19 and cytokine storm syndrome have both functionally depleted $CD4^+$ T cells and highly active ones. SARS-CoV2-specific $CD4^+$ T cells have been linked to reduced COVID-19 disease severity and a quick decrease in viral load [33]. A study found that having more $CD4^+$ T cells, especially naïve $CD4^+$ T cells, correlates with faster viral clearance and less critical illness [34].

Th17 cells, which produce IL-17 and IL-23, contribute significantly to tissue damage and inflammatory responses related to CS [35]. A recent study found a strong correlation between IL-17A expression and the severity of illness [36]. Studies indicate that a systemic neutrophil milieu may influence $CD4^+$ T cells to promote Th17 and IL-17A, leading to increased inflammation [36, 37].

Follicular helper T cells or Tfh contributes in humoral immunity by interacting with B cells in the germinal center. This allows B cells to differentiate into plasma cells, which has the capability of producing high-affinity neutralizing antibodies [38]. Tfh cells may potentially promote disease progression in some circumstances and in case of severe COVID-19 patients, higher number of cytotoxic Tfh cells is observed which in turn elevates different cytokine levels including IFNγ, IL-2, TNFα [39].

SARS-CoV-2-specific $CD8^+$ T-cell responses may correlate with disease severity and lower numbers found in severe patients and this inadequate $CD8^+$ T cell responses, may contribute to the development of acute COVID-19 and its related manifestations like CS [40]. Recent research suggests that severe COVID-19 patients have exhausted $CD8^+$ T cells with high expression of PD-1, TIM-3, LAG-3, CTLA-4, NKG2A, and CD39. Additionally, there is an inverse relationship between PD-1 and CD38 (an activation marker) in $CD8^+$ T cells [41] and it was also revealed by recent researches that acute COVID-19 patients have higher levels of PD-1 expression in their $CD8^+$ T cells [41]. A recent study found the role of SARS-CoV-2-specific $CD8^+PD\text{-}1^+$ T cells in the disease severity [42].

B cells are not generally responsible for the development of cytokine storm. However, B-cell depletion has been effective in treating different cytokine storm illnesses, including HHV-8-associated multicentric Castleman's disease. This implies that these cells can initiate cytokine storms, especially when infected with viruses including COVID-19 [22]. The involvement and implication of different immune cells in the cytokine storm are described in Fig. 9.1.

Inflammatory Signaling Cascade of Cytokine Storm

The precise mechanism involved in the initiation of COVID-CS induction and the primary cytokine responsible for COVID-CS immunopathology are still unclear. SARS-CoV-2 lacks a distinct pathogen-associated molecular pattern (PAMP) and possibly, upon cellular entry via its ACE2 receptor, viral genomic RNAs can be sensed as PAMPs by related PRRs, such as TLRs and RLRs in host cells. Subsequently, PAMP recognition activates the transcription factors IRF3/7 and NF-κB, resulting in the generation of IFN-I and pro-inflammatory cytokines [31–33]. Additionally, NLRP3 inflammasome produces mature IL-1β and IL-18. Proinflammatory cytokines such IL-6, IL-2, TNF-α, and IFN-γ stimulate JAK-STAT or NF-κB signaling by connecting to immune cell receptors. This leads to increased production of proinflammatory genes, creating a positive feedback loop

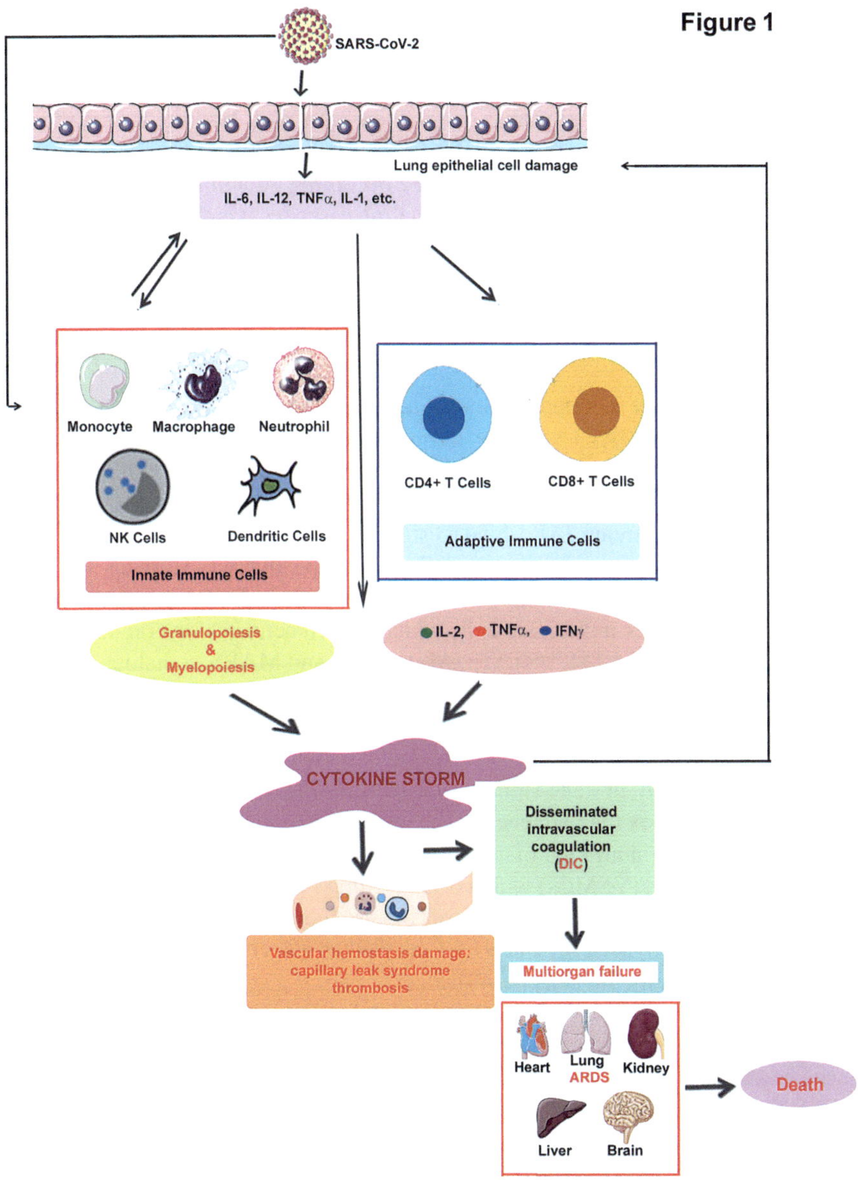

Fig. 9.1 Involvement and implication of immune cells in cytokine storm. Initially, SARS-CoV-2 infects lung epithelial cells and causes tissue damage, which in turn results in the secretion of several inflammatory cytokines like IL-1, IL-6, TNFα, etc. These inflammatory cytokines then recruit innate immune cells like monocytes, macrophages, dendritic cells and natural killer cells which subsequently activate adaptive immune cells ($CD4^+$ and $CD8^+$ T cells) and results in excessive and sustained production of circulating cytokines which can further augment epithelial cell damage. A severe cytokine storm can activate macrophages (Macrophage Activation Syndrome, MAS) and cause erythro-phagocytosis (hemophagocytic lymphohistiocytosis, HLH), leading to anemia and disrupting vascular hemostasis. This can lead to capillary leak syndrome, thrombosis, and DIC (Disseminated intravascular coagulation). These events collectively cause ARDS, multi-organ failure, and death

that triggers CS. In contrast, regulatory cytokines such as IL-10 are released to counteract immunological hyperactivity. SARS-CoV-2 specifically inhibits the protective IFN-I response through many mechanisms, and it causes an increase in proinflammatory cytokines, hindering viral clearance and leading to paradoxical hyperinflammation, such as CS. COVID-CS may be an unintended consequence of the host's immune response to SARS-CoV-2 becoming inflammatory type [12]. The major cytokines and the inflammatory signaling cascades which are involved in the COVID-19 associated severity and CS are briefly described below.

IL-6/IL-6R/JAK-STAT Signaling Pathways

IL-6 is predominantly secreted by monocytes, macrophages, and DCs, which in turn activates the JAK1/STAT3 pathway during inflammation. COVID-19 severity is linked to the IL-6-JAK1-STAT3 axis, with an elevated level of phosphorylated STAT3 in leukocyte subsets in COVID patients compared to healthy controls [2, 43] IL-6 activates downstream JAK1/STAT3 signaling via two signaling pathways: cis-signaling (classic) and trans-signaling (sIL-6R), respectively [43]. In cis signaling, at first, mIL-6Rs (membrane-bound IL-6 receptor) recruits IL-6 to generate an IL-6/IL-6R/gp130 complex in immune cells which in turn activates several downstream signaling pathways like JAK1/STAT3, Akt/mTOR, and MAPK signaling pathways. This also exerts a pleiotropic effect on immune cells by promoting differentiation of Th17, $CD8^+$ T, and B cells, increasing neutrophil migration, and decreasing T_{reg} formation [2]. This leads to increased IL-6 release, which worsens inflammation.

On the other hand, in trans-signaling, IL-6 binds to sIL-6Rs (soluble IL-6 receptor), and this complex then attaches to the gp130 dimer and activates JAK1/STAT3 signaling pathways in all kind of cells specifically in endothelial cells and vascular smooth muscle cells (VSMCs) which do not express mIL-6R. The activation of the IL-6/IL-6R/JAK1-STAT3 pathway causes the production of a variety of mediators, including IL-6, IL-8, MCP-1, and VEGF, as well as a decrease in the expression of E-cadherin on endothelial cells [2, 43].

Furthermore, IL-6 can stimulate the release of acute-phase proteins such as CRP, hepcidin, fibrinogen, thrombopoietin, complement C3, and ferritin. So, altogether, it can be stated that in COVID-CS, IL-6 contributes in immune cell hyperactivation, which leads to malfunction of different vital organs. This pathway has been depicted in Fig. 9.2a.

IFNγ/JAK-STAT Signaling

IFN-γ, mostly produced by macrophages, T cells, and NK cells, plays a significant role in immunological processes like inflammation. When NK cells fail to eradicate infections, IFN-γ plays a significant role in CS-related illnesses [2]. Due to impaired NK activity, excessive T cell activation and IFN-γ production also do not effectively attack infections, leading to immunopathology. Although it is unclear if IFN-γ plays

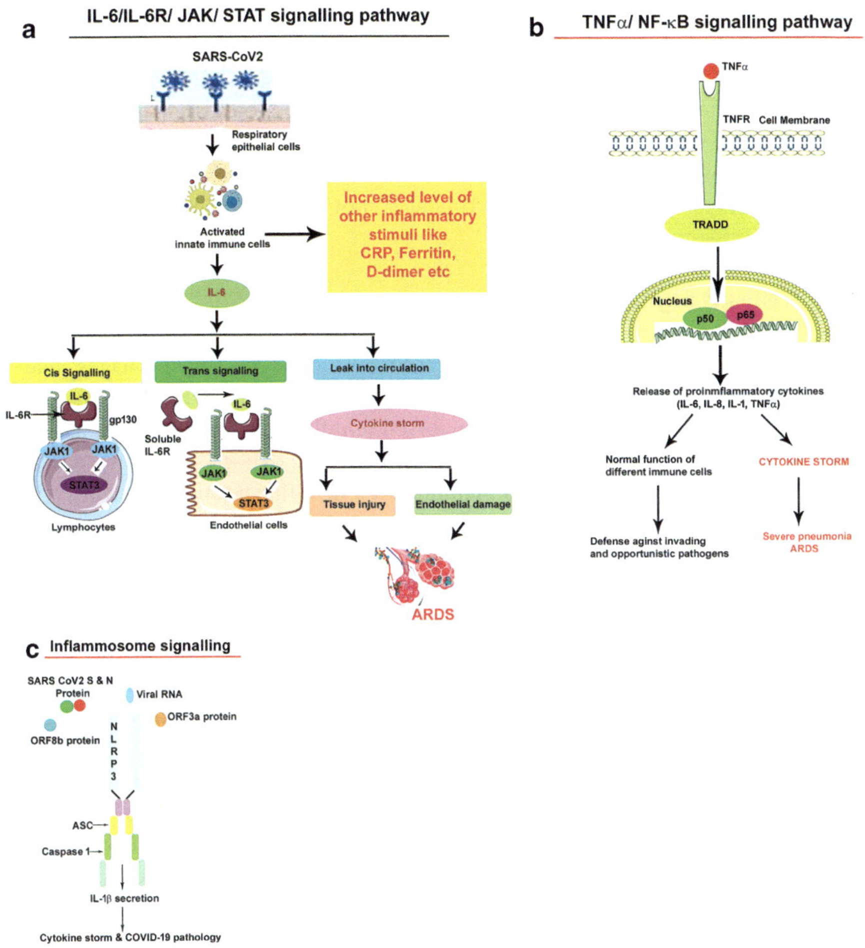

Fig. 9.2 Major signaling pathways involved in cytokine storm and inflammatory responses during SARS-CoV-2 infection. (**a**) IL-6/IL-6R/JAK/STAT (**b**) TNFα/NF-κB and (**c**) NLRP3/IL-1β inflammasome signaling cascade involved in cytokine storm during SARS-CoV-2 infection. (**a**) IL-6, a proinflammatory cytokine majorly operates via two signaling pathways: cis and trans. In cis signaling, IL-6 interacts with the membrane-bound IL-6 receptor (mIL-6R) and gp130, activating Janus kinases 1 (JAK1) and STAT3. In trans signaling, high levels of circulating IL-6 bind to the soluble version of the IL-6 receptor (sIL-6R), forming a complex with a gp130 dimer on almost all somatic cell types. (**b**) TNFα activates the NF-κB signaling cascade, leading to the production of proinflammatory and anti-apoptotic genes via TNFR1 and intermediate adapters. TNFα/NFκB signaling contributes to CS initiation and immune cell hyperactivation by causing epithelial cell death, leading to increased systemic inflammation. (**c**) The NLRP3 inflammasome is assembled after sensing spike (S) and nucleocapsid (N) proteins, viral RNA, and ORF3a. This assembly induces IL-1β secretion, which has been linked to COVID-19 disease pathology

a specific role in COVID-CS, its involvement in activating immune cells suggests that it may play a significant impact. Several investigations have found higher levels of IFN-γ in COVID-19 patients. One study found a decrease in IFN-γ production by $CD4^+$ T cells in severe COVID-19 individuals compared to those with mild illness. This could be attributed to reduced T cell numbers and functional exhaustion, as described above [2]. So, it could be stated that increased production of IFN-γ in COVID-19 is mostly because of macrophages, not by the T cells.

TNFα/NF-κB Signaling

TNFα, a proinflammatory cytokine released by monocytes, macrophages, and T cells, has been linked to many infections, autoimmune disorders, and cancer. NF-κB regulates the expression of cytokines involved in inflammation, a role that has been preserved throughout evolution. TNFα activates the NF-κB signaling cascade, resulting in the production of pro-inflammatory and antiapoptotic genes via TNFR1 and intermediate adapters. NF-κB can activate TNFα during inflammation, such as LPS stimulation. TNFα/NF-κB signaling may cause immune cell hyperactivation and epithelial cell death in CS, potentially leading to disease progression. The interaction of SARS-CoV-2 S protein with ACE2 leads to increased production of NF-κB. PBMC of COVID-19 patients had higher levels of inflammatory NF-κB pathway signaling mediators, including IL1R1, MYD88, IRAK1, TNF receptor-associated factor 6, NFκB1, and RELA indicating the involvement of NF-κB signaling pathway in COVID-19 pathogenesis [44].

Previous research indicates that high level of TNFα is associated with a poor outcome in SARS-CoV and MERS patients. Additionally, inhibiting NF-κB can ameliorate pulmonary symptoms in SARS-CoV-infected mice [45]. However, the relevance of TNFα in COVID-19 remains unclear. According to recent research, TNFα levels in serum have been found to be increased in severe COVID-19 cases. Further research is needed to understand the role of TNFα in detail in case of COVID-CS. This pathway has been depicted in Fig. 9.2b in this book chapter. A recent paper suggests that inhibiting the TNFα-NF-κB pathway may have protective effects in COVID-19, but care should be applied based on two considerations. [a] TNFα's significance in COVID-CS remains unclear. [b] Blocking NF-κB may affect its protective effects in cellular homeostasis, as demonstrated by a general example.

NLRP3/IL-1β Signaling Cascade

IL-1β plays a significant role in autoinflammatory illnesses like gout and chronic inflammatory arthritis. IL-1β is secreted by macrophages through apoptosis and pyroptosis. It promotes immune cell migration to inflamed tissues, Th17 cell differentiation, expression of cytokines and adhesion factors, and activation of the NF-κB pathway, resulting in positive feedback for self-production [2]. The NLRP3 protein forms a complex with ASC and caspase-1 to convert dormant IL-1β precursor to mature IL-1β. IL-1β plays a favorable impact in both initiating and maintaining inflammation [2, 11].

Previous research indicates that IL-1β may have a role in coronavirus-induced CS. Zhang et al. [46] found that higher levels of IL-1β, in COVID-19 subjects with severe symptoms. Huang et al. [47] found increased serum levels of IL-1β in COVID-19 patients. Furthermore, a prior study demonstrated that NLRP3 can be directly activated by SARS-CoV viral proteins such as ORF3a and ORF8b, implying that direct activation of NLRP3 by SARS-CoV-2 protein may have a similar effect. Relevant reviews examine the probable functions of the NLRP3 inflammasome in severe COVID-19. ROS is allegedly an initiator of NLRP3 activation. Excessive ROS production during severe COVID-19 infection may activate NLRP3 and cleave IL-1β precursors, worsening inflammation in COVID-CS. Detailed molecular mechanism related to NLRP3/IL-1β signaling cascade has been depicted in Fig. 9.2c.

IL-10 Signaling

IL-10, produced by immune cells such as Th2 cells, T_{regs}, $CD8^+$ T cells, B cells, DCs, macrophages, and NK cells, plays a crucial role in immunoregulation via the IL-10R/JAK/STAT3 pathway. Mechanistic details related to IL-10/JAK/STAT3 pathway and its contribution in tissue damage/protection during SARS-CoV2 infection have been described in Fig. 9.3a. IL-10 exerts its anti-inflammatory properties directly by inhibiting the innate immunological capabilities of macrophages and DCs or indirectly by promoting T_{reg} formation. Furthermore, IL-10 can activate mast cells and improve the activities of $CD8^+$ T, B, and NK cells. Excessive IL-10 production may operate as a negative feedback mechanism to balance the immune system's hyperactivity. In COVID-CS, IL-10's ability to regulate inflammatory mediator release and proinflammatory cell activation is limited, so external administration of IL-10 is often suggested [1, 2].

However, a new analysis suggests that IL-10 may be harmful during the early stages of SARS-CoV-2 infection as it promotes T cell fatigue. A group of researchers suggest that inhibition of IL-10 with a neutralizing antibody could be a promising treatment option for SARS-CoV-2 infection [2]. Clinical trial data of 102 COVID-19 patients and 45 control individuals from Wuhan, China, demonstrated that patients with critical illness ($n = 17$) had significantly higher serum IL-10 levels compared to those with moderate ($n = 42$) and severe ($n = 43$) illness. Additionally, IL-10 levels were positively correlated with serum CRP concentrations, suggesting that IL-10 could be used as a disease severity indicator. Huang *et al.* [47] found significantly higher plasma levels of IL-10 in COVID-19 patients admitted to the ICU. A follow-up clinical research with 71 COVID-19 patients (53 mild and 18 severe) from Beijing, China, and 18 controls found a consistent trend of increased IL-10 production in early COVID-19 patients [48].

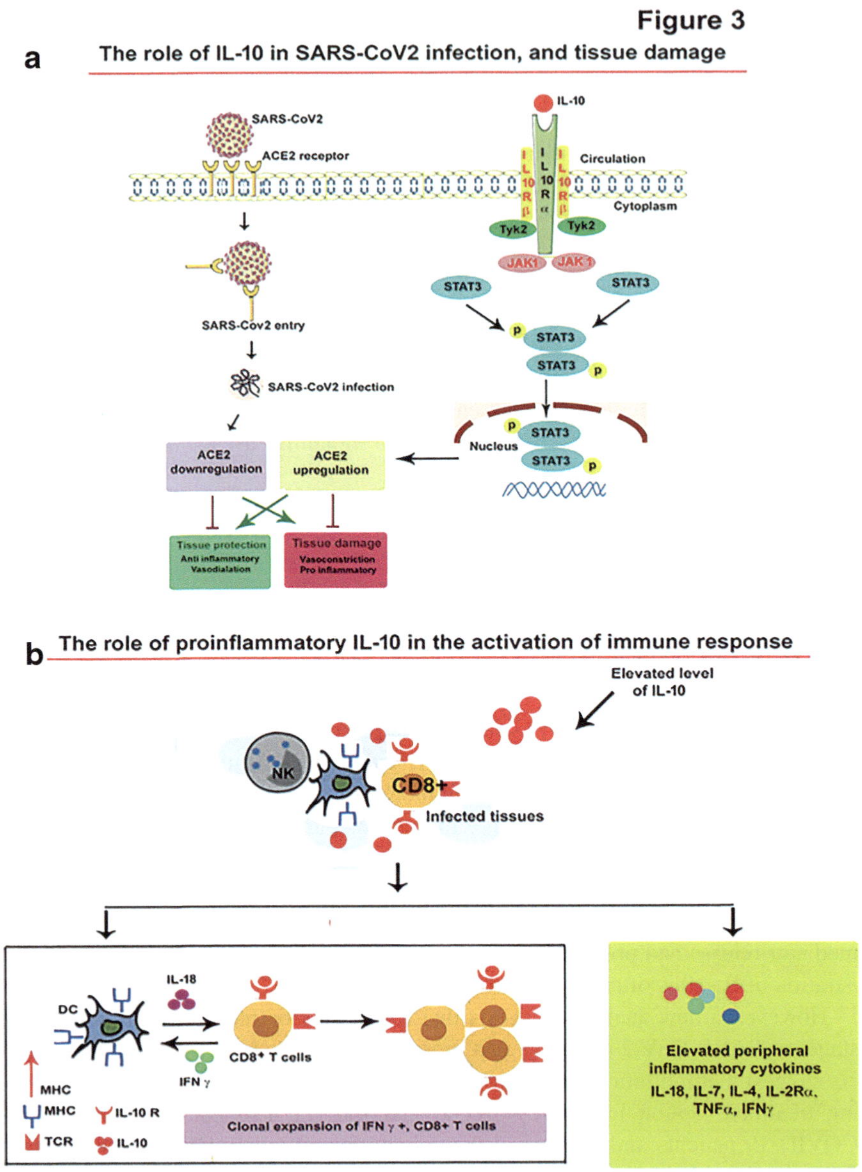

Fig. 9.3 Role of IL-10 in cytokine storm and SARS-CoV-2 pathogenesis. (**a**) Potential role of IL-10 in the regulation of ACE2 downregulation and counteracting tissue damage during severe COVID-19 infection. (**b**) In humans, IL-10 plays a proinflammatory and immune-activating role. IL-10 enhances inflammatory cytokine production, activating and promoting effector CD8⁺ T cell proliferation. The combined effect may be deleterious in COVID-19 pathogenesis

These clinical observations suggest that IL-10 may contribute to COVID-19 development by promoting inflammation and activating the immune system. Several cytokines, including IL-2Rα, IL-4, IL-7, IL-18, IFN-γ, GM-CSF, TNF-α, and chemokines IP-10 and CXCL9, have been found to be raised in the peripheral blood of severe/critically ill COVID-19 patients and COVID-19 patients with severe/critical illness had hyperactivated and proliferating cytotoxic $CD8^+$ T cells, despite a drop in peripheral $CD8^+$ T-cell count [2, 4]. Moreover, in severe/critically ill COVID-19 patients, elevated serum IL-10 levels can increase the fraction of IFN-γ-producing effector $CD4^+$ and $CD8^+$ T cells in peripheral blood [41]. This phenomenon has described in Fig. 9.3b. IL-10 may play a dual role in COVID-19, depending on when it is secreted. In the initial stage, after a few days of infection, IL-10 produced in the lungs may act through a negative feedback mechanism to counteract inflammation and restore tissue homeostasis. In later stages, IL-10 production becomes continuous and elevated, potentially promoting the release of proinflammatory cytokines and chemokines, hyperactivating cytotoxic effector $CD8^+$ T cells, and exacerbating systemic inflammation [2, 26].

IL-10 may exacerbate viral sepsis-related hyperinflammation in severe COVID-19 patients [1, 2]. Hyperactivation of adaptive immunity in COVID-19 patients may worsen the condition, as IL-10 has been shown to expand cytotoxic effector $CD8^+$ T cells in humans [26]. So, it could be hypothesized from the above observations that IL-10's role in boosting systemic inflammatory cytokine production and encouraging T cell activation and proliferation in COVID-19 patients may lead to a fatal immune pathology.

Apart from the above main cytokine and their associated pathways, there are several cytokines that contribute significantly in cytokine storm like IL-7, IL-17, GM-CSF, etc. According to recent findings, COVID-19 patients have higher levels of IL-7, which correlates with disease severity.

However, the consequence of elevated IL-7 levels in COVID-19 remains uncertain. Increased IL-7 levels may be a reaction to lymphopenia in severe/critical COVID-19 patients, given its protective effect [2]. On the other hand, IL-17 can increase lung injury and reduce survival by recruiting neutrophils, promoting proinflammatory factors, and preventing apoptosis in ARDS and LPS-induced acute lung injury models. Evidence suggests that COVID-19 patients, particularly those with severe and critical illnesses, have increased levels of IL-17. Moreover, recent research suggested that IL-17 promotes neutrophil recruitment and causes symptoms as fever, matrix degradation, tissue remodeling, and inflammatory infiltration during COVID-CS. Although IL-17 has been identified as a promising intervention target for COVID-CS, levels were found to be normal in 102 mild and 21 acute patients, respectively. So, more clinical trials and foundational studies are needed for greater clarity [2]. In hyperinflammatory circumstances like CS, GMCSF promotes emergency myelopoiesis and attracts myeloid cells to inflammatory areas, perpetuating inflammatory responses [2].

Cytokine Storm and PANoptosis

SARS-CoV-2 infection in the lungs causes innate immune cells to produce pro-inflammatory cytokines. The combination of TNF-α and IFN-γ promotes PANoptosis, resulting in a cytokine storm loop that amplifies cytokine release and continues the process. This positive feedback loop can cause systemic inflammation, multiorgan failure, and death. TNF-α and IFN-γ synergy increase STAT1 signaling and IRF1 expression, which regulates iNOS for NO generation. This route induces PANoptosis, a newly defined inflammatory programmed cell death pathway induced by a multiprotein PANoptosome complex which allows for interaction and activation of the machinery required for pyroptosis, apoptosis, and necroptosis [49]. TNF-α and IFN-γ activate PANoptosis through molecular pathways and this pathway have depicted in Fig. 9.4. Several medicines have been developed to address this. Transcriptomic analysis of patient-derived 'inflammatory macrophages' showed a high level of transcriptomic similarity to those macrophages triggered by TNF-α and IFN-γ. These cells have elevated amounts of STAT1, IFNGR1, IFNGR2, NFκB1, IL-1β, and other markers [50]. These findings suggest that TNF-α and IFN-γ synergism contribute to COVID-19-associated cytokine storm.

To treat severe COVID-19, clinical medicines have been repurposed. In addition to vaccinations, the FDA has licensed certain antivirals and immunomodulators for COVID-19 treatment. Remdesivir, a viral polymerase inhibitor, has been approved for the treatment of hospitalized patients [49].

Anti-TNF-α medication for rheumatic disorders has been linked to lower hospitalization rates from COVID-19, indicating a potential preventive effect in disease development. Targeting the JAK-STAT signaling pathway, which causes PANoptosis in response to TNF-α and IFN-γ, can reduce inflammation and ameliorate symptoms (Fig. 9.4). Baricitinib, a JAK1-JAK2 inhibitor, has been approved for COVID-19 treatment in conjunction with Remdesivir due to its effectiveness in shortening recovery time in hospitalized patients [49].

Conclusion and Future Perspective

Fundamental research leads to a better knowledge of COVID-19's immunopathology. A comprehensive COVID-CS scenario is forming, indicating a significant impact. This CS is more harmful due to its higher cytokine content compared to other known situations. However, lymphopenia is usually described in this state, indicating that innate cells are the primary producers of these cytokines. Although the exact role of different cytokines in the development of immunopathogenic parameters in COVID-19 remains unknown, it is clear that COVID-CS is intimately linked to the disease's key pathogenic alterations. Several biologic therapies targeting inflammatory cytokines or signaling pathways have shown promising outcomes in treating COVID-CS, with others under development. The therapeutic plan should aim to reduce inflammatory cytokine production and restore equilibrium. However, there are currently very few safe and effective medications to manage CS, making

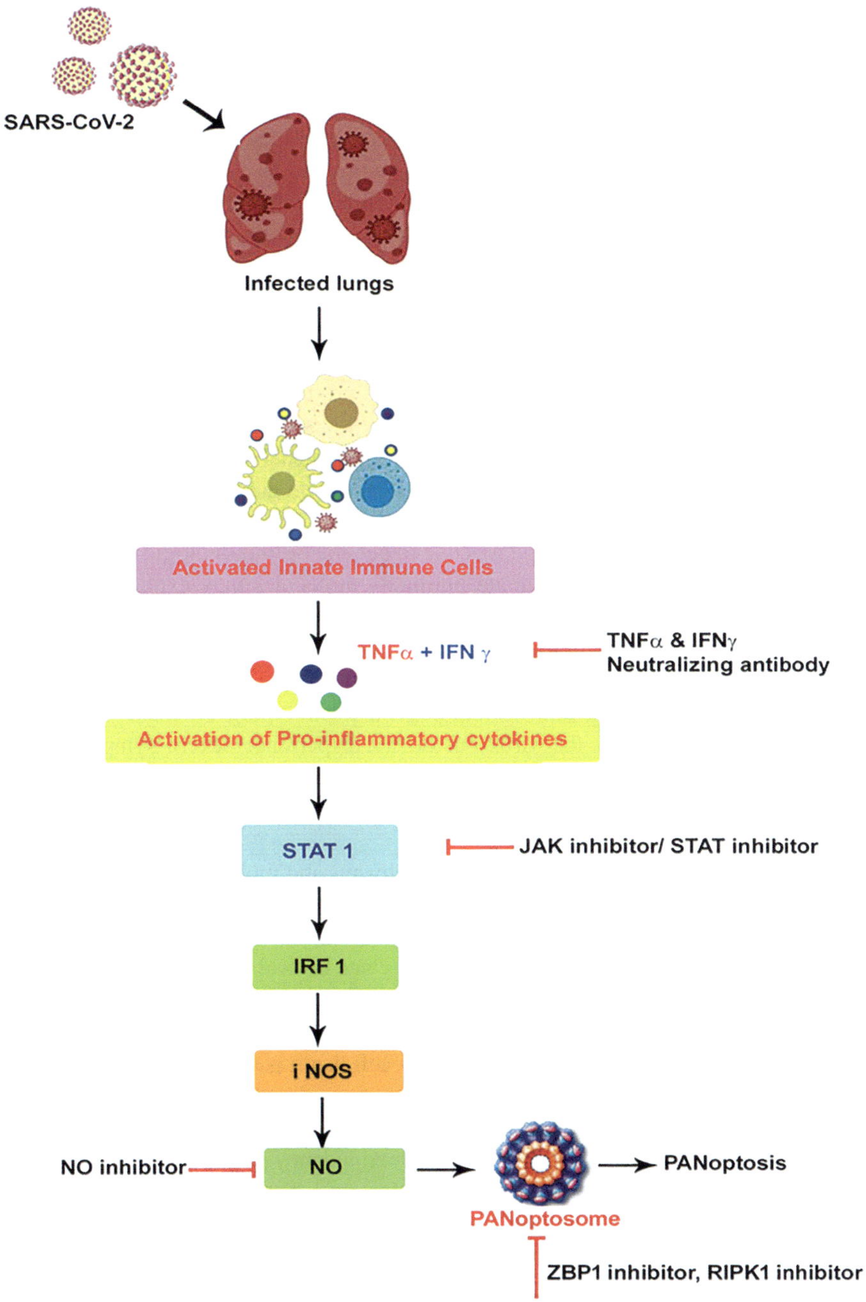

Fig. 9.4 Molecular mechanism, probable treatment options and the implication of PANoptosis in COVID-CS. SARS-CoV-2 lung infection triggers the production of proinflammatory cytokines by innate immune cells. TNF-α and IFN-γ synergistically activate STAT1 signaling, leading to the expression of IRF1, which regulates iNOS for NO production. PANoptosis is then induced through a multiprotein complex called PANoptosome. Based on the molecular mechanism, several treatment options including neutralizing antibody (anti-TNFα, anti-IFNγ antibody) and repurposing agents (like Baricitinib which is a JAK1/2 inhibitor and received EUA to treat COVID-19 in 2020) are indicated in the diagram

clinical treatment challenging for a variety of reasons. (1) Limited clinical diagnostic and treatment recommendations due to a lack of awareness among medical physicians. (2) Targeting many cytokines simultaneously is a pharmacological hurdle. To address cytokine production, it is crucial to target cytokines that are induced or directly target cells like monocytes and macrophages. Identification of more effective and safe anti-inflammatory factors/cytokines could be a better option for therapeutic intervention. Inappropriate targeting of inflammatory cytokines may result in acquired immunodeficiency and subsequent infections. Balancing CS and protective immunity in infectious diseases is challenging due to the protective role of inflammatory cytokines. Additionally, individual differences in age, immune status, and comorbidities can lead to disparities in CS and treatment. Thus, careful therapy is essential. Targeting dominant cytokines or molecular pathways in a certain CS scenario should be prioritized and implemented promptly. Although we are now in a preliminary state to understand the function of innate immunity, PANoptosis, and cytokine storm in COVID-19 pathogenesis, various issues still remain to be solved. For example, why do some individuals experience dysregulations of innate immune pathways after SARS-CoV-2 infection, whereas others do not? Individuals may experience varying immune responses due to genetic or immunological characteristics. Genetic heterogeneity in innate immunity molecules may explain why some SARS-CoV-2 patients are asymptomatic while others develop severe disease, given the importance of innate immune signaling in protecting against infection.

To better understand the immune response patterns associated with each patient's prognosis, genetic testing and immunological phenotyping on those individuals (who were infected with the virus but do not experience severe symptoms) are required. Moreover, monitoring innate immune responses in COVID-19 patients specifically of those patients with post-acute sequela of SARS-CoV-2 infection over time may provide additional insights.

Apart from identifying innate immune pathways and molecular targets related to COVID-CS-related consequences, determination of the proper time to clinical interventions also crucial for better prognosis and survival of COVID patients. In the early stages of infection, immune responses are critical for clearing pathogens. However, identification of proper timing of treatment is challenging since patients typically report to the hospital with significant clinical symptoms. To evaluate if targeting innate immunity and inflammatory cell death is beneficial, proper patient stratification and timing of treatment should be considered.

Pharmaceutical therapies targeting specific pathways may not effectively reduce inflammation and disease due to molecular redundancy within the key players of innate immune system. Cells' mechanisms for viral detection and cell death may vary by tissue and cell type. Although studies indicate that some cytokines are linked to COVID-CS-related complications and pathogenicity, further research is needed to understand the biological mechanisms that influence the COVID-CS etiology and discover potential treatment targets of COVID-CS. COVID-19 has taught us important lessons about dealing with natural pathogenic adversaries. COVID-CS research will provide vital knowledge and therapeutic possibilities for various diseases.

Acknowledgment The authors would like to acknowledge the Department of Biotechnology, NIT Durgapur for extending facilities for research work in the related field. CD was supported by DBT-RA fellowship.

References

1. Datta C, Bhattacharjee A. Cytokine storm and its implication in coronavirus diseases 2019 (COVID-19). J Immunol Sci. 2020;4(3):4–21.
2. Yang L, Xie X, Tu Z, Fu J, Wa D, Zhou Y. The signal pathways and treatment of cytokine storm in COVID-19. Signal Transduct Target Ther. 2021;6:255. https://doi.org/10.1038/s41392-021-00679-0.
3. Hadjadj J, Yatim N, Barnabei L, Corneau A, Boussier J, Pere H, et al. Impaired type I interferon activity and exacerbated inflammatory responses in severe COVID-19 patients. Science. 2020;369:718–24. https://doi.org/10.1126/science.abc6027.
4. Yang Y, Shen C, Li J, Yuan J, Yang M, Wang F, et al. Exuberant elevation of IP-10, MCP-3 and IL- 1ra during SARS-CoV-2 infection is associated with disease severity and fatal outcome. MedRxiv. 2020; https://doi.org/10.1101/2020.03.02.20029975.
5. Ni W, Yang X, Yang D, Bao J, Li R, Xiao Y, et al. Role of angiotensin-converting enzyme 2 (ACE2) in COVID-19. Crit Care. 24:422. https://doi.org/10.1186/s13054-020-03120-0.
6. He L, Ding Y, Zhang Q, Che X, He Y, Shen H, et al. Expression of elevated levels of pro-inflammatory cytokines in SARS-CoV-infected ACE21 cells in SARS patients: relation to the acute lung injury and pathogenesis of SARS. J Pathol. 2006;210:288–97. https://doi.org/10.1002/path2067.
7. Chang Y-J, Liu CY-Y, Chiang B-L, Chao Y-C, Chen C-C. Induction of IL-8 release in lung cells via activator protein-1 by recombinant baculovirus displaying severe acute respiratory syndrome-coronavirus spike p Proteins: identification of two functional regions. J Immunol. 2004;173:7602–14. https://doi.org/10.4049/jimmunol.173.12.7602.
8. Chen I-Y, Chang SC, Wu H-Y, Yu T-C, Wei W-C, Lin S. Upregulation of the chemokine (C-C motif) ligand 2 via a severe acute respiratory syndrome coronavirus spike-ACE2 signaling pathway. J Virol. 2010;84:7703–12. https://doi.org/10.1128/JVI.02560-09.
9. Park A, Iwasaki A. Type I and type III interferons – induction, signaling, evasion, and application to combat COVID-19. Cell Host Microbe. 2020;27:870–8. https://doi.org/10.1016/j.chom.2020.05.008.
10. Yue Y, Nabar NR, Shi C-S, Kamenyeva O, Xiao X, Hwang I-Y, et al. SARS-coronavirus open reading frame-3a drives multimodal necrotic cell death. Cell Death Dis. 2009;9:904. https://doi.org/10.1038/s41419-018-0917-y.
11. Shah A. Novel coronavirus-induced NLRP3 inflammasome activation: a potential drug target in the treatment of COVID-19. Front Immunol. 2020;11:1021. https://doi.org/10.3389/fimmu.2020.01021.
12. Mangalmurti N, Hunter CA. Cytokine storms: understanding COVID-19. Immunity. 2020;53:19–25.
13. Fajgenbaum DC, June CH. Cytokine storm. N Engl J Med. 2020;383:2255–73. https://doi.org/10.1056/NEJMra2026131.
14. Sayahinouri M, Mashayekhi Firouz S, Ebrahimi Sadrabadi A, Masoudnia M, Abdolahi M, Jafarzadeh F, et al. Functionality of immune cells in COVID-19 infection: development of cell-based therapeutics. BioImpacts. 2023;13(2):159–79.
15. Barnes BJ, Adrover JM, Baxter-Stoltzfus A, Borczuk A, Cools- Lartigue J, Crawford JM, et al. Targeting potential drivers of COVID-19: Neutrophil extracellular traps. J Exp Med. 2020;217:e20200652. https://doi.org/10.1084/jem.20200652.
16. Cicco S, Cicco G, Racanelli V, Vacca A. Neutrophil Extracellular Traps (NETs) and Damage-Associated Molecular Patterns (DAMPs): two potential targets for COVID-19 treatment. Mediat Inflamm. 2020;2020:7527953. https://doi.org/10.1155/2020/7527953.

17. Magro CM, Mulvey JJ, Laurence J, Sanders S, Crowson AN, Grossman M, et al. The differing pathophysiologies that underlie COVID-19-associated perniosis and thrombotic retiform purpura: a case series. Br J Dermatol. 2021;184:141–50. https://doi.org/10.1111/bjd.19415.
18. Manickam C, Sugawara S, Reeves RK. Friends or foes? The knowns and unknowns of natural killer cell biology in COVID-19 and other coronaviruses in July 2020. PLoS Pathog. 2020;16:e1008820. https://doi.org/10.1371/journal.ppat.1008820.
19. Kempuraj D, Selvakumar GP, Ahmed ME, Raikwar SP, Thangavel R, Khan A, et al. COVID-19, mast cells, cytokine storm, psychological stress, and neuroinflammation. Neuroscientist. 2020;26:402–14. https://doi.org/10.1177/1073858420941476.
20. Merad M, Martin JC. Pathological inflammation in patients with COVID-19: a key role for monocytes and macrophages. Nat Rev Immunol. 2020;20:355–62.
21. Al-Samkari H, Berliner N. Hemophagocytic lymphohistiocytosis. Annu Rev Pathol. 2018;13:27–49.
22. Suntharalingam G, Perry MR, Ward S, Brett SJ, Castello-Cortes A, et al. Cytokine storm in a phase 1 trial of the anti-CD28 monoclonal antibody TGN1412. N Engl J Med. 2006;355:1018–28.
23. Engelmann B, Massberg S. Thrombosis as an intravascular effector of innate immunity. Nat Rev Immunol. 2013;13:34–45.
24. Sakaguchi S. Naturally arising $CD4^+$ regulatory t cells for immunologic self tolerance and negative control of immune responses. Annu Rev Immunol. 2004;22:531–62.
25. Sakaguchi S, Ono M, Setoguchi R, Yagi H, Hori S, et al. $Foxp3^+CD25^+CD4^+$ natural regulatory T cells in dominant self-tolerance and autoimmune disease. Immunol Rev. 2006;212:8–27.
26. de Waal Malefyt R, Haanen J, Spits H, Roncorolo MG, Velde AT, Frigdor C, et al. Interleukin 10 (IL-10) and viral IL-10 strongly reduce antigen-specific human T cell proliferation by diminishing the antigen presenting capacity of monocytes via downregulation of class II major histocompatibility complex expression. J Exp Med. 1991;174:915–24.
27. Bogdan C, Vodovotz Y, Nathan C. Macrophage deactivation by interleukin. J Exp Med. 1991;174:1549–55.
28. Caricchio R, Gallucci M, Dass C, Zhang X, Galluci S, et al. Preliminary predictive criteria for COVID-19 cytokine storm. Ann Rheum Dis. 2021;80:88–95.
29. Chen LYC, Hoiland RL, Stukas F, Weillington CL, Mypinder SS. Confronting the controversy: interleukin-6 and the COVID-19 cytokine storm syndrome. Eur Respir J. 2020;56:2003006.
30. Mehta P, McAuley DF, Brown M, Sanchez E, Tattersall RS, et al. COVID-19: consider cytokine storm syndromes and immunosuppression. Lancet. 2020;395:1033–4.
31. Sallusto F. Heterogeneity of human $CD4^+$ T cells against microbes. Annu Rev Immunol. 2016;34:317–34.
32. Mosmann TR, Coffman RL. TH1 and TH2 cells: different patterns of lymphokine secretion lead to different functional properties. Annu Rev Immunol. 1989;7:145–73.
33. Zlei M, Sidorov IA, Joosten SA, Heemskerk MHM, Myeni SK, Pothast CR, et al. Immune determinants of viral clearance in hospitalised covid-19 patients: reduced circulating naïve $CD4^+$ T cell counts correspond with delayed viral clearance. Cells. 2022;11:2743. https://doi.org/10.3390/cells11172743.
34. Tan AT, Linster M, Tan CW, Le Bert N, Chia WN, Kunasegaran K, et al. Early induction of functional sars-Cov-2-Specific T cells associates with rapid viral clearance and mild disease in covid-19 patients. Cell Rep. 2021;34:108728. https://doi.org/10.1016/j.celrep.2021.108728.
35. Martonik D, Parfieniuk-Kowerda A, Rogalska M, Flisiak R. The role of Th17 response in covid-19. Cells. 2021;10:1550. https://doi.org/10.3390/cells10061550.
36. Choto TA, Makupe I, Cakana AZ, Sibanda EN, Mduluza T. Excessive neutrophil recruitment promotes typical T-helper 17 responses in coronavirus disease 2019 patients. PLoS One. 2022;17:e0273186. https://doi.org/10.1371/journal.pone.0273186.
37. De Biasi S, Meschiari M, Gibellini L, Bellinazzi C, Borella R, Fidanza L, et al. Marked T cell activation, senescence, exhaustion and skewing towards Th17 in patients with covid-19 pneumonia. Nat Commun. 2020;11:3434. https://doi.org/10.1038/s41467-020-17292-4.

38. Crotty S. Follicular helper Cd4 T cells (Tfh). Annu Rev Immunol. 2011;29:621–63. https://doi.org/10.1146/annurev-immunol-031210-101400.
39. Meckiff BJ, Ramírez-Suástegui C, Fajardo V, Chee SJ, Kusnadi A, Simon H, et al. Imbalance of regulatory and cytotoxic sars-Cov-2-reactive Cd4(+) T cells in covid-19. Cell. 2020;183:1340–53.e16. https://doi.org/10.1016/j.cell.2020.10.001.
40. Mazzoni A, Salvati L, Maggi L, Capone M, Vanni A, Spinicci M, et al. Impaired immune cell cytotoxicity in severe covid-19 is il-6 dependent. J Clin Invest. 2020;130:4694–703. https://doi.org/10.1172/jci138554.
41. Zhu Q, Xu Y, Wang T, Xie F. Innate and adaptive immune response in SARS-CoV-2 infection – current perspectives. Front Immunol. 2022;13:1053437. https://doi.org/10.3389/fimmu.2022.1053437.
42. Rha MS, Jeong HW, Ko JH, Choi SJ, Seo IH, Lee JS, et al. Pd-1-expressing sars-Cov-2-Specific Cd8(+) T cells are not exhausted, but functional in patients with covid-19. Immunity. 2021;54:44–52.e3. https://doi.org/10.1016/j.immuni.2020.12.002.
43. Kang S, Tanaka T, Narazaki M, Kishimoto T. Targeting interleukin-6 signaling in clinic. Immunity. 2019;50:1007–23.
44. Kircheis R, Haasbach E, Lueftenegger D, Heyken WT, Ocker M, Planz O. NF-κB pathway as a potential target for treatment of critical stage COVID-19 patients. Front Immunol. 2020;11:598444.
45. DeDiego ML, Nieato-Torres JL, Regla-Nava JA, Jimenez-Guardeno JM, Fernandez-Delgado R, et al. Inhibition of NF-kappaB-mediated inflammation in severe acute respiratory syndrome coronavirus-infected mice increases survival. J Virol. 2014;88:913–24.
46. Zhang W, Zhao Y, Zhang F, Wang Q, Li T, et al. The use of anti-inflammatory drugs in the treatment of people with severe coronavirus disease 2019 (COVID-19): the perspectives of clinical immunologists from China. Clin Immunol. 2020;214:108393.
47. Huang C, Wang Y, Li X, Ren L, Zhao J, et al. Clinical features of patients infected with 2019 novel coronavirus in Wuhan, China. Lancet. 2020;395:497–506.
48. Han H, Maa C, Li Q, Liu R, Zhao L, et al. Profiling serum cytokines in COVID-19 patients reveals IL-6 and IL-10 are disease severity predictors. Emerg Microbes Infect. 2020;9:1123–30.
49. Karki R, Kanneganti TD. Innate immunity, cytokine storm, and inflammatory cell death in COVID-19. J Transl Med. 2022;20:542. https://doi.org/10.1186/s12967-022-03767-z.
50. Zhang F, Mears JR, Shakib L, Beynor JI, Shanaj S, Korsunsky I, et al. IFN-γ and TNF-α drive a $CXCL10^{+}$ $CCL2^{+}$ macrophage phenotype expanded in severe COVID-19 lungs and inflammatory diseases with tissue inflammation. Genome Med. 2021;13:64.

COVID-19: Modulation of Signaling Pathways

10

Anupriya Sinha and Srinivas Patnaik

Abstract

The pathogenesis of SARS-CoV-2 infection involves intricate molecular pathways, with notable implications for host physiology and disease progression. Two key pathways, the Renin–Angiotensin system (RAS) and the Kallikrein-Kinin system (KKS), are central to understanding the pathophysiology of COVID-19. SARS-CoV-2 utilizes the angiotensin-converting enzyme 2 (ACE2) receptor for cell entry, influencing RAS dynamics and contributing to the dysregulation of blood pressure and inflammation. Similarly, activation of the KKS pathway, mediated by mast cells, exacerbates vascular permeability and inflammatory responses in severe COVID-19 cases. Furthermore, dysregulated immune responses, characterized by cytokine storms and aberrant signaling pathways, play pivotal roles in disease severity and multiorgan dysfunction. The interleukin-6 (IL-6) pathway, Toll-like receptor (TLR) signaling, and interferon-gamma (IFN-γ) responses are implicated in the dysregulated immune milieu observed in severe COVID-19. Additionally, signal transduction pathways such as PI3K/Akt/mTOR, NF-κB, MAPK, and HIF-1 pathways are intricately involved in orchestrating host responses to viral infection, highlighting potential therapeutic targets. Targeting these pathways holds promise for mitigating inflammation, viral replication, and tissue damage associated with SARS-CoV-2 infection. This comprehensive understanding of molecular pathways in SARS-CoV-2 infection is critical for effective therapeutic interventions and management strategies to combat COVID-19 disease.

A. Sinha · S. Patnaik (✉)
School of Biotechnology, KIIT Deemed to Be University, Bhubaneswar, Odisha, India
e-mail: srinivas.patnaik@kiitbiotech.ac.in

M. K. Ghosh, M. Basu (eds.), *Pathogenesis, Management and Socio-Economic Impact of COVID-19 Pandemic*, Medical Virology: From Pathogenesis to Disease Control, https://doi.org/10.1007/978-981-95-1426-7_10

Keywords

COVID-19 · SARS-CoV-2 · Inflammation · Cytokine storm · Signaling pathway

Introduction

In the course of SARS-CoV-2 infection, the interaction between the virus's spike (S) protein and the cell surface receptor ACE2 (Angiotensin-converting enzyme-2) likely triggers multiple signaling pathways. The potential harm of suppressing ACE2 in COVID-19 lies in the risk of overstimulating the adverse classic pathway, subsequently inducing various proinflammatory pathways, and causing damage to tissues and organs [12]. Acute inflammation in COVID-19 patients is marked by heightened activation of the NF-kβ and JAK/STAT signaling pathways. The exacerbated inflammatory responses in the host are likely a result of the excessive activity observed in these pathways [4]. Cytokine storm was identified as the primary factor contributing to COVID-19 severity. Toll-like receptors (TLRs) play a crucial role in recognizing viral particles and initiating the innate immune system. Tumor necrosis factor-α, interleukin-6 (IL-6), IL-1, and interferons (IFNs) are among the proinflammatory cytokines released upon activation of TLR pathways. Various TLRs are potentially significant in the case of COVID-19 [17]. Anti-inflammatory and anticytokine storm therapy are primary strategies used to treat COVID-19. Nevertheless, there are no particular medications that work well to treat viral infections. Understanding the molecular signaling during SARS-CoV-2 infection is therefore critical. This chapter covers potential molecular processes and dysregulated signaling pathways that may arise from infection with SARS-CoV-2.

Molecular Signalings Involved in COVID-19 Disease

Renin-Angiotensin System (RAS)

RAS comprises two distinct pathways. Both routes commence with the production of renin, which transforms angiotensinogen into angiotensin I (ANG I). Initially, angiotensin-converting enzyme (ACE) converts ANG I into ANG II. ANG II then binds to the receptor, known as ANG II type 1 receptor (AT1R), leading to an elevation in blood pressure through vasoconstriction and by retaining sodium. The second pathway operates independently of ACE, wherein the enzyme ACE2 converts ANG I into angiotensin-1-9 (ANG-1-9) and ANG II type 2 receptor (AT2R), working in opposition to the ACE-dependent pathway by inducing vasodilation. Consequently, this pathway contributes to the reduction of blood pressure and exerts other cardioprotective effects [31]. SARS-CoV-2 employs ACE2 as its cell entry receptor, potentially influencing the RAS pathway within the body. The respiratory tract serves as the primary entry route for SARS-CoV-2 virus entry, and the mucosa of the nose, mouth, and lungs' alveolar epithelium expresses ACE2 abundantly [16]. Studies have indicated fluctuations in the RAS and ACE2 levels during

SARS-CoV-2 infection, with a notable increase in ACE2 during the initial phases of the disease and a subsequent rapid decline in more advanced stages. The findings suggest a protective function of ACE2, and its decrease is associated with a deterioration of the inflammatory pulmonary condition [8]. The binding of ACE2 to the spike (S) protein of the virus reduces cellular ACE2 levels. This subsequent decrease in surface ACE2 expression raises the ANG II level, likely playing a role in the notable mortality rates associated with severe lung damage and SARS-related fibrosis. While ACE2 is known to lower the ANG II levels and reduces fibrosis and other related disorders, its diminished presence in SARS and COVID-19 may be linked to heightened inflammation [8, 39]. Furthermore, viral infection and the activation of RAS stimulate the sudden increase in oxygen radicals. Elevated reactive oxygen species (ROS) have deleterious effects on cellular macromolecules. Determining the involvement of ACE2 in the pathogenesis of SARS-CoV-2 is crucial because of the multiple-organ dysfunction that is seen in the course of infection, which includes breathing difficulties, heart diseases, damage to the kidney, and changes in gut and hepatic functions [13].

Kallikrein-Kinin System Pathway Involved in COVID-19

Mast cells (MCs) play a significant role in inflammatory and vascular hyperpermeability occurrences associated with activating the kallikrein–kinin system (KKS). These events impact individuals suffering from the severe manifestations of COVID-19. In the proteolytic cleavage of low-molecular-weight kininogens (LMWK) produced by tissues and high-molecular-weight kininogens (HMWK) produced by the liver, kallikrein, a polypeptide with degradative properties, plays a critical role in the KKS pathway [32]. Due to the influence of kallikrein, HMWKs undergo degradation to form the hormone bradykinin (BK), and LMWKs are metabolized into lys-BK. Lys-BK exhibits an equivalent physiological effect to BK, as both substances target the bradykinin receptor B2 (B2R). Kininase I facilitates the cleavage of BK and lys-BK into des-Arg9-bradykinin (DABK) and lys-des-Arg9-bradykinin (lysinated-DABK), respectively. Both DABK and Lys-DABK specifically interact with the B1R [30, 36]. Although B2R is consistently found in various healthy tissues, particularly within endothelial vasculature, B1R becomes selectively activated during localized inflammatory occurrences. This activation is caused by cytokines like TNF-α and IL-1β, as well as by DABK and lysinated-DABK. When these receptors interact, vascular tone and inflammatory events are modulated, which increases the permeability of blood vessels and pain perception [10]. Recent investigations into the involvement of the KKS in flavivirus infections have established a connection between microvascular dysfunctions and the signaling by BK through the B2R receptor, which is consistently found in cells of the endothelial system. The significance of activation of KKS as a natural defense mechanism to viral infections has garnered growing interest, especially in light of reports highlighting thrombogenic events associated with COVID-19. Recent discoveries elucidating the involvement of the KKS in viral pathogenesis indicate heightened activation of KKS, leading to increased bradykinin peptide in the bloodstream as well as tissues. Additionally, reports have shown that virus replication prompts the expression of BKR. Increased levels of kallikrein in tissues and metabolites generated from BK were observed in the bronchoalveolar lavage fluid of individuals affected by COVID-19. Hospitalized patients with COVID-19 exhibited notable elevations in plasma concentrations of both whole and cleaved kininogen. Additionally, the detection of increased levels of kallikrein: C1 and FXIIa: C1 complexes suggests heightened activation of the KKS. Additionally, in a somewhat larger cohort of 63 patients who developed serious pneumonia as a result of COVID-19, the KKS was activated [3]. A correlation was found between reduced BK and elevated levels of BK-(1–8) and carboxypeptidase N, suggesting that BK was being degraded. Notably, an upregulation of B1R mRNA expression in circulating cells was demonstrated, suggesting a potential preference for the signaling cascade of BK1-8/B1R. Moreover, a correlation was found between the elevated BK1 and coagulation markers (D-dimer and fibrinogen), cytokines that cause inflammation (IL-6, TNF, and IL1-β), and the severity of the condition [37]. Elevated KKS signaling and bradykinin accumulation may underlie the dry cough observed in COVID-19. When a person is infected with SARS-CoV-2, the activity of ACE2 decreases due to the mechanisms through which the virus enters cells. With the reduction in ACE2 function, the degradation of DABK is inhibited, leading to prolonged activation of B1 receptors. This prolonged activation exacerbates smooth muscle contraction, lung damage, and inflammation, which contribute to pulmonary symptoms, notably the dry cough. It is worth mentioning that numerous other symptoms of COVID-19, such as fatigue, vomiting, diarrhea, and headaches, are observed in conditions characterized by elevated levels of BK and increased vascular permeability, such as pulmonary angioedema [40]. The impact of heightened DABK levels will be compounded by the rise in B1 receptor expression triggered by cellular stress and inflammation. Researchers have identified an accumulation of BK in the bronchoalveolar lavage fluid of individuals with COVID-19, establishing a connection between this accumulation and the symptoms characteristic of the disease [9].

Immune Responses and Related Signaling Pathways in COVID-19

Heightened Cytokine Response

In order to combat viral infections within the host organism, the immune system is essential. However, an overactive immune response can precipitate pathological conditions. Clinical investigations have revealed higher concentrations of proinflammatory cytokines in the plasma of critically ill patients compared to those with mild to moderate symptoms, indicating a direct correlation between the severity of COVID-19 and the onset of the *cytokine storm*. Both innate and adaptive immune cells contribute to the initiation of the *cytokine storm*, with macrophages playing a significant role in various conditions, including SARS and MERS. Upon activation, macrophages can release multiple proinflammatory cytokines such as TNF-α and IL-6, which can potentially ignite the inflammatory cascade, thereby instigating a cytokine storm [17]. Cytokine release syndrome is a potential occurrence in the progression towards severe forms of COVID-19 and eventual mortality. Although the exact mechanisms driving the uncontrolled release of inflammatory agents remain unclear, various hypotheses have been proposed. The initial hypothesis revolves around virus replication, which instigates pyroptosis, an intensely inflammatory type of cell death akin to apoptosis. In individuals with COVID-19, pyroptosis prompts the discharge of cytokines that trigger inflammation and disrupts the functioning of innate immune system cells [44]. Increasing evidence suggests a disruption in innate immunity induced by IFN-1. IFN-1 plays a crucial role in viral replication and facilitates the functioning of the adaptive system. After a viral infection, macrophages, neutrophils, and dendritic cells initiate the immune response, serving as the body's primary defense. This concept is supported by the findings from lung autopsies of individuals who succumbed to COVID-19, which showed significant infiltration of macrophages into the bronchial mucosa [22]. Reduced adaptive immunological responses are the result of disruption of the innate immune system, as seen by a sharp drop in antibody production. In order to initiate the antiviral response in the host, innate immune cells must first detect the presence of the virus. First, the virus is recognised by the immune system's innate defences using receptors called pattern recognition receptors (PRRs), such as TLRs on dendritic cells. These receptors identify particular molecular patterns linked to viruses, also called molecular patterns associated with pathogens (PAMPs). When TLRs identify PAMPs in the pathogen, adaptor protein molecules mediate a variety of cellular responses. These reactions involve the activation of pathways, such as MAPKs and NF-kβ, which result in the expression of IFN and the generation of inflammatory molecules. Interferon-activated gene expression is promoted by downstream signaling mechanisms that become active upon IFN production. Interferon functions to stop the virus from spreading and influences the immune response by improving macrophages' capacity to phagocytose antigens [12]. Targeting the virus's surface antigen, neutralizing antibodies are produced in response to adaptive immunity, according to the second theory. Multiple studies in animals have suggested that IgG can bind to the S protein, initiating inflammatory pathways. This binding process can lead to the accumulation of inflammatory mediators like macrophages as well as monocytes in the lungs, facilitated by the secretion of IL-8 and monocyte chemoattractant protein (MCP)-1. The viral-anti-S-IgG complex and the Fc receptor on the macrophage's surface work together to promote the response of inflammation. This concept finds support in observations of reduced levels of proinflammatory cytokines following the blockade of macrophage receptors. Moreover, a number of lines of data suggested that in COVID-19 patients, the onset of severe respiratory illness corresponds with the existence of antiviral IgGs [26]. In the realm of acquired immunity against viral infections, T helper (Th1) cells are recognized as pivotal players. Both T cells (CD4+ and CD8+) exhibit a significant antiviral function by maintaining a delicate equilibrium between combating the virus and managing excessive inflammation. CD4+ T cells undertake the task of generating targeted antibodies against the virus by initiating B cell activation, which relies on T cell assistance. On the other hand, CD8+ T cells are tasked with the eradication of virus-infected cells. T cells extracted from individuals with COVID-19 exhibited significantly elevated levels of PD-1, a marker associated with T cell exhaustion. T cells are essential in eliminating viruses from the host. CD8+ cytotoxic T cells release various molecules like granzymes, perforin, and IFN to eradicate viruses from cells (Fig. 10.1). In parallel, CD4+ cells assist cytotoxic and B cells in clearing the pathogen [24]. The initial innate immune response serves as a rapid and efficiently coordinated frontline defense mechanism. However, dysregulated immune responses can potentially result in harm to the host's immune system. Increased cytokine levels have been associated with the development of acute respiratory distress syndrome (ARDS) and subsequent damage to multiple organs (Fig. 10.1). Despite advancements, numerous uncertainties persist regarding the signaling pathways underlying COVID-19. Understanding and managing the cytokine storm has become a key focus in the development of treatments for COVID-19. Therapies such as corticosteroids and monoclonal antibodies targeting specific cytokines aim to mitigate this hyperinflammatory response and improve patient outcomes. Additionally, research continues to explore the underlying mechanisms driving

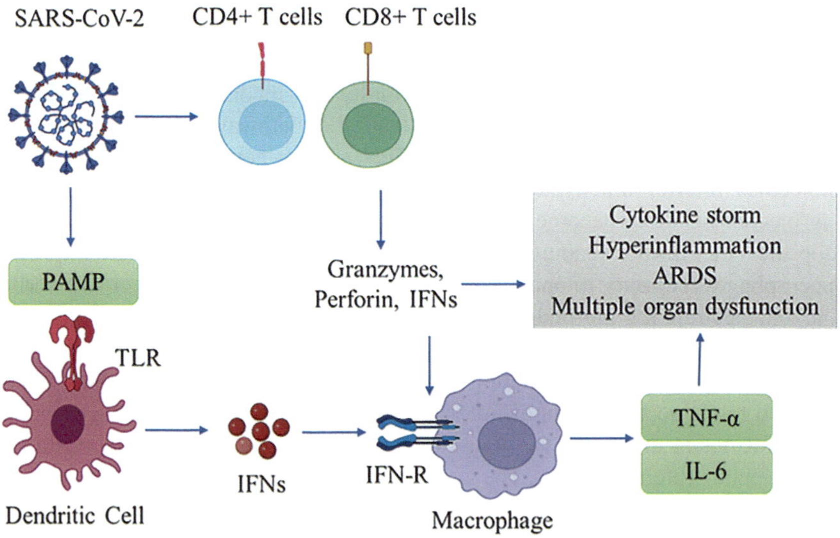

Fig. 10.1 *Cytokine storm*. The immune system's innate defenses recognize the virus through PRR, including TLRs on dendritic cells. These receptors detect PAMPs on the viruses. In response, dendritic cells secrete interferons (IFNs), which bind to interferon receptors (IFN-R) on macrophages. Activated macrophages can then release pro-inflammatory cytokines such as TNF-α and IL-6, potentially triggering an inflammatory cascade and initiating a *cytokine storm*. Additionally, CD8+ cytotoxic T cells release molecules like granzymes, perforin, and interferon to eliminate viruses from infected cells. Simultaneously, CD4+ helper T cells support cytotoxic T cells and B cells in clearing the pathogen

cytokine dysregulation in COVID-19 to develop more effective therapeutic interventions.

IL-6 Pathway

In the classical IL-6 signaling pathway, IL-6 binds to membrane-bound IL-6R to form a heterohexameric complex. This complex is the first step in the activation of the JAK/STAT signaling cascade, which in turn activates the transcription of genes that STAT3 is specifically targeting. The classical pathway orchestrates defensive immune responses vital for host defense and maintaining internal balance. When IL-6 trans-signaling conditions are met, membrane-bound IL-6R is cleaved by ADAM17 to produce soluble IL-6R (sIL-6R), which then binds to IL-6. Once IL-6 is associated with the soluble type, the resulting complex engages the glycoprotein receptor, facilitating downstream pathway activation akin to the classical mode [12]. This route has implications for the pathophysiology of various inflammatory diseases. Reduced amounts of membrane-bound IL-6R allow IL-6 to affect cells through soluble trans-signaling of IL-6R. In the process, a complex is formed between IL-6 and the sIL-6R. This complex then interacts with the gp130 receptor, which is present on the surface of many different types of cells. By this approach, IL-6 increases its cellular uptake and strengthens its ability to regulate a variety of physiological processes, including inflammation, immunological responses, and tissue regeneration (Fig. 10.2). IL-6 trans-signaling is a major contributor to COVID-19 pathogenesis.

IFN-γ Signaling

Preclinical research data indicate that interferons (IFNs) are important for protection from COVID-19. IFNs, a group of cytokines, serve as crucial mediators between cells in combating pathogens and exert pivotal functions in the immune system. Following natural infection with SARS-CoV-2, there is a rapid expansion of both CD8+ and CD4+ T cells targeting various regions of the viral proteome. Particularly, T cells recognizing epitopes within the spike, membrane, and nucleoprotein structural proteins tend to dominate the immune response in the majority of individuals. Variations in the SARS-CoV-2 reactions have been documented among individuals experiencing asymptomatic infection or those with mild or severe COVID-19. Notably, a higher incidence of delayed onset, diminished production of IFN-γ, aberrant activation, impaired function, or exhaustion of T cells was observed in severe cases compared to mild or asymptomatic ones [29]. IFNs activate natural killer (NK) cells and macrophages, which have immunomodulatory and antiviral qualities. Based on sequence homology and receptor specificity, they are divided into type I and type II. IFN-γ and IFN-β are examples of IFNs, but IFN-γ is the only form of type II IFN. Of them, IFN-γ is the main element that connects the innate and adaptive immune responses [35]. Furthermore, IFN-γ's ability to prevent viral

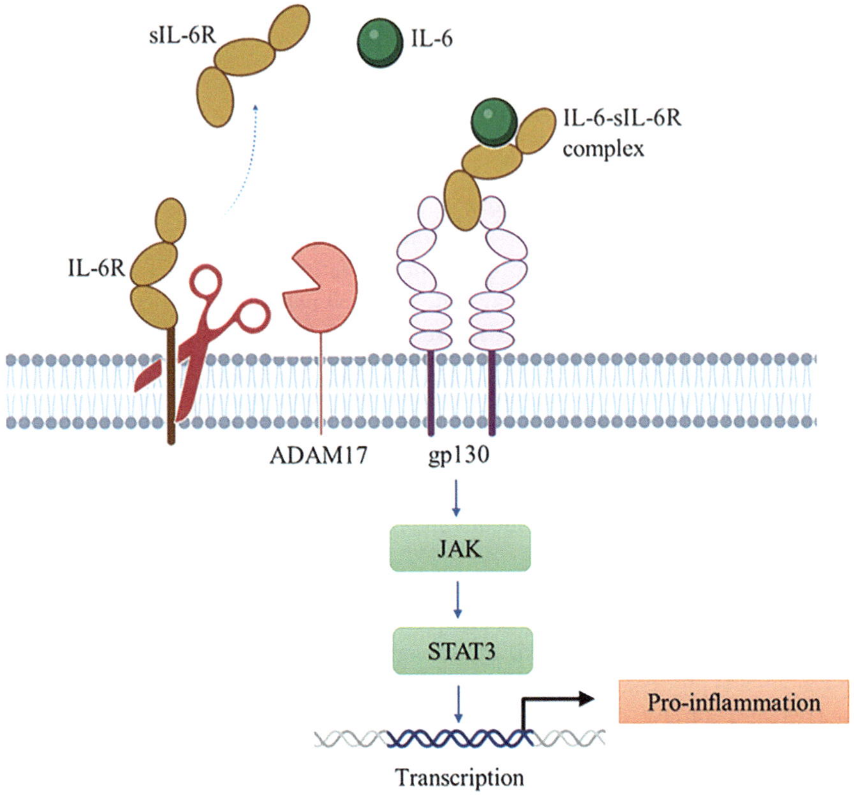

Fig. 10.2 IL-6 pathway. When IL-6 trans-signaling occurs, ADAM17 cleaves the membrane-bound IL-6R to generate sIL-6R, which subsequently binds to IL-6. The IL-6/sIL-6R complex attaches to gp130 on the cell surface. This IL-6/IL-6R/gp130 complex then starts cellular pathways via the JAK/STAT pathway, leading to the expression of downstream target genes of STAT3. Through this mechanism, IL-6 enhances its cellular uptake and augments its capacity to regulate inflammation

replication was first identified as a means of battling viral infections. The mechanism of action of IFN-γ involves the secretion of IFNs when cells are subjected to viral infections, facilitating intercellular communication against pathogens and microbes. Subsequently, IFN-γ-induced macrophages initiate the antiviral mechanism, attract leukocytes, and enhance the activity of NK cells [20]. Antigen-presenting cells (APCs) produce cytokines that control the synthesis of IFN-γ. These cytokines create a strong link between the innate immune system's release of IFN-γ and viral infection. One of the suggested defense mechanisms against COVID-19 is the JAK/STAT pathway triggered by IFN-γ. A crucial part of IFN-mediated immunity is played by STAT1 [33]. In this regard, studies have demonstrated that SARS-CoV-2 inhibits the STAT1 cascade by preventing IFN signaling. An alternative mechanism involves the presence of high concentrations of the S protein's receptor-binding domain, which interacts with host cell receptors, and nucleocapsid protein. Key proteases, particularly the main protease (M-pro or 3CLpro), are essential for COVID-19's replication within host cells. These enzymes are pivotal in processing viral polyproteins, critical for generating new viral particles and facilitating viral replication [18]. The presence of NP-specific T cells producing IFN-γ appears to correlate with a strengthened overall T cell response against SARS-CoV-2. This bolstered T cell response plays a vital role in controlling viral infection and eliminating the virus from the body. This phenomenon was noted in individuals who also exhibited elevated levels of neutralizing antibodies. The presence of S-RBD-specific T cells secreting IFN-γ indicates an immune response specifically targeting this critical region of the virus. On the basis of these findings, it is suggested that S-RBD could potentially serve as a beneficial component in the treatment of COVID-19 [28].

Toll-Like Receptor (TLR) Signaling

The genetic diversity within a population concerning PRR remains constrained as there are no chromosomal rearrangements observed during their expression. TLRs represent a significant subset of PRRs found in humans, being expressed across various cell types including phagocytes, leukocytes, and epithelial cells. Multiple homo- or heterodimers of TLRs can bind to different components of infections. Adaptor proteins are drawn into the cytoplasm of phagocytic cells when they bind to TLRs on those cells. MyD88 (myeloid differentiation primary response 88) serves as a pivotal adaptor protein in TLR signaling (TLR1, 2, 5, 6, 10). Upon recognition of their specific ligands, TLRs undergo conformational changes, enabling them to recruit adaptor proteins like MyD88 to their cytoplasmic domains. MyD88 then acts as a scaffold, facilitating the assembly of other signaling molecules, thereby activating downstream signaling pathways such as NF-κB and MAPK [23]. The MyD88-dependent pathway is employed by all TLRs except TLR3. When bound to a ligand, TLRs undergo a conformational change and subsequently recruit MyD88, an adaptor molecule belonging to the Toll-interleukin family. TLR3 and TLR4 use the TRIF-dependent pathway. In addition to MyD88, certain TLRs have the ability to recruit other adaptor proteins like TRIF (TIR-domain-containing adapter-inducing interferon-β) or TRAM (TRIF-related adaptor molecule). These additional adaptors initiate distinct signaling pathways, particularly in response to viral infections [2, 12]. Both MyD88 and TRIF-dependent pathways offer the most potent antiviral defense against SARS-CoV-2 disease. Removal of TLR signaling correlates with severe or lethal COVID-19. For instance, mice with TLR3/TLR4 adaptor TRIF defects are very susceptible to SARS-CoV and have a higher chance of dying from the virus; whereas, animals with TLR3/TLR4 adapter−/− are more susceptible to SARS-CoV but do not have a higher mortality rate. Additionally, following infection of TRIF−/− mice, aberrant signaling pathways and proinflammatory cytokine production were similar to those of patients with fatal results from MERS-CoV or SARS-CoV infection [42]. Numerous viruses stimulate the innate immune system via TLRs, facilitating virus elimination; however, this can lead to detrimental effects on the host, including tissue damage due to hyperinflammation. For example, the severity of COVID-19 patients has been linked to increased production of IL-6, which may result from TLR pathway induction. Activation of TLRs triggers the inflammasome and subsequently produces IL-1β, which in turn activates IL-6. Furthermore, the TLR4 signaling pathway triggers the activation of neutrophil extracellular traps (NETs). The formation of NETs in individuals with COVID-19 may result in prolonged inflammation, exacerbating patient outcomes. The elevated expression of TLR4 in males due to their higher testosterone levels potentially elucidates the increased levels of proinflammatory cytokines observed in males compared to females. Intriguingly, COVID-19 patients exhibit heightened levels of neutrophil myeloperoxidase, responsible for generating oxidized phospholipids, alongside activation of the TLR signaling during the pulmonary phase of infection, leading to oxidative damage. Consequently, TLR may play a role in the COVID-19 pathogenesis [6]. Detection of the surface S glycoprotein located on the virus envelope binding to ACE2 may trigger a response from TLR7. Numerous cytokines, such as interleukins and interferons, can be secreted in response to TLR7 activation. Furthermore, it has the potential to trigger the release of MCP-1, a chemokine that pulls monocytes and other immune cells to inflammatory areas [46]. Analysis of the entire genome sequences of SARS-CoV, MERS-CoV, and SARS-CoV-2 unveiled a potentially heightened involvement of TLR7 in the pathogenesis of SARS-CoV-2 compared to the two other forms of disease, as the former possesses more single-stranded RNA (ssRNA) motifs capable of binding to TLR7. It's worth noting that TLR7 is another potential stimulator for triggering the formation of NETs in patients with COVID-19. TLR7/8 hyperactivation can cause patients to have a robust proinflammatory response, which can result in acute pulmonary damage [27]. Additionally, the agonist of TLR2/6/9 is utilized to turn on pulmonary epithelial cells and intrinsic immune cells, prompting the production of diverse factors to combat infection. Intriguingly, numerous clinical trials are assessing the impact of anti-inflammatory agents in mitigating mortality resulting from persistent lung inflammation in individuals with COVID-19. TLRs may therefore exert both detrimental and advantageous impacts in cases of SARS-CoV-2 infection. While TLR7/8 is the primary recognizer of ssRNA, the genetic material of COVID-19, it's conceivable that other TLRs, including TLR3, TLR4, and TLR6, might also play a role in the infection (Fig. 10.3) [3]. Therefore, to determine the therapeutic and potentially damaging effects of TLRs in COVID-19 cases, it is essential to investigate both TLR protagonists and antagonists.

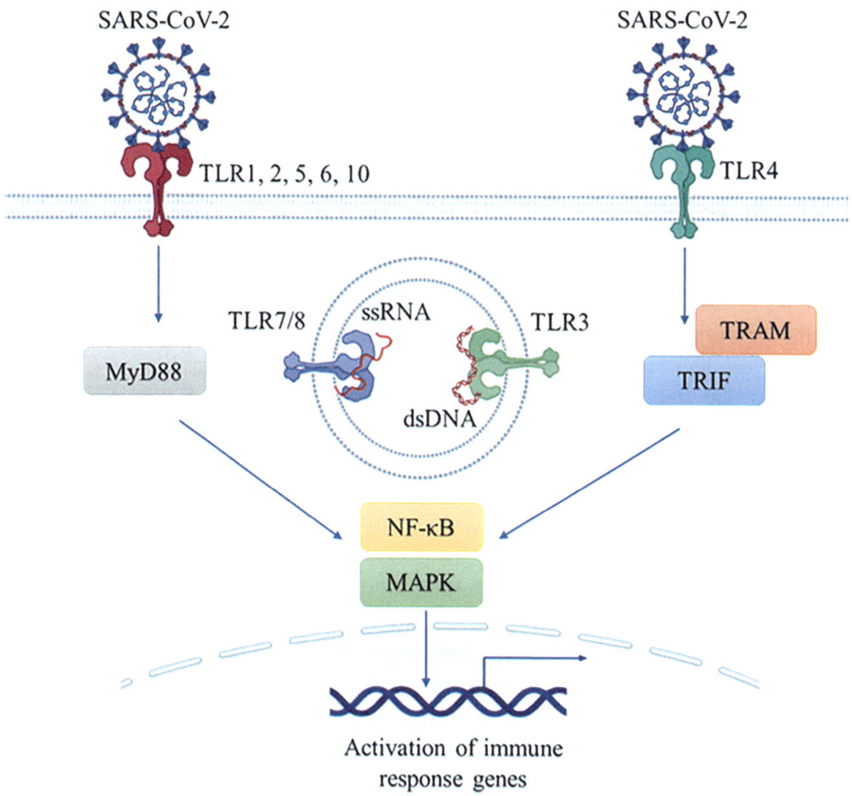

Fig. 10.3 Toll-like receptor (TLR) signaling. Following ligand-induced dimerization, adaptor molecules bind to the cytoplasmic signaling domain to relay the signal. All TLRs, except TLR3, interact with MyD88 proteins. In contrast, TLR3 and TLR4 engage a TRIF-dependent pathway. Both the MyD88-dependent and TRIF-dependent pathways activate downstream signaling components, including NF-κB and MAPK pathways, which are crucial for the expression of inflammatory genes mediated by TLRs

Signal Transduction in COVID-19

NF-κB Signaling

An important part of promoting the innate immune response against pathogens is the NF-κB signaling pathway. NF-κB activation is triggered by SARS-CoV-2 via multiple pathways. The activation leads to the translocation of the NF-κB transcription factor to the nucleus, where it activates multiple genes implicated in inflammation. Positive feedback loops can hyperactivate the NF-κB pathway in predisposed individuals, which in turn exacerbates an inflammatory response mediated by elevated cytokine levels [5]. NF-κB, triggers the production of IL-6, a cytokine that plays a major part in inflammation and is strongly associated with the severity of COVID-19 [19]. According to a study on SARS-CoV, the virus that caused the global SARS outbreak in 2003, the nucleocapsid (N) protein of SARS-CoV has been shown to activate NF-κB in kidney epithelial, Vero E6 cells [21]. Simultaneously, SARS-CoV with the Envelope (E) gene deleted displayed decreased proinflammatory cytokine production, decreased neutrophil infiltration, and decreased lung pathology, all of which improved BALB mice survival. Inhibitors of the NF-κB signaling increased the survival rate. Using the NF-κB pathway, the S protein triggers a potent cytokine response in infected mononuclear cells, according to in vitro research conducted during the previous SARS pandemic. NF-κB regulates immune responses, including cytokine production. TLR triggers NF-κB through a protein kinase C (PKC)-dependent pathway. PKC phosphorylates IκB, leading to NF-κB release and cytokine gene transcription. Blocking NF-κB can suppress TLR-induced cytokine responses, potentially mitigating inflammation in diseases like COVID-19 [11]. Numerous anti-inflammatory and antiviral medications prescribed for COVID-19 treatment, including N-acetylcysteine, hydroxychloroquine, dexamethasone, and macrolide antibiotics, are also linked to suppression of the NF-κB pathway. These results, however, demonstrated that inhibiting the NF-κB pathway has the potential to depress innate immunity and is not very specific in combating viral infections like SARS-CoV-2 [45]). Thus, it may be appropriate to directly target downstream effectors like CXCL2, TNFα, or certain receptor-stimulated NF-κB pathways. Thus, pharmacological inhibition of NF-κB signaling may serve as a potential therapeutic approach for severe COVID-19.

PI3K/Akt/mTOR Signaling Pathway

One important cell signaling system that controls many different aspects of cell life, including anabolic processes, nutrient uptake, growth, proliferation, survival, and motility, is the PI3K/Akt/mTOR pathway. This pathway's upregulated expression is seen in several malignancies, viral infections, and autoimmune disorders, and plays an important role in the pathophysiology of these conditions [47]. A proteomic analysis revealed that approximately 12.5% of the about 12,000 phosphorylation sites were examined for alterations in response to coronavirus infection. These modifications point out the participation of several kinases, including RPS6Ks, p38, Akt, ERK, and JNK, which are known to regulate numerous important cellular processes [41]. The mTOR, HIF-1, and TNF signaling pathways were substantially elevated following SARS-CoV-2 infection. SARS-CoV infection leads to severe pulmonary fibrosis through the activation of EGFR. Many proteins included in these pathways, like Akt, mTOR, and S6K1, were found to be phosphorylated at elevated levels. Notably, these proteins were activated 24 h after infection, indicating that SARS-CoV-2 initiates the PI3K/Akt/mTOR pathway at the early stages of infection. Activated mTOR and NLRP3 inflammasome pathway leads to the synthesis of IL-1β, a mediator of fever and inflammation [7, 12]. Memory B cell populations are decreased by mTORC1 inhibition. In individuals with heightened susceptibility, mitigating the onset of severe symptoms may involve targeted inhibition of memory B cell activation in the initial phase, thereby decreasing the generation of cross-reactive antibodies against SARS-CoV-2 (Fig. 10.4) [48]. The PI3K/Akt pathway modulates many immune cells implicated in the response to COVID-19; in particular, the function of CD8, mast cells, B cells, and neutrophils is thought to be controlled by the PI3K-δ isoform. Therefore, it should not be surprising that mutations in this isoform, such as a gain of function mutation that increases Akt activation and causes activation-induced death in lymphocytes, are linked to an increased risk of respiratory tract infections. Consequently, given that PI3K-δ was successful in reducing allergic lung inflammation, it was proposed that it could be a viable target in the treatment of COVID-19 [1]. By focusing on this pathway in the early stages of the illness, one can improve patient outcomes by preventing the virus from entering the body and multiplying, as well as by lowering the viral load. Targeting PI3K and Akt can also improve effector performance and reduce suppressor function during early phases of the immune response, assisting in the removal of the pathogen

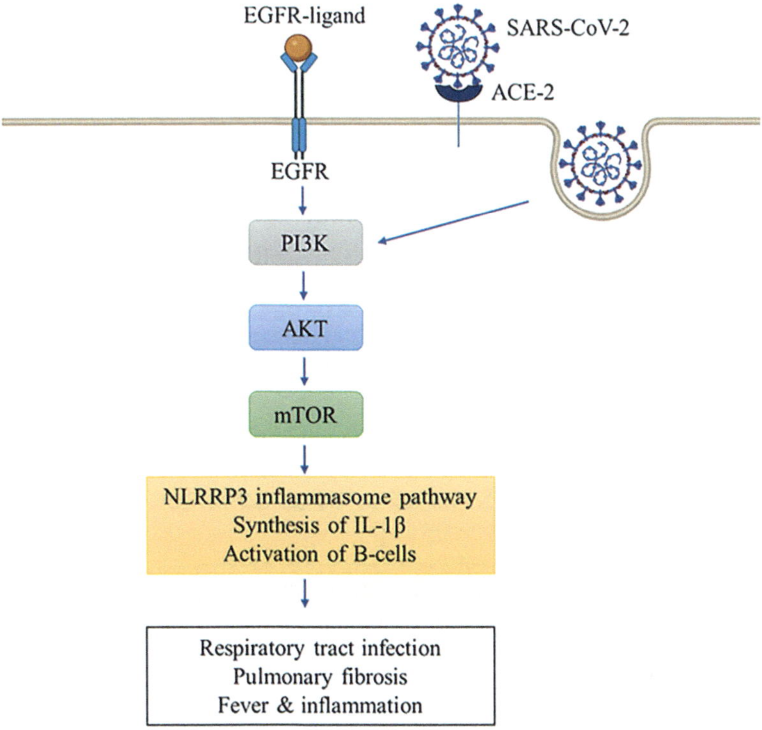

Fig. 10.4 PI3K/Akt/mTOR signaling pathways. SARS-CoV infection triggers the activation of EGFR. PI3K then activates mTORC1 by phosphorylating Akt. The activation of mTOR by SARS-CoV-2 subsequently stimulates the NLRP3 inflammasome pathway, leading to the production of IL-1β, which mediates fever, pulmonary inflammation, and fibrosis

before it causes immunological dysregulation. Targeting mTOR may help reduce cytokine storms, limit neutrophil recruitment, and promote regulatory T-cell function.

MAPK Signaling

Mitogen-activated protein kinases (MAPKs) are widely expressed serine/threonine kinases that regulate essential cellular processes. The p38 family of kinases, JNK, and ERK are the three primary subgroups of MAPKs (Fig. 10.5). Certain functions of the p38 MAPK molecular pathway are seen in cells infected by SARS-CoV-2.

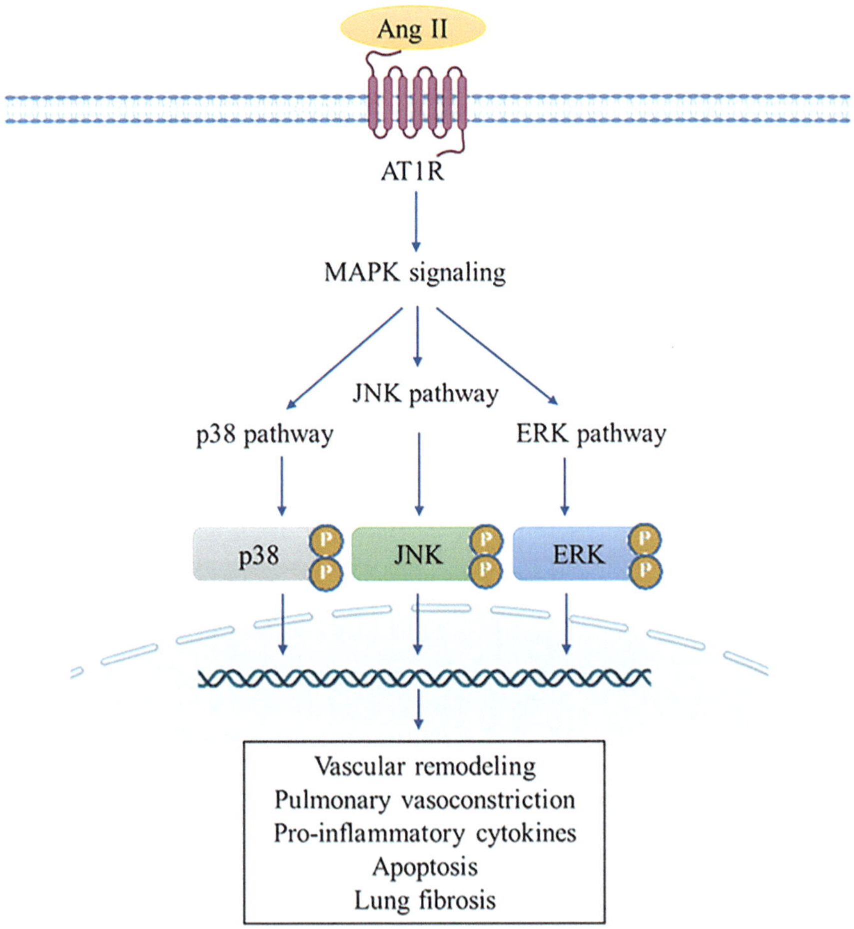

Fig. 10.5 MAPK Signaling. The reduction of ACE2 results in excessive production of Ang II, which binds to AT1R, thereby triggering the activation of MAPK signaling including p38, JNK, and ERK pathways

The studies reveal that SARS-CoV-infected cells have phosphorylated downstream targets of p38 MAPK, including hyper-phosphorylation of eIF4E, Tyr-705 STAT3 dephosphorylation, and p38 MAPK activation [25]. By activating p38 MAPK, overproduction of Ang II causes vasoconstrictive and inflammatory effects; in contrast, Ang 1–7 anti-inflammatory action inhibits p38 MAPK. After the viral invasion, ACE2 becomes inactive. However, to facilitate cell entrance, viral p38 MAPK activation causes endocytosis of viral receptors. Clinical trials are underway for certain inhibitors of p38 MAPK, like losmapimod, which is the most extensively studied in clinical settings. Another inhibitor, dilmapimod, was previously administered to patients with acute pulmonary damage after trauma in a clinical trial, where no significant safety concerns were identified. These inhibitors are being considered for repurposing in clinical trials involving COVID-19 patients [12]. Vascular remodeling and hypoxic pulmonary vasoconstriction are associated with elevated p38 MAPK levels, and these processes are not good indicators of how an acute lung injury would turn out. Cardiac ACE-2 expression suggests that this pathway is active in heart as well. The other pathway, the JNK, sometimes referred to as the death pathway, regulates how cells react to cytokines that cause inflammation. Apoptosis, inflammation, and tissue cytokine production are triggered when the JNK pathway is activated by a variety of cellular stressors, including DNA damage, growth factor generation, and pro-inflammatory cytokines. As per reports, SARS-CoV-2 exacerbates its pathogenicity by triggering JAK-STAT and JNK signaling pathways, leading to the release of cytokines, and eventual development of lung fibrosis [38]. ERK1 (p44 MAPK) and ERK2 (p42 MAPK) are two different types of ERK signaling, which is another pathway of MAPK. Proinflammatory factors COX-2 and IL-8 are produced as a result of the ERK pathway being activated by the SARS-CoV spike and nucleocapsid proteins through activation of ERK1/2. It has been shown in the investigation of SARS-CoV pathogenesis that pulmonary epithelial cells infected with the virus, phosphorylate ACE2 through the expression of the enzyme casein kinase, which initiates the pathways of activator protein 1 and ERK1/2 [15].

HIF-1 Signaling

Prolonged oxygen deprivation in individuals with COVID-19 could trigger the activation of the HIF-1_{α} pathways. This activation could lead to an increase in the transcription of ADAM17, a significant molecule involved in the activation of cytokine expression through HIF-1_{α}. ADAM17 plays a pivotal role in the signaling of IL-6. Furthermore, HIF-1_{α} might regulate crucial cell surface proteins, such as ACE2 and TMPRSS2, which are relevant to SARS-CoV-2 infection. Hence, targeting HIF-1 could be a strategy to mitigate SARS-CoV-2 infection. Previous research has elucidated that among the pivotal genes in the HIF-1 pathway, HMOX1 could potentially contribute to the attenuation of severe inflammation. The human HMOX1 protein exhibits a high-confidence interaction with the ORF3a protein of SARS-CoV-2 [34]. The eNOS enzyme is crucial for the cell's innate antiviral defenses, and variations in its genetic makeup may be linked to the severity of COVID-19 [14]. Recent reports have indicated elevated levels of the Plasminogen Activator Inhibitor 1 (PAI-1) protein in COVID-19 patients requiring supplemental oxygen [49]. Wing and colleagues emphasize the significance of HIF signaling in governing different facets of SARS-CoV-2 infection. Their research conducted on lung epithelial cells brings to light a significant finding—the promising therapeutic application of Roxadustat, a HIF inhibitor, and similar agents in addressing COVID-19. These inhibitors exhibit the ability to intervene at various phases of the life cycle of the viruses, encompassing its entry and replication, indicating their potential efficacy against evolving SARS-CoV-2 strains [43].

Conclusion

Currently, the global landscape is grappling with the multifaceted implications of COVID-19 on healthcare and economic sectors. Our understanding of the intricate molecular pathways facilitating SARS-CoV-2 infection and the interplay between the virus and its host remains incomplete. Obtaining thorough insights into the molecular processes underlying virus-triggered inflammation and cellular demise is imperative for devising efficacious treatments for SARS-CoV-2. This encompasses the repurposing of existing therapeutics as well as the innovation of novel treatment modalities. Various signaling pathways and molecular targets involved in COVID-19 are summarized in Table 10.1.

Table 10.1 The molecular pathways linked to SARS-CoV-2 infection and potential treatment targets based on these pathways

Signaling system	Molecular pathways	Potential target molecules	References
Endocrine System	Renin–Angiotensin system (RAS)	ACE2, ACE, AT1R	[16, 31]
	Kinin–kallikrein system (KKS)	BK, B1R, BB1R, DABK	[3, 9, 37, 40]
Immune responses and related signaling pathways	IL-6 pathway	IL-6, IL-6R, ADAM17	[12, 17]
	IFN-γ signaling	IFN-γ, JAK/STAT	[29, 33, 35]
	Toll-like receptor (TLR) signaling pathway	TLR7/8, TLR3, MCP-1, TLR4, and TLR6	[27, 42, 46]
Signal transduction	NF-κB signaling pathway	NF- κB, CXCL2, TNFα	[11, 45]
	PI3K/Akt/mTOR signaling pathways	mTOR, HIF-1, NLRP3, TNF, Akt, mTOR, and S6K1	[7, 41]
	MAPK signaling pathway	ERK1/2, p38 MAPK, JNK, COX-2, IL-8	[15, 25, 38]
	HIF-1 signaling pathway	HIF-1_{α}, ADAM17, TMPRSS2	[34, 43]

As any modification of the related molecular networks and the regulatory mechanisms may have pathogenic outcomes, directing attention towards dysregulated signaling pathways is an effective strategy. Prospective signaling pathways and potential treatment targets were identified and discussed in this chapter. The erratic pathways playing a crucial part in the response to SARS-CoV-2 infection share the greatest number of components with the cytokine-cytokine receptor interaction pathway. Despite the fact that the main routes were highlighted and cross-talks between pathways were identified. Targeting a single cytokine or signaling pathways regulating cytokines has led to the clinical evaluation of various therapeutic approaches thus far. Future research exploring virus–host interactions will likely reveal additional signaling nodes amenable to therapeutic targeting.

Acknowledgement We are thankful to the KIIT School of Biotechnology, KIIT Deemed to be University, which made this study possible.

Authors Contribution AP conceived the idea of structuring the manuscript, wrote the manuscript, drafted all the figures and tables. SP revised, edited, and approved the final draft of the manuscript.

Data Availability The software Biorender (https://www.biorender.com/) was used to create certain elements in the figures of the manuscript.

Declarations

Conflicts of Interests The authors whose names appear in this manuscript declare that they have no affiliations with or involvement in any organization or entity with any financial or nonfinancial interest in the subject matter discussed. There is no conflict of interest to disclose.

References

1. Abu-Eid R, Ward FJ. Targeting the PI3K/Akt/mTOR pathway: a therapeutic strategy in COVID-19 patients. Immunol Lett. 2021;240:1–8. https://doi.org/10.1016/J.IMLET.2021.09.005.
2. Akira S, Takeda K. Toll-like receptor signalling. Nat Rev Immunol. 2004;4(7):499–511. https://doi.org/10.1038/NRI1391.
3. Alfaro E, Díaz-García E, García-Tovar S, Zamarrón E, Mangas A, Galera R, Nanwani-Nanwani K, Pérez-de-Diego R, López-Collazo E, García-Río F, Cubillos-Zapata C. Impaired Kallikrein-Kinin system in COVID-19 patients' severity. Front Immunol. 2022;13 https://doi.org/10.3389/FIMMU.2022.909342.
4. Amini-Farsani Z, Yadollahi-Farsani M, Arab S, Forouzanfar F, Yadollahi M, Asgharzade S. Prediction and analysis of microRNAs involved in COVID-19 inflammatory processes associated with the NF-kB and JAK/STAT signaling pathways. Int Immunopharmacol. 2021;100:108071. https://doi.org/10.1016/J.INTIMP.2021.108071.
5. Attiq A, Yao LJ, Afzal S, Khan MA. The triumvirate of NF-κB, inflammation and cytokine storm in COVID-19. Int Immunopharmacol. 2021;101(Pt B) https://doi.org/10.1016/J.INTIMP.2021.108255.
6. Cicco S, Cicco G, Racanelli V, Vacca A. Neutrophil Extracellular Traps (NETs) and Damage-Associated Molecular Patterns (DAMPs): two potential targets for COVID-19 treatment. Mediat Inflamm. 2020;2020 https://doi.org/10.1155/2020/7527953.

7. Conti P, Ronconi G, Caraffa A, Gallenga CE, Ross R, Frydas I, Kritas SK. Induction of pro-inflammatory cytokines (IL-1 and IL-6) and lung inflammation by Coronavirus-19 (COVI-19 or SARS-CoV-2): anti-inflammatory strategies. J Biol Regul Homeost Agents. 2020;34(2):327–31. https://doi.org/10.23812/CONTI-E.
8. Cook JR, Ausiello J. Functional ACE2 deficiency leading to angiotensin imbalance in the pathophysiology of COVID-19. Rev Endocr Metab Disord. 2022;23(2):151. https://doi.org/10.1007/S11154-021-09663-Z.
9. Cooper SL, Boyle E, Jefferson SR, Heslop CRA, Mohan P, Mohanraj GGJ, Sidow HA, Tan RCP, Hill SJ, Woolard J. Role of the renin–angiotensin–aldosterone and kinin–kallikrein systems in the cardiovascular complications of COVID-19 and long COVID. Int J Mol Sci. 2021;22(15) https://doi.org/10.3390/IJMS22158255.
10. Dagnino APA, Campos MM, Silva RBM. Kinins and their receptors in infectious diseases. Pharmaceuticals. 2020;13(9):1–28. https://doi.org/10.3390/PH13090215.
11. DeDiego ML, Nieto-Torres JL, Regla-Nava JA, Jimenez-Guardeño JM, Fernandez-Delgado R, Fett C, Castaño-Rodriguez C, Perlman S, Enjuanes L. Inhibition of NF-κB-mediated inflammation in severe acute respiratory syndrome coronavirus-infected mice increases survival. J Virol. 2014;88(2):913. https://doi.org/10.1128/JVI.02576-13.
12. Farahani M, Niknam Z, Mohammadi Amirabad L, Amiri-Dashatan N, Koushki M, Nemati M, Danesh Pouya F, Rezaei-Tavirani M, Rasmi Y, Tayebi L. Molecular pathways involved in COVID-19 and potential pathway-based therapeutic targets. Biomed Pharmacother = Biomedecine & Pharmacotherapie. 2022;145 https://doi.org/10.1016/J.BIOPHA.2021.112420.
13. Gheblawi M, Wang K, Viveiros A, Nguyen Q, Zhong JC, Turner AJ, Raizada MK, Grant MB, Oudit GY. Angiotensin-converting enzyme 2: SARS-CoV-2 receptor and regulator of the renin-angiotensin system: celebrating the 20th anniversary of the discovery of ACE2. Circ Res. 2020;126(10):1456–74. https://doi.org/10.1161/CIRCRESAHA.120.317015.
14. Guan SP, Seet RCS, Kennedy BK. Does eNOS derived nitric oxide protect the young from severe COVID-19 complications? Ageing Res Rev. 2020;64 https://doi.org/10.1016/J.ARR.2020.101201.
15. Hemmat N, Asadzadeh Z, Ahangar NK, Alemohammad H, Najafzadeh B, Derakhshani A, Baghbanzadeh A, Baghi HB, Javadrashid D, Najafi S, Ar Gouilh M, Baradaran B. The roles of signaling pathways in SARS-CoV-2 infection; lessons learned from SARS-CoV and MERS-CoV. Arch Virol. 2021;166(3):675–96. https://doi.org/10.1007/S00705-021-04958-7.
16. Hoffmann M, Kleine-Weber H, Krueger N, Mueller MA, Drosten C, Poehlmann S. The novel coronavirus 2019 (2019-nCoV) uses the SARS-coronavirus receptor ACE2 and the cellular protease TMPRSS2 for entry into target cells. BioRxiv. 2020;2020(01):31.929042. https://doi.org/10.1101/2020.01.31.929042.
17. Khanmohammadi S, Rezaei N. Role of Toll-like receptors in the pathogenesis of COVID-19. J Med Virol. 2021;93(5):2735. https://doi.org/10.1002/JMV.26826.
18. Le Bert N, Tan AT, Kunasegaran K, Tham CYL, Hafezi M, Chia A, Chng M, Lin M, Tan N, Linster M, Chia WN, Chen MI-C, Wang L-F, Ooi EE, Kalimuddin S, Tambyah PA, Guek J, Low H, Tan Y-J, Bertoletti A. Different pattern of pre-existing SARS-COV-2 specific T cell immunity in SARS-recovered and uninfected individuals; 2020. https://doi.org/10.1101/2020.05.26.115832.
19. Leisman DE, Ronner L, Pinotti R, Taylor MD, Sinha P, Calfee CS, Hirayama AV, Mastroiani F, Turtle CJ, Harhay MO, Legrand M, Deutschman CS. Cytokine elevation in severe and critical COVID-19: a rapid systematic review, meta-analysis, and comparison with other inflammatory syndromes. Lancet Respir Med. 2020;8(12):1233–44. https://doi.org/10.1016/S2213-2600(20)30404-5.
20. Li G, De Clercq E. Therapeutic options for the 2019 novel coronavirus (2019-nCoV). Nat Rev Drug Discov. 2020;19(3):149–50. https://doi.org/10.1038/D41573-020-00016-0.
21. Liao QJ, Ye LB, Timani KA, Zeng YC, She YL, Ye L, Wu ZH. Activation of NF-κB by the full-length nucleocapsid protein of the SARS coronavirus. Acta Biochim Biophys Sin. 2005;37(9):607. https://doi.org/10.1111/J.1745-7270.2005.00082.X.

22. McGonagle D, Sharif K, O'Regan A, Bridgewood C. The role of cytokines including Interleukin-6 in COVID-19 induced pneumonia and macrophage activation syndrome-like disease. Autoimmun Rev. 2020;19(6) https://doi.org/10.1016/J.AUTREV.2020.102537.
23. Melvold RW, Sticca RP. Basic and tumor immunology: a review. Surg Oncol Clin N Am. 2007;16(4):711–35. https://doi.org/10.1016/J.SOC.2007.08.003.
24. Mescher MF, Curtsinger JM, Agarwal P, Casey KA, Gerner M, Hammerbeck CD, Popescu F, Xiao Z. Signals required for programming effector and memory development by CD8+ T cells. Immunol Rev. 2006;211:81–92. https://doi.org/10.1111/J.0105-2896.2006.00382.X.
25. Mizutani T. Signaling pathways of SARS-CoV in vitro and in vivo. Mol Biol SARS-Coronavirus. 2010;305 https://doi.org/10.1007/978-3-642-03683-5_19.
26. Montazersaheb S, Hosseiniyan Khatibi SM, Hejazi MS, Tarhriz V, Farjami A, Ghasemian Sorbeni F, Farahzadi R, Ghasemnejad T. COVID-19 infection: an overview on cytokine storm and related interventions. Virol J. 2022;19(1) https://doi.org/10.1186/S12985-022-01814-1.
27. Moreno-Eutimio MA, López-Macías C, Pastelin-Palacios R. Bioinformatic analysis and identification of single-stranded RNA sequences recognized by TLR7/8 in the SARS-CoV-2, SARS-CoV, and MERS-CoV genomes. Microbes Infect. 2020;22(4–5):226–9. https://doi.org/10.1016/J.MICINF.2020.04.009.
28. Ni L, Ye F, Cheng ML, Feng Y, Deng YQ, Zhao H, Wei P, Ge J, Gou M, Li X, Sun L, Cao T, Wang P, Zhou C, Zhang R, Liang P, Guo H, Wang X, Qin CF, Dong C. Detection of SARS-CoV-2-specific humoral and cellular immunity in COVID-19 convalescent individuals. Immunity. 2020;52(6):971. https://doi.org/10.1016/J.IMMUNI.2020.04.023.
29. Olea B, Albert E, Giménez E, Torres I, Amat P, Remigia MJ, Alberola J, Carbonell N, Ferreres J, Blasco ML, Navarro D. SARS-CoV-2-reactive IFN-γ-producing CD4+ and CD8+ T cells in blood do not correlate with clinical severity in unvaccinated critically ill COVID-19 patients. Sci Rep. 2022;12(1) https://doi.org/10.1038/S41598-022-18659-X.
30. Oschatz C, Maas C, Lecher B, Jansen T, Björkqvist J, Tradler T, Sedlmeier R, Burfeind P, Cichon S, Hammerschmidt S, Müller-Esterl W, Wuillemin WA, Nilsson G, Renné T. Mast cells increase vascular permeability by heparin-initiated bradykinin formation in vivo. Immunity. 2011;34(2):258–68. https://doi.org/10.1016/J.IMMUNI.2011.02.008.
31. Santos RAS, Oudit GY, Verano-Braga T, Canta G, Steckelings UM, Bader M. The renin-angiotensin system: going beyond the classical paradigms. Am J Physiol Heart Circ Physiol. 2019;316(5):H958–70. https://doi.org/10.1152/AJPHEART.00723.2018.
32. Schmaier AH. The contact activation and kallikrein/kinin systems: pathophysiologic and physiologic activities. J Thromb Haemost JTH. 2016;14(1):28–39. https://doi.org/10.1111/JTH.13194.
33. Schroder K, Hertzog PJ, Ravasi T, Hume DA. Interferon-γ: an overview of signals, mechanisms and functions. J Leukoc Biol. 2004;75(2):163–89. https://doi.org/10.1189/JLB.0603252.
34. Serebrovska ZO, Chong EY, Serebrovska TV, Tumanovska LV, Xi L. Hypoxia, HIF-1α, and COVID-19: from pathogenic factors to potential therapeutic targets. Acta Pharmacol Sin. 2020;41(12):1539–46. https://doi.org/10.1038/S41401-020-00554-8.
35. Shahabinezhad F, Mosaddeghi P, Negahdaripour M, Dehghani Z, Farahmandnejad M, Moghadami M, Nezafat N, Masoompour SM. Therapeutic approaches for COVID-19 based on the dynamics of interferon-mediated immune responses. 2020; https://doi.org/10.20944/PREPRINTS202003.0206.V1.
36. Sheikh IA, Kaplan AP. Studies of the digestion of bradykinin, lysyl bradykinin, and kinin-degradation products by carboxypeptidases A, B, and N. Biochem Pharmacol. 1986;35(12):1957–63. https://doi.org/10.1016/0006-2952(86)90727-6.
37. Shibayama Y, Skoner D, Suehiro S, Konishi JE, Fireman P, Kaplan AP. Bradykinin levels during experimental nasal infection with rhinovirus and attenuated influenza virus. Immunopharmacology. 1996;33(1–3):311–3. https://doi.org/10.1016/0162-3109(96)00051-3.
38. Singh Y, Gupta G, Satija S, Pabreja K, Chellappan DK, Dua K. COVID-19 transmission through host cell directed network of GPCR. Drug Dev Res. 2020;81(6):647. https://doi.org/10.1002/DDR.21674.

39. Sinha A, Vaggu RG, Swain R, Patnaik S. Repurposing of RAS-pathway mediated drugs for intestinal inflammation related diseases for treating SARS-CoV-2 infection. Curr Microbiol. 2023;80(6) https://doi.org/10.1007/S00284-023-03304-1.
40. Sodhi CP, Wohlford-Lenane C, Yamaguchi Y, Prindle T, Fulton WB, Wang S, McCray PB, Chappell M, Hackam DJ, Jia H. Attenuation of pulmonary ACE2 activity impairs inactivation of des-Arg9 bradykinin/BKB1R axis and facilitates LPS-induced neutrophil infiltration. Am J Physiol Lung Cell Mol Physiol. 2018;314(1):L17–31. https://doi.org/10.1152/AJPLUNG.00498.2016.
41. Stukalov A, Girault V, Grass V, Karayel O, Bergant V, Urban C, Haas DA, Huang Y, Oubraham L, Wang A, Hamad MS, Piras A, Hansen FM, Tanzer MC, Paron I, Zinzula L, Engleitner T, Reinecke M, Lavacca TM, Pichlmair A. Multilevel proteomics reveals host perturbations by SARS-CoV-2 and SARS-CoV. Nature. 2021;594(7862):246–52. https://doi.org/10.1038/S41586-021-03493-4.
42. Totura AL, Whitmore A, Agnihothram S, Schäfer A, Katze MG, Heise MT, Baric RS. Toll-like receptor 3 signaling via TRIF contributes to a protective innate immune response to severe acute respiratory syndrome coronavirus infection. MBio. 2015;6(3):1–14. https://doi.org/10.1128/MBIO.00638-15.
43. Wing PAC, Keeley TP, Zhuang X, Lee JY, Prange-Barczynska M, Tsukuda S, Morgan SB, Harding AC, Argles ILA, Kurlekar S, Noerenberg M, Thompson CP, Huang KYA, Balfe P, Watashi K, Castello A, Hinks TSC, James W, Ratcliffe PJ, McKeating JA. Hypoxic and pharmacological activation of HIF inhibits SARS-CoV-2 infection of lung epithelial cells. Cell Rep. 2021;35(3) https://doi.org/10.1016/J.CELREP.2021.109020.
44. Yang M. Cell pyroptosis, a potential pathogenic mechanism of 2019-nCoV infection. SSRN Electron J. 2020; https://doi.org/10.2139/SSRN.3527420.
45. Yang L, Xie X, Tu Z, Fu J, Xu D, Zhou Y. The signal pathways and treatment of cytokine storm in COVID-19. Signal Transduct Target Ther. 2021;6(1) https://doi.org/10.1038/S41392-021-00679-0.
46. Yazdanpanah F, Hamblin MR, Rezaei N. The immune system and COVID-19: friend or foe? Life Sci. 2020;256 https://doi.org/10.1016/J.LFS.2020.117900.
47. Zhang Y, Yan H, Xu Z, Yang B, Luo P, He Q. Molecular basis for class side effects associated with PI3K/AKT/mTOR pathway inhibitors. Expert Opin Drug Metab Toxicol. 2019;15(9):767–74. https://doi.org/10.1080/17425255.2019.1663169.
48. Zheng Y, Li R, Liu S. Immunoregulation with mTOR inhibitors to prevent COVID-19 severity: a novel intervention strategy beyond vaccines and specific antiviral medicines. J Med Virol. 2020;92(9):1495–500. https://doi.org/10.1002/JMV.26009.
49. Zuo Y, Warnock M, Harbaugh A, Yalavarthi S, Gockman K, Zuo M, Madison JA, Knight JS, Kanthi Y, Lawrence DA. Plasma tissue plasminogen activator and plasminogen activator inhibitor-1 in hospitalized COVID-19 patients. MedRxiv: The Preprint Server for Health Sciences; 2020. https://doi.org/10.1101/2020.08.29.20184358.

Immunological Persistence and Risk of SARS-CoV-2 Reinfection: Insights and Implications

11

Sunny Kumar and Malini Basu

Abstract

The SARS-CoV-2 virus, responsible for the COVID-19 pandemic, presents ongoing challenges to global public health, notably through the persistence of immunity and the risk of reinfection. This chapter delves into the mechanisms of immune response to SARS-CoV-2, examining both the innate and adaptive immune responses, with a focus on humoral and cellular immunity. It explores the duration of immunity following natural infection and vaccination, highlighting the roles of neutralizing antibodies, memory B cells and T cells. Factors influencing the risk of reinfection, such as waning immunity, the emergence of variants of concern (VOCs), immune compromise and demographic variables, are thoroughly analyzed. The chapter underscores the importance of booster vaccinations, continuous surveillance and adaptive public health measures in mitigating reinfection risks. Future directions emphasize the need for longitudinal studies, identifying immune correlates of protection and developing personalized immunization strategies. This comprehensive exploration aims to inform and guide public health strategies in managing the long-term impact of COVID-19. This chapter explores the mechanisms of immune response to SARS-CoV-2, the longevity of immune protection, factors influencing reinfection risk and implications for public health strategies.

S. Kumar (✉)
Cancer Biology and Inflammatory Disorder Division, Council of Scientific and Industrial Research-Indian Institute of Chemical Biology (CSIR-IICB), TRUE Campus, CN-6, Sector–V, Salt Lake, Kolkata, India

M. Basu (✉)
Department of Microbiology, Dhruba Chand Halder (DCH) College - University of Calcutta, Jaynagar Mazilpur, West Bengal, India

M. K. Ghosh, M. Basu (eds.), *Pathogenesis, Management and Socio-Economic Impact of COVID-19 Pandemic*, Medical Virology: From Pathogenesis to Disease Control, https://doi.org/10.1007/978-981-95-1426-7_11

Keywords

SARS-CoV-2 reinfection · Immune response · Immunological persistence

Introduction

The COVID-19 pandemic, driven by SARS-CoV-2, has highlighted the need to understand the duration of immunity and the risk of reinfection. Immunological persistence refers to how long the immune system retains the ability to combat the virus after initial exposure, whether through natural infection or vaccination. The immune response involves both immediate, nonspecific actions by the innate immune system and targeted, long-lasting responses by the adaptive immune system, which includes B cells and T cells. Antibody levels may decrease over time, memory B cells and T cells can provide lasting immunity. This suggests that the body can respond quickly to subsequent exposures, potentially reducing the severity of reinfections. However, factors such as the strength of the initial immune response, the emergence of viral variants and individual health can influence reinfection risk. Variants like alpha, beta, delta and omicron have mutations that can partially evade immune responses, though memory T cells often still recognize these variants. Vaccination remains crucial in combating SARS-CoV-2, with various vaccines showing high efficacy. Booster doses enhance immunity, particularly against new variants. Breakthrough infections can occur but are generally less severe, emphasizing the importance of vaccines and boosters. Hybrid immunity, resulting from both infection and vaccination, offers robust and durable protection. Public health strategies must consider these factors, promoting vaccination and booster programs and maintaining measures like mask-wearing and social distancing. Ongoing research is vital to adapt policies to evolving viral landscapes, ensuring effective management of the pandemic and protection of global health [1–6].

This book chapter aims to synthesize current knowledge on immunological persistence following SARS-CoV-2 infection, examining both humoral and cellular immunity aspects. It will explore the mechanisms underlying the immune response to the virus, the longevity of antibodies and T-cell responses and factors influencing the durability of immunity. Additionally, the review will address the risk of SARS-CoV-2 reinfection, discussing the evidence of reinfection cases, the implications of emerging variants and the role of waning immunity. By providing a comprehensive overview of the immune response's persistence and the risk of reinfection, this article seeks to inform ongoing public health efforts and guide future research priorities in the fight against COVID-19.

Immune Response to SARS-CoV-2

Currently, SARS-CoV-2 infection is a major concern for public health due to insufficient availability of efficacious therapeutic options. Development of immune-pathological changes (hyper-inflammation) due to this infection is a major determinant of infection severity. Antiviral immune responses against this infection were generated through the collaborative role of innate as well as adaptive immune responses. Immune system attempt to clean the unwanted material from host cell and reduce the viral load [7]. Mechanistic pathways of both immune responses are explained in the following sections (Fig. 11.1).

Innate Immune Responses

The innate immunity acknowledges the quick response (within few hours of infection) and stimulates the "alarm bells" for production of type 1 IFN and other inflammatory molecules. It restricts further viral replication, permits the recruitment of effector cells and finally governs the priming of adaptive immunity. In first two steps, it slows down the replication and spreading of the virus while the next step triggers the activation of adaptive immunity.

Recently, a study of biochemical profiling of 452 SARS-CoV-2 patients from Wuhan documented higher neutrophil:lymphocyte ratio and an increase in neutrophil count and cytokines (IL-6, TNF-α), while the reduction of basophils, eosinophils and monocytes was noted [7, 8]. However, SARS-CoV-2 virus selectively prefers ACE-2 receptor for entry into host cells, which is highly expressed in the lung epithelium and the GUT. Lung tissue damage is primarily noticed due to the destruction of lung macrophages and epithelial cells [3, 9, 10]. After viral entry, host innate immune cells react immediately and express the various pathogen-recognition receptors (PRRs) (*viz.*, CTLRs, TLRs, RLRs and NLRs) and play crucial roles in sensing pathogen-associated molecular pattern (PAMP) [11, 12]. However, the RNA-based SARS-CoV-2 virus is specifically recognized by RIG-1 in the cytosol, and TLR-2/3/7 helps to recognize endosomal virus [13–15]. Polyinosinic-polycytidylic acid-based activation of TLR3 can inhibit the infection rate [16]. After recognition, transcription factors (*viz.*, NF-kB and IRF3) gets activated in the cytosol and translocated to the nucleus which induces several pro-inflammatory cytokines, chemokines and type-1 IFN. These type-1 INF molecules are documented as a first-line defensive innate immune response [17]. Further, these interferon molecules activate the JAK-STAT pathway by stimulating the IFN-α/β receptor. Subsequently, JAK1 and TYK2 phosphorylate the STAT1 & 2, which form a new complex with IFN9 and are translocated into nucleus for the activation of interferon-stimulated genes (ISGs) that result in the generation of antiviral immune responses [18].

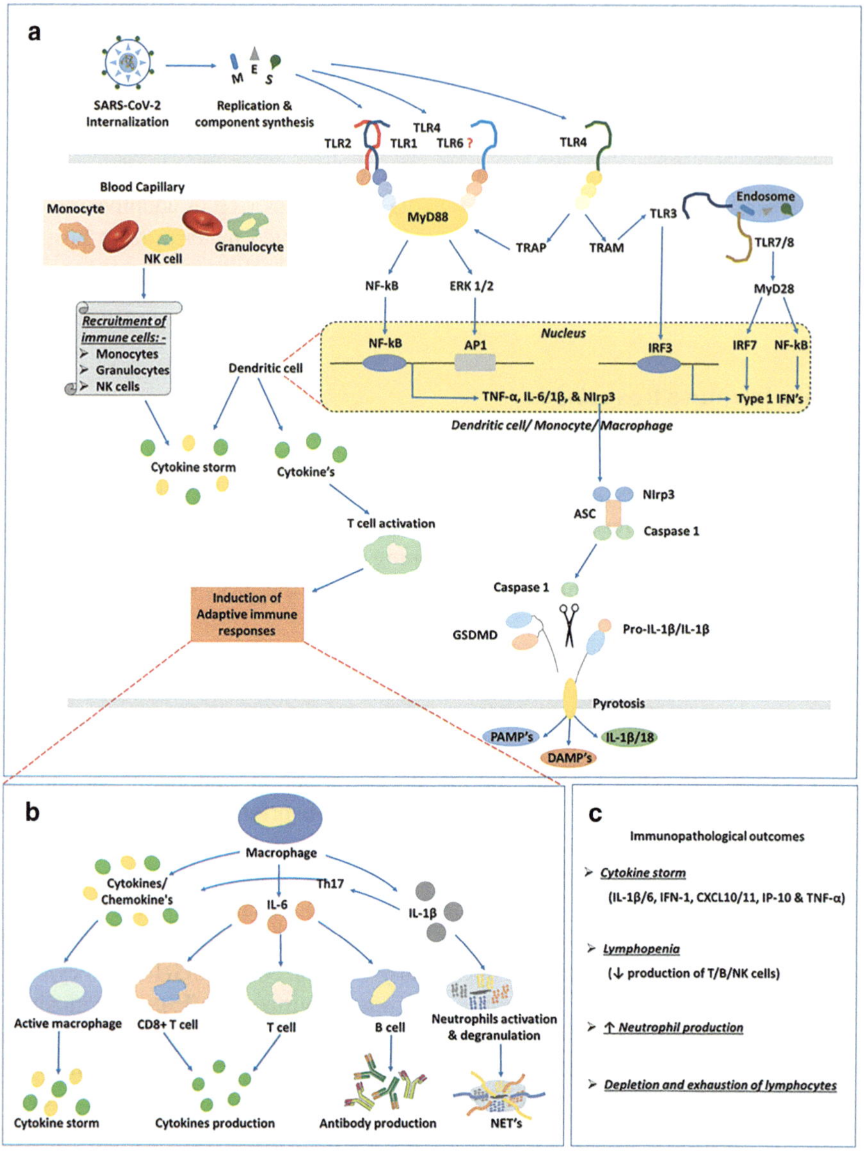

Fig. 11.1 Mechanisms of innate as well as adaptive immunity. (**a**) Upper panel shows the mechanism of induction of innate immunity, (**b**) lower left panel shows the mechanism of adaptive immunity and (**c**) lower right panel shows the occurrence of various immunopathological changes

The infectious viral particles keep on generating several strategies to lower down the interferon-mediated antiviral immune response. In SARS-CoV-2 infected patients, low-level IFN-α/β expression was observed in dendritic cells and macrophages [19]. Preclinical studies state that lung immunopathological changes primarily occur due to the huge production of interferons in SARS-CoV-2 infection. Plasmacytoid-dendritic cells (pDCs) are the primary source of type-1 interferons while the nonstructural proteins regulate the response of interferons in pDCs and other inflammatory cells [19–21]. Exceptionally, nonstructural proteins (*viz.*, ORF6 and NSP1) interfere with the interferon signaling by obstructing the STAT1 phosphorylation and nuclear translocation [22, 23].

SARS-CoV-2-infected host cell macrophages induce the production of chemokines (*viz.*, chemokine C–C ligand 2 (CCL2), monocyte chemotactic protein 1 (MCP1) and C-X-C chemokine (CXCL10), IFN-γ-inducible protein 10 (IP-10)) and several cytokines (*viz.*, IL-2/7, TNF-α, IFN-γ, IP-10, MIP-1α and MCP-1) in severe COVID-19 patients that plays important roles in lung pathology [24, 25]. Increased expressions of CCL7/MCP-3, CCL8, MCP-2 and CCL3, macrophage inflammatory protein 1α (MIP-1α) were also documented in infected hosts, having chemotactic activity for macrophages [26, 27].

Eosinophils and natural killer (NK) cells are crucial players in the development of innate antiviral immune responses. Eosinophil-mediated antiviral immune response is due to the increased production of nitric oxide through the activation of nitric oxide synthase-2 (NOS-2) [28]. On the other hand, NK cells express several receptors for the MHC-1 molecule that may either activate or obstruct the cytotoxicity or cytokine storm. Natural-killer group 2 member D (NKG2D) is an NK cell type that is responsible for the production of cytokines and chemokines to initiate the cytotoxic activity [29].

However, to date, the clear-cut role and mechanism of innate immunity in COVID-19 patients are not well understood. Unfit or physically ill individuals with weak immunity are more susceptible to infection than healthy individuals. Due to these concerns, it can be concluded that innate immune responses acts like a decisive factor for the disease outcome or severity [1]. Furthermore, severely infected patients are positively correlated with the increased expressions of several cytokines (*viz.*, granulocyte colony-stimulating factor, IP-10, TNF-α, MIP-1α and MCP-1) [25]. Moreover, the higher expression of pro-inflammatory cytokines is a sign of infection related progression, severity and deaths. Initially, the microbial pathogens get access to hijack the lungs or guts, then several innate immunological processes begin such as production of cytokines, chemokines, recruitment of neutrophils, macrophages and other inflammatory cells to the site of injury and finally induce the cytokine storm [30].

Adaptive Immune Responses

Prime players of adaptive immune response are B cells, $CD4^+$ T cells and $CD8^+$ T cells. B cells generate antigen-specific antibodies, $CD4^+$ T cells act as a helper and effector, and $CD8^+$ T cells kill the foreign viral particles. However, these immune mechanisms are important to control and mitigate almost all foreign viral particles and develop after ~6–10 days of priming. During this period, naive cells are extensively proliferated and differentiated into effector cells (helper and cytotoxic T cells and B cells). Moreover, post-initiation, memory cells of the respective effector cells are retained in the host for a longer duration, which can activate again to lower the chances of reinfection. These immunological responses and developed memory cells are important key factors that determine the success of all vaccines [4].

Immune cells (*viz.*, CD4+ and CD8+ T cells) play crucial antiviral roles by promoting the release of antigen-specific antibodies from B cells and killing the virus-infected cells [31]. The critical responsibility of CD4+ T cells in maintaining SARS-CoV-2 propagation and disease severity is well documented in T-cell-deficient BALB/c mice model [32]. CD4+ T cells regulate overall viral clearance while memory CD8+ T cells protects the host cell from fatal SARS-CoV-2 infection by generating several cytokines (*viz.*, IFN-γ, TNF-α and IL-2) and cytotoxic molecules (granzyme B) [33]. Additionally, the memory CD8+ T cells against M and N proteins of SARS-CoV infection are documented to be retained in the infected host cells for 11 years with a capability to proliferate and to produce interferons [34]. However, the count of total T cells, CD4+ and CD8+ T cells in COVID-19 infected individuals is lower than that of healthy individuals [35]. Recent reports have shown that reduced levels of T cells are a result of their apoptosis and necrosis due to higher production of multiple cytokines [35]. Similarly, the c-terminal BH3 domain of SARS-CoV E protein is documented to interact with BcL-xL, an anti-apoptotic protein which leads to the apoptosis of T cells [36]. Additionally, the generation of T_H-17 cytokine (IL-17) is reported in SARS/MERS-CoV infections which plays a role in the recruitment of neutrophils and monocytes to the site of injury; further, it also helps in the activation of downstream chemokine or cytokine cascade. The count of CCR6+ Th17 is also increased in SARS-CoV-2 infected patients that cause lung injury or hyper-inflammation lead by cytokine storm [37]. Furthermore, higher expression of Th1 cells has also been reported in the lungs of the infected patients which damage the organ. CD4 + T cells quickly converted into pathogenic Th1 cells induce the production of granulocyte colony-stimulating factor (G-CSF), cytokine that initiates tissue damage. It concludes that count of T cells is reduced in SARS-CoV-2 infected patients and is responsible for more severe symptoms [3] (Fig. 11.2).

Adaptive immunity is responsible and crucial to control the SARS-CoV-2 infection, but extensive investigations are still required. In SARS-CoV-2 infected patients, antibody profiling indicates a typical pattern of IgM and IgG secretions [38]. The SARS-CoV-2 selective IgG antibodies can be retained for a longer duration than IgM antibodies which indicates that IgG antibodies are responsible for defensive role [38].

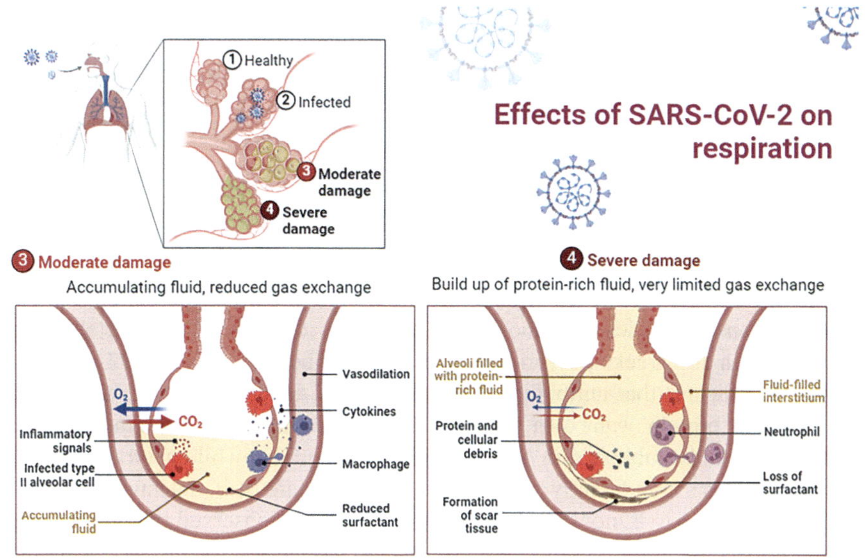

Fig. 11.2 Effect of SARS-CoV-2 on the respiratory system. (This image is adopted from Biorender. https://www.biorender.com/)

Immunological Persistence

Overall risk of SARS-CoV-2 reinfection is relatively low, but the chances are still high. To determine these probabilities, a longitudinal study has been performed that compares the chances of reinfection in previously infected patients versus the population with no infection history. To perform this study, the authors have collected data from MEDLINE (Ovid), the World Health Organization global literature database, ClinicalTrials.gov, COVID19reviews.org and reference lists. This study showed that the overall chances of reinfection ranged from 0% to 2.2% and after the first time symptomatic infection with wild-type or alpha variants, the rate of reinfection was lowered by 80% to 98%. Another meta-analysis study shows the probability of reinfection was found to be 87% less than the first infection. Taken together, protection against the chances of reinfection is approximately more than 80% which may be retained/persists for at least 7 months in SARS-CoV-2 (wild-type or alpha variants) infected individuals. To date, no reinfection study is documented regarding infection with the Delta and Omicron variants in the respective infected patients. The duration for which immunological protection persists in Delta and Omicron-infected individuals has not been studied yet. Results for the patients from the elderly age group, immunocompromised, racial and ethnic subgroups are conflicting and remain unclear [39, 40]. Understanding the duration of immunity conferred by natural infection and vaccination is crucial for managing public health strategies, including vaccination programs and predicting the course of the pandemic. This literature review synthesizes current findings on the duration and nature of immunity following natural infection and vaccination against SARS-CoV-2.

Immunity After Natural Infection

Natural infection with SARS-CoV-2 induces a multifaceted immune response, including the production of neutralizing antibodies (NAbs), memory B cells and T-cell responses. Studies have shown that most individuals develop detectable antibodies within 1–3 weeks post-infection, with levels peaking at around 3–4 weeks. Research indicates that antibodies, particularly IgG against the spike protein and nucleocapsid protein, wane over time but remain detectable for several months post-infection. A study published in *Nature Medicine* found that while neutralizing antibodies decrease significantly after 6 months, they are still detectable in most individuals for up to 8 months. Another study in *Science* observed that neutralizing antibodies and memory B cells persisted for up to 8 months post-infection, although the levels of antibodies declined gradually. Memory T cells, both CD4+ and CD8+, play a crucial role in long-term immunity. Evidence suggests that T-cell responses are robust and can last for several months. Research in *Cell* demonstrated that SARS-CoV-2-specific memory T cells persist for at least 6–8 months post-infection, even as antibody levels decline. This cellular immunity provides a layer of protection against severe disease upon re-exposure to the virus. Cases of reinfection with SARS-CoV-2 have been documented, though they appear to be relatively rare within the first year post-infection. The risk of reinfection is higher with emerging variants that partially escape immune recognition, such as the Beta and Delta variants. However, the severity of reinfections tends to be lower, suggesting some degree of protective immunity remains [41–43].

Immunity After Vaccination

Vaccination against SARS-CoV-2, using various platforms (mRNA, vector-based, protein subunit), induces a strong immune response. mRNA vaccines like Pfizer-BioNTech (BNT162b2) and Moderna (mRNA-1273) have shown high efficacy in eliciting robust antibody and T-cell responses. Post-vaccination antibody levels are generally higher than those observed after natural infection. Studies indicate that vaccine-induced antibodies peak within a few weeks after the second dose and remain high for at least 6 months. Research published in *The New England Journal of Medicine* reported that antibodies elicited by mRNA vaccines persist at high levels for at least 6 months, though a gradual decline is observed. Given the observed decline in antibody levels over time, booster doses have been recommended to sustain immunity. Studies have shown that booster doses significantly elevate antibody titers and enhance protection, particularly against variants of concern. A study in

The Lancet reported that a booster dose of the Pfizer-BioNTech vaccine increased neutralizing antibody levels by more than 25-fold against the Omicron variant compared to the initial two-dose series. Similar to natural infection, vaccines also induce strong T-cell responses. Memory T cells have been detected for up to 8 months following vaccination. These T-cell responses are crucial for long-term immunity and may provide protection against severe disease even when antibody levels wane. Research in *Nature* highlighted that T-cell responses elicited by mRNA vaccines are durable and recognize multiple viral epitopes, including those from variants of concern. Breakthrough infections, where vaccinated individuals contract COVID-19, have been reported, particularly with the emergence of variants like Delta and Omicron. However, these infections are generally less severe, indicating that vaccination provides substantial protection against severe outcomes. A study in *The New England Journal of Medicine* showed that vaccinated individuals have a significantly reduced risk of severe disease and hospitalization compared to unvaccinated individuals [5, 44, 45].

Comparative Analysis

Comparing natural infection and vaccination, both provide substantial immunity, but the duration and robustness vary. Vaccination generally induces higher initial antibody levels compared to natural infection. However, the decline in antibody levels over time is a common feature in both scenarios. Booster doses are more common in the context of vaccination, aimed at maintaining high levels of neutralizing antibodies and enhancing protection against emerging variants. In contrast, natural infection typically does not have a "booster" equivalent unless through reinfection, which is less desirable due to potential health risks. Both natural infection and vaccination elicit durable T-cell responses, critical for long-term immunity. The T-cell responses appear to be broad and recognize multiple viral components, providing cross-protection against variants. Variants of concern pose challenges to both naturally acquired and vaccine-induced immunity. Vaccines, especially mRNA types, have been shown to be adaptable through updated formulations and booster doses, which enhance protection against new variants. Natural infection provides some degree of variant cross-protection, but the extent can vary depending on the variant and the individual's immune response [41, 46].

Understanding the duration and quality of immunity after natural infection and vaccination against SARS-CoV-2 is crucial for public health strategies. Both natural infection and vaccination confer substantial immunity, with vaccination generally providing higher initial antibody levels and more controlled booster interventions. Long-term protection appears to be mediated significantly by T-cell responses in both cases. However, the emergence of variants underscores the need for ongoing monitoring and potential vaccine updates to ensure continued protection against COVID-19. Public health policies should continue to emphasize vaccination, including booster doses, to maintain robust immunity across populations and mitigate the impact of emerging variants [46].

Factors Influencing Risk of Reinfection

COVID-19 reinfection occurs when an individual who has recovered from a previous infection with SARS-CoV-2 becomes infected again. Understanding the factors that influence the risk of reinfection is crucial. Elderly individuals generally have a higher risk of reinfection compared to younger people. Aging is associated with a decline in immune function, known as immunosenescence, which can impair the body's ability to mount an effective immune response against reinfections. Individuals with underlying health conditions, such as diabetes, cardiovascular diseases and immunosuppressive conditions, are at a higher risk of reinfection. These conditions can weaken the immune system, making it more difficult to respond effectively to a subsequent exposure to the virus. People with weakened immune systems, whether due to medical treatments (e.g., chemotherapy and organ transplantation) or chronic illnesses (e.g., HIV/AIDS), are more susceptible to reinfections. Their impaired immune responses might not produce sufficient levels of antibodies or maintain memory cells effectively [39, 41].

Individuals with high exposure risk, such as healthcare workers or those living in densely populated areas, are more likely to encounter the virus multiple times. Repeated exposure increases the probability of reinfection, especially if protective measures (e.g., mask-wearing, social distancing) are not consistently followed. Vaccination significantly reduces the risk of severe disease and reinfection. However, disparities in vaccine coverage and uptake can influence reinfection rates. Areas with low vaccination rates may see higher incidences of reinfections as the population remains more vulnerable. The timing of the initial infection in relation to the emergence of new variants plays a role in reinfection risk. Individuals infected early in the pandemic with the original strain may have a higher reinfection risk when exposed to newer variants, which might evade the immunity developed against the original strain. There is evidence suggesting that the severity of the initial infection can impact the strength and duration of the immune response. Individuals who experienced mild or asymptomatic infections may develop a weaker immune response compared to those who had severe disease. Consequently, those with mild initial infections might be at higher risk for reinfection [39, 41].

Different vaccines have varying efficacy rates, and the type of vaccine received can influence reinfection risk. mRNA vaccines (e.g., Pfizer-BioNTech and Moderna) have shown high efficacy in preventing symptomatic infection and severe disease, while other vaccines might offer varying levels of protection. Booster doses have been shown to enhance immunity, increasing antibody titers and strengthening memory cell responses. Individuals who receive booster doses are likely to have a reduced risk of reinfection compared to those who only received the primary series of the vaccine. The natural decline in antibody levels over time is a primary factor influencing the risk of reinfection. Although memory B and T cells provide a layer of protection, reduced circulating antibodies can lead to a higher susceptibility to reinfection, especially with high exposure levels. SARS-CoV-2 has undergone significant mutations, leading to the emergence of variants of concern (VOCs) such as

Alpha, Beta, Gamma, Delta and Omicron. These variants may possess mutations in the S protein, potentially evading neutralizing antibodies generated from previous infections or vaccinations. This antigenic variation can increase the risk of reinfection and reduce vaccine efficacy [38, 41, 47].

Understanding these factors helps inform public health strategies, including the importance of booster vaccinations, continued surveillance of emerging variants and targeted protection for high-risk groups. Ongoing research is essential to further elucidate the mechanisms behind reinfection and to optimize strategies for preventing and managing reinfections in the population.

Implications for Public Health

To counteract waning immunity and enhance protection against emerging variants, booster vaccinations have become a critical strategy. Booster doses can reinvigorate immune responses, increasing antibody titers and strengthening T-cell responses. The timing and frequency of booster doses are continually evaluated based on epidemiological data and vaccine performance against new variants. Ongoing surveillance of SARS-CoV-2 variants and breakthrough infections is essential for assessing reinfection risks and guiding public health interventions. Genomic sequencing helps track the emergence and spread of VOCs, while serological studies monitor population immunity levels. Adapting vaccines to target emerging variants is a proactive approach to maintaining vaccine efficacy. mRNA vaccines, in particular, offer the flexibility for rapid updates. Research into pan-coronavirus vaccines, which aim to provide broad protection against multiple coronaviruses, is also underway. Despite vaccination efforts, nonpharmaceutical interventions (NPIs) such as mask-wearing, physical distancing and hand hygiene remain important, especially in settings with high transmission rates or vulnerable populations. These measures help reduce the risk of reinfection and transmission [5, 48–50].

Future Directions

Longitudinal studies are crucial for understanding the duration of immunity post-infection and post-vaccination. These studies will help determine the longevity of immunological memory and inform booster vaccination schedules. Identifying immune correlates of protection—specific immune markers associated with immunity—is vital for evaluating vaccine efficacy and guiding public health strategies. These correlates can help predict the risk of reinfection and optimize vaccine formulations. As our understanding of individual variability in immune responses grows, personalized immunization strategies could be developed. Tailoring vaccination schedules and doses based on factors such as age, health status and genetic background may enhance immune protection and reduce reinfection risk [39, 41].

Conclusion

The persistence of immunity and the risk of reinfection with SARS-CoV-2 are indeed complex and multifaceted issues that require ongoing attention and adaptation. Both natural infection and vaccination have been shown to provide substantial protection against COVID-19. However, the phenomenon of waning immunity, coupled with the emergence of new variants, continues to pose significant challenges. The durability of immunity following natural infection varies among individuals and can be influenced by factors such as age, health status and the severity of the initial infection. Similarly, the immune response elicited by vaccines, while robust, is not permanent and tends to diminish over time. This necessitates continuous monitoring of immunity levels in populations to understand the duration of protection and to identify groups at increased risk of reinfection. The emergence of SARS-CoV-2 variants has further complicated the landscape. Variants with mutations in key regions of the virus can partially evade the immune response, leading to reduced effectiveness of both natural and vaccine-induced immunity. This ongoing viral evolution underscores the need for adaptive public health strategies and the potential for regular updates to vaccine formulations, similar to the annual adjustments made for influenza vaccines. Booster vaccinations have emerged as a critical tool in sustaining immunity levels and extending protection, particularly in vulnerable populations such as the elderly and immunocompromised. Evidence suggests that booster doses can significantly enhance the immune response, restoring high levels of protection against severe disease and reducing the likelihood of reinfection. In addition to vaccination, nonpharmaceutical interventions such as masking, social distancing and improved ventilation continue to play an important role in controlling the spread of the virus. Public health measures must be adaptable and responsive to the evolving understanding of the virus and the immune response it elicits. Understanding the dynamics of immune responses to SARS-CoV-2 is crucial for developing robust and long-term vaccination strategies. This includes not only determining optimal timing for booster doses but also identifying the most effective vaccine platforms and formulations. Research into cross-protective immunity and the development of vaccines that target multiple viral antigens or variants will be key to enhancing and prolonging immunity. In conclusion, managing the long-term impact of COVID-19 requires a multifaceted approach that integrates continuous monitoring of immunity, adaptive vaccination strategies and flexible public health measures. By staying vigilant and responsive to new scientific insights and viral developments, we can better protect global health and mitigate the risk of reinfection and widespread transmission of SARS-CoV-2. The collective effort of the scientific community, public health officials and policymakers will be essential in navigating the ongoing challenges posed by this pandemic.

Author's Contributions SK and MB conceived the idea, review structure and wrote the manuscript. All authors have read, agreed and approved the final draft of the manuscript.

Funding This work is supported by CSIR (OLP121 & HCP40) and AcSIR.

Data Availability The software Biorender (https://www.biorender.com/) is used to draw some elements of the few figures.

Declarations

Conflicts of Interests There is no competing and conflict of interest.

References

1. Prompetchara E, Ketloy C, Palaga T. Immune responses in COVID-19 and potential vaccines: lessons learned from SARS and MERS epidemic. Asian Pac J Allergy Immunol. 2020;38(1):1–9.
2. Nile SH, Nile A, Qiu J, Li L, Jia X, Kai G. COVID-19: pathogenesis, cytokine storm and therapeutic potential of interferons. Cytokine Growth Factor Rev. 2020;53:66–70.
3. Zhou Y, Fu B, Zheng X, et al. Pathogenic T-cells and inflammatory monocytes incite inflammatory storms in severe COVID-19 patients. Natl Sci Rev. 2020;7(6):998–1002.
4. Sette A, Crotty S. Adaptive immunity to SARS-CoV-2 and COVID-19. Cell. 2021;184(4):861–80.
5. Kumar S, Basu M, Ghosh P, Ansari A, Ghosh MK. COVID-19: clinical status of vaccine development to date. Br J Clin Pharmacol. 2022;
6. Ghosh MK, Kumar S, Ganguly KK, et al. COVID-19 and cancer: insights into their association and influence on genetic and epigenetic landscape. Epigenomics. 2023;15(4):227–48. https://doi.org/10.2217/epi-2023-0052.
7. Lagunas-Rangel FA. Neutrophil-to-lymphocyte ratio and lymphocyte-to-C-reactive protein ratio in patients with severe coronavirus disease 2019 (COVID-19): a meta-analysis. J Med Virol. 2020;
8. Qin C, Zhou L, Hu Z, et al. Dysregulation of immune response in patients with coronavirus 2019 (COVID-19) in Wuhan, China. Clin Infect Dis. 2020;71(15):762–8.
9. Kumar S, Chatterjee M, Ghosh P, Ganguly KK, Basu M, Ghosh MK. Targeting PD-1/PD-L1 in cancer immunotherapy: an effective strategy for treatment of triple-negative breast cancer (TNBC) patients. Genes Dis. 2023;10(4):1318–50.
10. Kumar S, Basu M, Ghosh MK. Chaperone-assisted E3 ligase CHIP: a double agent in cancer. Genes Dis. 2021;9(6) https://doi.org/10.1016/j.gendis.2021.08.003.
11. Ishii KJ, Koyama S, Nakagawa A, Coban C, Akira S. Host innate immune receptors and beyond: making sense of microbial infections. Cell Host Microbe. 2008;3(6):352–63.
12. Tartey S, Takeuchi O. Pathogen recognition and Toll-like receptor targeted therapeutics in innate immune cells. Int Rev Immunol. 2017;36(2):57–73.
13. Rogers NC, Slack EC, Edwards AD, et al. Syk-dependent cytokine induction by Dectin-1 reveals a novel pattern recognition pathway for C type lectins. Immunity. 2005;22(4):507–17.
14. Dosch SF, Mahajan SD, Collins AR. SARS coronavirus spike protein-induced innate immune response occurs via activation of the NF-κB pathway in human monocyte macrophages in vitro. Virus Res. 2009;142(1-2):19–27.
15. Totura AL, Whitmore A, Agnihothram S, et al. Toll-like receptor 3 signaling via TRIF contributes to a protective innate immune response to severe acute respiratory syndrome coronavirus infection. MBio. 2015;6(3):e00638-15.

16. Mazaleuskaya L, Veltrop R, Ikpeze N, Martin-Garcia J, Navas-Martin S. Protective role of Toll-like receptor 3-induced type I interferon in murine coronavirus infection of macrophages. Viruses. 2012;4(5):901–23.
17. Koop A, Lepenies I, Braum O, et al. Novel splice variants of human IKKε negatively regulate IKKε-induced IRF3 and NF-kB activation. Eur J Immunol. 2011;41(1):224–34.
18. Hosseini A, Gharibi T, Marofi F, Javadian M, Babaloo Z, Baradaran B. Janus kinase inhibitors: a therapeutic strategy for cancer and autoimmune diseases. J Cell Physiol. 2020;235(9):5903–24.
19. Channappanavar R, Perlman S. Pathogenic human coronavirus infections: causes and consequences of cytokine storm and immunopathology. In: Seminars in immunopathology, vol. 39. Springer; 2017. p. 529–39.
20. Cervantes-Barragan L, Züst R, Weber F, et al. Control of coronavirus infection through plasmacytoid dendritic-cell–derived type I interferon. Blood. 2007;109(3):1131–7.
21. Channappanavar R, Fehr AR, Vijay R, et al. Dysregulated type I interferon and inflammatory monocyte-macrophage responses cause lethal pneumonia in SARS-CoV-infected mice. Cell Host Microbe. 2016;19(2):181–93.
22. Frieman M, Yount B, Heise M, Kopecky-Bromberg SA, Palese P, Baric RS. SARS-CoV ORF6 antagonizes STAT1 function by sequestering nuclear import factors on the rER/Golgi membrane. J Virol. 2007;81(18):9812–24.
23. Wathelet MG, Orr M, Frieman MB, Baric RS. Severe acute respiratory syndrome coronavirus evades antiviral signaling: role of nsp1 and rational design of an attenuated strain. J Virol. 2007;81(21):11620–33.
24. Zhou J, Chu H, Li C, et al. Active replication of Middle East respiratory syndrome coronavirus and aberrant induction of inflammatory cytokines and chemokines in human macrophages: implications for pathogenesis. J Infect Dis. 2014;209(9):1331–42.
25. COVID-19: consider cytokine storm syndromes and immunosuppression – PubMed. https://pubmed.ncbi.nlm.nih.gov/32192578/. Accessed 1 Mar 2022.
26. Cheung CY, Poon LL, Ng IH, et al. Cytokine responses in severe acute respiratory syndrome coronavirus-infected macrophages in vitro: possible relevance to pathogenesis. J Virol. 2005;79(12):7819–26.
27. Van Damme J, Proost P, Lenaerts JP, Opdenakker G. Structural and functional identification of two human, tumor-derived monocyte chemotactic proteins (MCP-2 and MCP-3) belonging to the chemokine family. J Exp Med. 1992;176(1):59–65.
28. Phipps S, Lam CE, Mahalingam S, et al. Eosinophils contribute to innate antiviral immunity and promote clearance of respiratory syncytial virus. Blood J Am Soc Hematol. 2007;110(5):1578–86.
29. Lodoen MB, Lanier LL. Viral modulation of NK cell immunity. Nat Rev Microbiol. 2005;3(1):59–69.
30. Mangalmurti N, Hunter CA. Cytokine storms: understanding COVID-19. Immunity. 2020;53(1):19–25.
31. Maloir Q, Ghysen K, Von Frenckell C, Louis R, Guiot J. Acute respiratory distress revealing antisynthetase syndrome. Rev Med Liege. 2018;73(7–8):370–5.
32. Chen J, Lau YF, Lamirande EW, et al. Cellular immune responses to severe acute respiratory syndrome coronavirus (SARS-CoV) infection in senescent BALB/c mice: CD4+ T cells are important in control of SARS-CoV infection. J Virol. 2010;84(3):1289–301.
33. Channappanavar R, Fett C, Zhao J, Meyerholz DK, Perlman S. Virus-specific memory CD8 T cells provide substantial protection from lethal severe acute respiratory syndrome coronavirus infection. J Virol. 2014;88(19):11034–44.
34. Ng OW, Chia A, Tan AT, et al. Memory T cell responses targeting the SARS coronavirus persist up to 11 years post-infection. Vaccine. 2016;34(17):2008–14.

35. Diao B, Wang C, Tan Y, et al. Reduction and functional exhaustion of T cells in patients with coronavirus disease 2019 (COVID-19). Front Immunol. 2020:827.
36. Yang Y, Xiong Z, Zhang S, et al. Bcl-xL inhibits T-cell apoptosis induced by expression of SARS coronavirus E protein in the absence of growth factors. Biochem J. 2005;392(1):135–43.
37. Mubarak A, Alturaiki W, Hemida MG. Middle East respiratory syndrome coronavirus (MERS-CoV): infection, immunological response, and vaccine development. J Immunol Res. 2019;
38. Li X, Geng M, Peng Y, Meng L, Lu S. Molecular immune pathogenesis and diagnosis of COVID-19. J Pharm Anal. 2020;10(2):102–8.
39. Helfand M, Fiordalisi C, Wiedrick J, et al. Risk for reinfection after SARS-CoV-2: a living, rapid review for American College of Physicians Practice Points on the Role of the antibody response in conferring immunity following SARS-CoV-2 infection. Ann Intern Med. 2022; https://doi.org/10.7326/M21-4245.
40. Tillett RL, Sevinsky JR, Hartley PD, et al. Genomic evidence for reinfection with SARS-CoV-2: a case study. Lancet Infect Dis. 2021;21(1):52–8.
41. Sciscent BY, Eisele CD, Ho L, King SD, Jain R, Golamari RR. COVID-19 reinfection: the role of natural immunity, vaccines, and variants. J Community Hosp Intern Med Perspect. 2021;11(6):733–9.
42. Spellberg B, Nielsen TB, Casadevall A. Antibodies, immunity, and COVID-19. JAMA Intern Med. 2021;181(4):460–2.
43. Stephens DS, McElrath MJ. COVID-19 and the path to immunity. JAMA. 2020;324(13):1279–81.
44. Kamińska D, Dęborska-Materkowska D, Kościelska-Kasprzak K, et al. Immunity after COVID-19 recovery and vaccination: similarities and differences. Vaccines. 2022;10(7):1068.
45. Tiyo BT, Schmitz GJ, Ortega MM, et al. What happens to the immune system after vaccination or recovery from COVID-19? Life. 2021;11(11):1152.
46. Goldberg Y, Mandel M, Bar-On YM, et al. Protection and waning of natural and hybrid immunity to SARS-CoV-2. N Engl J Med. 2022;386(23):2201–12. https://doi.org/10.1056/NEJMoa2118946.
47. Ciotti M, Ciccozzi M, Terrinoni A, Jiang WC, Wang CB, Bernardini S. The COVID-19 pandemic. Crit Rev Clin Lab Sci. 2020;57(6):365–88. https://doi.org/10.1080/10408363.2020.1783198.
48. Del Rio C, Collins LF, Malani P. Long-term health consequences of COVID-19. JAMA. 2020;324(17):1723–4.
49. Ambalavanan R, Snead RS, Marczika J, Malioukis A. Epidemiological contemplation for a currently pragmatic COVID-19 health passport: a perspective. Front Public Health. 2024;12:1347623.
50. Jebril N. World Health Organization declared a pandemic public health menace: a systematic review of the coronavirus disease 2019 "COVID-19." Available SSRN 3566298. 2020.

12 Convalescent Plasma Therapy in COVID-19

Ranit D'Rozario, Jafar Sarif, and Dipyaman Ganguly

Abstract

Passive antibody therapy, utilizing plasma or serum, has been employed successfully to combat infectious diseases for over a century by transferring antibodies from recovered individuals to recipients, potentially mitigating disease progression. During the COVID-19 pandemic, convalescent plasma emerged as a promising therapeutic option, receiving emergency use authorization and widespread adoption. Its historical efficacy in previous pandemics, biological plausibility, and initial lack of targeted antivirals made it a cornerstone response. While initial studies suggested benefits, later research showed mixed results and faced challenges in determining its effectiveness. However, recent evidence indicates that when administered early and with high antibody levels, convalescent plasma can be effective in preventing disease progression and reducing mortality in COVID-19 patients. This chapter provides a comprehensive overview of the principles, and mechanisms of convalescent plasma therapy, summarizes current evidence for its use in COVID-19, and highlights its efficacy, importance, and associated challenges.

Keywords

COVID-19 · Plasma therapy · Convalescent plasma

R. D'Rozario · J. Sarif · D. Ganguly (✉)
IICB-Translational Research Unit of Excellence, CSIR-Indian Institute of Chemical Biology, Kolkata, India
e-mail: dipyaman@iicb.res.in

M. K. Ghosh, M. Basu (eds.), *Pathogenesis, Management and Socio-Economic Impact of COVID-19 Pandemic*, Medical Virology: From Pathogenesis to Disease Control, https://doi.org/10.1007/978-981-95-1426-7_12

Introduction

Since its initial detection in Wuhan, China, in December 2019, the SARS-CoV-2 virus, responsible for COVID-19, has triggered a global health and social crisis of unmatched proportions. Over the past four years, it has infected more than 770 million people worldwide, with confirmed deaths exceeding 7 million [1]. Current treatment strategies for COVID-19 management include a combination of therapies with varying effectiveness. These include oxygen supplementation, antiviral medications like remdesivir and lopinavir/ritonavir, anti-inflammatory drugs such as corticosteroids and tocilizumab, and anticoagulants like heparins [2]. In the early stages of the COVID-19 pandemic, one of the first treatment approaches explored was the passive transfer of anti-SARS-CoV-2 neutralizing antibodies, which involved administering convalescent plasma of blood obtained from individuals who had recently recovered from COVID-19, to patients with severe illness [3, 4]. Despite the widespread use of neutralizing anti-spike monoclonal antibodies for COVID-19 management, the emergence of resistant viral variants and increased pathogenicity/transmissibility in new strains pose challenges. Conversely, COVID-19 convalescent plasma appears to retain clinical effectiveness against these variants due to its extensive array of neutralizing antibodies. This led to a resurgence of interest in its use, particularly in immunocompromised patients who may not elicit a robust antibody response, or in those who experience undesirable reactions to antiviral drugs [5–8]. Prior to the widespread availability of commercially produced vaccines, such as the Pfizer-BioNTech (BNT162b2) or Moderna (mRNA-1273) formulations, and the adoption of antivirals and immunomodulatory drugs, convalescent plasma (CP) therapy emerged as a critical intervention for severe and critical COVID-19 cases. This approach aimed to provide immediate immune support to patients with critical illness manifested by dyspnoea, hypoxemia, respiratory failure, septic shock, or multiple organ dysfunction. Convalescent plasma can exert a dual effect against SARS-CoV-2 infection. Not only does it neutralize the invading virus, but it also provides passive immunomodulatory properties, potentially aiding in regulating the systemic cytokine response triggered by the viral infection [9, 10]. Despite widespread clinical use, the effectiveness of convalescent plasma therapy remains a topic of debate among clinicians and researchers. While observational studies conducted worldwide have shown promising signs of efficacy, there is a lack of robust data from large randomized controlled trials (RCTs) to conclusively confirm the benefits [11]. The current chapter provides a critical analysis of the effectiveness of convalescent plasma therapy against COVID-19 analysing its entire process from donor selection and plasma collection to clinical application and treatment outcomes.

Historical Background

For more than a century, convalescent plasma has been a successful passive vaccination technique for the prevention and treatment of epidemic diseases. Prior to the discovery of monoclonal antibodies and gamma globulin products, immunological or convalescent blood products (such as serum or plasma) from donors or animals that had recovered were used in passive immunization therapy. These were employed as therapeutic agents to treat or prevent particular infections in persons who were at-risk or infected. By transferring pre-formed antibodies, passive antibody injection differs from active immunization therapy (vaccination), which can have different clinical results among patients and requires time to elicit an immune response. This strategy is the only one that works well for giving vulnerable people instant protection. A possible approach was to use it in the wake of the most recent COVID-19 epidemic. CP was initially used during the Spanish influenza A (H1N1) pneumonia pandemic of 1918–1920, according to most accounts. During that time, a number of studies suggested that convalescent plasma might be an effective way to lower the risk of mortality. A meta-analysis carried out by Luke et al. in 2006, which examined data from 1703 patients across eight pertinent papers, later confirmed this. The use of serum from convalescent patients dates back many years, even before the Spanish influenza pandemic. For example, it was investigated as a medication for acute paralysis during the 1916 New York poliomyelitis outbreak. In 1916, Nicolle and Conseil employed serotherapy once more to contain a small-scale measles outbreak in Tunis. Hess employed the same treatment strategy in 1915 to cure mumps and avoid its consequences related to the testicles. Lastly, it is believed that Francesco Cenci of Italy was the first to employ convalescent serum as a therapeutic method to shield children from contracting measles [3, 12–15].

The Principle of Convalescent Plasma Therapy

In short, convalescent plasma is a form of passive antibody therapy that uses the antibodies specific to the pathogen from a patient who has recovered to treat or prevent disease. Unlike immunizations and monoclonal antibodies, convalescent plasma therapy only requires a traditional blood collection infrastructure to collect and distribute convalescent plasma, as well as the availability of disease survivors who are willing to donate plasma [16, 17]. Since there were no known vaccines or monoclonal antibodies at the beginning of the COVID-19 pandemic, convalescent plasma was an easily deployable option. CP is also widely used in low-resource environments worldwide because of this. CP must follow the three main tenets of passive antibody treatment in order to neutralize SARS-CoV-2 and provide a therapeutic benefit. The requirements for convalescent plasma are as follows [3, 17, 18]:

1. It must have particular antibodies against the pathogen, the SARS-CoV-2 virus.
2. It must have enough anti-SARS-CoV-2 antibody.
3. It must be transfused either as a preventative measure or early in the course of the illness.

Patients receiving convalescent plasma therapy mostly benefit from immunity or therapeutic effects due to neutralizing antibodies that target a particular infectious pathogen. Through complement activation, antibody-dependent phagocytosis, and antibody-dependent cellular toxicity, these antibodies stop the infection from invading host cells and aid in its removal. Additional bioactive compounds, including as pentraxin, natural antibodies, anti-inflammatory cytokines, endosomes, defensins, and other agents, may further reduce the burden of disease in the convalescent plasma [19]. CP also combats infections through direct and indirect cellular and humoral immune pathways. The major goals of convalescent plasma therapy are to enhance the course of the disease, decrease its rate of advancement, stimulate or fortify the humoral immune response, and reduce the likelihood of a cytokine storm [18–21].

Mechanisms of Action of Convalescent Plasma in COVID-19

The spike protein on the SARS-CoV-2 virus acts like a key, unlocking the door to enter human cells. Neutralizing antibodies produced by the body in response to infection target a specific region on this spike protein, the receptor-binding domain (RBD). In vitro studies utilizing animal models have demonstrated that plasma derived from individuals immunized against the receptor-binding domain (RBD) of SARS-CoV can inhibit SARS-CoV-2 infection of Vero 6 cells [22]. This finding aligns with previous observations made with SARS-CoV and MERS coronaviruses. Just like with SARS-CoV-2, antibodies that target the spot on the spike protein that binds to receptors were able to block the virus from entering cells and multiplying [23].

While convalescent plasma shares the basic building blocks of regular plasma (water, proteins, and salts), it has a key advantage: specific antibodies (immunoglobulins) tailored to fight a particular infection. These antibodies act like specialized warriors, targeting the virus in multiple ways to suppress its spread in the bloodstream (viremia). Here are some of their battle strategies:

Blocking Entry: Antibodies have the ability to bind to certain proteins on the surface of viruses, such as glycoproteins, and so function as a barrier to stop the virus from adhering and infecting human cells.

Fusion Blockade: Some antibodies might hinder the virus's ability to fuse with endosomes, cellular compartments where it normally sheds its outer shell to gain access to the cell's machinery.

Offspring Blockade: Antibodies can target the virus within infected cells, hindering the production and release of new viral particles (progeny virions).

Protein Disruption: Certain antibodies might interfere with enzymes outside the cell that help the virus cleave (cut) its proteins, potentially disrupting its ability to function properly.

By employing these diverse strategies, convalescent plasma containing the right antibodies can significantly weaken the virus and aid the patient's immune response [9, 24].

CP contains not only neutralizing antibodies (NAbs) that block the virus, but also other protective antibodies. These include IgG and IgM, which are different types of immune globulins. Even if they do not directly stop the virus from multiplying, these non-neutralizing antibodies can still play a role in preventing infection (prophylaxis) or speeding up recovery. For example, in SARS-CoV infections, the body starts producing IgG antibodies targeting the virus's nucleoprotein (N) as early as 4 days after symptoms appear. By day 14, most patients have detectable levels of these antibodies (seroconversion). Studies have shown that 2 years after recovering from SARS, 89% of patients still have both IgG antibodies and NAbs in their blood. IgM antibodies, another type involved in the early immune response, peak around day 9 of illness. Soon after, in the second week, the body switches to producing more IgG antibodies [25, 26].

Convalescent Plasma Donor Selection and Collection

The selection criteria for CP donors for COVID-19 treatment prefer healthy adults with complete recovery from the infection, confirmed by a minimum asymptomatic period of 14 days. This timeframe is consistent with estimations of antibody development and the potential infectious window. The obligatory neutralizing antibody test is essential for the initial screening of donors. Ideally, selection is based on the level of these antibodies, measured using the plaque reduction neutralization test (PRNT). However, due to the time and resource limitations of PRNT, enzyme-linked immunosorbent assays (ELISAs) targeting the virus's receptor-binding domain (RBD) have been used as an alternative in several studies, which has been supported by reports suggesting a good correlation between ELISA ratios/indexes and PRNT titres. To minimize the potential impact of mutations in circulating viral strains, prioritizing CP donors from the same geographic region as the intended recipient(s) is recommended. In theory, the selection of CP donors should prioritize individuals with high neutralizing antibody (NAb) titres to maximize therapeutic efficacy. However, during urgent scenarios, expediency may necessitate accepting donors without high-titre NAb screening or utilizing low-titre donations. Once immediate needs are met, subsequent collections should strategically target high-titre donors to replenish stock for optimal treatment [27, 28].

In India, the potential CP donor must meet the usual criteria for blood donation as outlined in the Drugs and Cosmetics Rules (1945) and the National Blood Transfusion Council (NBTC) COVID-19 guidelines, and also necessitates informed consent. While whole blood donation is an option in resource-limited settings, plasmapheresis is preferred due to the larger plasma volume collected. To ensure safety and extended shelf life, donated plasma undergoes rapid freezing (ideally −40 °C to −80 °C) within 8–24 h of collection. Storage duration can reach up to 12 months. Additional post-donation processing options exist to further enhance safety and quality. These options include pathogen inactivation techniques like photochemical inactivation or solvent detergent treatment, and pooling plasma from multiple donors [28].

Case Studies and Real-World Applications

Early reports on CP therapy for COVID-19 emerged during the initial stages of the pandemic. Case reports, case series, and larger non-randomized studies originating from China, Italy, and other regions documented its use. These studies consistently described the administration of CP collected from donors who had recovered from a confirmed SARS-CoV-2 infection to patients with a positive clinical diagnosis. However, significant heterogeneity was observed across studies regarding key characteristics. This variation included donor and product features, administered dose (both volume and neutralizing antibody titre), dosage timing, and recipient characteristics, particularly disease severity. The following summarizes the critical aspects of each included study:

A prospective case series from Shenzhen, China, published in March 2020 by Shen et al. [29], investigated the use of CP in critically ill COVID-19 patients with ARDS. Five patients received 400 ml of CP within 10–22 days of ICU admission. The study reported positive outcomes, including improved clinical parameters, reduced viral load, and increased antibody titres, with no fatalities at 60 days. Zhang et al. (2020) described positive clinical and radiological outcomes in a small study of four patients with COVID-19 who received CP [30]. All patients demonstrated remission, with a median duration of treatment of 19 days and a median dose of 300 ml per dose. Notably, the study included a 73-year-old man with chronic renal failure and a 31-year-old pregnant woman at 35 + 2 weeks of gestation. In Korea, a study by Ahn et al. (2020) found that two critically ill COVID-19 patients with severe pneumonia received CP and demonstrated positive clinical outcomes [31]. Both patients showed significant clinical improvement, including improved oxygenation (Pao2/FiO2 ratio) and a decrease in inflammatory markers. Additionally, both patients survived the study period. A study conducted by Duan et al. in China investigated the effects of CP in ten COVID-19 patients. Following a single 200 mL dose, patients exhibited improvements in clinical presentation and several parameters such as lymphocyte count and C-reactive protein (CRP). Chest X-ray findings showed variable degrees of improvement. Notably, seven out of ten patients achieved viral clearance. The median time for CP administration in this study was 16.5 days after the first symptoms appeared [32]. Ye et al. (2020) reported a case series investigating CP in six patients with moderate to severe COVID-19, conducted in China [33]. Notably, patients selected in this study did not require intensive care unit admission. Following CP infusion, all patients demonstrated clinical improvement and resolution of radiological observations. Additionally, 75% patients achieved negative viral load, and no mortality or adverse events were noted. A retrospective observational study by Zeng et al. [34] explored the putative benefits of CP therapy in critically ill COVID-19 patients requiring intensive care admission. The study included 21 patients, with six receiving 300 ml of CP after a median of 21.5 days. While mortality rates were high in both groups (83.3% in the CP group and 93.3% in the control group), a noteworthy finding emerged: all patients who received CP achieved viral clearance compared to only 26.7% in the control group. The study summarized that CP may be efficacious in eliminating the virus, but did not appear to have a significant impact on mortality in critically ill patients. The inclusion of a comparison arm allowed for a more robust assessment of the effectiveness of CP in COVID-19.

Later investigations extended the initial small-scale case studies by recruiting larger cohorts in single-arm studies and randomized controlled trials. These studies, conducted in the United States, Italy, Israel, Spain and other countries, have substantiated the concept of CP therapy for COVID-19 and demonstrated its potential and safety. Single-arm intervention studies conducted across various centres in the USA, Italy, and Israel reported positive clinical outcomes in patients receiving early CP therapy, including improved discharge rates and reduced mortality. However, the lack of standardization in donor characteristics, antibody titres, dosing regimens and timing of administration restricted the extent to which these results could be generalized [35].

In a seminal study by Li et al. (2020) [36], the first randomized controlled trial (RCT) investigating CP therapy for COVID-19 was conducted in Wuhan, China. The study included 103 severely ill adults who received either CP with standard care or standard care alone. While the trial was stopped early due to low enrollment, it revealed a significantly higher viral clearance rate in the CP group compared to controls. However, the study was limited by a lack of statistical significance, lower doses of CP than in later studies, and a long time lag between symptom onset and plasma administration. Rasheed et al. (2020) [37] found that in critically ill COVID-19 patients in Iran, the addition of CP (400 mL) to standard care significantly reduced mortality (4.8% vs. 28.5% in controls) and shortened the course of the disease. Another RCT from Netherlands by Gharbharan et al. (2020) [38] terminated early due to high pre-existing neutralizing antibody titres in patients (79% comparable to CP donors), enrolled only 86 participants (20% of target) and found no significant differences in mortality, hospital stay, or disease severity at day 15, despite being underpowered. Similar results with no significant differences in primary and secondary outcomes were also observed and reported in an RCT performed in Spain by Avendano-Sola et al. (2020) [39]. In support of the potential benefits of CP therapy, two separate randomized controlled trials were conducted in India. Agarwal et al. (2020) [40] investigated the efficacy of CP therapy in an open-label multicentre RCT. The study enrolled 464 hospitalized adults in India, randomizing them to receive either two 200 mL doses of CP (median neutralizing antibody titre 1:40) or standard care alone. While CP did not significantly reduce mortality (13.6% vs. 14.6%) or prevent severe disease progression, it demonstrated benefits in symptom resolution, oxygen reduction, and viral clearance. However, subgroup analysis revealed no advantage for CP use within three days of symptom onset or with higher titre antibodies (>1:80). An open-label, single-centre RCT by Ray et al. (2020) [41] investigated CP therapy in moderate to severely ill COVID-19 patients ($n = 80$) in Kolkata, India. While CP significantly reduced systemic cytokine abundance, it did not yield clinically significant improvements in mortality, oxygen support duration, or hospital discharge period. Subgroup analysis in patients younger than 67 years suggested potential benefits, including reduced mortality and shorter duration of hypoxia. Other than these two, RCTs from Argentina have been reported. Simonovich et al. (2020) [42] conducted an open-label, multicentre RCT in Argentina ($n = 333$) to assess CP therapy in mild-to-moderate COVID-19 patients. Patients were randomized 2:1 to receive CP ($n = 208$) or placebo ($n = 105$). The study found no significant differences in mortality or clinical outcomes at day 30 according to the WHO ordinal scale. However, a substantial proportion (46.5%) of patients lacked detectable baseline antibodies, and the median titre was low (1:50). Libster et al. (2021) [43] conducted a multicentre, double-blind, placebo-controlled trial ($n = 160$) to evaluate CP for mild COVID-19 in elderly patients (>75 years). A single 250 mL unit of high-titre CP was administered within 72 h of diagnosis, but recruitment was halted at 76% due to logistical challenges. The trial showed a trend towards reduced progression to severe disease (16% CP vs. 31% placebo), with a potential dose-dependent effect.

Anti-inflammatory Role of Convalescent Plasma

Initial case studies and reports suggested the potential benefits of convalescent plasma (CP) infusion in treating severe COVID-19 patients [29–34]. However, larger-scale randomized controlled trials failed to substantiate these early findings [36–44]. As alternative therapeutic strategies for COVID-19 emerged, the initial promise of CP waned. As a result, on 8 February 2022, the Infectious Diseases Society of America (IDSA) issued a statement firmly advising against the use of convalescent plasma in hospitalized COVID-19 patients [45].

Limited research suggests a potential immediate impact of convalescent plasma (CP) infusion on severe COVID-19 patients. Casadevall and colleagues demonstrated a strong inverse correlation between CP utilization and subsequent mortality. They postulated that the efficacy of CP therapy might be contingent on early administration during the course of infection. In a multicentre RCT co-authored by Casadevall, involving patients with COVID-19 symptoms onset within eight days, it was demonstrated that early administration of high-titre SARS-CoV-2 convalescent plasma reduced hospitalization rates by 54% over a 28-day period compared to a control group receiving plasma from COVID-19-negative donors [46, 47]. A concurrent study conducted post-completion of an RCT in Kolkata yielded comparable findings. Analysis of circulatory cytokine profiles among patient COVID-19 cohorts revealed a marked reduction in several key cytokines predominantly within the CP therapy treatment group, suggesting its potential anti-inflammatory properties.

Further exploratory analyses identified a significant clinical benefit in younger (<67 years) patients experiencing severe COVID-19. Apolipoprotein D and members of the SERPIN family, specifically SERPINA1 and heparin cofactor 2, were identified as key anti-inflammatory proteins within the convalescent plasma proteome, exhibiting a widespread cytokine-attenuating effect [41, 48, 49].

These studies collectively conclude that CP therapy effectively mitigates severe inflammation by directly addressing its underlying cause: an overactive immune response. Interestingly, the therapy's efficacy is not primarily attributed to its antiviral antibodies. Rather, CPT harnesses a complex interplay of immune components, including general antibodies, complement proteins, and clotting factors, to attenuate excessive inflammatory responses. By suppressing the release of harmful cytokines, key drivers of the detrimental "cytokine storm" associated with severe COVID-19, CP therapy safeguards vital organs from damage.

Challenges of Convalescent Plasma Therapy in COVID-19

CP has been widely used to enhance survival rates in patients during various viral epidemics, including SARS, MERS, influenza, and Ebola virus disease. Recently, some COVID-19 patients in China have also shown improvement with CP therapy. However, to fully capitalize on this promising treatment, several critical issues still need to be addressed. According to studies, viremias like SARS often peak during the first week of illness. Patients typically experience an immunological response in the second week following the onset of symptoms, which increases the risk of a potentially fatal cytokine storm. Based on the premise that CP therapy is passive immunity by giving patients pathogen-specific antibodies, CP infusion still carries a latent risk of exacerbating hyperimmune attacks, despite one study suggesting that CP therapy may reduce serum cytokine response with uncertain implementation time. The exact therapeutic window for convalescent plasma's antiviral effects is unknown, but convalescent plasma has likely limited efficacy in the later stages of COVID-19, such as in patients requiring mechanical ventilation. Some studies have found no benefit from convalescent plasma in patients with advanced COVID-19 respiratory failure, attributing this to an excessive inflammatory response [36–44]. These findings should not be surprising and may be seen as evidence of the limitations of the therapeutic window for convalescent plasma, rather than as evidence of its ineffectiveness.

The optimal therapeutic dose of convalescent plasma that effectively neutralizes the virus and improves the clinical condition of recipients is currently unknown and a complex concept. Moreover, there is no established "gold standard" serological assay for detecting SARS-CoV-2 antibodies, and the assays being used are still under investigation. As a result, some studies may not have observed a clinical benefit from convalescent plasma due to low antibody levels. Additionally, variations in antibody levels and the different assays used between trials may contribute to the inconsistent outcomes observed in different studies.

CP therapy also comes with risks. Besides the typical side effects of transfusions like chills, fever, and allergic reactions, it can also lead to more serious issues like lung injury, fluid overload, and red blood cell breakdown [50]. There's also a small but concerning risk of catching infections like HIV, hepatitis B and C, or syphilis from the plasma itself. Therefore, careful consideration and proper use of CP therapy are crucial.

Conclusion

The current findings on CP therapy in COVID-19 present a complex picture. While RCTs have largely failed to show a survival benefit in severe COVID-19 patients, some trials have highlighted clinical improvements such as faster viral resolution and reduced oxygen requirements following transfusion. However, when the WHO published updated guidelines advising against CPT for COVID-19 patients, Casadevall and colleagues submitted a formal request for reconsideration. They contended that the majority of earlier RCTs involved patients in advanced disease stages and CP was administered in later disease stages, potentially limiting its impact on clinical outcomes. Their findings, in contrast, suggested that early CP administration in patients with high SARS-CoV-2 antibody titres correlated with accelerated hospital discharge and reduced mortality rates [46]. They also highlighted a substantial reduction in hospitalization rates among immunocompetent outpatients and the potential benefits of CP in immunosuppressed individuals who often exhibit inadequate antibody responses following vaccination or infection [7, 8]. These observations, together with the historical use of convalescent plasma in other outbreaks, support its continued exploration as a potential armament against infectious diseases. Despite recognized limitations and uncertainties, CP remains a readily accessible therapeutic option in regions with established blood banking infrastructure, especially those lacking financial resources for monoclonal antibodies or antiviral drugs.

However, a crucial gap remains, the lack of standardized regulatory guidance for future pandemic preparedness on the implementation of CP and the design of RCTs. Poor trial design in the initial stages of the COVID-19 pandemic, exemplified by inadequate neutralizing antibody titres in CP units compared to recipients, highlights the importance of avoiding such shortcomings in the future. CP can serve as a valuable interim therapy until definitive vaccine solutions are developed. The lessons learned from the COVID-19 pandemic may provide a valuable model for future responses to novel disease outbreaks.

Author's Contributions R. D. and J. S. contributed equally to writing the manuscript, D. G. supervised and edited the manuscript.

Funding The authors acknowledge grant no. MLP129 from Council of Scientific and Industrial Research, India, for this manuscript.

Data Availability Not applicable.

Declarations

Conflicts of Interests Authors declare absence of any conflict of interest.

References

1. WHO. Coronavirus disease (COVID-19) pandemic. https://www.who.int/emergencies/diseases/novel-coronavirus-2019. Accessed on 20 June 2024.
2. Yuan Y, Jiao B, Qu L, Yang D, Liu R. The development of COVID-19 treatment. Front Immunol. 2023;14:1125246. Published 26 Jan 2023
3. Casadevall A, Pirofski LA. The convalescent sera option for containing COVID-19. J Clin Invest. 2020;130(4):1545–8.
4. Franchini M, Corsini F, Focosi D, Cruciani M. Safety and efficacy of convalescent plasma in COVID-19: an overview of systematic reviews. Diagnostics (Basel). 2021;11(9):1663. Published 11 Sept 2021
5. Focosi D, McConnell S, Casadevall A, Cappello E, Valdiserra G, Tuccori M. Monoclonal antibody therapies against SARS-CoV-2 [published correction appears in Lancet Infect Dis. 2022;22(9):e239]. Lancet Infect Dis. 2022;22(11):e311–26. https://doi.org/10.1016/S1473-3099(22)00468-6.
6. Senefeld JW, Franchini M, Mengoli C, et al. COVID-19 convalescent plasma for the treatment of immunocompromised patients: a systematic review and meta-analysis. JAMA Netw Open. 2023;6(1):e2250647. Published 3 Jan 2023
7. Focosi D, Franchini M. Potential use of convalescent plasma for SARS-CoV-2 prophylaxis and treatment in immunocompromised and vulnerable populations. Expert Rev Vaccines. 2022;21(7):877–84.
8. Senefeld JW, Klassen SA, Ford SK, et al. Use of convalescent plasma in COVID-19 patients with immunosuppression. Transfusion. 2021;61(8):2503–11.
9. Rojas M, Rodríguez Y, Monsalve DM, et al. Convalescent plasma in Covid-19: possible mechanisms of action. Autoimmun Rev. 2020;19(7):102554.
10. Noor R, Tasnim N, Saha C. COVID-19 pandemic and the convalescent plasma therapy: possible benefits and risks. Curr Clin Microbiol Rep. 2021;8(3):194–8.
11. Casadevall A, Grossman BJ, Henderson JP, et al. The assessment of convalescent plasma efficacy against COVID-19. Med. 2020;1(1):66–77.
12. Brown BL, McCullough J. Treatment for emerging viruses: convalescent plasma and COVID-19. Transfus Apher Sci. 2020;
13. Cao H, Shi Y. Convalescent plasma: possible therapy for novel coronavirus disease 2019. Transfusion. 2020;60:1078–83.
14. Tiberghien P, de Lamballiere X, Morel P, Gallian P, Lacombe K, Yazdanpanah Y. Collecting and evaluating convalescent plasma for COVID-19 treatment: why and how? Vox Sang; 2020.
15. Ripoll JG, van Helmond N, Senefeld JW, et al. Convalescent plasma for infectious diseases: historical framework and use in COVID-19. Clin Microbiol Newsl. 2021;43(4):23–32.
16. Casadevall A, Scharff MD. Return to the past: the case for antibodybased therapies in infectious diseases. Clin Infect Dis. 1995;21:150–61.
17. Bloch EM, Shoham S, Casadevall A, Sachais BS, Shaz B, Winters JL, et al. Deployment of convalescent plasma for the prevention and treatment of COVID-19. J Clin Invest. 2020;130:2757–65.
18. Casadevall A, Pirofski L, Joyner MJ. The principles of antibody therapy for infectious diseases with relevance for COVID-19. MBio. 2021:12.

19. Garraud O, Heshmati F, Pozzetto B, Lefrere F, Girot R, Saillol A, et al. Plasma therapy against infectious pathogens, as of yesterday, today and tomorrow. Transfus Clin Biol. 2016;23:39–44.
20. Naranjo-Gomez M, Lambour J, Piechaczyk M, Pelegrin M. Neutrophils are essential for induction of vaccine-like effects by antiviral monoclonal antibody immunotherapies. JCI. Insight. 2018:3.
21. Nasser R, Pelegrin M, Plays M, Gros L, Piechaczyk M. Control of regulatory T cells is necessary for vaccine-like effects of antiviral immunotherapy by monoclonal antibodies. Blood. 2013;121:1102–11.
22. Walls AC, Park YJ, Tortorici MA, Wall A, McGuire AT, Veesler D. Structure, function, and antigenicity of the SARS-CoV-2 spike glycoprotein. Cell. 2020;181:281–92.
23. Du L, He Y, Zhou Y, Liu S, Zheng B-J, Jiang S. The spike protein of SARS-CoV–a target for vaccine and therapeutic development. Nat Rev Microbiol. 2009;7:226–36.
24. Franchini M, Glingani C, Liumbruno GM. Potential mechanisms of action of convalescent plasma in COVID-19. Diagnosis (Berl). 2021;8(4):413–20.
25. Hsueh P-R, Huang L-M, Chen P-J, Kao C-L, Yang P-C. Chronological evolution of IgM, IgA, IgG and neutralisation antibodies after infection with SARS-associated coronavirus. Clin Microbiol Infect. 2004;10:1062–6.
26. Gorse GJ, Donovan MM, Patel GB. Antibodies to coronaviruses are higher in older compared with younger adults and binding antibodies are more sensitive than neutralizing antibodies in identifying coronavirus-associated illnesses. J Med Virol. 2020;92:512–7.
27. Franchini M, Marano G, Velati C, Pati I, Pupella S, Maria LG. Operational protocol for donation of anti-COVID-19 convalescent plasma in Italy. Vox Sang. 2021;116(1):136–7.
28. Focosi D, Anderson AO, Tang JW, Tuccori M. Convalescent plasma therapy for COVID-19: state of the art. Clin Microbiol Rev. 2020;33(4):e00072-20. Published 12 Aug 2020
29. Shen C, Wang Z, Zhao F, et al. Treatment of 5 critically ill patients with COVID-19 with convalescent plasma. JAMA. 2020;323(16):1582–9.
30. Zhang B, Liu S, Tan T, et al. Treatment with convalescent plasma for critically ill patients with severe acute respiratory syndrome coronavirus 2 infection. Chest. 2020;158(1):e9–e13.
31. Ahn JY, Sohn Y, Lee SH, et al. Use of convalescent plasma therapy in Two COVID-19 patients with acute respiratory distress syndrome in Korea. J Korean Med Sci. 2020;35(14):e149. Published 13 Apr 2020
32. Duan K, Liu B, Li C, et al. Effectiveness of convalescent plasma therapy in severe COVID-19 patients. Proc Natl Acad Sci U S A. 2020;117(17):9490–6.
33. Ye M, Fu D, Ren Y, et al. Treatment with convalescent plasma for COVID-19 patients in Wuhan. China. J Med Virol. 2020;92(10):1890–901.
34. Zeng QL, Yu ZJ, Gou JJ, et al. Effect of convalescent plasma therapy on viral shedding and survival in patients with coronavirus disease 2019. J Infect Dis. 2020;222(1):38–43.
35. Wood EM, Estcourt LJ, McQuilten ZK. How should we use convalescent plasma therapies for the management of COVID-19? Blood. 2021;137(12):1573–81.
36. Li L, Zhang W, Hu Y, et al. Effect of convalescent plasma therapy on time to clinical improvement in patients with severe and life-threatening COVID-19: A Randomized Clinical Trial [published correction appears in JAMA. 2020;324(5):519]. JAMA. 2020;324(5):460–70.
37. Rasheed AM, Fatak DF, Hashim HA, et al. The therapeutic potential of convalescent plasma therapy on treating critically-ill COVID-19 patients residing in respiratory care units in hospitals in Baghdad, Iraq. Infez Med. 2020;28(3):357–66.
38. Gharbharan A, Jordans CCE, GeurtsvanKessel C, et al. Effects of potent neutralizing antibodies from convalescent plasma in patients hospitalized for severe SARS-CoV-2 infection. Nat Commun. 2021;12(1):3189. Published 27 May 2021
39. Avendaño-Solá C, Ramos-Martínez A, Muñez-Rubio E, et al. A multicenter randomized open-label clinical trial for convalescent plasma in patients hospitalized with COVID-19 pneumonia. J Clin Invest. 2021;131(20):e152740.
40. Agarwal A, Mukherjee A, Kumar G, et al. Convalescent plasma in the management of moderate covid-19 in adults in India: open label phase II multicentre randomised con-

trolled trial (PLACID Trial) [published correction appears in BMJ. 2020;371:m4232]. BMJ. 2020;371:m3939. https://doi.org/10.1136/bmj.m4232. Published 22 Oct 2020
41. Ray Y, Paul SR, Bandopadhyay P, et al. A phase 2 single center open label randomised control trial for convalescent plasma therapy in patients with severe COVID-19. Nat Commun. 2022;13(1):383. Published 19 Jan 2022
42. Simonovich VA, Burgos Pratx LD, Scibona P, et al. A randomized trial of convalescent plasma in Covid-19 severe pneumonia. N Engl J Med. 2021;384(7):619–29.
43. Libster R, Pérez Marc G, Wappner D, et al. Early high-titer plasma therapy to prevent severe Covid-19 in older adults. N Engl J Med. 2021;384(7):610–8.
44. RECOVERY Collaborative Group. Convalescent plasma in patients admitted to hospital with COVID-19 (RECOVERY): a randomised controlled, open-label, platform trial. Lancet. 2021;397(10289):2049–59.
45. https://www.idsociety.org/globalassets/idsa/practice-guidelines/covid-19/treatment/idsa-covid-19-gl-tx-and-mgmt%2D%2D-convalescent-plasma-2022-02-03.pdf
46. Sullivan DJ, Gebo KA, Shoham S, et al. Early outpatient treatment for Covid-19 with convalescent plasma. N Engl J Med. 2022;386(18):1700–11.
47. Habtehyimer F, Zhu X, Redd AD, et al. COVID-19 convalescent plasma therapy decreases inflammatory cytokines: a randomized controlled trial. Microbiol Spectr. 2024;12(1):e0328623.
48. Bandopadhyay P, D'Rozario R, Lahiri A, et al. Nature and dimensions of systemic hyperinflammation and its attenuation by convalescent plasma in severe COVID-19. J Infect Dis. 2021;224(4):565–74.
49. Raychaudhuri D, Bandopadhyay P, D'Rozario R, et al. Clinical trial subgroup analyses to investigate clinical and immunological outcomes of convalescent plasma therapy in severe COVID-19. Mayo Clin Proc Innov Qual Outcomes. 2022;6(6):511–24.
50. Samad N, Sodunke TE, Banna HA, et al. Convalescent plasma therapy for management of COVID-19: perspectives and deployment in the current global pandemic. Risk Manag Healthc Policy. 2020;13:2707–28. Published 23 Nov 2020

Part IV

COVID-19 Drug Repurposing: Mode of Action, Clinical Trials, and Future Perspectives

Repurposing of Antiviral Agents in COVID-19

13

Shailendra Dwivedi, Vijay Singh, Akash Bansal, Radhieka Misra, Alka Tripathi, and Sanjeev Misra

Abstract

Coronavirus disease (COVID-19) has been caused by the recent Severe Acute Respiratory Syndrome Coronavirus 2 (SARS-CoV-2) outbreak, which has had a global impact on public health. Researchers are diligently striving despite numerous obstacles to develop an efficacious treatment for COVID-19, given the absence of a well-established therapeutic approach. Limited empirical treatment protocols were substantiated by scientific evidence from previous viral epidemics of a similar nature. An investigation into therapeutic approaches, such as the repurposing of pre-existing antiviral agents, has become imperative in light of the COVID-19 pandemic. This chapter provides an analysis of the potential of repurposed pharmaceuticals to combat SARS-CoV-2, with a specific emphasis on their efficacy, safety profiles, and mechanisms of action. Remdesivir, an RNA polymerase inhibitor initially designed for the treatment of the Ebola virus, and favipiravir, a broad-spectrum antiviral utilised against influenza, are among the agents under discussion. Furthermore, this chapter provides an in-depth analysis of the repurposing capabilities of pharmaceuticals such as lopinavir/ritonavir and

S. Dwivedi (✉) · V. Singh · A. Bansal
Department of Biochemistry, All India Institute of Medical Sciences, Gorakhpur, Uttar Pradesh, India

R. Misra
Undergraduate Medical Student, Era's Lucknow Medical College and Hospital, Lucknow, Uttar Pradesh, India

A. Tripathi
Department of Ophthalmology, All India Institute of Medical Sciences, Gorakhpur, Uttar Pradesh, India

S. Misra
Vice Chancellor, Atal Bihari Vajpayee Medical University, Lucknow, Uttar Pradesh, India

M. K. Ghosh, M. Basu (eds.), *Pathogenesis, Management and Socio-Economic Impact of COVID-19 Pandemic*, Medical Virology: From Pathogenesis to Disease Control, https://doi.org/10.1007/978-981-95-1426-7_13

hydroxychloroquine, illuminating the contentious issues that encircle their safety and efficacy in the management of COVID-19. Novel methods including protease inhibitors like Camostat mesylate and molnupiravir, an oral antiviral that promotes deadly RNA mutagenesis, are being investigated to reduce viral proliferation and illness severity. Researchers and medical professionals have been investigating innovative medication prophylaxis and therapy using drug repurposing techniques. Thus, novel and effective medications that may provide a fresh perspective on the management of the current SARS-CoV-2 pandemic have become a critical need. By conducting an extensive examination of these repurposed agents, the objective of this chapter is to offer valuable perspectives on their therapeutic efficacy, obstacles, and potential future developments in the fight against COVID-19.

Keywords

Repurpose drugs · Immunomodulatory · COVID-19 · SARS-CoV-2 · Antiviral drugs · Drug action mechanisms

Introduction

The emergence of the lethal coronavirus was evident from the start of the twenty-first century. The discussion about a viral pandemic began in 2003 with the outbreak of severe acute respiratory syndrome caused by the coronavirus known as SARS-CoV. The virus was quickly identified as a highly contagious pathogen that tends to spread rapidly. The initial incidence was documented in China, and thereafter, the virus spread to encompass five continents, with a recorded death rate of 9.6% during the duration of the pandemic. Due to the limited number of medications available, the primary approach employed was to restrict the spread of the infection, which yielded some degree of success. In 2012, a second epidemic of the Middle East respiratory syndrome-related coronavirus (MERS-CoV) emerged on the Arabian Peninsula, resulting in a death rate of approximately 34.4%. Once again, a prevention approach was implemented to control the disease's spread. Additionally, numerous treatment regimens were tested. As a result, the disease was successfully contained within a limited geographical area. In December 2019, a new and highly dangerous strain of the virus emerged in Wuhan, Hubei Province, China, causing the COVID-19 sickness. The virus rapidly disseminated and caused a global cessation of activities when the World Health Organization (WHO) officially designated COVID-19 as a pandemic on March 11, 2020 [33, 35]. Previously, extensive research was conducted in various fields such as immunology, genetics, medical biotechnology, molecular engineering, nutrition, nanotechnology, and regenerative medicine to address the issues related to the origin, diagnosis, treatment, and management of coronaviruses. However, these efforts were scattered and pursued in diverse directions. Therefore, since there was no definitive explanation, the primary course of action available was to regulate the spread. This time, the significant global impact was due to the extensive commercial aviation sector, which facilitated international connections, and the highly contagious nature of the virus. As a result, before specialised preventive measures could be effectively implemented, the virus had already caused widespread disruption worldwide. Given the current COVID pandemic, it has become crucial to incorporate expertise from other disciplines.

At present, a treatment plan that combines symptomatic medication and supportive measures is helpful in managing mild diseases. However, for moderate-to-severe diseases and patients at high risk, a targeted therapy tailored to the disease is necessary. The de novo drug design process, together with its subsequent approval, is a lengthy and time-consuming endeavour. As a result, an alternative approach is to repurpose or re-profile a medicine that has already been approved for human use. Drug repurposing is a methodical approach in drug discovery that helps identify new uses for existing drugs. Having a well-established safety profile, understanding of pharmacokinetics, knowledge of adverse effects, and awareness of drug interactions significantly decreases the time and cost involved in drug development. This makes it a favourable option, especially during urgent situations such as a pandemic.

We conducted a comprehensive search for relevant scientific articles in PubMed, Google Scholar, and EMBASE using specific keywords related to coronavirus, COVID-19, and SARS-CoV-2, as well as terms related to treatment, therapy, antiviral drugs, and specific medications such as hydroxychloroquine, chloroquine, lopinavir, ritonavir, Arbidol, azithromycin, remdesilvir, oseltamivir, steroids, and interferon. We also included terms related to clinical trials, novel antiviral therapy, and stem cell therapy in the context of COVID-19.We incorporated the studies into our analysis that satisfied the following criteria: (1) Preclinical and clinical studies conducted on verified COVID-19 cases in humans, cell lines, and animal models; (2) Original papers including clinical trials, case series, and case reports; (3) papers that reference a specific therapeutic agent (drug) and its associated clinical or virological outcomes.

Various Repurposed Drugs and Their Characteristics Utilised Against COVID-19

Drug repurposing typically involves two concepts: either a one drug interacts with numerous targets, allowing for the exploration of new target sites of action for the existing compound, or disease-associated targets are often linked to various biological processes involved in the development of the disease. Hence, the quest for repurposing medications has focused on the etiological agents, namely the coronavirus, and typical manifestations encompass inflammation, Acute Respiratory Distress Syndrome (ARDS), among various others [39]. Figure 13.1 summarise the repurpose drugs in COVID 19 patients. Remdesivir has been determined to be a secure and efficient therapeutic choice for expectant ladies diagnosed with COVID-19. A study involving 150 pregnant women with moderate to severe COVID-19 infection found that administering remdesivir at an early stage resulted in improved outcomes. These benefits included reduced hospitalisation duration and decreased rates

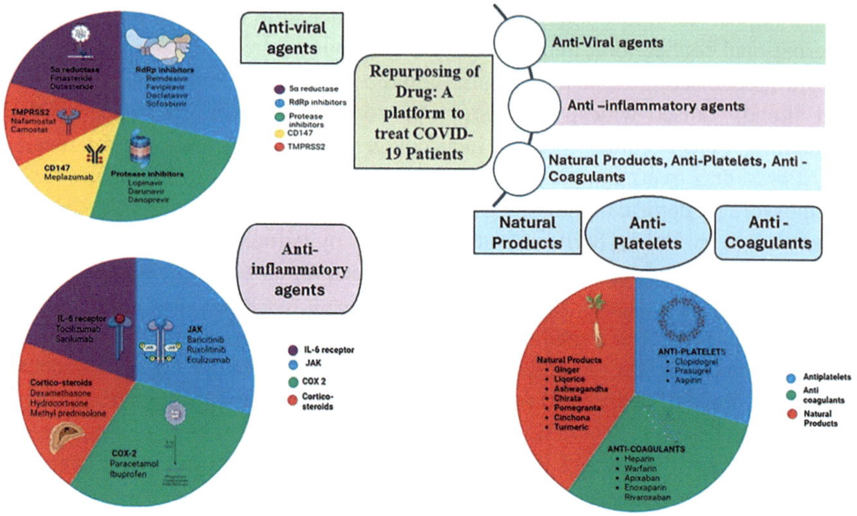

Fig. 13.1 Repurpose drugs in COVID-19

of admission to the intensive care unit (ICU). In addition, the administration of remdesivir was discovered to decrease the likelihood of negative outcomes for both the mother and the newborn, including placental abruption, pregnancy-induced hypertension, and chorioamnionitis. The study indicates that remdesivir can be an advantageous therapy choice for pregnant women who have COVID-19, especially those who have moderate to severe illness [20, 28].

Hydroxychloroquine (HCQ)/Azithromycin (AZ)

Chloroquine (CQ) and its derivative hydroxychloroquine (HCQ) are aminoquinoline chemicals that are utilised for the management of malaria and systemic lupus erythematosus (SLE). Experiments were conducted on human immunodeficiency virus (HIV), hepatitis A, and influenza viruses to gather data on the antiviral capabilities of CQ [17]. Hydroxychloroquine does not competitively block the virus at the ACE-2 inhibitor site. Instead, its antiviral activity is hypothesised to be a result of a pH alteration. As acidity levels rise, the viral spike protein breaks down and loses its ability to attach to ACE-2 receptors or remain viable to infect others. In vitro efficacy of chloroquine (CQ) against SARS-CoV-2 was demonstrated as early as February 2020. It was noted that the recommended dosage of CQ might achieve an effective concentration (EC) 90 value of 6.90 μM. These data were later reinforced by other writers who demonstrated that HCQ is more powerful than CQ, achieving lower EC 50 values. This hypothesis was proposed based on the elevated intracellular concentration of HCQ. A combination therapy of azithromycin, which competes for attachment to ACE-2 receptors on the host cell, was suggested as a prophylactic and treatment option for mild-to-moderate COVID-19 infections, in addition to HCQ [25].

During the early stages of the COVID-19 outbreak in France, two hospitals carried out a clinical trial that was not randomised and had no blinding, which indicated that the combination of hydroxychloroquine (HCQ) and azithromycin (AZ) was effective in treating patients [18]. The intervention arm of the study administered oral HCQ sulphate at a dosage of 200 mg, three times daily for a duration of 10 days. Additionally, AZ was given at a dosage of 500 mg on the first day, followed by 250 mg per day for the subsequent 4 days. This treatment was provided to individuals aged 12 years and older who had confirmed COVID-19 cases as determined by Polymerase Chain Reaction (PCR) testing. On the sixth day after starting treatment, all patients who received a combination of HCQ and AZ were completely cured of the virus, compared to just 57.1% of patients who received HCQ alone, and 12.5% of patients in the control group. This difference was statistically significant ($P < 0.001$). Despite being categorised as a clinical trial, this study appeared to be more of an experimental endeavour with a retrospective investigation of treatment within difficult circumstances. The study had a limited sample size, with a total of 36 participants, divided into 20 in the intervention arm and 16 in the control arm. Although the study was not randomised, it nonetheless offered a preliminary direction for further research and therapy during the initial stages of the pandemic. Another randomised control experiment [9] was undertaken in China, involving 62 patients (31 in each group), Chen et al.'s study evaluated the effectiveness of hydroxychloroquine (HCQ) in treating 62 COVID-19 patients. The patients were divided into a control group and an experimental group. The experimental group received standard care and HCQ at a daily dosage of 400 mg. The study found that HCQ reduced the duration of clinical recovery and improved pneumonia resolution in the experimental group (80.6%) compared to the control group (54.8%). However, two patients in the experimental group experienced moderate adverse effects [10]. Nevertheless, this article did not undergo peer review. A multicenter case series comprising 100 patients from 10 hospitals in China demonstrated that the administration of chloroquine (CQ) phosphate at the recommended therapeutic dosage effectively prevented worsening of pneumonia, reduced lung damage as seen in radiographic imaging, and expedited the clearance of the virus, leading to a shorter duration and less severe course of the disease, with no reported negative side effects [16].

A study on hydroxychloroquine (HCQ) in early COVID-19 patients found no significant difference in symptom severity between HCQ and a placebo. The study, which included 491 adult outpatients from the US and Canada, found no significant difference in symptom severity over a 14-day period. HCQ was also linked to a higher incidence of drug adverse effects. The study's limitations, such as limited access to testing, did not alter the conclusion that HCQ is ineffective as an oral treatment for early COVID-19 in outpatients. [42].

Subsequent research, however, disproved numerous early assertions. A multicenter, open-label, randomised controlled trial [16] was conducted across multiple nations to investigate the effects of hydroxychloroquine (HCQ) with or without a macrolide. This study did not find any substantial advantage of intervention compared to conventional care in terms of intubation, mortality, survival, and poor virological outcome. In several other investigations [21], the HCQ and non-HCQ groups were carefully selected to have similar age and concomitant illnesses, such as hypertension, diabetes, and chronic lung disorders. HCQ groups experienced adverse effects such as arrhythmia, cardiac arrest, other electrocardiogram (ECG) alterations, and diarrhoea. In other investigations, comparable findings indicated a bias in the administration of HCQ to patients with a higher prevalence of comorbidities. A multicentric randomised controlled clinical study (RCT) conducted on the American continent reported treatment failure of hydroxychloroquine (HCQ) among 1309 asymptomatic COVID-19 patients. A study conducted in China found that there was no notable enhancement when using HCQ compared to a placebo. Therefore, the definitive conclusion is that the effectiveness of HCQ or CQ in treating COVID-19 remains uncertain to this day and requires further clinical research for official clearance [2].

Lopinavir/Ritonavir

A protease inhibitor is a substance that inhibits the activity of proteases, enzymes that break down proteins. Lopinavir and ritonavir have been authorised as antiretroviral medications for the treatment of HIV. Lopinavir has been found to exhibit inhibitory effects on both SARS-CoV and Middle East respiratory syndrome coronavirus (MERS-CoV) in a limited number of in vitro investigations. The combination therapy with ritonavir competitively binds to the cytochrome P450-3A4 enzyme, which delays the metabolism of lopinavir and extends its plasma half-life, thus prolonging the inhibitory activity [34, 43].

A trial conducted in Wuhan involved 94 adult patients who were proven to have COVID-19. Out of them, 100 patients in the control group received conventional therapy only. The remaining patients received lopinavir/ritonavir, with a dosage of 400 mg and 100 mg respectively, twice a day for a duration of 14 days. The intervention arm did not show any meaningful advantage compared to the control arm (hazard ratio for median time to clinical improvement: 1.39; 95% CI: 1.00–1.91). A recent study conducted in China found that there was a significant reduction in the duration of viral shedding when lopinavir/ritonavir treatment was initiated within 10 days of the beginning of symptoms. The median duration of viral shedding was 19 days compared to 28.5 days in the control group. This difference was statistically significant with a log-rank P value of less than 0.001. This study is a retrospective analysis conducted by observational methods on a sample of 78 patients [49]. In randomised controlled research conducted in China, Li et al. [50] examined 86 individuals and concluded that there was no discernible therapeutic advantage of LPV/r over umifenovir in patients with mild-to-moderate COVID-19. A limited retrospective case series documented positive outcomes in terms of viral load reduction, radiographic improvement, and decrease in eosinophil count after administering lopinavir. It is worth noting that these patients were also receiving a comprehensive treatment regimen consisting of IFN- α2b atomisation inhalation, Arbidol, methylprednisolone, and immunoglobulin [32].

The LOTUS China trial was a clinical study that compared the effectiveness and safety of lopinavir-ritonavir and conventional therapy in hospitalised adult COVID-19 patients. The trial involved 199 patients with proven SARS-CoV-2 infection and severe respiratory disease, divided into two groups: one group received lopinavir-ritonavir, and the other received conventional therapy alone. The study found no significant difference in the duration of clinical improvement between the two groups. The mortality rate and viral RNA detection frequencies were comparable. Despite lopinavir-ritonavir having a higher incidence of gastrointestinal adverse events, the standard-care group experienced a greater frequency of severe adverse events. The trial concluded that lopinavir-ritonavir did not offer any extra advantage compared to usual therapy for severe COVID-19 patients admitted to the hospital. The study also found that hydroxychloroquine (HCQ) did not significantly reduce the intensity of symptoms compared to a placebo. These findings provide crucial guidance for treatment techniques and emphasise the need for ongoing research to identify effective therapeutics for COVID-19 [6].

In January 2020, the enigma of WHO's ambitious global "mega-trial" called solidarity was resolved. The experiment randomly assigns confirmed cases of COVD-19 to either usual care or one of four active therapy options: remdesivir, chloroquine or hydroxychloroquine, lopinavir/ritonavir, or lopinavir/ritonavir with interferon beta-1a. The administration of hydroxychloroquine, chloroquine, or lopinavir/ritonavir to hospitalised patients was stopped in early July 2020. The interim results, published in mid-October 2020, indicated that the four repurposed antiviral agents mentioned earlier had minimal or no impact on COVID-19 patients who were hospitalised. This was determined by assessing overall mortality, the need for ventilation, and the length of hospital stay [26]. Due to insufficient proof and pharmacodynamics that are potentially unfavourable, the National Institutes of Health (NIH) Panel on COVID-19 Treatment Guidelines and the Infectious Diseases Society of America (IDSA) guidelines do not endorse the use of lopinavir/ritonavir or any other HIV protease inhibitors in patients with COVID-19. The IDSA further notes that it has the property of inhibiting CYP3A, which can lead to severe cutaneous responses, QT prolongation, and the possibility of medication interactions. Currently, there are ongoing studies investigating the potential benefits of combining methylprednisolone or umifenovir with LPV/r, which have lately shown promising results in improving treatment outcomes [2].

Umifenovir (Arbidol) Versus Favipiravir

Umifenovir is a potent antiviral medication that specifically targets the hemagglutinin protein. It is currently employed in both China and Russia for the treatment of influenza. Vankadari's studies on structural and molecular dynamics confirmed that the spike glycoproteins of SARS-CoV-2, like those of H3N2, serve as the therapeutic target for umifenovir [38, 45]. A retrospective study conducted in China examined 81 non-ICU hospitalised patients with COVID-19. The study found no evidence of improvement in viral clearance or prognosis. A separate trial ($n = 86$) conducted to assess the effectiveness of lopinavir/ritonavir or umifenovir monotherapy with conventional care in patients with mild-to-moderate COVID-19 found no significant variation between the treatment groups [2]. Favipiravir, also known as Avigan and developed by Appili Therapeutics, is an orally administered antiviral medication that has been authorised for the treatment of influenza in Japan. In Russia, it has received approval for the treatment of COVID-19. Favipiravir specifically hinders the activity of RNA polymerase, an essential enzyme for the reproduction of the virus. A clinical trial was undertaken in Russia to compare the efficacy of favipiravir with the standard of therapy in hospitalised patients with mild COVID-19. The trial was adaptive, multicenter, open-label, randomised, and conducted in phases 2 and 3. Both dosage regimens of favipiravir had comparable virologic response. On Day 5, viral clearance was achieved in 62.5% of patients (25/40) in the favipiravir group, compared to 30% of patients (6/20) in the usual treatment group ($p = 0.018$). On Day 10, 92.5% of patients using favipiravir achieved viral clearance, compared to 80% in the standard treatment group ($p = 0.155$) [2].

A clinical experiment was conducted to examine the effectiveness of umifenovir (Arbidol) with favipiravir in treating confirmed instances of COVID-19 in 240 adult patients. The trial was prospective, randomised, controlled, open-label, and multicentric. Each patient was administered either umifenovir (Arbidol) at a dosage of 200 mg three times a day or favipiravir at a dosage of 1600 mg twice a day on the first day, followed by a dosage of 600 mg twice a day for a duration of 10 days. On the seventh day of observation, there was no significant difference in the rate of clinical recovery between the group treated with favipiravir (61.2%) and the group treated with Arbidol (51.6%). Favipiravir was linked to reduced time for fever recovery (1.70 days difference, $P < 0.0001$) and cough recovery (1.75 days difference, $P < 0.0001$). No significant differences were observed in the need for auxiliary oxygen or mechanical ventilation. Favipiravir's most common side effect was increased serum uric acid (13.8%, $p = 0.0014$). Favipiravir did not improve the 7-day recovery rate but accelerated symptom relief with manageable side effects [6].

An increase in the concentration of serum uric acid was also observed in the group treated with favipiravir. However, both medications have been determined to be equally effective thus far. Therefore, a clinical decision can be made by optimising the patient's illness condition and comorbidity when choosing between the two options [8].

Remdesivir

Remdesivir is a broad-spectrum antiviral drug that works by inhibiting RNA polymerases. It is a nucleoside analogue that was initially created to treat Ebola but has also shown effectiveness against other viruses such as filoviruses, paramyxoviruses, pneumoviruses, and coronaviruses [15]. This medicine has undergone testing through both in vitro and in vivo trials, namely using mice and Rhesus monkeys, to evaluate its effectiveness against SARS-CoV-2. In addition, clinical experiments have been conducted on adults and children infected with SARS-CoV-2 using various dosage levels, and they have shown minimal harm. In addition, double-blind, randomised, multicenter clinical trials have shown a notable decrease in viral load during the illness. However, there was no significant decrease in the death rate when compared to patients who received a placebo during the same time frame. Furthermore, in vitro cell culture Vero-E6 experiments demonstrated that this medication had antiviral effects during the post-entry phase of the cells, with an EC50 value of 1.76 μM, as shown in EC50 [4].

The Food and Drug Administration (FDA) has approved remdesivir (Veklury) for hospitalised COVID-19 patients after evaluating evidence from three randomised controlled trials: ACTT-1 and SIMPLE. The ACTT-1 trial showed Veklury reduced recovery time from 15 days to 10 days and improved clinical results by Day 15. The SIMPLE trials confirmed the benefits of 5-day and 10-day treatment durations for moderate and severe cases. The 5-day group showed significant improvements for moderate cases. The FDA's decision ensures the safety and effectiveness of Veklury, but does not address medicine pricing [41]. Remdesivir may be considered for the treatment of patients with mild-to-moderate COVID-19 disease. In contrast, the earliest trials of remdesivir did not show any therapeutic advantage compared to a placebo. Specifically, in several studies, when patients were given a 10-day course of remdesivir, there was no statistically significant improvement in their health compared to those receiving normal care through random selection. Nevertheless, individuals who underwent a 5-day treatment regimen demonstrated enhanced results compared to those who received normal care. However, the precise clinical significance remained ambiguous. However, a study conducted by Grein et al. [21] revealed that 68% of the 61 patients with severe COVID-19 who participated in the study experienced notable improvement when treated with remdesivir [21]. A recent study undertaken by the National Institutes of Health (NIH) in the United States, involving 1062 COVID-19 patients, demonstrated promising findings. These findings included a decrease in the time it took for patients to recover and a tendency towards reduced fatality rates. As a consequence, the US Food and Drug Administration granted emergency use and authorisation, while the European Medicines Agency and the National Health Service in the United Kingdom provided their endorsement. Recent research has led to a change in WHO guidelines, which no longer suggest the use of remdesivir in hospitalised patients with COVID-19. According to the reports, remdesivir has shown no reduction in death rates, the requirement for mechanical ventilation, or the length of hospital stay. Therefore, additional extensive randomised controlled trials (RCTs) are necessary to provide a more comprehensive understanding of the role of remdesivir in the treatment of individuals with COVID-19 [2].

Remdesivir in Children

The paediatric dose for remdesivir emergency use authorisation was determined based on pharmacokinetic data obtained from healthy people. Since February 2020, remdesivir has been accessible for compassionate use in children with severe cases of COVID-19. In June 2020, a Phase 2/3 trial called CARAVAN was started to evaluate the safety, tolerability, pharmacokinetics, and effectiveness of remdesivir in children with moderate-to-severe COVID-19. CARAVAN is a clinical trial that investigates the use of remdesivir in children aged from birth to 18 years. It is an open-label study, meaning that all participants will receive the treatment and there will be no placebo group. The findings were unveiled on July 10–11, 2020, demonstrating enhancements in the majority of the 77 children afflicted with severe COVID-19. 80% of children receiving ventilator or ECMO treatment and 87% of those not requiring invasive oxygen support demonstrated clinical recovery [5].

Remdesivir Use in Pregnant Women

A study has been conducted on the results of the first 86 pregnant women who underwent treatment with remdesivir between March 21 and June 16, 2020. The results showed that the rates of recovery were high among pregnant women who received remdesivir, with 67 recovering while pregnant and 19 recovering within the first three days after giving birth. No new safety concerns were identified. Initially, 40% of pregnant women (with a median gestational age of 28 weeks) required invasive ventilation, while 95% of postpartum women (with a median gestational age at delivery of 30 weeks) needed this intervention. Among those who required mechanical ventilation, 93% of pregnant women and 89% of postpartum women were successfully extubated and recovered [5]. Remdesivir is a safe and effective treatment option for pregnant women with COVID-19. A study involving 150 women found that early administration of remdesivir reduced hospitalisation duration and ICU admissions. It also reduced the risk of negative outcomes for both the mother and newborn, such as placental abruption, pregnancy-induced hypertension, and chorioamnionitis. This suggests remdesivir as an advantageous therapy choice for pregnant women with moderate-to-severe COVID-19 [22].

Molnupiravir

Molnupiravir, also known as MK-4482 (formerly EIDD-2801), is an orally administered antiviral medication that acts as a prodrug of the nucleoside derivative N4-hydroxycytidine. It induces antiviral effects by causing mistakes in the replication process of viral RNA in the SARS-CoV-2 virus [24]. The initial findings of the Phase 2 dose-ranging MOVe-OUT study ($n = 2020$) revealed that, on average, 24% of patients in the placebo group were still carrying infectious SARS-CoV-2 10 days after their first episode of signs and symptoms. In contrast, all outpatients treated with balapiravir were able to recover non-infectious virus as early as day 5. The inpatient study of molnupiravir (MOVe-IN) has been stopped, but the Phase 3 trial for outpatients with at least one risk factor for negative outcomes (such as advanced age, obesity, and diabetes) is continuing. Patients in this trial are getting 800 mg of the drug orally twice daily. We are now waiting for the final results [47].

Miscellaneous Strategies Against COVID-19 Viral Disease

COVID-19 prevention and treatment involve a multifaceted approach, including diagnostic methods like RT-PCR, isothermal amplification techniques, CRISPR-based detection, and immunoassays. Treatment strategies include repurposing existing drugs and using convalescent plasma from recovered patients. Prevention involves standard precautions like hand hygiene, mask use, eye protection, ventilation, and chemical air disinfection [40]. Vaccination is crucial for preventing severe illness and death. Further research and clinical trials are needed to refine these strategies. This situation resembles a perplexing problem found in a literature. Since the initial documented incidence of COVID-19 infection, various antiviral medications have been employed to combat the virus through drug repurposing. However, none of these treatments has demonstrated total effectiveness [13]. Scientists are diligently researching new methods to cure the current viral sickness, which are promising and could be utilised to handle future viral outbreaks (Table 13.1).

Stem Cell, Synthetic Nano Stem Cells-Based Therapy, DNA/RNA Vaccines

Stem cell therapy has recently become a promising approach for treating various diseases, including viral diseases [7]. However, the progress in developing and advancing this type of therapy, particularly with pluripotent stem cells, has been slow due to limited sources for isolating these cells and ethical limitations. The decision to choose mesenchymal stem cells (MSCs) may be influenced by factors such as their source and availability, their capacity to multiply rapidly, their non-invasive method, and their lack of ethical concerns.

One of the advantages and features of utilising mesenchymal stem cell-based therapy is their accessibility and ability to be obtained from multiple tissues including bone marrow, peripheral blood, and adipose tissues such as abdominal fat, infrapatellar fat pad, and buccal fat pad. Additional sites encompass neonatal birth-related tissues such as the placenta, umbilical cord, Wharton's jelly, amniotic fluid, and cord blood. Isolated MSCs can be stored for future applications. They are multipotent stem cells that can self-renew and differentiate into specialised cells. MSCs can be expanded to the desired volume within a suitable timeframe. They can also be preserved for repeated therapeutic use. No severe adverse reactions to allogeneic MSCs have been reported in any clinical trials so far. The safety and effectiveness of MSCs have been clearly demonstrated in numerous clinical trials [19, 27].

Furthermore, it is imperative to analyse the justifications for utilising MSCs in the context of viral infections. It is commonly acknowledged that viral infections that have a high rate of reproduction and the propensity to cause significant harm might impair the body's natural and acquired immune defences. A dysfunction has occurred in the coordinated immune response, specifically in the capacity of antigen-presenting cells (APCs) like dendritic cells and macrophages, to efficiently capture, process, and present antigens to the adaptive immunological components,

Table 13.1 Key repurpose drugs used and their mechanism of action on COVID-19 infection

Drugs	Class/type	Mechanism of action	Indications	Mechanism of action on COVID-19 infection
Chloroquine and hydroxychloroquine	4-aminoquinoline	Glycosylation inhibition refers to the modification of newly produced proteins after translation.	Malaria, rheumatoid arthritis, lupus	Inhibits viral replication by increasing pH levels in endosomes, making it difficult for the virus to enter host cells
Lopinavir/ritonavir	Protease inhibitors	Inhibits viral penetration into cells.	HIV	Inhibits viral replication by blocking the protease enzyme, preventing the virus from replicating
Zanamivir	Neuraminidase inhibitor	Zanamivir functions by inhibiting the activity of neuraminidase, which consequently impacts the aggregation and release of viral particles. The medicine immobilises the influenza virus, preventing it from exiting its host cell and spreading to neighbouring cells.	Influenza	Inhibits viral release from infected cells, preventing the spread of the virus
Amprenavir	Protease inhibitor (HIV)	It attaches to the active site of the protease and hinders the functioning of the enzyme. This inhibition hinders the process of splitting the viral polyproteins, which leads to the creation of immature viral particles that are not capable of causing infection.	HIV	Inhibits viral replication by blocking the protease enzyme, preventing the virus from replicating
Darunavir	Second generation protease inhibitor	Darunavir exhibits molecular flexibility, enabling it to accommodate variations in the conformation of a protease enzyme. It inhibits the reproduction of HIV.	HIV	

Oseltamivir	Neuraminidase inhibitor	A selective inhibitor of influenza virus neuraminidase enzymes which is important for viral entry into uninfected cells, for the release of recently formed virus particles from infected cells.	Influenza	Inhibits viral release from infected cells, preventing the spread of the virus
Favipiravir	RNA-dependent RNA polymerase inhibitors	Favipiravir is a prodrug that undergoes ribosylation and phosphorylation inside cells to produce the active metabolite favipiravir ibofuranosyl-5′-triphosphate. This active metabolite competes with purine nucleosides and disrupts replication by integrating into the RNA of the virus. As a result, it has the potential to inhibit the RNA-dependent RNA polymerase (RdRP) of RNA viruses.	Influenza	Inhibits viral replication by disrupting the RNA-dependent RNA polymerase, preventing the virus from replicating
Faldaprevir	Protease inhibitor	The precise process is yet unidentified, despite its application in treating chronic hepatitis C virus.	Hepatitis C	Inhibits viral replication by blocking the protease enzyme, preventing the virus from replicating
Galidesivir	Protease inhibitor—adenosine analogue	Originally designed for Zaire ebolavirus, it attaches to the viral RNA polymerase instead of the normal nucleotide, causing the viral RNA polymerase to stop working properly and resulting in the premature ending of the RNA strand being produced.	Ebola	Inhibits viral replication by blocking the RNA-dependent RNA polymerase, preventing the virus from replicating

(continued)

Table 13.1 (continued)

Drugs	Class/type	Mechanism of action	Indications	Mechanism of action on COVID-19 infection
Baloxavir marboxil	RNA polymerase inhibitor (endonuclease inhibitor)	CAP endonuclease inhibitor. A crucial part of viral RNA polymerase is the influenza endonuclease. Cap endonuclease converts host pre-mRNAs into viral mRNA primers, making it a common target for anti-influenza medication research. Inhibiting endonuclease prevents mRNA transcription and inactivates influenza.	Influenza	Inhibits viral replication by disrupting the RNA-dependent RNA polymerase, preventing the virus from replicating
Indinavir	Protease inhibitor	Indinavir hinders the activity of the HIV viral protease enzyme, hence impeding the cleavage of the gag-pol polyprotein. This leads to the production of immature viral particles that are incapable of causing infection.	HIV	Inhibits viral replication by blocking the protease enzyme, preventing the virus from replicating
Ribavirin	Nucleoside analogue (guanine)—inhibits viral RNA-dependent RNA polymerase	It inhibits viral mRNA polymerase by attaching to its nucleotide binding site, reducing viral replication, and it inhibits de novo guanine nucleotide synthesis and intracellular GTP pools, reducing viral protein synthesis.	Hepatitis C	Inhibits viral replication by disrupting the RNA-dependent RNA polymerase, preventing the virus from replicating
Sofosbuvir	Nucleotide analogue inhibitor and NS5B RNA-dependent RNA polymerase inhibitor	The compound is a nucleotide analogue inhibitor that specifically targets the HCV NS5B protein, which is an RNA-dependent RNA polymerase. Following metabolism, it undergoes conversion into uridine analogue triphosphate (GS-461203). This compound then integrates into HCV RNA through the NS5B polymerase and functions as a chain terminator.	Hepatitis C	Inhibits viral replication by disrupting the RNA-dependent RNA polymerase, preventing the virus from replicating

Tipranavir	Nonpeptidic protease inhibitor	It hinders the processing of the viral gag and gag-pol polyproteins in cells infected with HIV-1, hence delaying the generation of fully developed virions.	HIV	Inhibits viral replication by blocking the protease enzyme, preventing the virus from replicating
Nelfinavir	Protease inhibitor	Nelfinavir hinders the activity of the HIV viral proteinase enzyme, hence impeding the cleavage of the gag-pol polyprotein. This leads to the production of immature viral particles that are incapable of causing infection.	HIV	Inhibits viral replication by blocking the protease enzyme, preventing the virus from replicating
Remdesivir	Adenosine nucleotide analogues	It undergoes metabolism to transform into an active form called GS-441524, which is an analogue of adenosine. GS-441524 inhibits the activity of viral RdRP and avoids being corrected by viral exoribonuclease (ExoN), resulting in a reduction in viral RNA synthesis.	Ebola	Inhibits viral replication by disrupting the RNA-dependent RNA polymerase, preventing the virus from replicating
Acyclovir	Nucleotide analogue	Acyclovir is transformed into acyclovir monophosphate by viral thymidine kinase, and then further changed into acyclovir triphosphate. Acyclovir triphosphate has a stronger attraction to viral DNA polymerase compared to cellular DNA polymerase. It binds to the DNA and leads to the termination of the DNA chain.	Herpes	Inhibits viral replication by disrupting the DNA-dependent DNA polymerase, preventing the virus from replicating

(continued)

Table 13.1 (continued)

Drugs	Class/type	Mechanism of action	Indications	Mechanism of action on COVID-19 infection
Entecavir	Guanine analogue (HCV)	Anti-hepatitis B agent. Entecavir inhibits all three HBV polymerase (reverse transcriptase) activities by competing with the natural substrate deoxyguanosine triphosphate: (1) base priming, (2) reverse transcription of the negative strand from the pre-genomic messenger RNA, and (3) positive strand synthesis of HBV DNA.	Hepatitis B	Inhibits viral replication by disrupting the RNA-dependent RNA polymerase, preventing the virus from replicating
GS-441524	Adenosine nucleoside analogue	GS-441524 undergoes phosphorylation three times to produce the active nucleoside triphosphate. This nucleotide is then integrated into the genetic material of the virions, thereby stopping their replication process.	Ebola	Inhibits viral replication by disrupting the RNA-dependent RNA polymerase, preventing the virus from replicating
Tenofovir	Acyclic nucleoside analogue adenosine monophosphate	It is an antiviral acyclic nucleoside phosphonate after activation. A powerful inhibitor of viral reverse transcriptase, viral polymerase, and viral synthesis, it terminates chains.	HIV	Inhibits viral replication by disrupting the RNA-dependent RNA polymerase, preventing the virus from replicating
Umifenovir	Fusion inhibitor	The targeted membrane effectively prevents viral fusion and hence hinders the passage of the virus into the cell.	Influenza	Inhibits viral entry into host cells, preventing the virus from replicating

Thalidomide	Immunomodulatory	Its anti-inflammatory effect is attributed to its capacity to accelerate the degradation of messenger RNA in blood cells, resulting in a reduction of tumour necrosis factor-α (TNF-α). Thalidomide has the ability to enhance the production of interleukins (IL), namely IL-12, and stimulate the activity of natural killer cells.	Multiple myeloma	Enhances the immune response by stimulating the production of interleukins and natural killer cells, which helps in combating the virus
Umifenovir	Direct-acting antiviral (DAA), host-targeting agent (HTA) and hemagglutinin inhibitor (influenza)	Furthermore, its capacity to engage in interactions with viral proteins and lipids potentially disrupts subsequent phases of the viral life cycle.	Influenza	Inhibits viral entry into host cells, preventing the virus from replicating
Amantadine	Antiviral used in the prophylactic or symptomatic treatment of influenza A, and also used as an anti-parkinsonian agent.	It facilitates the release of dopamine, in addition to stimulating the norepinephrine response. Additionally, it exhibits antagonistic effects on NMDA receptors. The medicine disrupts the function of a viral protein called M2, which acts as an ion channel. This protein is necessary for the viral particle to shed its protective coating when it enters the cell through endocytosis.	Influenza	Inhibits viral replication by disrupting the M2 protein, preventing the virus from replicating
Pleconaril	Viral capsid inhibitor	It attaches to a hydrophobic pocket in viral protein 1, the primary component of the capsid (outer shell) of picornaviruses. This causes the viral capsid to become inflexible and compacted, thereby inhibiting the removal of its RNA. Consequently, the virus is prevented from binding to the host cell and initiating infection.	Picornaviruses	Inhibits viral replication by disrupting the capsid structure, preventing the virus from replicating

Source: https://go.drugbank.com/drugs

T cells and B cells. The compromised function of both the innate and adaptive immune systems facilitates viral replication and exacerbates the severity of infections. Studies have shown that an inequilibrium in the innate and adaptive immune system results in the emergence of a severe cytokine storm, similar to what is observed in various infectious diseases and cancer. This identical phenomenon is initiated by viral infections caused by COVID-19. The primary factor responsible for this phenomenon is largely ascribed to a heightened presence of inflammatory compounds, particularly interleukins and immune cells [7, 12]. The cytokine storm produced during this process ultimately impacts the antigen-presenting cells (APCs) and adaptive immune cells. Presently, scientists are engaged in the development of MSCs therapy employing either natural APCs or artificial APCs to provoke the intended immune response, as specified in Table 13.2 [48].

Previous investigations on viral infections, such as COVID-19, have demonstrated that mesenchymal stem cells (MSCs) has the capacity to modulate the immune system by altering immunological responses. They achieve this by enhancing the proliferation and activity of diverse immune cells, while inhibiting the differentiation of monocytes into dendritic cells (DCs). MSCs have the ability to alter the cytokine profiles of DCs by enhancing the production of regulatory cytokines and decreasing the production of inflammatory cytokines. Mesenchymal stem cells (MSCs) promote the acquisition of tolerant traits in both naïve and effector T cells, restrict the generation of antibodies by B cells, and diminish the proliferation and cytotoxicity of natural killer (NK) cells [36]. The immunomodulatory activities occur through cellular interactions and the secretion of specialised signalling molecules known as cytokines. The cytokines mentioned are interferon- (IFN-) γ, indoleamine 2,3-dioxygenase (IDO), transforming growth factor- (TGF-) β, interleukin (IL-) 6, IL-10, and prostaglandin E2 [44]. Therefore, many experiments are currently being conducted on MSCs by countries like as the United States, China, Israel, Iran, Italy, and Iraq, in order to address the covid-19 pandemic.

Moreover, a recent case report recorded a pivotal diagnosis of COVID-19 in a 65-year-old female patient. The patient's neutrophil count increased by 87% but their lymphocyte count decreased by 9.8%. Despite the administration of antiviral drugs including lopinavir / ritonavir, IFN-α, and oseltamivir, the patient's vital signs worsened. Afterwards, the patient was treated solely with umbilical cord mesenchymal stem cells (UC-MSCs) and α1 thymosin at a dosage of 5 × 107 cells, given three times. The trial results showed a progressive decline in serum albumin, CRP, and ALT / AST levels after the second injection. Furthermore, other crucial measures of health also demonstrated enhancement. Afterwards, the patient's life-sustaining device was disconnected, allowing her to walk. The patient's leukocyte and neutrophil counts decreased to a normal level, but the number of lymphocytes climbed to their normal level. Significantly, there was a considerable increase in the numbers of CD3+ T cells, CD4+ T cells, and CD8+ T cells. In addition, the CT scans showed a significant improvement in pneumonia after three injections of cord stem cells were given. Ultimately, the patient was discharged from the intensive care unit. This discovery indicates that umbilical cord mesenchymal stem cells may be an excellent

Table 13.2 Nucleic Acid candidates under preclinical and clinical trials

S. no.	Platform	Candidate vaccine	Developer	Target diseases	Current stage of clinical evaluation
1.	DNA	DNA	Takis/applied DNA sciences/ Evvivax	COVID-19	Clinical trial
2.	DNA	DNA vaccine	Alvea Holdings, LLC	Sars-CoV-2 infection	Clinical trial Phase 1
3.	DNA	SARS-CoV-2 DNA vaccine (ICCOV)	Immuno cure holding (HK) LIMITED	COVID-19	Clinical trial
4.	DNA	DNA vaccine	Matti Sällberg	COVID-19 respiratory infection COVID-19 vaccine adverse reaction	Clinical trial
5.	DNA	DNA vaccine	GeneOne Life Science, Inc.	COVID-19 Healthy	Clinical trial
6.	DNA	DNA vaccine	AnGes, Inc.	COVID-19	Clinical trial
7.	DNA	DNA vaccine	AnGes, Inc	COVID-19	Clinical trial
8.	DNA	DNA vaccine (COVIGEN)	University of Sydney	Sars-CoV-2 COVID-19	Clinical trial
9.	DNA	Plasmid DNA, needle-free delivery	Immunomic Therapeutics, Inc./EpiVax, Inc./PharmaJet, Inc.	COVID-19	Preclinical (update not available or published)
10.	DNA	DNA plasmid vaccine	Zydus Cadila	COVID-19	Clinical- phase III
11.	DNA	DNA vaccine	BioNet Asia	COVID-19	Preclinical, Phase I trial

(continued)

Table 13.2 (continued)

S. no.	Platform	Candidate vaccine	Developer	Target diseases	Current stage of clinical evaluation
12.	DNA	DNA vaccine	University of Waterloo	COVID-19	Preclinical (update not available or published)
13.	RNA	RNA vaccine	Vinbiocare Biotechnology Joint Stock Company	COVID-19 vaccines	Clinical trial Phase 1 Phase2
14.	RNA	RNA vaccine	The University of Texas Health Science Center at San Antonio	COVID-19 vaccine Cancer	Preclinical trial (completed)
15.	RNA	mRNA vaccine	Gaziosmanpasa Research and Education Hospital	COVID-19 virus infection Ovarian function insufficiency Fertility disorders Vaccine adverse reaction	Preclinical trial (completed)
16.	RNA	RNA-based vaccine	BioNTech SE	SARS-CoV-2 infection COVID-19	Clinical trial Phase 2 Phase 3
17.	RNA	LNP-encapsulated mRNA cocktail encoding VLP	Fudan University/ Shanghai JiaoTong University/RNACure Biopharma	COVID-19	Preclinical (update not available or published)
18.	RNA	LNP-encapsulated mRNA encoding RBD	Fudan University/ Shanghai JiaoTong University/RNACure Biopharma	COVID-19	Preclinical (update not available or published)
19.	RNA	Replicating defective SARS-CoV-2 derived RNAs	Centro Nacional Biotecnología (CNB-CSIC), Spain	COVID-19	Preclinical (update not available or published)
20.	RNA	LNP-encapsulated mRNA	University of Tokyo/ Daiichi-Sankyo	COVID-19	Preclinical (update not available or published)

21.	RNA	Liposome-encapsulated mRNA	BIOCAD	COVID-19	Preclinical (update not available or published)
22.	RNA	mRNA	China CDC/Tongji University/ Stermina	COVID-19	Preclinical (update not available)
23.	RNA	mRNA	Arcturus/Duke-NUS	COVID-19	Third phase trials completed
24.	RNA	mRNA	BioNTech/Fosun Pharma/ Pfizer	COVID-19	#Available in market and completed third phase trial
25.	RNA	saRNA	Imperial College London	COVID-19	Preclinical, phase II trials completed
26.	RNA	mRNA	Curevac	COVID-19	Shall be available in May to June 2021, as third phase trials completed

Source: https://www.who.int/docs/default-source/blue-print/draft-landscape-of-covid-19-candidate-vaccines.pdf
World Health Organization [48], October 15. *DRAFT landscape of COVID-19 candidate vaccines*

treatment option, either alone or in combination with other immune modulators, for persons with acute conditions. COVID-19 [31].

A recent study conducted by researchers from China, specifically Beijing, Shanghai, and Hubei, involved the successful transplantation of human MSCs. The details of this trial may be seen in Table 13.2 [48]. The study observed a positive change in the condition of seven patients who were enrolled and treated with COVID-19 pneumonia at Beijing YouAn Hospital in China, between January 23, 2020, and February 16, 2020. Following the administration of MSC injections, the clinical outcomes, changes in inflammatory and immune function levels, and potential negative effects were assessed in a group of seven registered patients over a period of 14 days. The functional outcomes of all seven patients showed a significant improvement following the transplantation of MSCs, with no reported side effects. The pulmonary function tests and symptoms of these seven patients shown a notable improvement within a span of 48 hours following the donation of mesenchymal stem cells (MSC). Following the transplantation, there was an observed rise in the number of peripheral lymphocytes, accompanied by a drop in the levels of C-reactive protein, which is a marker for infection and inflammation. Furthermore, during a span of 3–6 days, the cytokine-secreting immune cells CXCR3 + CD4+ T cells, CXCR3 + CD8+ T cells, and CXCR3+ NK cells ceased to exist.There was also an observed increase in the population of regulatory dendritic cells that express CD14, CD11c, and have low levels of CD11b. In addition, the MSC-based treatment group showed a significant decrease in the pro-inflammatory cytokine TNF-α, and a notable increase in the anti-inflammatory cytokine IL-10, compared to the control placebo group. The initial investigation into the intravenous transplantation of MSCs has revealed promising prospects, demonstrating both safety and efficacy in treating patients with COVID-19 pneumonia [30].

Pluristem Therapeutics Inc. (Nasdaq: PSTI) (TASE:PSTI) stated doing another trial on MSCs in March–April 2020. A prominent regenerative medicine business based in HAIFA, Israel has reported successful treatment of individuals experiencing difficulties with COVID-19 in the United States. The therapy was administered in accordance with the US Food and Drug Administration's (FDA) Single Patient Expanded Access Programme, which is also known as a compassionate use programme. This is a component of the US Coronavirus Treatment Acceleration Programme (CTAP), which is an emergency initiative aimed at promoting medicines by utilising all available approaches to expedite the delivery of novel medications to patients.

Prior to receiving PLX treatment, the patient's condition was described as critically sick, experiencing respiratory failure caused by acute respiratory distress syndrome (ARDS). The patient required life-saving interventions such as mechanical ventilation and was admitted to the intensive care unit (ICU) for a duration of three weeks. PLX cells are allogeneic mesenchymal-like cells that possess immunomodulatory features and have the ability to stimulate the natural regulatory T cells and M2 macrophages of the immune system. Using this, they can potentially stop or reverse the dangerous exacerbation of the immune system. PLX cells have the ability to decrease the occurrence and intensity of COVID-19 pneumonia and pneumonitis, which could improve the outlook for patients. The company intends to collaborate with the BIH Centre for Regenerative Therapy (BCRT) and the Berlin Centre for Advanced Therapies (BeCAT) at Charite' University of Medicine Berlin to investigate PLX cell product options for Covid-19. According to a recent article on May 8, 2020, the company registered a phase II trial after receiving clearance from the FDA. The title of this trial is "A Randomized, Double-Blind, Placebo-Controlled, Multicenter, Parallel-Group phase II Study to Assess the Effectiveness and Safety of Intramuscular Injections of PLX-PAD for the Treatment of severe COVID-19". This research aims to treat 140 adult patients who are intubated and mechanically ventilated and are experiencing respiratory failure and acute respiratory distress syndrome (ARDS) caused by COVID-19. The main measure of effectiveness in this trial is the total number of days that patients are able to breathe without the assistance of a ventilator, specifically over the 28-day period from day 1 to day 28 of the study [46].

The stem cell trials described above, as well as a few others, are summarised in Table 13.3.

The outcomes of these trials including therapy with MSCs may establish it as an optimal choice for the treatment approach in the event of future outbreaks of comparable new viruses. Additionally, it has demonstrated potential as a viable treatment option for patients with COVID-19.

Synthetic Nano Stem Cells Based Therapy

LIFNano, being an advanced technology, has the potential to be used as a substitute for cell-based therapy, reducing the requirement for large quantities of therapeutic chemicals that are readily available. Therefore, this therapy has the potential to revitalise damaged tissues and mitigate the cytokine storm associated with pneumonia [1, 14]. As per the specifications, MSCs are acknowledged for their capacity to secrete LIF. Nevertheless, the expense associated with processing and keeping them is exorbitant for ordinary persons. The use of LIFNano offers a feasible solution to this issue. The approach employs nanotechnology to administer synthetic stem cells called "LIFNano," which has a potency that is one thousand times higher than soluble LIF [37]. LIFNano, a revolutionary technology, has the potential to serve as an alternative to cell-based therapy, hence decreasing the need for vast amounts of medicinal chemicals that are easily accessible. Hence, this therapy possesses the capacity to rejuvenate impaired tissues and alleviate the cytokine storm linked to pneumonia.

DNA/RNA Vaccines and COVID-19

Over the past 30 years, a clever method of vaccination has been developed. Instead of using a whole inactivated virus or pathogen, a plasmid containing the DNA sequence that codes for the antigen(s) is directly injected into the necessary tissues

Table 13.3 Stem Cell Therapy based trials on COVID-19 (Updated list on 30th April 2024 and removed the cancelled trials on trials registry)

S. no.	First submitted date/ date of registration	Study Phase	ID	Project title	Cell type	Participants (Number, age, sex)	Intervention
1	2020-03-15	Phase 1	NCT04313322	Treatment of COVID-19 patients using Wharton's jelly-mesenchymal stem cells	WJ -MSCs	Enrolment: 5 Age: 18 years to older Sex: All	WJ-MSCs will be derived from cord tissue of newborns, screened for HIV1/2, HBV, HCV, CMV, mycoplasma, and cultured to enrich for MSCs. WJ-MSCs will be counted and suspended in 25 ml of saline solution containing 0.5% human serum albumin, and will be given to patient intravenously.
2.	2020-07-16	Phase1 Phase 2	NCT04473170	Study evaluating the safety and efficacy of autologous non-Haematopoietic peripheral blood stem cells in COVID-19 (SENTAD-COVID)	Autologous non-haematopoietic peripheral blood stem cells (NHPBSC)	146	Biological: Autologous non-Haematopoietic peripheral blood stem cells (NHPBSC) Drug: COVID-19 standard care
3.	2021-08-30	Phase 1	NCT04486001	COVID-19 stem cell therapy: A phase I study of intravenous administration of allogeneic adipose stem cells	Adipose-derived allogeneic mesenchymal stem cell	10	Biological: PSC-04

4.	2020-09-01	Phase 1	NCT04573270	A pilot phase study evaluating the effects of a single mesenchymal stem cell injection in patients with suspected or confirmed COVID-19 infection and healthcare providers exposed to coronavirus patients	Mesenchymal stem cell	40	Biological: PrimePro
5.	2021-11-14	Phase 3	NCT05122234	Effectiveness and safety profile of mesenchymal stem cell Secretomes as a treatment for severe cases of COVID-19	Mesenchymal stem cell Secretome	40	Biological: Injection of secretome—mesenchymal stem cell Other: Placebo Drug: Standard treatment of Covid-19
6.	2024-02-02	Phase 2	NCT04361942	Treatment of severe COVID-19 pneumonia with allogeneic mesenchymal stem cells (MSV-COVID)	Allogeneic mesenchymal stem cells	24	Biological: mesenchymal stem cells (MSCs) Intravenous injection of one million MSCs (MSV cells)/kg suspended in 100 ml physiological saline solution. Other names: MSV (GMP-compliant MSCs manufactured by IBGM in Valladolid) Other: Placebo Intravenous injection of 100 ml physiological saline solution containing no cells Other names: Saline

(continued)

Table 13.3 (continued)

S. no.	First submitted date/ date of registration	Study Phase	ID	Project title	Cell type	Participants (Number, age, sex)	Intervention
7.	2020/09–07	Phase 1	NCT04276987	A pilot clinical study on inhalation of mesenchymal stem cells exosomes treating severe novel coronavirus pneumonia	allogenic adipose mesenchymal stem cells	24 participants	Biological: MSCs-derived exosomes 5 times aerosol inhalation of MSCs-derived exosomes (2.0*10E8 nano vesicles/3 ml at day 1, day 2, day 3, day 4, day 5).
8.	2020-01-27	Phase 1	NCT04252118	Mesenchymal stem cell treatment for pneumonia patients infected with COVID-19	MSCs	Enrolment: 20 Age: 18 years to 70 years Sex: All	Biological: MSCs 3 times of MSCs (3.0*10E7 MSCs intravenously at day 0, day 3, day 6).
9.	2020-04-02	Phase 1 Phase 2	NCT04339660	Clinical research of human mesenchymal stem cells in the treatment of COVID-19 pneumonia	UC-MSCs	Enrolment: 30 Age: 18 years to 75 years Sex: All	Biological: UC-MSCs 1*10E6 UC-MSCs /kg body weight suspended in 100 mL saline Other: Placebo 100 mL saline intravenously

10.	2020-02-24	Phase 2	NCT04288102	Treatment With Mesenchymal Stem Cells for Severe Corona Virus Disease 2019(COVID-19)	MSCs	Enrolment: 90 Age: 18 years to 75 years Sex: All	Biological: MSCs three times of MSCs (4.0*10E7 cells per time) intravenously at day 0, day 3, day 6. Biological: Saline containing 1% human serum albumin solution of MSC three times of placebo intravenously at day 0, day 3, day 6
11.	2020-02-07	Phase 2	NCT04269525	Umbilical cord (UC)-derived mesenchymal stem cells (MSCs) treatment for the 2019-novel coronavirus (nCOV) pneumonia	UC-MSCs	Enrolment: 10 Age: 18 years to 75 years Sex: All	Biological: UC-MSCs After enrolment, each subject will receive UC-MSCs infusion intravenously on day 1, day 3, day 5, and day 7.

(continued)

Table 13.3 (continued)

S. no.	First submitted date/ date of registration	Study Phase	ID	Project title	Cell type	Participants (Number, age, sex)	Intervention
12.	2020-04-13	Phase 1 Phase 2	NCT04355728	Use of UC-MSCs for COVID-19 patients	UC-MSCs	Enrolment: 24 Age: 18 years or older Sex: All	Biological: Umbilical cord mesenchymal stem cells UC-MSC will be administered at 100x10^6 cells/infusion administered intravenously in addition to the standard of care treatment. Other: Standard of care Standard of care treatment per the treating hospital protocol.
13.	2017-02-01	Phase 1 Phase 2	NCT03042143	Repair of acute respiratory distress syndrome by stromal cell administration (REALIST) (COVID-19)	Human UC-MSCs CD362 enriched	Enrolment: 75 Age: 16 years or older Sex: All	Biological: Human umbilical cord-derived CD362 enriched MSCs infusion of human umbilical cord-derived CD362 enriched MSCs Biological: Placebo (plasma-Lyte 148) infusion of placebo

14.	2020-04-20	Phase 2 Phase 3	NCT04366063	Mesenchymal stem cell therapy for SARS-CoV-2-related acute respiratory distress syndrome	MSCs	Enrolment: 60 Age: 16 years to 65 years Sex: All	Biological: Cell therapy protocol 1 Cell therapy protocol 1(n = 20). Patients will receive two doses of MSCs 100 × 10e6 (±10%) at day 0 and day 2 plus conventional treatment. Biological: Cell therapy protocol 2 Patients will receive two doses of MSCs 100 × 10e6 (±10%)at day 0 and day 2, intravenously plus two doses of EVs at day 4 and day 6 plus conventional treatment.

(continued)

Table 13.3 (continued)

S. no.	First submitted date/ date of registration	Study Phase	ID	Project title	Cell type	Participants (Number, age, sex)	Intervention
15.	2020-03-29	Phase 1 Phase 2	NCT04333368	Cell therapy using umbilical cord-derived mesenchymal stromal cells in SARS-CoV-2-related ARDS	UC-MSCs	Enrolment: 60 Age: 18 years or older Sex: All	Biological: Umbilical cord Wharton's jelly-derived human umbilical cord Wharton's jelly-derived human MSC (at the dose of one million / kg) will be administered via a peripheral or central venous line over 30–45 minutes, using tubing with a 200-μm filter. Cells, in a 150 mL volume, will be delivered at D1–D3–D5. Other: NaCl 0.9% NaCl 0.9% (150 mL) given via an intravenous route at D1–D3–D5

16	2020-02-14	(no information on the trial registry)	NCT04273646	Study of human umbilical cord mesenchymal stem cells in the treatment of severe COVID-19	Human UC-MSCs	Estimated enrolment: 48 Age: 18 years to 65 years Sex: All	Biological: UC-MSCs 4 times of UC-MSCs (0.5*10E6 UC-MSCs/kg body weight intravenously at day 1, day 3, day 5, day 7). Drug: Placebo 4 times of cell-free stem cell suspension (saline containing 1% human albumin) intravenously at day 1, day 3, day 5, day 7).
17	2020-02-27	Phase 1	NCT04302519	Novel coronavirus-induced severe pneumonia treated by dental pulp mesenchymal stem cells	Dental pulp mesenchymal stem cells	Enrolment:24 Age: 18 years to 75 years Sex: All	Biological: Dental pulp mesenchymal stem cells On the basis of clinical standard treatment, the injection of dental mesenchymal stem cells was increased on day 1, 3 and 7 of the trial.
18	2020-02-18	Phase 1 Phase 2	ChiCTR2000029990	Clinical trials of mesenchymal stem cells for the treatment of pneumonitis caused by novel coronavirus pneumonia (COVID-19)	MSCs	Enrolment: 120 Age: 18 years to 95 years Sex: All	Experimental group: MSCs Control group: Saline

(continued)

Table 13.3 (continued)

S. no.	First submitted date/ date of registration	Study Phase	ID	Project title	Cell type	Participants (Number, age, sex)	Intervention
19	2020-04-08	- (not available on trial registry)	ChiCTR2000031735	Clinical study for natural killer (NK) cells from umbilical cord blood in the treatment of novel coronavirus pneumonia (COVID-19)	NK cells	Enrolment:20 Age: 18 years and older Sex: All	Experimental group: NK cells were injected by intravenous drip once a day for 2 to 3 times in total, each dose was 4 10^7 pieces /kg body weight,100 ml normal saline suspension Control group: 100 ml normal saline intravenous drip
20.	2020-03-27	- (not available on trial registry)	ChiCTR2000031319	Safety and efficacy study of allogeneic human dental pulp mesenchymal stem cells to treat severe novel coronavirus pneumonia (COVID-19) patients	hDPSC	Enrolment: 20 Age: 18 years to 65 years Sex: All	hDPSCs group: Routine treatment + intravenous injection of human dental pulp stem cells Control group: Routine treatment + placebo

21	2020-03-22	- (not available on trial registry)	ChiCTR2000031139	Safety and effectiveness of human embryonic stem cell-derived M cells (CAStem) for pulmonary fibrosis correlated with novel coronavirus pneumonia (COVID-19)	Human embryonic stem cell-derived M cells	Enrolment: 20 Age: 18 years to 80 years Sex: All	Case series: The cell dose was $3*10^6$ cells / kg. It was intravenously infused twice in a row, and the interval between each infusion was 1 week (+/−2 days). If the investigator considered it necessary, an additional infusion could be performed. Infusion interval 1 week (+/−2 days) from the last time.
22	2020-03-18	Phase 1	ChiCTR2000030944	Clinical study of human NK cells and MSCs transplantation for severe novel coronavirus pneumonia (COVID-19)	NK cells and MSCs	Enrolment: 20 Age: 4 years to 80 years Sex: All	Experimental group: On the basis of the current clinical treatment of SNCP, NK cells and MSCs were increased Control group: current clinical treatment of SNCP

(continued)

Table 13.3 (continued)

S. no.	First submitted date/ date of registration	Study Phase	ID	Project title	Cell type	Participants (Number, age, sex)	Intervention
23	2020-02-24	- (not available on trial registry)	ChiCTR2000030173	Key techniques of umbilical cord mesenchymal stem cells for the treatment of novel coronavirus pneumonia (COVID-19) and clinical application demonstration	UC-MSCs	Enrolment: 60 Age: 18 years to 70 years Sex: All	Experimental group: umbilical cord mesenchymal stem cells Control group: convention treatment
24	2020-02-24	Phase 2	ChiCTR2000030138	Clinical trial for human mesenchymal stem cells in the treatment of severe novel coronavirus pneumonia (COVID-19)	UC-MSCs	Enrolment: 60 Age: 16 years to 75 years Sex: All	Experimental group: Intravenous injection of human umbilical cord mesenchymal stem cells (UC-MSC) Control group: Routine treatment + placebo
25	2020-02-23	- (not available on trial registry)	ChiCTR2000030116	Safety and effectiveness of human umbilical cord mesenchymal stem cells in the treatment of acute respiratory distress syndrome of severe novel coronavirus pneumonia (COVID-19)	MSCs	Enrolment: 16 Age: 16 years to 75 years Sex: All	Two groups: Different stem cell doses

26	2020-02-22	- (not available on trial registry)	ChiCTR2000030088	Umbilical cord Wharton's jelly-derived mesenchymal stem cells in the treatment of severe novel coronavirus pneumonia (COVID-19)	Wharton's jelly MSCs	Enrolment: 40 Age: 18 years to 80 years Sex: All	Experimental group: Iv injection of Wharton's jelly-mesenchymal stem cells (1×10^6/kg), cell suspension volume: 40 ml Control group: Iv 40 ml saline
27	2020-02-20	- (not available on trial registry)	ChiCTR2000030020	The clinical application and basic research related to mesenchymal stem cells to treat novel coronavirus pneumonia (COVID-19)	MSCs	Enrolment: 20 Age: 18 years to 70 years Sex: All	Case series: Mesenchymal stem cells therapy
28	2020-02-07	- (not available on trial registry)	ChiCTR2000029606	Clinical study for human menstrual blood-derived stem cells in the treatment of acute novel coronavirus pneumonia (COVID-19)	Human menstrual blood-derived stem cells preparations	Enrolment: 63 Age: 1 year to 99 years Sex: All	experimental group A: Conventional treatment followed by intravenous infusion of human menstrual blood-derived stem cells preparations control gorup A: Conventional treatment Experimental group B1: Artificial liver therapy + conventional treatment Experimental group B2: Artificial liver therapy followed by intravenous infusion of human menstrual blood-derived stem cells preparations + conventional treatment Control Gorup A: Conventional treatment

(continued)

Table 13.3 (continued)

S. no.	First submitted date/ date of registration	Study Phase	ID	Project title	Cell type	Participants (Number, age, sex)	Intervention
29.	2020-02-05	- (not available on trial registry)	ChiCTR2000029580	Severe novel coronavirus pneumonia (COVID-19) patients treated with ruxolitinib in combination with mesenchymal stem cells: a prospective, single blind, randomised controlled clinical trial	Mesenchymal stem cell	Enrolment: 70 Age: 18 years to 75 years Sex: All	Experimental group: Ruxolitinib combined with mesenchymal stem cell Control group: Routine treatment
30.	2020-02-05	- (not available on trial registry)	ChiCTR2000029572	Safety and efficacy of umbilical cord blood mononuclear cells in the treatment of severe and critically novel coronavirus pneumonia(COVID-19): a randomised controlled clinical trial	Umbilical cord blood mononuclear cells	Enrolment: 30 Age: 18 years or older Sex: All	Experimental group: conventional treatment combined with umbilical cord blood mononuclear cells group Control group: Conventional treatment
31.	2020-02-04	- (not available on trial registry)	ChiCTR2000029569	Safety and efficacy of umbilical cord blood mononuclear cells conditioned medium in the treatment of severe and critically novel coronavirus pneumonia (COVID-19): a randomised controlled trial	Umbilical cord mesenchymal stem cell	Enrolment: 30 Age: 18 years or older Sex: All	Experimental group: conventional treatment combined with umbilical cord mesenchymal stem cell conditioned medium group Control group: Conventional treatment

to elicit an immune response. This methodology offers numerous potential advantages compared to traditional methods, such as the ability to stimulate both B- and T-cell responses, enhanced vaccine durability, the absence of a complete infectious entity (ensuring safety), and the potential for large-scale production [29].

Animal studies have examined the immune responses to DNA vaccination using genes from different infectious agents, such as human immunodeficiency virus, influenza virus, hepatitis B virus, rabies virus, lymphocytic chorio-meningitis virus, malarial parasites, and mycoplasmas. Various gene-based vaccines have been authorised for veterinary use to specifically target diseases like canine melanoma (Merial), West Nile virus (Wyeth), fish haematopoietic necrosis virus (Novartis), and swine growth hormone-releasing hormone (Inovio).

Currently, vaccines utilise the DNA of the primary antigen and the accessory segment to facilitate the entry of DNA into the specific cells. Bacterial plasmid DNA is the most commonly approved adjuvant. However, there is now a growing interest in using RNA complexes, which consist of nucleic acid molecules and other components of pathogens, as vaccine antigens. The effectiveness of the vaccine ultimately depends on the type and characteristics of the agent, the antigen, the carrier, and their interaction with the host. Nonetheless, a few promising studies have laid the groundwork for future research, and the pandemic has significantly increased the demand for such vaccines.

Amidst the COVID-19 pandemic, many research institutions and pharmaceutical businesses have conducted studies for DNA and RNA vaccines, as seen in Tables 13.2 and 13.3. These treatments are currently undergoing different phases of clinical testing, and only a few have been licensed for use in different countries. In February 2020, Applied DNA Sciences subsidiary LineaRx and Takis Biotech formed a collaborative venture to produce a linear DNA vaccine as a preventive measure against the coronavirus. The vaccine is developed using Polymerase Chain Reaction (PCR)-based DNA synthesis technology [11, 35].

Inovio Pharmaceuticals and the Coalition for Epidemic Preparedness Innovations (CEPI) have jointly launched another clinical trial for DNA vaccines. An open-label Phase I trial is now underway with 40 participants. The purpose of the trial is to assess the safety, tolerability, and immunological profile of INO-4800. The vaccine is administered through intradermal injection followed by electroporation utilising the CELLECTRA® 2000 device. The trial is being conducted with healthy adult volunteers.

Alvea Holdings has jointly launched another clinical trial for DNA Vaccines. The study aims to evaluate the safety, tolerability, and immunogenicity of the Alveavax-v1.2 vaccine in healthy individuals. The vaccine is optimised for the Omicron/BA.2 variant, which is currently unavailable due to high costs, cold chain issues, and manufacturing issues. A total of 130 participants aged between 18 and 65 years will be enrolled in five groups and administered the vaccine according to their dose arm. The primary outcome measures include safety and tolerability of the Alveavax-v1.2 vaccine compared to a control booster vaccine, incidence of local and systemic adverse events within seven days of dose administration, overall number of participants and incidence proportion of unsolicited adverse events within

28 days of dose administration, safety and tolerability of Alveavax-v1.2 in healthy participants compared to a control booster vaccine, and the overall number of participants and incidence of serious adverse events [3].

Moreover, the National Institute of Allergy and Infectious Diseases (NIAID) is currently undertaking a Phase I clinical trial on an RNA-based vaccine. The study is officially titled "Safety and Immunogenicity Study of the mRNA-1273 Vaccine for Preventing SARS-CoV-2 Infection (COVID-19)" and involves 105 participants. Entos Pharmaceuticals is currently developing the Fusogenix DNA vaccine. It employs a distinctive drug delivery system to mitigate COVID-19 infections. The Fusogenix drug delivery platform is a proteo-lipid vehicle that delivers genetic material directly into cells. Entos is currently focused on creating a highly efficient payload that includes various protein epitopes obtained from SARS-COV-2 proteins. This payload aims to stimulate a strong immune response in the body, thereby preventing COVID-19 infection. In a similar vein, Zydus Cadila has declared the initiation of preclinical trials aimed at developing a vaccine for COVID-19 utilising two innovative methodologies. The initial strategy will entail creating a DNA vaccine targeting the viral membrane protein of the virus. The subsequent strategy will require developing a live attenuated recombinant measles virus (rMV) vectored vaccine. Various prestigious institutions and universities, such as Karolinska Institute, Waterloo University, Fudan University/Shanghai JiaoTong University, as well as pharmaceutical companies like Zydus Cadila and BioNet Asia Immunomic Therapeutics, have initiated numerous preclinical trials worldwide for DNA/RNA-based and nucleic acid-based vaccines. Table 13.2 presents a summary of various trials, some of which are currently available in markets [23]. Several miscellaneous interventions for COVID-19 are being evaluated in clinical trials to determine their safety, tolerability, and efficacy. These studies include RNA vaccine candidates like BNT162a1, BNT162b1, BNT162b2, and BNT162c2, tested in a large Phase 1/2 trial involving 7600 participants aged 18 to 85. Another Phase 1 study is assessing the mRNA-1273 vaccine in 105 participants aged 18 to 99. The safety and immunity of aAPC vaccines are being investigated in a Phase 1 trial with 100 participants aged 6 months to 80 years. Additionally, the INO-4800 DNA vaccine is being tested for safety, tolerability, and immunogenicity in 40 healthy volunteers aged 18 to 50. These trials are crucial for developing effective COVID-19 vaccines and treatments. Table 13.4 presents a summary of Miscellaneous trials, some of which are currently available in markets.

Table 13.4 Miscellaneous Interventions for COVID-19

S. no.	First submitted date/ date of registration	Study phase/ recruitment status	ID	Project title		Participants (Number, age, sex)	Intervention
1	2020-04-27	Phase 1 Phase 2	NCT04368728	Study to describe the safety, tolerability, immunogenicity, and potential efficacy of RNA vaccine candidates against COVID-19 in healthy adults	RNA	Enrolment: 7600 Age: 18 years to 85 years Sex: All	Biological BNT162a1, BNT162b1, BNT162b2, BNT162c2, and placebo are available for intramuscular injections in a 0.5 mL volume.
2.	2020-06-27	Phase 3	IRCT20111224008507N3	Effectiveness of Ivermectin in the treatment of coronavirus infection in patients admitted to educational hospitals of Mazandaran in 2020	Ivermectin	Enrolment: 60 patients Age: 5 years Sex: Both	The intervention group will use a single dose of Ivermectin 3 mg oral tablet with 0.2 mg/kg of Tadbir kala jam company in weight groups 15–24, 25–3, 36–50, 51–80, and over 80.
3	2020-02-21	Phase 1	NCT04283461	Safety and immunogenicity study of 2019-nCoV vaccine (mRNA-1273) for prophylaxis of SARS-CoV-2 infection (COVID-19)	mRNA	Enrolment: 105 Age: 18 years to 99 years Sex: All	mRNA-1273 is a lipid nanoparticle dispersion containing an mRNA drug substance, composed of proprietary lipid SM-102 and three commercially available lipids, cholesterol, DSPC, and PEG2000 DMG.

(continued)

Table 13.4 (continued)

S. no.	First submitted date/ date of registration	Study phase/ recruitment status	ID	Project title		Participants (Number, age, sex)	Intervention
4	2020-03-05	Phase 1	NCT04299724	Safety and immunity of Covid-19 aAPC vaccine	Artificial antigen-presenting cells (aAPC)	Enrolment: 100 Age: 6 months to 80 years Sex: All	Subjects will receive three subcutaneous injections of Covid-19/aAPC vaccine, INO-4800, on day 0 and week 4, administered via biological route.
5	2020-04-03	Phase 1	NCT04336410	Safety, tolerability and immunogenicity of INO-4800 for COVID-19 in healthy volunteers	INO-4800 (DNA)	Enrolment: 40 Age: 18 years to 50 years Sex: All	Drug: INO-4800 INO-4800 will be administered ID on day 0 and week 4.
6.	2023-07-26	Phase 2	NCT05595369	RECOVER-VITAL: Platform protocol to measure the effects of antiviral therapies on long COVID symptoms (RECOVER-VITAL)	Drug:Paxlovid 25 day dosing	Enrolment: 900 Age: 18 years and older (adult, older adult) Sex: All	Drug: Experimental: Paxlovid 25-day dosing Drug: Experimental: Paxlovid 15-day dosing Drug: Placebo comparator: Control

Conclusion

In conclusion, this comprehensive chapter of clinical data, in vivo, and in vitro evidence on over 25 medicines and substances, aims to assist in determining the appropriate clinical treatment for COVID patients and also determine their mechanism of action and indications in COVID-19 patients. Figure 13.2 summarise the mechanism of repurposed drugs in COVID-19 patients. Furthermore, we anticipate that this could serve as a guiding principle for future prophylactic research with a substantial number of participants. We have incorporated a wide range of chemicals, including FDA-approved pharmaceuticals, drug-like molecules, and approved medications with newly repurposed modes of action. The integration of clinical trials data, computational modelling, laboratory-based experiments, and live organism testing can aid in the logical application of novel antiviral medications at different stages of managing the ongoing pandemic. Moreover, this chapter aims to prevent the simultaneous use of pharmaceuticals linked to polypharmacy, which can result in notable harmful effects, occasionally involving toxic and degenerative interactions between drugs in people. The role of mesenchymal stem cells (MSCs) in reducing inflammation by activating antigen-presenting cells is a significant breakthrough in developing a novel biological treatment against COVID-19 and other

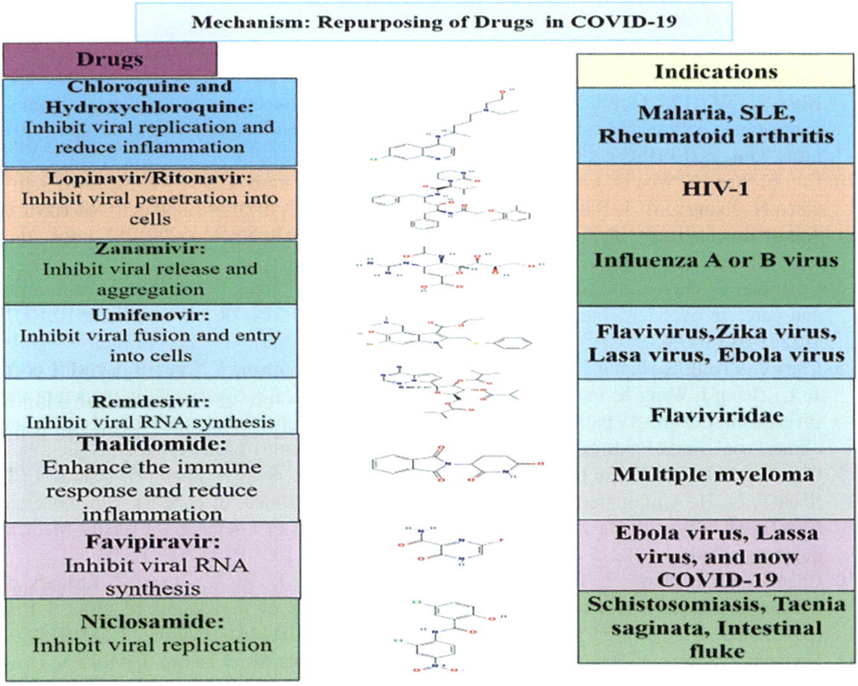

Fig. 13.2 Mechanism: repurposing of drugs in COVID-19

viral outbreaks. Additionally, the synergistic action of DNA and RNA-based vaccines plays a crucial role in achieving the desired immune modification. This innovative approach requires further development.

Acknowledgements I am thankful to the department staff, LDC Rohit Nigam, office attendants Daulat Singh, and Shubham Kumar for their help while preparing this chapter.

References

1. Abdelgawad M, Bakry NS, Farghali AA, Abdel-Latif A, Lotfy A. Mesenchymal stem cell-based therapy and exosomes in COVID-19: current trends and prospects. Stem Cell Res Ther. 2021;12:469. https://doi.org/10.1186/s13287-021-02542-z.
2. Abubakar AR, Sani IH, Godman B, Kumar S, Islam S, Jahan I, Haque M. Systematic review on the therapeutic options for COVID-19: clinical evidence of drug efficacy and implications. Infect Drug Resist. 2020;13:4673–95. https://doi.org/10.2147/IDR.S289037.
3. Alvea Holdings LLC. A Phase 1 Open-label, Active-controlled, Randomized Dose-finding Study to Evaluate Safety, Tolerability, and Immunogenicity of Intradermal and Subcutaneous Application of the Plasmid DNA SARS-CoV-2 Omicron BA.2 Vaccine Alveavax-v1.2 in Primary Ad26.COV2.S Vaccinated Healthy Individuals (Clinical Trial Registration NCT05844202). 2023. clinicaltrials.gov. https://clinicaltrials.gov/study/NCT05844202
4. Andrade BS, Rangel F d S, Santos NO, Freitas ADS, Soares WR d A, Siqueira S, Barh D, Góes-Neto A, Birbrair A, Azevedo VA d C. Repurposing approved drugs for guiding COVID-19 prophylaxis: A systematic review. Front Pharmacol. 2020;11:590598. https://doi.org/10.3389/fphar.2020.590598.
5. Burwick RM, Yawetz S, Stephenson KE, Collier A-RY, Sen P, Blackburn BG, Kojic EM, Hirshberg A, Suarez JF, Sobieszczyk ME, Marks KM, Mazur S, Big C, Manuel O, Morlin G, Rose SJ, Naqvi M, Goldfarb IT, DeZure A, et al. Compassionate use of Remdesivir in pregnant women with severe coronavirus disease 2019. Clin Infect Dis. 2021;73(11):e3996–4004. https://doi.org/10.1093/cid/ciaa1466.
6. Cao B, Wang Y, Wen D, Liu W, Wang J, Fan G, Ruan L, Song B, Cai Y, Wei M, Li X, Xia J, Chen N, Xiang J, Yu T, Bai T, Xie X, Zhang L, Li C, et al. A Trial of Lopinavir–Ritonavir in Adults Hospitalized with Severe Covid-19. N Engl J Med. 2020:NEJMoa2001282. https://doi.org/10.1056/NEJMoa2001282.
7. Chambial S, Dwivedi S, Shukla KK, John PJ, Sharma P. Vitamin C in disease prevention and cure: an overview. Indian J Clin Biochem. 2013;28(4):314–28. https://doi.org/10.1007/s12291-013-0375-3.
8. Chen C, Zhang Y, Huang J, Yin P, Cheng Z, Wu J, Chen S, Zhang Y, Chen B, Lu M, Luo Y, Ju L, Zhang J, Wang X. Favipiravir versus Arbidol for clinical recovery rate in moderate and severe adult COVID-19 patients: a prospective, multicenter, open-label, randomized controlled clinical trial. Front Pharmacol. 2021;12:683296. https://doi.org/10.3389/fphar.2021.683296.
9. Chen J, Liu D, Liu L, Liu P, Xu Q, Xia L, Ling Y, Huang D, Song S, Zhang D, Qian Z, Li T, Shen Y, Lu H. A pilot study of hydroxychloroquine in treatment of patients with moderate COVID-19. Zhejiang Da Xue Bao. Yi Xue Ban = Journal of Zhejiang University Medical Sciences. 2020a;49(2):215–9. https://doi.org/10.3785/j.issn.1008-9292.2020.03.03.
10. Chen Z, Hu J, Zhang Z, Jiang S, Han S, Yan D, Zhuang R, Hu B, Zhang Z. Efficacy of hydroxychloroquine in patients with COVID-19: Results of a randomized clinical trial (p. 2020.03.22.20040758). medRxiv. 2020b. https://doi.org/10.1101/2020.03.22.20040758.
11. Dwivedi S, Rawat A, Ranjan A, Agrawal R, Misra R, Gupta SK, Kishore S, Misra S. Drug repurposing and novel antiviral drugs for COVID-19 management. In: COVID-19. CRC Press; 2022.

12. Dwivedi S, Singh S, Goel A, Khattri S, Mandhani A, Sharma P, Misra S, Pant KK. Pro-(IL-18) and anti-(IL-10) inflammatory promoter genetic variants (intrinsic factors) with tobacco exposure (extrinsic factors) may influence susceptibility and severity of prostate carcinoma: a prospective study. Asian Pacific J Cancer Prev. 2015;16(8):3173–81. https://doi.org/10.7314/apjcp.2015.16.8.3173.
13. Dwivedi S, Yadav SS, Singh MK, Shukla S, Khattri S, Pant KK. Pharmacogenomics of viral diseases. In: Omics for Personalized Medicine; 2013. p. 637–76. https://doi.org/10.1007/978-81-322-1184-6_28.
14. Foronjy RF, Dabo AJ, Cummins N, Geraghty P. Leukemia inhibitory factor protects the lung during respiratory syncytial viral infection. BMC Immunol. 2014;15:41. https://doi.org/10.1186/s12865-014-0041-4.
15. Frediansyah A, Nainu F, Dhama K, Mudatsir M, Harapan H. Remdesivir and its antiviral activity against COVID-19: a systematic review. Clin Epidemiol Glob Health. 2021;9:123–7. https://doi.org/10.1016/j.cegh.2020.07.011.
16. Gao J, Tian Z, Yang X. Breakthrough: chloroquine phosphate has shown apparent efficacy in treatment of COVID-19 associated pneumonia in clinical studies. Biosci Trends. 2020;14(1):72–3. https://doi.org/10.5582/bst.2020.01047.
17. Gatignol A, Dubuisson J, Wainberg MA, Cohen ÉA, Darlix J-L. New pandemics: HIV and AIDS, HCV and chronic hepatitis, Influenza virus and flu. Retrovirology. 2007;4(1):8. https://doi.org/10.1186/1742-4690-4-8.
18. Gautret P, Lagier J-C, Parola P, Hoang VT, Meddeb L, Mailhe M, Doudier B, Courjon J, Giordanengo V, Vieira VE, Tissot Dupont H, Honoré S, Colson P, Chabrière E, La Scola B, Rolain J-M, Brouqui P, Raoult D. Hydroxychloroquine and azithromycin as a treatment of COVID-19: results of an open-label non-randomized clinical trial. Int J Antimicrob Agents. 2020;56(1):105949. https://doi.org/10.1016/j.ijantimicag.2020.105949.
19. Golchin A, Farahany TZ, Khojasteh A, Soleimanifar F, Ardeshirylajimi A. The clinical trials of mesenchymal stem cell therapy in skin diseases: an update and concise review. Curr Stem Cell Res Ther. 2019;14(1):22–33. https://doi.org/10.2174/1574888X13666180913123424.
20. Gottlieb Robert L, Vaca Carlos E, Roger P, Jorge M, Webb Brandon J, Gilberto P, Godson O, Pablo R, Nielsen Bibi U, Michael B, Ausberto H, Yessica S, Shilpi M, Olayemi O, Jacek S, Kavita J, Hyland Robert H, Anu O, Shuguang C, et al. Early Remdesivir to prevent progression to severe Covid-19 in outpatients. N Engl J Med. 2022;386(4):305–15. https://doi.org/10.1056/NEJMoa2116846.
21. Grein J, Ohmagari N, Shin D, Diaz G, Asperges E, Castagna A, Feldt T, Green G, Green ML, Lescure F-X, Nicastri E, Oda R, Yo K, Quiros-Roldan E, Studemeister A, Redinski J, Ahmed S, Bernett J, Chelliah D, et al. Compassionate use of Remdesivir for patients with severe Covid-19. N Engl J Med. 2020;382(24):2327–36. https://doi.org/10.1056/NEJMoa2007016.
22. Gutierrez R, Mendez-Figueroa H, Biebighauser JG, Bhalwal A, Pineles BL, Chauhan SP. Remdesivir use in pregnancy during the SARS-CoV-2 pandemic. J Matern Fetal Neonatal Med. 2022;35(25):9445–51. https://doi.org/10.1080/14767058.2022.2041595.
23. Home | ClinicalTrials.gov. (n.d.). Retrieved June 6, 2024, from https://clinicaltrials.gov/
24. Jayk Bernal A, Gomes da Silva MM, Musungaie DB, Kovalchuk E, Gonzalez A, Reyes VD, Martín-Quirós A, Caraco Y, Williams-Diaz A, Brown ML, Du J, Pedley A, Assaid C, Strizki J, Grobler JA, Shamsuddin HH, Tipping R, Wan H, Paschke A, Butterton JR, Johnson MG, et al. Molnupiravir for oral treatment of Covid-19 in nonhospitalized patients. N Engl J Med. 2022;386(6):509–20. https://doi.org/10.1056/NEJMoa2116044.
25. Kálai T, Pongrácz JE, Mátyus P. Recent advances in influenza, HIV and SARS-CoV-2 infection prevention and drug treatment—the need for precision medicine. Chemistry. 2022;4(2):Article 2. https://doi.org/10.3390/chemistry4020019.
26. Khan SA, Al-Balushi K. Combating COVID-19: the role of drug repurposing and medicinal plants. J Infect Public Health. 2021;14(4):495–503. https://doi.org/10.1016/j.jiph.2020.10.012.
27. Krawczenko A, Klimczak A. Adipose tissue-derived mesenchymal stem/stromal cells and their contribution to Angiogenic processes in tissue regeneration. Int J Mol Sci. 2022;23(5):2425. https://doi.org/10.3390/ijms23052425.

28. Kumar S, Basu M, Ghosh P, Pal U, Ghosh MK. COVID-19 therapeutics: clinical application of repurposed drugs and futuristic strategies for target-based drug discovery. Genes Dis. 2023;10(4):1402–28. https://doi.org/10.1016/j.gendis.2022.12.019.
29. Leitner WW, Ying H, Restifo NP. DNA and RNA-based vaccines: principles, progress and prospects. Vaccine. 1999;18(9–10):765–77.
30. Leng Z, Zhu R, Hou W, Feng Y, Yang Y, Han Q, Shan G, Meng F, Du D, Wang S, Fan J, Wang W, Deng L, Shi H, Li H, Hu Z, Zhang F, Gao J, Liu H, et al. Transplantation of ACE2- mesenchymal stem cells improves the outcome of patients with COVID-19 pneumonia. Aging Dis. 2020;11(2):216–28. https://doi.org/10.14336/AD.2020.0228.
31. Liang B, Chen J, Li T, Wu H, Yang W, Li Y, Li J, Yu C, Nie F, Ma Z, Yang M, Xiao M, Nie P, Gao Y, Qian C, Hu M. Clinical remission of a critically ill COVID-19 patient treated by human umbilical cord mesenchymal stem cells: a case report. Medicine (Baltimore). 2020;99(31):e21429. https://doi.org/10.1097/MD.0000000000021429.
32. Liu F, Xu A, Zhang Y, Xuan W, Yan T, Pan K, Yu W, Zhang J. Patients of COVID-19 may benefit from sustained Lopinavir-combined regimen and the increase of eosinophil may predict the outcome of COVID-19 progression. Int J Infect Dis. 2020;95:183–91. https://doi.org/10.1016/j.ijid.2020.03.013.
33. Mackenzie JS, Smith DW. COVID-19: A novel zoonotic disease caused by a coronavirus from China: what we know and what we don't. Microbiol Australia. 2020;MA20013 https://doi.org/10.1071/MA20013.
34. Magro P, Zanella I, Pescarolo M, Castelli F, Quiros-Roldan E. Lopinavir/ritonavir: repurposing an old drug for HIV infection in COVID-19 treatment. Biom J. 2021;44(1):43–53. https://doi.org/10.1016/j.bj.2020.11.005.
35. Maurya CK, Misra R, Sharma P, Singh N, Awasthi H, Agrawal R, Misra S, Dwivedi S. Novel stem cells and nucleic acid-based vaccine trials against viral outbreak: A systematic evaluation during COVID-2019 pandemic. Indian J Clin Biochem. 2020;35(4):397–409. https://doi.org/10.1007/s12291-020-00907-4.
36. Mehta P, McAuley DF, Brown M, Sanchez E, Tattersall RS, Manson JJ, HLH Across Speciality Collaboration, UK. COVID-19: consider cytokine storm syndromes and immunosuppression. Lancet (London, England). 2020;395(10229):1033–4. https://doi.org/10.1016/S0140-6736(20)30628-0.
37. Metcalfe SM, Strom TB, Williams A, Fahmy TM. Multiple sclerosis and the LIF/IL-6 Axis: use of nanotechnology to harness the Tolerogenic and reparative properties of LIF. Nano. 2015;2:5. https://doi.org/10.5772/60622.
38. Proskurnina EV, Izmailov DY, Sozarukova MM, Zhuravleva TA, Leneva IA, Poromov AA. Antioxidant potential of antiviral drug Umifenovir. Molecules. 2020;25(7):Article 7. https://doi.org/10.3390/molecules25071577.
39. Pushpakom S, Iorio F, Eyers PA, Escott KJ, Hopper S, Wells A, Doig A, Guilliams T, Latimer J, McNamee C, Norris A, Sanseau P, Cavalla D, Pirmohamed M. Drug repurposing: Progress, challenges and recommendations. Nat Rev Drug Discov. 2019;18(1):41–58. https://doi.org/10.1038/nrd.2018.168.
40. Rahmani AM, Mirmahaleh SYH. Coronavirus disease (COVID-19) prevention and treatment methods and effective parameters: A systematic literature review. Sustain Cities Soc. 2021;64:102568. https://doi.org/10.1016/j.scs.2020.102568.
41. Research, C. for D. E. and. Remdesivir (Veklury) approval for the treatment of COVID-19—the evidence for safety and efficacy. FDA. 2024. https://www.fda.gov/drugs/spotlight-cder-science/remdesivir-veklury-approval-treatment-covid-19-evidence-safety-and-efficacy
42. Skipper CP, Pastick KA, Engen NW, Bangdiwala AS, Abassi M, Lofgren SM, Williams DA, Okafor EC, Pullen MF, Nicol MR, Nascene AA, Hullsiek KH, Cheng MP, Luke D, Lother SA, MacKenzie LJ, Drobot G, Kelly LE, Schwartz IS, et al. Hydroxychloroquine in nonhospitalized adults with early COVID-19. Ann Intern Med. 2020:M20-4207. https://doi.org/10.7326/M20-4207.
43. Song Y, Peng W, Tang D, Dai Y. Protease inhibitor use in COVID-19. Sn Comprehensive Clin Med. 2020;2(9):1436–43. https://doi.org/10.1007/s42399-020-00448-0.

44. Spaggiari GM, Capobianco A, Becchetti S, Mingari MC, Moretta L. Mesenchymal stem cell-natural killer cell interactions: evidence that activated NK cells are capable of killing MSCs, whereas MSCs can inhibit IL-2-induced NK-cell proliferation. Blood. 2006;107(4):1484–90. https://doi.org/10.1182/blood-2005-07-2775.
45. Umifenovir—An overview | ScienceDirect Topics. n.d.. Retrieved June 8, 2024, from https://www.sciencedirect.com/topics/chemistry/umifenovir
46. U.S. FDA Clears Pluristem's IND Application for Phase II COVID-19 Study. n.d.. Retrieved June 6, 2024, from https://edition.cnn.com/business/newsfeeds/globenewswire/7917412.html
47. Wahl A, Gralinski LE, Johnson CE, Yao W, Kovarova M, Dinnon KH, Liu H, Madden VJ, Krzystek HM, De C, White KK, Gully K, Schäfer A, Zaman T, Leist SR, Grant PO, Bluemling GR, Kolykhalov AA, Natchus MG, et al. SARS-CoV-2 infection is effectively treated and prevented by EIDD-2801. Nature. 2021;591(7850):451–7. https://doi.org/10.1038/s41586-021-03312-w.
48. World Health Organization. 2020, October 15. DRAFT landscape of COVID-19 candidate vaccines.
49. Yan D, Liu X-Y, Zhu Y-N, Huang L, Dan B-T, Zhang G-J, Gao Y-H. Factors associated with prolonged viral shedding and impact of lopinavir/ritonavir treatment in hospitalised non-critically ill patients with SARS-CoV-2 infection. Eur Respir J. 2020;56(1):2000799. https://doi.org/10.1183/13993003.00799-2020.
50. Li Y, Xie Z, Lin W, Cai W, Wen C, Guan Y, Mo X, Wang J, Wang Y, Peng P, Chen X, Hong W, Xiao G, Liu J, Zhang L, Hu F, Ding J, Zheng Z, Wang S, Zhang Y, Xu L. An exploratory randomized, controlled study on the efficacy and safety of lopinavir/ritonavir or arbidol treating adult patients hospitalized with mild/moderate COVID-19 (ELACOI). MedRxiv. 2020. https://doi.org/10.1101/2020.03.19.20038984.

The Role of Anti-inflammatory Agents in the Management of COVID-19

14

Sibani Sarkar, Sunny Kumar, and Malini Basu

Abstract

The COVID-19 pandemic caused by SARS-CoV-2 has created unprecedented challenges for worldwide public health, necessitating the quick development of effective treatment methods. Systemic inflammation and sometimes hyper-inflammatory responses with cytokine storm are associated with the severity of the disease and poor prognosis and even death. Several inflammatory hallmarks including C-reactive protein. Ferritin, D-dimers and different cytokines such as IL-6, IL-10 and TNF-α have been reported. The complications are managed by the use of anti-inflammatory drugs to alleviate the dysregulated immune response and cytokine-mediated inflammation that increased the survival rate and lesser hospitalization. Moreover, several customized medicine techniques must be taken that depend on patient variables, disease severity and inflammatory biomarkers in order to adjust treatment regimens accordingly. Despite these achievements, there are still many hurdles and open issues to fight against this disease. Thus, the optimized use of anti-inflammatory drugs avoids possible risks and bad effects. The effectiveness and safety of anti-inflammatory medicines in COVID-19, including corticosteroids, IL-6 inhibitors, JAK inhibitors and colchicine, among others, have been looked upon. Various research approaches investigate the possible synergistic effects of anti-inflammatory drugs by identifying novel targets and pathways implicated in inflammation after COVID-19 infection.

S. Sarkar (✉) · S. Kumar
Cancer Biology and Inflammatory Disorder Division, Council of Scientific and Industrial Research-Indian Institute of Chemical Biology (CSIR-IICB), TRUE Campus, Kolkata, West Bengal, India

M. Basu (✉)
Department of Microbiology, Dhruba Chand Halder (DCH) College - University of Calcutta, Jaynagar Mazilpur, West Bengal, India

M. K. Ghosh, M. Basu (eds.), *Pathogenesis, Management and Socio-Economic Impact of COVID-19 Pandemic*, Medical Virology: From Pathogenesis to Disease Control, https://doi.org/10.1007/978-981-95-1426-7_14

Keywords

SARS-CoV-2 · Anti-inflammatory agents · Cytokine storm

Introduction

The new coronavirus SARS-CoV-2 which produced the COVID-19 pandemic, had quickly spread over the world, posing a threat to healthcare systems and igniting a desperate hunt for effective treatment solutions. Inflammatory mechanisms, dysregulated immunological responses and viral replication all play intricate roles in the pathophysiology of COVID-19. An overactivation of the virus triggers a cytokine storm consisting of pro-inflammatory cytokines like IL-6, TNF-α and IL-1β after severe COVID-19 infections. Since inflammation plays a major role in the pathophysiology of COVID-19, inflammatory pathways are the major focus for a sensible treatment strategy. Anti-inflammatory drugs have been studied for their ability to modify the immune response and lessen the hyper-inflammatory state linked to the severity of disease. These drugs include corticosteroids, colchicine, JAK inhibitors, IL-6 inhibitors and others. It was noted that controlled trials have shown that corticosteroids, including dexamethasone, are effective in lowering mortality and improving clinical outcomes in hospitalized COVID-19 patients with severe respiratory disease. Similarly, IL-6 inhibitors like tocilizumab and sarilumab slow disease progression and increase survival in critically sick patients with increased inflammatory markers [16, 24].

The effectiveness of anti-inflammatory medicines suggests future avenues for clinical study and treatment with evaluation for the safety and long-term effects of anti-inflammatory drugs. Drugs are implemented in these patients by investigating the possible synergistic benefits of combination therapy. Even with these developments, there are still a number of problems and unsolved issues. Research is still ongoing to determine the best patient demographics, minimize risks and side effects and optimize the time of anti-inflammatory drug initiation. Furthermore, a tailored medicine strategy takes into account unique patient characteristics, the severity of the disease, and inflammatory biomarkers, which may improve therapy results even more [8]. Moreover, it is critical to make steps to guarantee that all patients, even those in remote with limited resources, have fair access to these drugs. To sum up, anti-inflammatory drugs have great potential to help control the disease since they can lessen the cytokine storm and also the severity of the illness.

General Features of COVID-19 Pandemic

COVID-19 was identified first in December 2019 in the city of Wuhan, Hubei Province, China. Here is a brief overview of the COVID-19 pandemic.

Emergence and Origin

Prior to the twenty-first century, coronaviruses included as a pathogen in veterinary medicine but had little impact on human physiology. But global concern started in 2002, when the SARS dreaded infectious disease in humans was identified. Ten years after that, new atypical respiratory diseases caused concern worldwide, named as Middle East Respiratory Syndrome (MERS), more fatal than SARS (2002). At the end of 2019, a new epidemic created a historic repeat called COVID-19, a novel coronavirus SARS-CoV-2 was identified [15].

Coronaviridae family includes enveloped viruses which has a positive-sense single-stranded RNA genome of α-coronavirus, β-coronavirus, γ-coronavirus and δ-coronavirus. SARS-CoV and MERS-CoV belong to the genus of β-coronavirus. Recently, SARS-CoV-2 was also included in the same genus after full genome sequencing and phylogenetic analysis. These all were caused by the coronavirus that caused flu-like symptoms but the clinical outcome was more severe. SARS-CoV-2's exact origins are unknown; however, it was thought to have originated in bats and may spread to people via an intermediate animal host. SARS-CoV-2 was first detected in December 2019 in Wuhan, Hubei Province, China. Then the virus swiftly spread from Wuhan to other parts of China, and eventually to other countries around the world. Human-to-human transmission was mostly by respiratory droplets produced when an infected person coughs, sneezes or speaks, but it can also spread through contact with contaminated surfaces [13, 17].

Clinical Presentation, Severity and Risk Factors

On 11 March 2020, WHO considered COVID-19 as a pandemic and it crossed the milestone of five million deaths by Oct 2021 in spite of restrictions on movements, social distancing and lockdown. COVID-19 symptoms can range from mild to severe and occur between 2 and 14 days after infection. It was an acute respiratory disease caused by SARS-CoV-2. Patient diagnosed with COVID-19 had fever, cough, shortness of breath, exhaustion, muscle or body aches, loss of taste or smell, sore throat, congestion or runny nose, nausea or vomiting and diarrhoea. The infection was asymptomatic at the earlier ages until the emergence of pneumonia, dyspnoea, organ dysfunction and even death. The clinical spectrum was wide and can result in ARDS and death, especially in older persons and those with underlying medical disorders. The infection spread across all ages but children had lower clinical impacts than adults and better recovery. There was also a variation in infections, hospitalization and deaths within different regions in the country and worldwide as raising questions about protection and risk factors of the COVID-19 pandemic. These concerned factors will be helpful to prevent disease infection, progression and adverse outcomes of infected patients [23].

Hypertension was the most common co-morbid condition. ACE2 protects the cardiovascular system from the action of angiotensin II but it is altered in diabetes and hypertensive renal disease. SARS-CoV-2 infection downregulates ACE2 by binding the virus to the ACE2 receptor, and thus, clinical deterioration occurs by releasing inflammatory cytokines. Previously certain characteristics, such as older age, underlying health disorders (viz., cardiac, diabetes, obesity and immunosuppression) and demographic factors (viz., male gender), have been linked to an increased risk of severe COVID-19 but it had been shown that the rate of infection was same but even higher in adolescents than adults and women were more prone than male. Differences in the immune systems may be partly explained by the severity of the pathogenesis of the disease. Based on reports, risk factors to develop COVID-19 in adults had ranged demographically by older age, male sex and different diseases such as cardiovascular diseases, hypertension and chronic obstructive pulmonary diseases (COPD). Younger children and specific comorbidities such as obesity had a higher risk of infection and more severe consequences of COVID-19. Additionally, the major risk factors for severity and mortality of COVID-19 include laboratory indices, pro-inflammatory cytokine levels and complications. But healthy diet/nutrition, COVID-19 vaccine and atopic conditions may be protected from diseases and also avoid progression [1, 4, 25].

Impact of COVID-19 Pandemic

Recently, our world is facing the worst effects of COVID-19 not only in economic sectors but also in the health of people, the environment, education and human psychology. Certainly, virus spreading was more than the hospital capacity, and thus, infected people were at home care, which caused a source of waste contamination with the virus and disruption of municipal waste management. The COVID-19 rapidly expanded beyond China, resulting in a global epidemic. WHO have proclaimed COVID-19 a public health emergency of international concern (PHEIC) on 30 January 2020 and then a pandemic on 11 March 2020, as instances increased internationally [9].

COVID-19 has had far-reaching consequences on public health, economies and cultures globally. Lockdowns, travel restrictions, quarantine and isolation protocols, mask mandates, social distancing rules and vaccine campaigns were among the steps implemented by governments and health authorities around the world to combat COVID-19 spread. These policies have disturbed daily life, resulted in widespread job loss, taxed healthcare systems and worsened existing social and economic disparities. COVID-19 has put enormous strain on worldwide healthcare systems, resulting in shortages of medical supplies, hospital beds, and healthcare workers. The pandemic also had major social, economic, and psychological consequences, affecting daily life and causing economic downturns, exacerbating health disparities and straining social cohesion [21, 27, 33].

Preventive Measures

Different prevention measures were followed in the COVID-19 period like frequent handwashing with hand sanitizer, wearing face masks in public places to avoid spreading. Besides, some rules must be followed for preventive measures such as a physical distance from others where social distancing was difficult to maintain, avoiding large gatherings and staying at home when sick. Vaccination initiatives were also vital in preventing the spread of the virus. Basic hygiene habits can help limit the spread of COVID-19. These include regularly washing hands with soap and water for at least 20 seconds, using hand sanitizer with at least 60% alcohol if soap and water are not available. One should avoid touching the face, covering coughs and sneezes with a tissue or elbow, cleaning, and disinfecting the objects and surfaces which were frequently touched (Fig. 14.1).

Maintaining physical distance from others, particularly indoors, can help to limit the likelihood of COVID-19 transmission. Health officials normally advise to remain separated at least 6 feet away from persons who were not in your family.

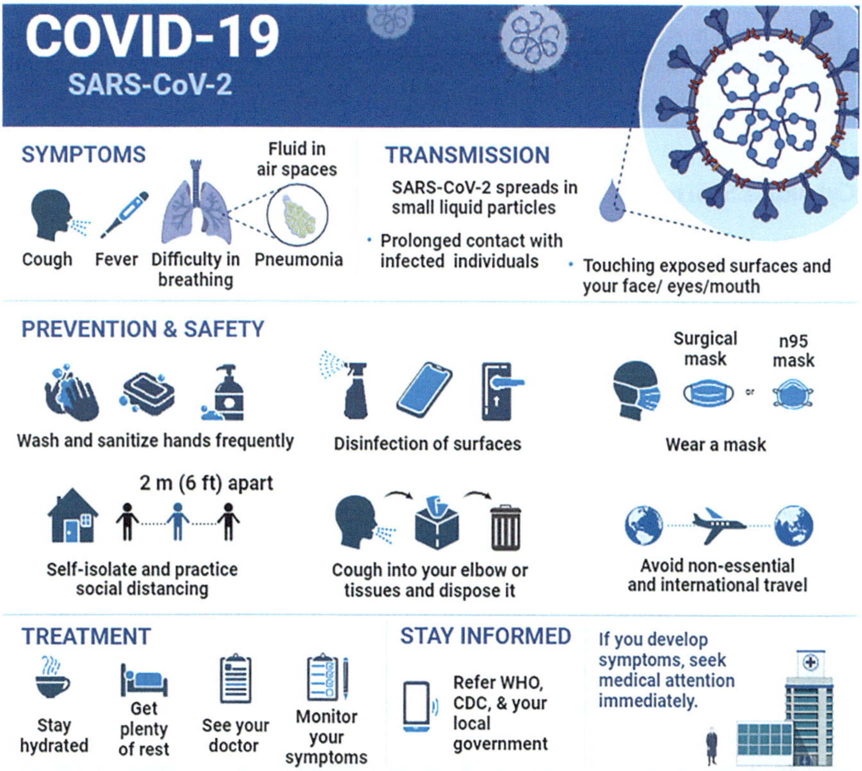

Fig. 14.1 Preventive measures to follow during SARS-CoV-2 infection. (This image is adopted from Biorender (https://www.biorender.com/))

Avoiding crowded places and events can also reduce the risk of contracting the virus. Masks serve to minimize the spread of respiratory droplets that could carry the virus. Wearing masks was a vital preventive step, especially indoors or when physical distance was difficult to maintain. N95 respirators or surgical masks were typically advised for increased protection, particularly in high-risk areas.

Ensuring adequate ventilation in indoor places can help to lower the concentration of virus-containing respiratory droplets. Opening windows and doors, utilizing fans and upgrading HVAC systems can all help to improve air circulation and ventilation. One would stay up-to-date with COVID-19 developments by adhering to public health authorities. Vaccination was an important tool in controlling COVID-19. Thus, combining vaccination with other preventive measures can provide extra benefits, especially in areas where transmission rates were higher with the coming of new variants [27].

Development of Inflammation in COVID-19

Inflammation plays an important role in the COVID-19 disease. In severe cases, the immune system is dysregulated, leading to the overproduction of inflammatory cytokines, such as IL-6, TNF-α, and IL-1β. Excessive inflammation can aggravate the severity of the disease. There are some points to be discussed regarding the inflammation in COVID-19.

Cytokine Storm

COVID-19 can trigger an excessive immune response that leads to the development of a cytokine storm. Thus, it is necessary for the proper functioning of the immune system. Cytokine plays a double-edged role in the human system. Several pathophysiological processes are necessary for survival, such as tissue repair, inflammation, fibrosis, and coagulation but excessive production of cytokines may become harmful, leading to hyper-inflammation called cytokine storm. COVID-19 patients had increased levels of inflammatory cytokines that include IL-1β, IL-2, IL-6, IL-10, TNF-α and IFN-γ inducible protein 10 (IP-10) and were also linked with disease severity. This cytokine storm can cause influenza-like illness and multiorgan damage, though it varies depending on the underlying pathology. In severe cases, fever was almost persistent with some common symptoms like fatigue, headache, diarrhoea, lymphadenopathy, hepato-splenomegaly, sensory alterations and skin rash. Tachypnoea and hypoxemia were common leading to ARDS. In severe cases, renal disease, liver damage, stress-related cardiomyopathy and neurological involvement, including encephalopathy, also happened. Coagulation leads to the possibility of thrombotic events in diffuse intravascular coagulopathy. Thus, researchers were running trials for the treatment of COVID-19 patients using anti-inflammatory agents to mitigate cytokine storm [26] (Fig. 14.2).

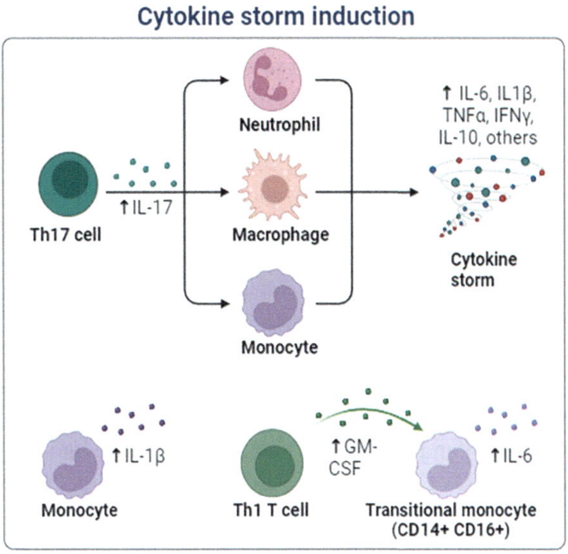

Fig. 14.2 Cytokine storm induction. (This image is adopted from Biorender (https://www.biorender.com/))

Pulmonary Inflammation

COVID pneumonia was an infection in the lungs that developed flu-like symptoms and suffering from type-II pneumocytes by severely infecting lung cells and can cause rich inflammatory cell infiltration consisting of neutrophils, macrophages, CD4+, CD8+ T-lymphocytes and increased production of cytokines that leads to bilateral pneumonia, ARDS, and multiorgan damage. As the immune system attacked the immune system in the lungs, they became inflamed and filled with fluid and difficult to breathe. This caused lung damage in both sides of the lungs and usually occurred after the initial phase of infected people who have had COVID-19 for a long time, and COVID-19 pneumonia spread across the lungs. It had a tendency to last longer and inflict damage in other locations. The severity of the disease was lessened by using different anti-inflammatory drugs that will be discussed later in this chapter [3, 25].

Endothelial Dysfunction

SARS-CoV-2 virus targeted both the pulmonary system and extrapulmonary systems as well and manifested myocardial injury, arterial stiffness, kidney injury, and liver injury. Thus, endothelial dysfunction occurred as a result of pathogenesis in COVID-19. The pulmonary capillary endothelium served as a fertile soil for viral entry and proliferation, making viral particles more easily transported through the bloodstream. SARS-CoV-2 infects ECs and epithelial cells in lung tissues using ACE2 and host cell alternative receptors. Endothelial integrity was essential for the pulmonary capillary alveolar barrier and lung homeostasis. COVID-19 caused broad endothelial injury, increased capillary permeability, inflammatory cell infiltration into perivascular tissue, interstitial oedema, and alveolar fluid retention. Furthermore, cytokine storm, disruption of the tight junction barrier, pulmonary hypertension, and lung fibrosis occurred in the patients as a result of SARS-CoV-2 infection. Severe COVID-19-induced cytokine storm increased multiorgan harm by destroying ECs, producing inflammation and thrombosis. Thus, future treatments that target endothelial dysfunction will be predicted to reduce COVID-19-related lung harm by using anti-inflammatory drugs [2, 31].

Immune Dysregulation

The mechanism of concurrent hyper-inflammatory damage in the presence of systemic immune suppression was undeniably complex. However, the production of IL-6, IL-8 and IL-10 may explain both innate and adaptive immune dysregulation. During COVID-19, macrophages generated higher IL-6, followed by a rise in IL-8. IL-8 attracted neutrophils to the site of infection, whereas IL-6 activated them, causing inflammation and epithelial cell damage. However, IL-6-activated neutrophils suppressed T cells through programmed death ligand (PD-1/PDL-1) signalling. T cell PD-1 expression was significantly higher in COVID-19 patients, resulting in apoptosis and possibly severe lymphopenia in critical COVID-19 patients [25, 30]. Up-regulation of STAT3 led to decreased myeloid cell activation and IFN-γ production from CD4+ and CD8+ T cells suggesting that STAT3 signalling may be a common mechanism for COVID-19 dysregulation. It was known that COVID-19 is largely characterized by a hyper-inflammatory cytokine storm. Thus, the immunological dysregulation exhibited in COVID-19 had serious consequences for treatments. Many immune suppressive medications such as anti-IL-6 receptor antibodies, IL-1 receptor antagonist and JAK/STAT inhibitors have been used for treatments [24, 29] (Fig. 14.3).

Immune Cells and Inflammatory Markers

In patients with severe COVID-19, immune cell numbers were abnormal and pro-inflammatory indicators were significantly elevated. Qin et al. observed that patients with COVID-19 had reduced levels of B cells, T cells and NK cells. The neutrophil lymphocyte ratio (NLR), leukocytes and neutrophil counts were all higher in severe patients, whereas the percentage of basophils, eosinophil and monocytes was lower. Severe cases showed elevated levels of infection-related inflammatory markers such as pro-calcitonin, CRP, serum ferritin, IL-2R, IL-6, IL-8 and TNF-α. These inflammatory indicator levels, lactate dehydrogenase and ferritin level indicate the death chance in patients with COVID-19. Elevated levels of these blood soluble markers were observed in critically ill patients. IL-6 and LDH were independent predictors of disease severity, whereas an early reduction in inflammatory markers led to better outcomes [7, 22, 35].

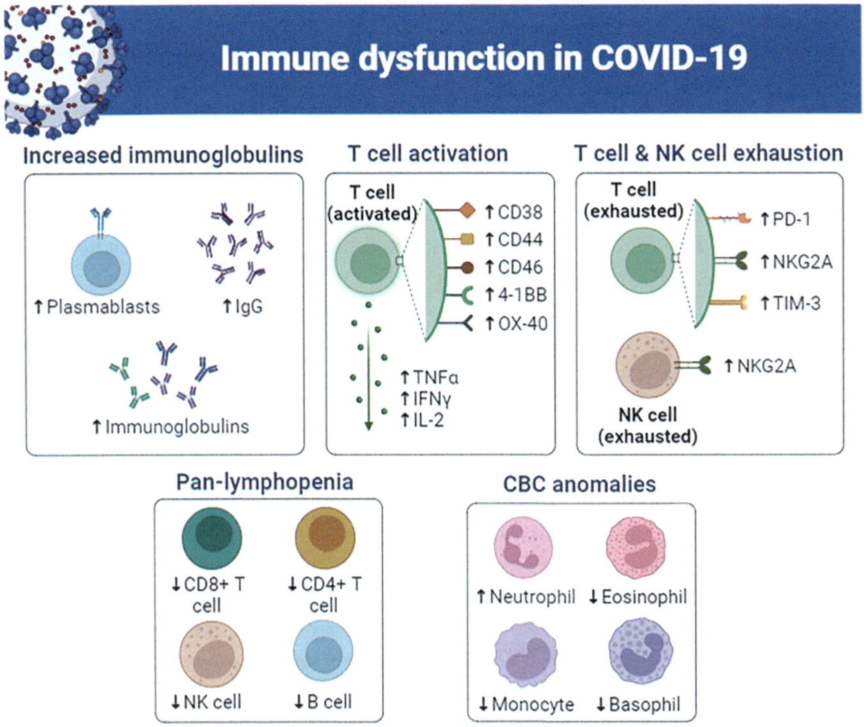

Fig. 14.3 Immune dysfunction in COVID-19 infection. (This image is adopted from Biorender (https://www.biorender.com/))

Management of COVID-19

Repurposing of anti-inflammatory agents in the context of COVID-19 has been an area of significant interest since the onset of the pandemic. This virus can lead to a range of symptoms from mild respiratory illness to severe pneumonia, acute respiratory syndrome and systemic inflammation. In severe cases, an excessive inflammatory response leads to cytokine storm associated with multiorgan failure. Several anti-inflammatory medications have been examined or proposed for the treatment of COVID-19. It would go over the mechanisms of action of various anti-inflammatory drugs and how they might reduce the inflammatory response associated with COVID-19. Suppressing cytokine synthesis, regulating immune cell activity and reducing tissue inflammation may play an important role. Timing of administration, patient heterogeneity, potential side effects and drug interactions were the problems and limitations of using anti-inflammatory drugs in COVID-19 management. The current evidence from clinical trials on the efficacy and safety of several anti-inflammatory medications for COVID-19 came from randomized controlled trials, observational research and meta-analyses [12, 20, 24]. Thus, the anti-inflammatory drugs were used to design therapeutic strategies in COVID-19 cases to identify biomarkers for patient selection with new pharmacological targets.

Anti-inflammatory Agents

Several immunological abnormalities were shown in severe COVID-19 patients. Thus, many immunomodulatory medicines had been used to these patients which mostly comprise of the following categories: i) Inhibition of systemic or local excessive immune and inflammatory responses (cytokine storm) such as glucocorticoids, inflammatory cytokine inhibitors, kinase inhibitors and nonsteroidal anti-inflammatory drugs ii) biological agents that can enhance the antiviral immune response, such as interferon and iii) drugs such as hydroxyl-chloroquine, that may have both antiviral and immune modulatory properties [22, 24]. In COVID-19, the research on the use of anti-inflammatory agents were also ongoing (Table 14.1).

Corticosteroids Glucocorticoids, one of the types in the corticosteroid family, are steroid hormones and have strong anti-inflammatory properties that can reduce inflammation by numerous factors. They work by reducing the immune response, specifically the inflammatory cascade and thus preventing cytokine storms. It could reduce mortality in severe COVID-19 patients, especially those who require oxygen supplementation or mechanical ventilation. However, it was noted that glucocorticoids should be used cautiously by following strict prescriptions and closely monitoring side effects. More evidence was required to use glucocorticoids judiciously, including dosage, course of treatment, time of administration and combination of drugs [10, 34].

Colchicine The systemic corticosteroids were commonly used to treat gout and Familial Mediterranean fever (FMF). It had anti-inflammatory properties by stabilizing polymorph nuclear cells and had been studied to lessen inflammation in COVID-19 patients. It inhibited NLRP3 and blocked the activation of IL-1β, IL-6, 8 by reducing and managing excessive cytokine storm. Colchicine appear to have considerable effects on improving the symptoms, course and fatality rates caused by novel corona virus disease. Some trials have been done to encourage advantages in minimizing the risk of complications in COVID-19 patients in the hospital [14, 28, 32].

Nonsteroidal Anti-inflammatory Agents Ibuprofen is a commonly prescribed medication that reduces inflammation and relieves symptoms in a number of conditions. The use of NSAIDs in the treatment of COVID-19 remained controversial, though they are generally used to treat fever and rheumatic disorders, due to several reasons. Firstly, the most serious consequences of COVID-19 were sepsis, cardiovascular and respiratory failure in the aged and persons with comorbidities. NSAIDs

Table 14.1 Effect of multiple anti-inflammatory agents

Sl. no.	Category	Drug	Mechanism	Effect	Clinical evidence
1	Corticosteroids	Dexamethasone	Inhibits cytokines	Decrease mortality	Reduced mortality
2	NSAIDS	Ibuprofen	Inhibits COX enzymes	Controversial	Use with caution
3	IL-6 inhibitor	Tocilizumab	Reduces inflammation	Reduces mortality	REMAP-CAP & RECOVERY
4	Antimalarial	Hydroxy-chloroquine	Anti-inflammatory and antiviral effects	Not clear	Potential risk
5	JAK inhibitors	Baricitinib	Inhibit JAK pathway	Reduces mortality	ACTT-2
6	Monoclonal antibodies	Casirivimab & imdevimab	Neutralization	Reduces hospitalization	Phase 3 trial
7	Anti-fibrotic agent	Pirfenidone	Reduces fibroblast proliferation, collagen synthesis, & cytokines	Reduces pulmonary fibrosis	Limited evidence
8	Traditional medicine	Curcumin	Inhibits NF-kB	Reduces inflammation	Limited evidence
9	Traditional medicine	Colchicine	Reduces neutrophils & inflammatory mediators	Reduces hospitalization	COLCORONA
10	Vitamins	Vitamin D	Reduces cytokines	Reduces severity	Ongoing
11	IL-6 inhibitors	Anakinra	Reduces inflammation	Improve serum IL-1β levels	Improves mortality

had adverse effects such as myocardial infarction, heart failure and stroke, limiting their use in COVID-19 treatment. Secondly, it could cause renal toxicity, which was exacerbated in patients who were dehydrated due to fever, gastrointestinal symptoms and dyspnoea. Thirdly, sometimes NSAIDs might reduce defence against viruses. It may delay the inflammatory response and alleviate infection in bacterial pneumonia by reducing polymorphonuclear cell recruitment. Sometimes NSAIDs may aggravate the disease by up-regulating ACE2 levels in the target cells. Thus, it had been suggested a risk of worsening outcomes, while others found no discernible influence. NSAIDs should be taken with caution and in the advice of a healthcare professional [12, 13, 22, 23].

IL-6 inhibitors (Tocilizumab) IL-6 is a cytokine that contributes to the inflammatory response. Current research suggests that IL-6 receptor monoclonal antibody (tocilizumab) in critically unwell COVID-19 patients was advantageous as serum IL-6 levels were critically linked with the severity of the disease, which can predict the necessity of ventilation. Thus, tocilizumab, an IL-6 receptor antagonist, has been used to treat COVID-19 individuals with severe inflammation, particularly those experiencing cytokine storms. Reducing the need for mechanical ventilation and improving survival in critically ill patients were observed, though it may be biased [16, 24, 31].

IL-6 inhibitors (Anakinra) Anakinra is a human recombinant IL-1 receptor antagonist, generally used to treat rheumatoid arthritis, NLRP3-related auto-inflammatory disease, severe cytokine release syndrome mediated by chimeric antigen receptor T cells (CAR-T cells) and macrophage activation syndrome. It had been suggested that anakinra may reduce inflammation and improve prognosis in patients, as serum IL-1β levels had been elevated in COVID-19 individuals [19].

JAK Inhibitors (Baricitinib) Sometimes, Janus kinase inhibitors interfere with the signal transducer and activator of the STAT pathway. These inhibitors reduce immune activation and inflammation in cells by reducing inflammatory cytokines. These drugs inhibit the activity of Janus kinases, which are implicated in the signalling pathways of several cytokines that cause inflammation. They are used to treat haematological disorders, rheumatoid arthritis and psoriasis. Baricitinib in combination with remdesivir was licensed for emergency use in hospitalized COVID-19 patients who required supplemental oxygen, invasive mechanical ventilation or extracorporeal membrane oxygenation [6, 11].

Sometimes, monoclonal antibodies, such as tocilizumab and sarilumab, are used to target specific inflammatory pathways, though they are not anti-inflammatory. They showed efficacy in lowering the risk of disease progression in hospitalized COVID-19 patients, particularly those with elevated inflammatory markers. The efficacy and safety of the anti-inflammatory agents can differ based on the stage and severity of the disease, patient characteristics and potential drug interactions. Treatment decisions should be based on the most recent evidence and recommendations from healthcare professionals. Furthermore, ongoing research is investigating new anti-inflammatory approaches and repurposed medications to treat COVID-19 [5, 18].

Mechanistic Study: Anti-inflammatory Agents in COVID-19

Anti-inflammatory medicines are essential for severe COVID-19 patients, as when overactive immune response can cause considerable tissue damage and organ failure. Anti-inflammatory drugs have an important role in COVID-19 management because they reduce the strong inflammatory response in people with this condition.

The mechanism is quite complicated and multifaceted. Dexamethasone is quietly involved in minimizing inflammation by blocking various inflammatory pathways, thus involving apoptosis of some immune cells such as: i) NF-κβ and IAP-1 are inhibited by suppressing pro-inflammatory genes. ii) Leucocyte migration to inflammatory areas is reduced. iii) Decreased phospholipase A2 reduces the generation of inflammatory mediators such as prostaglandin and leukotrienes [34]. Ibuprofen, as NSAID drug, has become more controversial than corticosteroids in COVID-19 management. The mechanism of action of NSAID drugs is in this way: (i) It reduces inflammation by inhibiting cyclooxygenase enzymes (COX) that lead to decreased synthesis of prostaglandins which are powerful mediators of inflammation. (ii) Decreased prostaglandin reduces vasodilation, vascular permeability and leucocyte migration [24]. Antagonist of IL-6, tocilizumab, was used for the treatment of COVID-19 patients. It suppresses IL-6, blocks the downstream signalling pathways, and thus lowers acute phase protein synthesis, immune cell recruitment and overall inflammation [24]. JAK inhibitors, as for example baricitinib, treat patients by blocking JAK-STAT signalling pathway which is crucial for various cytokine signalling molecules. These drugs reduce the phosphorylation and activation of STAT proteins by inhibiting JAK1, JAK2 that leads to decreased transcription of pro-inflammatory genes. Thus, inflammation was reduced in COVID-19 [24]. Colchicine is used to treat gout as well as in COVID-19 patient treatments by disrupting microtubule formation which is essential for the function of many inflammatory cells. It also impairs motility and function of neutrophils by inhibiting microtubule assembly and inflammasomes, particularly NLRP3 production are reduced. Thus, colchicine was used for the inhibition of inflammatory pathways in COVID-19 cases [28].

Providing anti-inflammatory drugs early in high-risk patients may help to reduce disease progression and its consequences. They can prevent mild or moderate COVID-19 instances but when it progresses to severe cases that require hospitalization or intensive care. COVID-19 pneumonia can cause ARDS, characterized by inflammation and decreased gas exchange in the lungs. Thus, anti-inflammatory medicines like corticosteroids have been shown to enhance oxygenation and lessen the need for mechanical ventilation until and unless the case is not severe. Early treatment with the anti-inflammatory agents can enhance survival chances in the severe patients where inflammation and cytokine storm lead to multiorgan failure and death. In the clinical study, it has been shown that corticosteroids like dexamethasone reduce mortality in hospitalized patients who require oxygen supplementation. Though anti-inflammatory medicines are clinically helpful to reduce the severity and mortality of the patients but they can also cause immunosuppression, hyper-glycemia, secondary infections and gastrointestinal problems. Thus, close monitoring of patients is essential to identify the critical side effects of using these types of drugs. But still they are used because they can moderate immune response, reduce inflammation and improve clinical outcomes, especially in severe cases where uncontrolled inflammation causes death [24].

Future Perspectives

COVID-19 has triggered a global health crisis. As the COVID-19 pandemic starts, several management measures must be adapted, including the critical use of new anti-inflammatory medicines in reducing the strong anti-inflammatory response. These approaches are based on the development of new anti-inflammatory drugs, the improvement of existing therapies, combination therapy, targeted delivery systems and customized medicines. Research is ongoing to create medications involving the target to inflammatory pathways that lead to complications such as ARDS, multiorgan failure and cytokine storm linked to severe COVID-19. The inflammatory response is driven by the immune system's overreaction to the virus that leads to excessive production of pro-inflammatory cytokines. We have already discussed some anti-inflammatory drugs to nullify cytokine storm, those are also currently used in managing COVID-19 such as dexamethasone, tocilizumab, baricitinib, etc. They have to target specific cytokine signalling pathways, reduce inflammation and lower mortality rates. But these treatments have several challenges. For example, corticosteroids can cause immunosuppression that increases the risk of secondary infection, and broad immunosuppressive agents can dampen the overall immune response, affecting the body's ability to fight against infection.

Scientists are focused on developing new anti-inflammatory drugs that specifically target key inflammatory pathways without broadly suppressing the immune system and also minimize adverse effects to make them safer for longer use. Combining antiviral agents with anti-inflammatory drugs may improve outcomes by controlling both viral replication and anti-inflammatory response. Combining therapy with drugs targeting different inflammatory pathways also may offer a more comprehensive strategy to manage severe COVID-19. As a drug discovery approach, genomics and proteomic approaches have been used to identify new molecular targets involved in the inflammatory response to SARS-CoV-2. High-throughput screening of chemical libraries is used to find out the potential anti-inflammatory compounds. These compounds must be validated in vitro cell cultures to assess their efficacy in reducing inflammatory markers and also in vivo COVID-19 animal models to study the effects of the new drug on inflammatory response and disease progression. Then, a clinical trial must be done to assess the safety of the dosage compared with current standard treatments in a larger, diverse patient population.

Conclusion

COVID-19 management with anti-inflammatory medicines is protected and becoming more sophisticated in future with a focus on novel drug discovery, therapy optimization, combinatorial approaches, targeted delivery systems and personalized medicines. Chronic inflammation and global accessibility are crucial to the efficacy of these strategies in reducing the effects of COVID-19 patient outcomes. Overall, the article emphasizes the significance of regulating the inflammatory response in COVID-19 patients. While several anti-inflammatory medicines show promise for further high-quality research and is required to validate their efficacy and safety. These agents must be used in accordance with the demands of each patient and the clinical situation.

References

1. Acosta MAT, Singer BD. Pathogenesis of COVID-19-induced ARDS: implications for an ageing population. Eur Respir J. 2020;56(3) https://erj.ersjournals.com/content/56/3/2002049.short
2. Acosta Saltos F, Acosta Saltos AD. Entry of SARS-CoV2 through the basal surface of alveolar endothelial cells–a proposed mechanism mediated by CD147 in Covid-19. 2020.
3. Ahmad S, Manzoor S, Siddiqui S, Mariappan N, Zafar I, Ahmad A, Ahmad A. Epigenetic underpinnings of inflammation: connecting the dots between pulmonary diseases, lung cancer and COVID-19. Semin Cancer Biol. 2022;83:384–98.
4. Barnes PJ. Role of HDAC2 in the pathophysiology of COPD. Annu Rev Physiol. 2009;71:451–64.
5. Benucci M, Giannasi G, Cecchini P, Gobbi FL, Damiani A, Grossi V, Infantino M, Manfredi M. COVID-19 pneumonia treated with sarilumab: a clinical series of eight patients. J Med Virol. 2020;92:2368.
6. Cantini F, Niccoli L, Nannini C, Matarrese D, Di Natale ME, Lotti P, Aquilini D, Landini G, Cimolato B, Di Pietro MA. Beneficial impact of Baricitinib in COVID-19 moderate pneumonia; multicentre study. J Infect. 2020;81(4):Article 4.
7. Chen J, Lau YF, Lamirande EW, Paddock CD, Bartlett JH, Zaki SR, Subbarao K. Cellular immune responses to severe acute respiratory syndrome coronavirus (SARS-CoV) infection in senescent BALB/c mice: CD4+ T cells are important in control of SARS-CoV infection. J Virol. 2010;84(3):Article 3.
8. Chen JS, Alfajaro MM, Chow RD, Wei J, Filler RB, Eisenbarth SC, Wilen CB. Nonsteroidal anti-inflammatory drugs dampen the cytokine and antibody response to SARS-CoV-2 infection. J Virol. 2021;95(7):e00014-21. https://doi.org/10.1128/JVI.00014-21.
9. Clemente-Suárez VJ, Navarro-Jiménez E, Moreno-Luna L, Saavedra-Serrano MC, Jimenez M, Simón JA, Tornero-Aguilera JF. The impact of the COVID-19 pandemic on social, health, and economy. Sustainability. 2021;13(11):6314.
10. Dixit KS, Kumthekar PU. Optimal Management of Corticosteroids in patients with intracranial malignancies. Curr Treat Options in Oncol. 2020;21(9):77. https://doi.org/10.1007/s11864-020-00771-7.
11. Favalli EG, Biggioggero M, Maioli G, Caporali R. Baricitinib for COVID-19: a suitable treatment? Lancet Infect Dis. 2020;20(9):Article 9.
12. Ghosh MK, Kumar S, Begam S, Ghosh S, Basu M. GBM immunotherapy: exploring molecular and clinical frontiers. Life Sci. 2024:123018. https://doi.org/10.1016/j.lfs.2024.123018.
13. Ghosh MK, Kumar S, Ganguly KK, Ghosh P, Tabassum S, Basu B, Basu M. COVID-19 and cancer: insights into their association and influence on genetic and epigenetic landscape. Epigenomics. 2023;15(4):227–48. https://doi.org/10.2217/epi-2023-0052.
14. Group CH, Mikolajewska A, Fischer A-L, Piechotta V, Mueller A, Metzendorf M-I, Becker M, Dorando E, Pacheco RL, Martimbianco ALC. Colchicine for the treatment of COVID-19. Cochrane Database Syst Rev. 1996;2021(10) https://www.cochranelibrary.com/cdsr/doi/10.1002/14651858.CD015045/abstract
15. Idris I, Adesola RO. Emergence and spread of JN.1 COVID-19 variant. Bull Natl Res Cent. 2024;48(1):27. https://doi.org/10.1186/s42269-024-01183-5.
16. Jiang Y, Zhao T, Zhou X, Xiang Y, Gutierrez-Castrellon P, Ma X. Inflammatory pathways in COVID-19: mechanism and therapeutic interventions. MedComm. 2022;3(3):e154. https://doi.org/10.1002/mco2.154.

17. Kadali RA, Janagama R, Peruru S, Malayala SV. Side effects of BNT162b2 mRNA COVID-19 vaccine: a randomized, cross-sectional study with detailed self-reported symptoms from healthcare workers. Int J Infect Dis. 2021;106:376–81.
18. Khan FA, Stewart I, Fabbri L, Moss S, Robinson K, Smyth AR, Jenkins G. Systematic review and meta-analysis of anakinra, sarilumab, siltuximab and tocilizumab for COVID-19. Thorax. 2021;76(9):907–19.
19. Khani E, Shahrabi M, Rezaei H, Pourkarim F, Afsharirad H, Solduzian M. Current evidence on the use of anakinra in COVID-19. Int Immunopharmacol. 2022;111:109075.
20. Kritas SK, Ronconi G, Caraffa AL, Gallenga CE, Ross R, Conti P. Mast cells contribute to coronavirus-induced inflammation: new anti-inflammatory strategy. J Biol Regul Homeost Agents. 2020;34(1):Article 1.
21. Kumar S, Basu M, Ghosh MK. Chaperone-assisted E3 ligase CHIP: a double agent in cancer. Genes Dis. 2021;9(6):Article 6. https://doi.org/10.1016/j.gendis.2021.08.003.
22. Kumar S, Basu M, Ghosh MK. E3 ubiquitin ligases and deubiquitinases in colorectal cancer: emerging molecular insights and therapeutic opportunities. Biochimica et Biophysica Acta (BBA) Mol Cell Res. 2024;119827
23. Kumar S, Basu M, Ghosh P, Ansari A, Ghosh MK. COVID-19: clinical status of vaccine development to date. Br J Clin Pharmacol. 2023a;89(1):114–49. https://doi.org/10.1111/bcp.15552.
24. Kumar S, Basu M, Ghosh P, Pal U, Ghosh MK. COVID-19 therapeutics: clinical application of repurposed drugs and futuristic strategies for target-based drug discovery. Genes Dis. 2023b;10(4):1402–28. https://doi.org/10.1016/j.gendis.2022.12.019.
25. Kumar S, Chatterjee M, Ghosh P, Ganguly KK, Basu M, Ghosh MK. Targeting PD-1/PD-L1 in cancer immunotherapy: an effective strategy for treatment of triple-negative breast cancer (TNBC) patients. Genes Dis. 2023c;10(4):1318–50.
26. Mangalmurti N, Hunter CA. Cytokine storms: understanding COVID-19. Immunity. 2020;53(1):Article 1.
27. McBride O, Murphy J, Shevlin M, Gibson-Miller J, Hartman TK, Hyland P, Levita L, Mason L, Martinez AP, McKay R. Monitoring the psychological, social, and economic impact of the COVID-19 pandemic in the population: context, design and conduct of the longitudinal COVID-19 psychological research consortium (C19PRC) study. Int J Methods Psychiatr Res. 2021;30(1):e1861.
28. Montealegre-Gómez G, Garavito E, Gómez-López A, Rojas-Villarraga A, Parra-Medina R. Colchicine: a potential therapeutic tool against COVID-19. Experience of 5 patients. *Reumatologia*. Clinica. 2021;17(7):371–5.
29. Qin C, Zhou L, Hu Z, Zhang S, Yang S, Tao Y, Xie C, Ma K, Shang K, Wang W. Dysregulation of immune response in patients with coronavirus 2019 (COVID-19) in Wuhan, China. Clin Infect Dis. 2020;71(15):Article 15.
30. Qiu Y, Yang Y, Yang R, Liu C, Hsu J-M, Jiang Z, Sun L, Wei Y, Li C-W, Yu D, Zhang J, Hung M-C. Activated T cell-derived exosomal PD-1 attenuates PD-L1-induced immune dysfunction in triple-negative breast cancer. Oncogene. 2021;40(31):Article 31. https://doi.org/10.1038/s41388-021-01896-1.
31. Sarkar S, Kumar S, Saha G, Basu M, Ghosh MK. PLGA-based dual-loaded nanoformulation of DIM and TMZ-an advanced clinical strategy for brain cancer treatment in a combinatorial approach. BioRxiv, 2023–05. 2023;
32. Schlesinger N, Firestein BL, Brunetti L. Colchicine in COVID-19: an old drug, new use. Curr Pharmacol Rep. 2020;6(4):137–45. https://doi.org/10.1007/s40495-020-00225-6.
33. Sidhu GS, Rai JS, Khaira KS, Kaur S. The impact of COVID-19 pandemic on different sectors of the Indian economy: a descriptive study. Int J Econ Financ Issues. 2020;10(5):113–20.
34. Sterne JA, Murthy S, Diaz JV, Slutsky AS, Villar J, Angus DC, Annane D, Azevedo LCP, Berwanger O, Cavalcanti AB. Association between administration of systemic corticosteroids and mortality among critically ill patients with COVID-19: a meta-analysis. JAMA. 2020;324(13):Article 13.

35. Van Berckelaer C, Rypens C, Van Dam P, Pouillon L, Parizel M, Schats KA, Kockx M, Tjalma WAA, Vermeulen P, Van Laere S. Infiltrating stromal immune cells in inflammatory breast cancer are associated with an improved outcome and increased PD-L1 expression. Breast Cancer Res. 2019;21(1):Article 1.

Application of Anti-platelet Drugs and Anti-coagulants in COVID-19

15

Bindu Kumari, Usha Kumari, Sandhya Singh,
Ravi Bhushan Singh, and Gireesh Kumar Singh

Abstract

Outbreak of COVID-19 was one of the deadliest health hazards encountered in the history of modern civilisation that claimed at least three million lives within a span of two years. The mortality and morbid conditions caused by COVID-19 were found to be associated with several underlying pathological mechanisms like hyper-inflammatory-immune response and coagulopathy. After infection, interaction of COVID-19 spike proteins with angiotensin-converting enzyme-2 receptors on cells initiates a cascade of events of biomolecular interaction in the affected person which culminates into fatal conditions like acute respiratory distress syndrome and multiple organ dysfunction syndrome. COVID-19-associated coagulopathy has been identified as one of the main culprits behind the symptoms and poor prognosis of the COVID-19, and its control through concerned medications became one of the determinants of the treatment outcome of the COVID-19. COVID-19-associated coagulopathy is a condition characterised by

B. Kumari · U. Kumari · S. Singh
Department of Pharmacy, Central University of South Bihar, Gaya, Bihar, India
e-mail: bindu@cusb.ac.in

R. B. Singh
Centre of Excellence in Pharmaceutical Sciences (CEPS), Guru Gobind Singh Indraprastha University, Dwarka, New Delhi, India

G. K. Singh (✉)
Department of Pharmacy, School of Health Science, Central University of South Bihar, Panchanpur, Gaya, Bihar, India
e-mail: gireesh@cusb.ac.in

M. K. Ghosh, M. Basu (eds.), *Pathogenesis, Management and Socio-Economic Impact of COVID-19 Pandemic*, Medical Virology: From Pathogenesis to Disease Control, https://doi.org/10.1007/978-981-95-1426-7_15

activation of microvascular blood clotting mediated by thrombin–fibrin interaction and involvement of activated platelets. This chapter discusses the molecular mechanism involved in the onset and progression of coagulopathy, biomarkers used to identify the severity level of coagulopathy and resulting complications. Treatment strategy specifically using anti-platelet drugs and anti-coagulants to control the coagulopathy has also been discussed in this chapter. Results of several observational studies, clinical trials and meta-analyses were thoroughly examined and have been included in this chapter. It is expected that this chapter will help clinicians and researchers to design and develop new treatment strategies for coagulopathy and related disease in general, and for COVID-19-associated coagulopathy in particular.

Keywords

COVID-19 · Coagulopathy · Biomarkers · Anti-platelet · Anti-coagulants · Treatment

Abbreviations

ACE-2	Angiotensin-converting enzyme-2 receptor
aPLAb	Anti-phospholipid antibodies
aPTT	Activated partial thromboplastin time
CAC	COVID-19-associated coagulopathy
COX	Cyclo-oxygenase
CRP	C-reactive protein
CRS	Cytokine release syndrome
DIC	Disseminated intravascular coagulation
ESR	Erythrocyte sedimentation rate
FDP	Fibrinogen degradation product
ICU	Intensive care unit
IFN	Interferon
IL	Interleukins
LC	Lymphocyte count
LDH	Lactate dehydrogenase
LMWH	Low molecular weight heparin
NLR	Neutrophil-to-lymphocyte ratio
PCT	Procalcitonin
PT	Prothrombin time
SARS-CoV-2	Severe acute respiratory syndrome coronavirus-2
TF	Tissue factor
TNF	Tumour necrosis factor
VTE	Venous thromboembolism
vWF	Von Willebrand Factor

Introduction

The causative agent of COVID-19 is a virus belonging to the family *Coronaviridae,* and it is referred to as Severe Acute Respiratory Syndrome Coronavirus 2 (SARS-CoV-2). The viral structure is characterised by a lipid bilayer envelope studded with spike (S) glycoproteins, enclosing a nucleocapsid composed of RNA and associated proteins. The SARS-CoV-2 virus is known to induce multiple clinical symptoms at a time, ranging from mild flu-like symptoms to death, particularly in older adults and those with additional health conditions like type 2 diabetes and cardiovascular disease. COVID-19-associated coagulopathy (CAC) is an acute health emergency condition induced by widespread thrombotic events that leads to several complications and, in many cases, even death. This coagulopathy can manifest at various anatomical sites, with venous thromboembolism (VTE), including deep vein thrombosis and pulmonary embolism, being common in severely ill COVID-19 patients [1]. The acute health crisis caused by COVID-19 infection is due to an increase in pro-coagulant (blood clotting) factors, such as von Willebrand Factor (VWF), factor VIII and fibrinogen, which contribute to an elevated risk of thrombosis. Several clinical studies have revealed the relationship between elevated levels of D-dimer and poor disease outcome in affected patients. D-dimer is a degradation product of stabilised fibrin polymer, i.e. fibrin cross-linked with factor XIII, by activated plasminogen. Thus, elevated levels of D-dimer in blood indicate thrombus production in blood by activation of the coagulation pathway and subsequent degradation by fibrinolytics. In severe cases of COVID-19, a condition termed as cytokine storm, there is overwhelming generation of pro-inflammatory cytokines such as tumour necrosis factor (TNF) and interleukins (IL), IL-6 and IL-1, which triggers the activation of coagulation and thrombin generation leading to life-threatening pulmonary embolism [2]. TNF and IL-1 are the key mediators of the endogenous coagulation cascade. This cytokine storm can contribute to a hyper-coagulable state. Figure 15.1 summarises the consequences of SARS-CoV-2 infection on the blood coagulation pathway.

There are many treatment strategies to combat the health menace presented by COVID-19 infection. Owing to the complexity of the pathophysiological machinery engaged by the SARS-CoV-2 virus, treatment of the associated symptoms and underlying causes is a challenging task. Like other viral infections, approaches to control the replication of the virus through anti-viral agents have been tried to control COVID-19. Several vaccines using conventional, i.e. attenuation of viral particles and innovative approaches like mRNA vaccines, have also been developed for prophylactic purposes and to reduce the severity of further infections. However, the use of anti-platelet drugs and anti-coagulant agents to control the coagulopathy during SARS-CoV-2 infection is the main strategy to reduce the severity and mortality. Platelet aggregation and initiation of the cascade of thrombus formation are one of the key inflammatory responses against SARS-CoV-2 infection. Spike proteins of SARS-CoV-2 are the main trigger factor behind these inflammatory responses. As

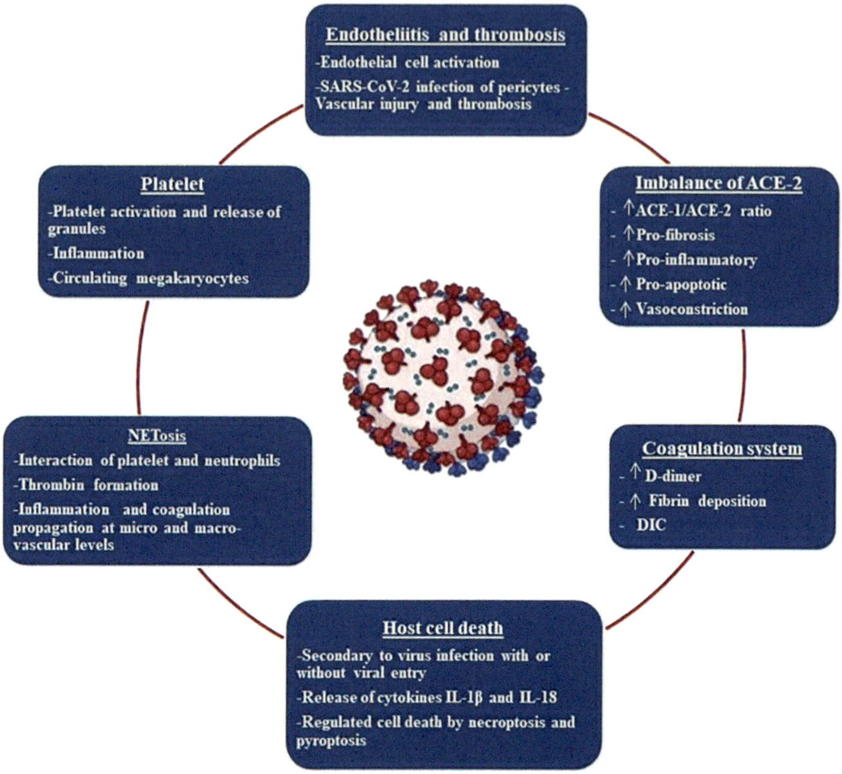

Fig. 15.1 Effects of SARS-CoV-2 on the coagulation system. ACE: angiotensin-converting enzyme, DIC: disseminated intravascular coagulation, IL: interleukin, and SARS-CoV-2: severe acute respiratory syndrome coronavirus-2

the name suggests, anti-platelet drugs inhibit the aggregation of platelets, and anti-coagulants are the drugs which prevent thrombus formation. Thus, these two drugs serve to check the coagulopathy induced by the COVID-19 virus. Since COVID-19-associated coagulopathy is the key pathological aspect that engages multiple organs during SARS-CoV-2 infection, it is pertinent to control the same using anti-platelet drugs and anti-coagulant agents. In the next part of this chapter, we will discuss the virus structure and coagulopathy induced by the same, followed by the use of anti-platelet drugs and anti-coagulant agents in the treatment of associated symptoms and pathologies.

COVID-19: SARS-CoV-2 Virus

Organisation of the SARS-CoV-2 Viral Particle

SARS-CoV-2 virus, commonly known as coronavirus, derived its name from its protein coat, which resembles the outer surface of the sun. The protein coat of

SARS-CoV-2 comprises multiple structural and functional proteins, each contributing to the virion's stability, infectivity and replication. SARS-CoV-2 possesses a single-stranded RNA of approximately 26–32 kilobases in length. Among the structural proteins, envelope (E), membrane (M), nucleocapsid (N) and the spike (S) proteins are the primary constituents, while functional proteins include RNA-dependent RNA polymerase (RdRp), papain-like protease (PLpro), 3-chymotrypsin-like protease (3CLpro) and helicase (H). Structural proteins play a diverse role in viral particle assembly and maintenance of the shape and structure. For example, E protein is essential for virus assembly, envelope formation and host cell membrane permeability, whereas M protein supports the viral assembly and is responsible for the shape of the viral membrane and virulence. N proteins serve to encapsulate and protect the nucleic acid of the virus. S proteins are essential for adherence to host cell receptors and are responsible for immune response generation in host [3].

The spike protein, protruding from the viral surface, mediates viral entry into host cells by binding to the angiotensin-converting enzyme-2 (ACE-2) receptors. It consists of two subunits, S1 and S2. S1 subunit recognises ACE-2 receptor, whereas S2 subunit is responsible for membrane fusion. The trimeric spike glycoprotein adopts a prefusion conformation, characterised by a receptor-binding domain (RBD) in the 'up' position, facilitating ACE-2 engagement. Structural studies have revealed conformational changes in the spike protein upon receptor binding, leading to membrane fusion and viral entry. S protein is the primary antigen that triggers an immune response in the host [4]. ACE-2 is a widely distributed zinc-dependent metalloenzyme [5]. However, organs composed of smooth muscle, chiefly the cardiovascular system, lungs and intestine are rich in ACE-2 receptors. It has been observed that strong binding between ACE-2 receptors and S protein leads to increased disease severity.

Mechanisms of Infection of SARS-CoV-2: Insights into COVID-19 Virus Entry and Replication

SARS-CoV-2 is the causative agent of COVID-19. It employs sophisticated mechanisms to infect host cells, evade immune surveillance and propagate within the host organism. The following are the key steps in the infection of SARS-CoV-2.

Viral Entry

SARS-CoV-2 enters host cells primarily through attachment of spike protein to ACE-2 receptors of the host cell. Some studies have also revealed the facilitatory action of heparin sulphate, a long, negatively charged polysaccharide expressed by all animal cells, which brings open conformation of the spike protein. This conformational change is essential for the binding of the spike protein to ACE-2 receptors [6]. The spike protein undergoes proteolytic cleavage by host cell proteases, such as TMPRSS2 or furin, to promote membrane fusion and viral entry [7]. Additionally, other host factors, including neuropilin-1 (NRP1) and basigin (BSG), have been implicated in SARS-CoV-2 entry [8].

Viral Replication

Once the viral RNA genome is released into the cytoplasm of host cell, it serves as a template for viral replication and transcription. The replication of SARS-CoV-2 is mediated by viral functional proteins, particularly the RNA-dependent RNA polymerase (RdRp), which catalyses the synthesis of viral RNA intermediates. Viral replication occurs in specialised membrane-bound compartments known as replication organelles, which shield viral RNA from host immune sensors and facilitate efficient replication [9].

Immune Evasion

SARS-CoV-2 employs various strategies to evade host immune responses, including the inhibition of type I interferon (IFN) signalling, modulation of antigen presentation pathways, and suppression of innate immune effectors. Viral proteins such as ORF3b, ORF6 and ORF9b interfere with host immune signalling pathways, facilitating viral replication and spread. Understanding these immune evasion mechanisms is crucial for developing immunotherapeutic approaches to enhance host anti-viral defences [10].

COVID-19-Associated Coagulopathy—Role of Platelets and Coagulation Factors

Blood coagulation is a complex process that completes in a cascading fashion with stepwise involvement of several clotting factors. Diverse triggering mechanisms such as tissue injury or viral infection may initiate the cascade of blood coagulation [11]. CAC is referred to as an acute illness characterised by thrombosis engaging multiple organs and consequent morbidity and mortality after SARS-CoV-2 infection [1]. In CAC, platelets have been identified as one of the main contributory factors. Platelets are component of blood which triggers the blood coagulation pathway in response to injury or inflammation of blood vessels via three major steps: activation, adhesion and aggregation. Platelet activation and aggregation have been observed in SARS-CoV-2 infection.

CAC present some unique feature which makes it different, albeit having some overlapping features as seen in other coagulopathies like disseminated intravascular coagulation (DIC). The thrombotic events as seen in COVID-19 infections are not fully established, but available data indicates that the interaction of the endothelium, the immune system and the coagulation system is the key factor behind CAC. CAC also involve Virchow's triad, i.e. endothelial dysfunction, stasis of blood flow and a hyper-coagulability state. The vascular endothelium, the innermost lining of blood vessels, plays an important role in maintaining haemostasis within the body through regulating numerous pathways like immune system, inflammation and blood vessel constriction. The endothelium exhibits a wide range of functions across different parts of the circulatory system. Dysfunction of the endothelium has multiple aspects, including its impact on coagulation. Endothelial cells control the coagulation system by keeping the inner surface of blood vessels non-thrombogenic and facilitating an adequate blood flow without causing systemic clotting or haemorrhage [12]. In case of SARS-CoV-2 infection, the virus binds with the ACE-2 receptor on endothelial cells. This leads to activation of endothelial cells and onset of inflammation response in order to neutralise the virus. Releases of VWF are one of the key inflammatory responses [1]. Once released into circulation, VWF stimulates the interaction between platelets and the vessel wall, ultimately leading to platelet aggregation. This process is accompanied by the release of factor VIII, which might increase coagulability [13]. Patients with severe COVID-19 often show increased levels of D-dimer, fibrinogen and VWF, indicating hyper-coagulability and endothelial activation. Up-regulation of TF triggers the extrinsic coagulation pathway, leading to thrombin generation and fibrin clot formation [12]. The pro-inflammatory condition promotes thrombus development by impairing the anti-coagulant machinery, which in turn is mediated by TF pathway inhibitor, thrombomodulin and endothelial cell protein C. Finally, during the inflammation endothelial cells release plasminogen activator inhibitor-1 (PAI-1) which inhibits fibrinolysis [14].

Both coagulation and inflammation are important host responses to COVID-19 [11]. In case of COVID-19, the platelets are activated towards a hyper-immune response, leading to an increase in platelet-monocyte and platelet-neutrophil aggregates, as well as enhanced platelet activity, TF activity and neutrophil extracellular trap (NET) formation (NETosis) [15]. Pro-inflammatory cytokines are attributed to the activation of the innate immune system and manifestation of symptoms of severe infection. Dysfunction of the immune system is a potential risk factor for unfavourable outcomes in COVID-19. The primary cause of morbidity and mortality in COVID-19 patients seems to be the cytokine storm, which results in an immune dysfunction in the whole body and in the lungs, particularly. Cytokine storm, often called cytokine release syndrome (CRS), is a systemic hyper-inflammatory response, triggered by SARS-CoV-2 infection. IL-6, TNF-α and IL-1 play a significant role in CRS and potentially cause vascular leakage and induce a hyper-coagulable state [16].

CAC has significant clinical consequences including increasing risk of thrombosis, cytokine storm and systemic inflammation, endothelial cells damage, abnormal coagulation markers and multiorgan failure. Overall, the clinical evidences suggest that thrombotic coagulopathy is very common in case of severe COVID-19 patients and significantly contribute to morbidity and mortality. Figure 15.2 illustrates the pathophysiology of CAC, highlighting the relationship between the endothelium, immune system and the coagulation system induced by SARS-CoV-2.

Biomarkers Used to Assess the Severity of CAC

Assessment of the level of severity of CAC is crucial to plan the treatment strategy for CAC and associated health conditions. Several biomarkers have been identified, validated and are frequently used to ascertain the status of CAC in the context of diagnostic and prognostic value. For convenience, these biomarkers are classified into two categories—biomarkers related to coagulation pathways and biomarkers belonging to inflammatory status.

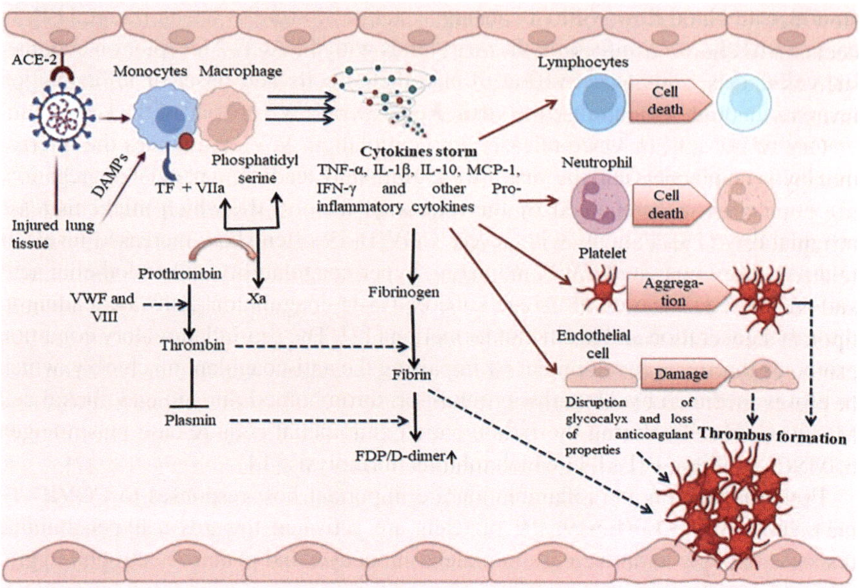

Fig. 15.2 Pathophysiology of CAC is a complex phenomenon. Damaged and inflamed endothelium alongwith activation of coagulation are two distinct axes of CAC. SARS-CoV-2 infects macrophage and endothelial cells via binding to the ACE-2 receptor. Tissue factor (TF) expressed on immune cells and excess production of cytokines stimulate the coagulation systems and damage the endothelial cells, leading to thrombus formation. ACE-2: angiotensin-converting enzyme-2, VWF: von Willebrand factor, FDP: fibrinogen degradation product, IL: interleukin, TNF-α: tumour necrosis factor, IFN-γ: interferon gamma, MCP-1: monocyte chemotactic protein-1, DAMPs: damage-associated molecular patterns

Coagulation-Related Biomarkers

These biomarkers are either components of the coagulation pathway or end products of the coagulation process.

D-Dimer

D-dimer is a degradation product of fibrin, and it may be used to identify the magnitude of blood clots. Several studies have confirmed that elevated D-dimer levels indicate poor prognosis and a higher chance of mortality in COVID-19 patients [17]. High level of D-dimer in COVID-19 patients correlates with high mortality chances.

Platelet Count

Platelet count is one of the most significant determinants for CAC. Activated platelets contribute to thrombus formation [18]. Data from 1476 COVID-19 patients indicate that thrombocytopenia significantly increased the risk of mortality [19].

Activated Partial Thromboplastin Time (aPTT) and Prothrombin Time (PT)

Prothrombin, a precursor to thrombin, is essential for converting fibrinogen to fibrin during blood coagulation. Elevated levels of prothrombin associated with higher plasma thrombin lead to coagulation activation and thrombosis [20]. PT and aPTT are coagulating factors that regulate blood coagulation. Patients with COVID-19 had elevated PT and aPTT levels.

Factor VIII and von Willebrand Factor (vWF)

Factor VIII and VWF are both the glycoproteins produced by megakaryocytes and vascular endothelial cells. They play an active role in platelet adhesion to injured and activated vascular endothelial cells during inflammation and immune activation [21]. Several studies have found the elevated levels of vWF antigen, enhanced VWF activity and higher binding to collagen in COVID-19 patients, indicating fatal CAC [22].

Ferritin

Ferritin is an iron-binding protein found both in the intracellular and extracellular space. It is an indirect marker for CAC. Qin et al., [23] showed an increased level of ferritin in 63% of COVID-19 patients, which was associated with poor clinical outcomes [23].

Inflammation-Related Biomarkers

Inflammation is the main precursor of CAC. Several mediators of inflammation serve as catalyst for CAC. An assessment of inflammatory markers gives insight into the status of CAC. The followings are the main inflammatory biomarkers:

Erythrocyte Sedimentation Rate

Elevated erythrocyte sedimentation rate (ESR) reveals the presence of inflammation, which is frequent in COVID-19 patients. The SARS-CoV-2 causes an intense inflammatory response, leading to an increase in the acute phase reactants such as fibrinogen, which contributes to the increase in ESR. It is also a part of systemic inflammatory response that indicates endothelial damage and hyper-coagulability. Measuring the ESR can aid in diagnosing CAC and in predicting adverse outcomes. It serves as a simple and effective marker to monitor the inflammatory and coagulative state of the COVID-19 patient [24].

C-Reactive Protein (CRP)

C-reactive protein is an acute phase protein primarily produced by the liver in response to inflammation, and it is significantly elevated in severe COVID-19 patients. CRP is considered an important biomarker of COVID-19 [25]. Several studies have shown that CRP can predict the likelihood of disease worsening in patients with mild COVID-19 symptoms. A study by Wang showed that the increase in CRP is strongly associated with COVID-19 severity and critical condition [26, 27].

Pro-Inflammatory Cytokines

COVID-19 patients have been observed to overexpress pro-inflammatory cytokines: IFN-γ, IL-1β, IL-2, IL-4, IL-6 and IL-10. Cytokine storm refers to elevated serum cytokine levels accompanied by an uncontrollable inflammatory response, leading to clinical worsening of symptoms and mortality in COVID-19 patients [28]. There is a direct connection between the generation of cytokine storm and the development of acute respiratory distress syndrome (ARDS) with a higher mortality rate in COVID-19 [29].

Neutrophil-to-Lymphocyte Ratio

The neutrophil-to-lymphocyte ratio (NLR) is a diagnostic marker for assessing the severity and prognosis of COVID-19, particularly in relation to coagulopathy, and the elevated NLR found in COVID-19 patients indicates an augmented inflammatory response [30].

Lactate Dehydrogenase

Elevated levels of lactate dehydrogenase (LDH) are significant risk factors for exacerbation in mild COVID-19 patients [31]. Numerous studies have found that LDH levels are higher in cases of severe COVID-19 compared with non-severe cases.

Table 15.1 summarises the different biomarkers associated with CAC and their relevance and importance for diagnostic purposes.

Application of Anti-platelet Drugs and Anti-coagulants in COVID-19

As discussed earlier, platelet activation, followed by its aggregation and adhesion, serves as an initiating factor for the thrombotic events during COVID-19 infection. Thus, preventing platelet activation and use of drugs to inhibit the cascade of thrombotic events is rational approach to minimise the severity and mortality caused by CAC.

Anti-platelet Drugs and Their Application in COVID-19

As the term suggests, anti-platelet drugs are those drugs which inhibit the activation, aggregation and adhesion of platelets. These drugs have been primarily used to treat cardiovascular and cerebral disease. Aspirin, ticlopidine, clopidogrel, prasugrel, cangrelor, ticagrelor, abciximab, eptifibatide, tirofiban and vorapaxar are clinically used anti-platelet drugs. These drugs interfere with specific steps involved in platelet aggregation. For example, aspirin inhibits thromboxane (TxA2) synthesis by acetylating cyclo-oxygenase enzyme (COX-1) irreversibly. Reduced TxA2 synthesis hampers platelet activation and their recruitment to the site of inflammation and injury. P2Y12, the key ADP receptor on the platelet surface, are irreversibly blocked by ticlopidine, clopidogrel, and prasugrel, whereas abciximab, eptifibatide and tirofiban inhibit the final common pathway of platelet aggregation by blocking fibrinogen and VWF from binding to activated GPIIb/IIIa. Vorapaxar inhibits

Table 15.1 Diagnostic markers with their significance in COVID-19-associated coagulopathy

Coagulation-related biomarkers			
Markers	levels	Significances of markers	Reference
D-dimer	Increased	There are significant evidence proving elevated D-dimer levels in COVID-19 patients, and it is associated with the severity of the disease and adverse outcomes, including mortality.	[17]
Platelets	Decreased	A lack of control of pre-coagulation platelets may contribute to the development of thrombotic complications caused by SARS-CoV-2. Threefold increase in mortality rate in COVID-19 patients who suffer from thrombocytopenia.	[19]
LC	Decreased	The decrease in lymphocytes, along with higher inflammatory markers such as IL-6, contributes to hyper-coagulability and increases the risk of thrombotic events. On the other hand, a decrease in lymphocytes, along with an increase in leucocyte count, can indicate severe disease and has been associated with a high mortality rate.	[32]
PT	Increased	Prolonged PT indicates the underlying coagulopathy associated with COVID-19, which is characterised by increased thrombotic and bleeding risk.	[20]
aPTT	Increased	Elevated levels of aPTT in COVID-19 patients were indicating a disturbance in the normal coagulation process. It is an important marker in the overall evaluation of coagulation status in affected patients.	[18]
Fibrinogen and FDP	Increased	Monitoring fibrinogen and FDP levels in COVID-19 patients helps in measuring the severity of coagulopathy	[33]
Factor VIII and VWF	Increased	Elevated levels of VWF increase factor VIII stability and activity, which promotes clot formation. This mechanism is implicated in the increased risk of thrombosis in COVID-19 patients.	[22]
Inflammation-related biomarkers			
ESR	Increased	ESR is frequently high in COVID-19 patients with coagulopathy, indicating that the infection has generated an inflammatory response. This is also linked to severe clinical manifestations and adverse outcomes.	[24]
NLR	Increased	Elevated NLR levels have been linked to higher disease severity and a poorer prognosis in COVID-19 patients. Monitoring NLR can help to identify high-risk patients primarily, allowing for faster intervention and better management of COVID-19 complications, particularly those associated with coagulopathy.	[30]

(continued)

Table 15.1 (continued)

Coagulation-related biomarkers			
Markers	levels	Significances of markers	Reference
aPLAb	Increased	aPLAb is associated with inflammatory markers and coagulation, and it has been shown that aPLAb has a pivotal role in worsening the coagulopathy in COVID-19 patients. Identification of these markers in patients to determine the risk and control thrombotic problems in that patient.	[34]
LDH	Increased	Increased LDH levels can indicate DIC and pulmonary thrombotic microangiopathy caused by CAC. LDH marker helps in early diagnosis of the extent of coagulopathy, identification and management of patients at risk of severe complications.	[35]
C-reactive protein	Increased	CRP level is elevated in COVID-19 patients, and it is used for the assessment of disease severity and is also useful to infer the risk of VTE.	[25]
IL-6	Increased	IL-6 is an important cytokine in the cytokine storm associated with severe COVID-19. Elevated levels of IL-6 are closely linked with increased inflammation and coagulopathy in COVID-19 patients and are useful for assessing disease severity and prognosis.	[36]
IL-10	Increased	Elevated levels of IL-10 to counteract the pro-inflammatory and pro-coagulant effects caused by other cytokines.	[36]

aPLAb anti-phospholipid antibodies, *aPTT* activated partial thromboplastin time, *ESR* erythrocyte sedimentation rate, *FDP* fibrinogen degradation product, *IL* interleukin, *LC* lymphocyte count, *LDH* lactate dehydrogenase, *NLR* neutrophil-to-lymphocyte ratio, *PCT* procalcitonin, *PT* prothrombin time

thrombin-mediated platelet activation by targeting PAR-1, the major thrombin receptor on platelets. Aspirin has been primarily used to prevent platelet aggregation for the management of CAC. Aspirin and other anti-platelet therapy may provide protection from severe COVID-19 infection by virtue of their anti-inflammatory effect and inhibition of platelet aggregation [37].

In a clinical study, Santoro et al. evaluated the safety and efficacy of anti-platelet therapy in 7824 hospitalised patients with COVID-19 and its effect on survival. These patients were enrolled under a multicentre international prospective registry. Clinical data and in-hospital complications were recorded. They found that anti-platelet therapy having aspirin or other drugs during hospitalisation for COVID-19 resulted in a lower death rate and minor requirements of mechanical ventilation. Additionally, the risk of bleeding was also minimised [38]. In another clinical study, Sokolski et al. investigated the effect of anti-platelet therapy on the in-hospital stay and medium-term outcomes in 2170 hospitalised COVID-19 patients. They found that patients already receiving anti-platelet therapy were less frequently hospitalised in the intensive care unit, developed less often shock and required less aggressive forms of therapy [37]. Additionally, the patients from the anti-platelet therapy group had lower 3-month mortality. Thus, overall a protective effect of anti-platelet therapy was observed on COVID-19 patients.

A retrospective study comprising 35,370 patients with and without active aspirin medications before contracting SARS-CoV-2 infection found that aspirin considerably decreased the risk of death by 32%. In the propensity-matched groups, mortality dropped from 6.3 to 2.5% at 14 days and from 10.5 to 4.3% at 30 days following matching of propensity scores and confounding correlate adjustments [39]. A meta-analysis assessing 12 retrospective investigations comprising patients with SARS-CoV-2 positive status showed that aspirin therapy was clearly beneficial in reducing a fatal progression of COVID-19 disease [40].

Despite promising results shown by anti-platelet drugs in COVID-19 treatment, some authors have reported conflicting outcomes where no efficacy or nonsignificant changes have been observed. After analysing four different randomised clinical trials, Zong et al. concluded that adding anti-platelet therapy to the standard care had no positive outcome irrespective of the disease severity and existing anti-coagulation intensity [41]. Similar results on the use of anti-platelet therapy in COVID-19 were also found by Duo et al. They conducted a systematic review and meta-analysis on different studies comprising total of 17,950 patients with COVID-19 to investigate the efficacy of anti-platelet therapy. Their findings revealed that there were no significant changes in disease outcome when anti-platelet drugs were given to patients. Moreover, a meta-analysis has reported the risk of developing major bleeding disorders when anti-platelet drugs were used for COVID-19 treatment [42].

Anti-coagulants and their Application in COVID-19

Anti-coagulants are the drugs which inhibit the coagulation cascade of the blood and prevent thrombus formation. Jay McLean discovered unfractionated heparin as the first anti-coagulant drug in 1916 [43]. Later, warfarin was discovered in 1948, which is the first oral anti-coagulant drug also [44]. Nowadays, a variety of anti-coagulants are available which are used for various clinical purposes including atrial fibrillation, venous thromboembolism, post-heart valve replacement and pregnancy-related complications. Anti-coagulants are classified into two categories—injectable and oral. Injectable anti-coagulants include heparin and its derivatives like—low molecular weight heparin, fondaparinux. These anti-coagulants have no intrinsic anti-coagulant activity, but they bind to antithrombin and augment the rate at which it inhibits various coagulation proteases. Oral anti-coagulants are those drugs which are administered orally and are subclassified according to their mechanism of action. For example, Vitamin K Antagonist—warfarin, Direct thrombin inhibitor—Dabigatran and Direct Factor Xa inhibitors—rivaroxaban, apixaban, and edoxaban.

Besides having anti-coagulant properties, heparin has been found to show anti-inflammatory and anti-viral activities which justify its application in the adjuvant therapy of COVID-19 [45, 46]. The anti-inflammatory activity of heparin is attributed to its ability to neutralise the inflammatory mediators like cytokines, neutrophil chemotaxis and leucocyte migration and complement factors [45]. Heparin is postulated to restrict viral entry into cells as it competes with heparan sulphate, an adherence molecule for COVID-19 spike proteins, and ultimately inhibits its interaction with ACE-2 receptors [46]. In various clinical studies, it has been observed that application of heparin for anti-coagulant therapy in COVID-19 yields protective effects on patients in terms of survival, hospital stay and severity of related complications. Heparins, which include both low molecular weight heparin (LMWH) and unfractionated heparin, have effects related to non-anti-coagulation by lowering neutrophil and platelet activation and reducing the response of inflammation [47]. The International Society for Thrombosis and Haemostasis recommends prophylactic doses for all patients, regardless of disease severity, who need to be admitted to the hospital because of a SARS-CoV-2 infection [48].

Apart from heparin, directly acting oral anti-coagulants like rivaroxaban, apixaban or dabigatran and warfarin have also been tried to manage CAC. They are the drug of choice for outpatient and mainly used to avoid drug–drug interaction of heparin with dexamethasone and anti-viral agents used during treatment of COVID-19. The protective effect of anti-coagulants in CAC also depends on the stage of administration. In order to assess such factors, Noh et al. conducted an observational clinical study to see whether pre-admission anti-coagulants use may have benefits by preventing COVID-19-induced thromboembolism. They conducted a study on the cohort of 4349 patients aged above 40 years and used mortality, mechanical ventilation and ICU admission as primary endpoints. They found that pre-admission anti-coagulants were not useful to extend the protective effect against severe COVID-19 outcomes [49].

Combination of Anti-platelet and Anti-coagulant Drugs in COVID-19

Since platelet aggregation and the thrombin cascade work in tandem to precipitate CAC, it is pertinent to administer a therapeutic regimen to control both at the same time. Various clinical studies have evaluated the prophylactic and curative potential of co-administered anti-platelet and anti-coagulant drugs in COVID-19 patients. Matil et al. in a propensity-matched cohort study have reported the improved outcome in hospitalised patients with COVID-19 who received combined anti-coagulant and anti-platelet therapy. They conducted a retrospective observational study comprising of 242 patients and found that combining both anti-platelet and anti-coagulant yields better results as compared to either one [50]. Despite better results in COVID-19 patients shown by anti-platelet and anti-coagulant combination, the data on similar studies is scarce and needs to be explored extensively.

Conclusion and Future Prospects

It is beyond any doubt that anti-platelet and anti-coagulants drugs are of definite therapeutic value for the management of CAC. Various clinical and analytical studies have proved their protective value for COVID-19 patients in terms of different determinants like duration of hospital stay, need for mechanical ventilation and mortality. However, some divergent results have also been seen where anti-platelet and anti-coagulants drugs could not produce the desired clinical outcome. This contradiction may be due to a myriad of factors like—age of the patients in the cohort, gender, ethnicity, associated health conditions like—type 2 diabetes, cardiovascular disease, etc. and most importantly, the strain of virus which infected the particular patients. As it was observed that no study considered which mutant of COVID-19 virus had infected the cohort subjects, there is a lack of uniformity in terms of the strain of COVID-19 virus which certainly has an important bearing on the severity of the disease and outcome of anti-platelet and anti-coagulants drugs treatment. Owing to the complexity of the factors associated with the COVID-19 disease as mentioned above, the use of anti-platelet and anti-coagulants drugs requires a highly individualised approach. Although some guidelines have been prescribed which clearly describe the usage of anti-platelet and anti-coagulants drugs as per the patient's profile, more detailed research is required to obtain optimal benefits of these two classes of drugs in case of CAC.

Declarations

Authors' contribution—All authors have contributed equally in the planning and drafting of the manuscript.

Funding—There was no financial support to conduct this literature-based study.

Conflict of interest—Authors declare no conflict of interest.

Data availability—No data was generated during this study.

References

1. Seo JW, Kim DY, Yun N, Kim DM. Coronavirus disease 2019-associated coagulopathy. Microorganisms. 2020;10(8):1556.
2. Wolf A, Khimani F, Yoon B, et al. The mechanistic basis linking cytokine storm to thrombosis in COVID-19. Thromb Update. 2022;8:100110.
3. Adhikari B, Sahu N. COVID-19 into chemical science per spective: chemical preventive measures and drug development. ChemSelect. 2021;6(9):2010–28.
4. Walls AC, Park YJ, Tortorici MA, Wall A, McGuire AT, Veesler D. Structure, function, and antigenicity of the SARS-CoV-2 spike glycoprotein. Cell. 2020;181(2):281–92.
5. Karges J, Cohen SM. Metal complexes as antiviral agents for SARS-CoV-2. Chembiochem. 2021;22(16):2600–7.
6. Kearns FL, Sandoval DR, Casalino L, Clausen TM, Rosenfeld MA, Spliid CB, Amaro RE, Esko JD. Spike-heparan sulfate interactions in SARS-CoV-2 infection. Curr Opin Struct Biol. 2022;76:102439.
7. Shang J, Ye G, Shi K, Wan Y, Luo C, Aihara H, Li F. Structural basis of receptor recognition by SARS-CoV-2. Nature. 2020;581(7807):221–4.
8. Cantuti-Castelvetri L, Ojha R, Pedro LD, Djannatian M, Franz J, Kuivanen S, Simons M. Neuropilin-1 facilitates SARS-CoV-2 cell entry and infectivity. Science. 2020;370(6518):856–60.

9. Hillen HS, Kokic G, Farnung L, Dienemann C, Tegunov D, Cramer P. Structure of replicating SARS-CoV-2 polymerase. Nature. 2020;584(7819):154–6.
10. Konno Y, Kimura I, Uriu K, Fukushi M, Irie T, Koyanagi Y, Sauter D. SARS-CoV-2 ORF3b is a potent interferon antagonist whose activity is further increased by a naturally occurring elongation variant. Cell Rep. 2020;32(12):108185.
11. Iba T, Levy JH, Connors JM, Warkentin TE, Thachil J, Levi M. The unique characteristics of COVID-19 coagulopathy. Crit Care. 2020;24:1–8.
12. Levi M, ten Cate H, van der Poll T. Endothelium: interface between coagulation and inflammation. Crit Care Med. 2002;30(5):S220–4.
13. Goshua G, Pine AB, Meizlish ML, Chang CH, Zhang H, Bahel P, Lee AI. Endotheliopathy in COVID-19-associated coagulopathy: evidence from a single-Centre, cross-sectional study. Lancet Haematol. 2020;7(8):e575–82.
14. Pons S, Arnaud M, Loiselle M, Arrii E, Azoulay E, Zafrani L. Immune consequences of endothelial cells' activation and dysfunction during sepsis. Crit Care Clin. 2020;36(2):401–13.
15. Manne BK, Denorme F, Middleton EA, Portier I, Rowley JW, Stubben C, Campbell RA. Platelet gene expression and function in patients with COVID-19. Blood J ASH. 2020;136(11):1317–29.
16. Lazzaroni MG, Piantoni S, Masneri S, Garrafa E, Martini G, Tincani A, Franceschini F. Coagulation dysfunction in COVID-19: the interplay between inflammation, viral infection and the coagulation system. Blood Rev. 2021;46:100745.
17. Cho ES, McClelland PH, Cheng O, Kim Y, Hu J, Zenilman ME, D'Ayala M. Utility of d-dimer for diagnosis of deep vein thrombosis in coronavirus disease-19 infection. J Vasc Surg Venous L. 2021;9(1):47–53.
18. Zhang J, Ding Y, Jiang D, Xie J, Liu Y, Ma J, Zhang Y. Deficiency of platelet adhesion molecule CD226 causes megakaryocyte development and platelet hyperactivity. FASEB J. 2020;34(5):6871–87.
19. Yang X, Yang Q, Wang Y, Wu Y, Xu J, Yu Y, Shang Y. Thrombocytopenia and its association with mortality in patients with COVID-19. J Thromb Haemost. 2020;18(6):1469–72.
20. Hong LZ, Shou ZX, Zheng DM, Jin X. The most important biomarker associated with coagulation and inflammation among COVID-19 patients. Mol Cell Biochem. 2021;476:2877–85.
21. Ruggeri ZM. Role of von Willebrand factor in platelet thrombus formation. Ann Med. 2000;32:2–9.
22. Goshua G, Pine AB, Meizlish ML, Chang CH, Zhang H, Bahel P, Lee AI. Endotheliopathy in COVID-19-associated coagulopathy: evidence from a single-centre, cross-sectional study. Lancet Haematol. 2020;7(8):e575–82.
23. Qin Z, Zhang X, Chen Z, Liu N. Establishment and validation of an immune-based prognostic score model in glioblastoma. Int Immunopharmacol. 2020;85:106636.
24. Long X, Zhang Z, Zou W, Ling J, Li D, Jing L, Fang M. Coagulopathy of patients with COVID-19 is associated with infectious and inflammatory markers. Risk Manag Healthc Policy. 2020:1965–75.
25. Brosnahan SB, Smilowitz NR, Amoroso NE, Barfield M, Berger JS, Goldenberg R, Maldonado TS. Thrombosis at hospital presentation in patients with and without coronavirus disease 2019. J Vasc Surg Venous L. 2021;9(4):845–52.
26. Wang G, Wu C, Zhang Q, Wu F, Yu B, Lv J, Zhong Y. C-reactive protein level may predict the risk of COVID-19 aggravation. Open Forum Infect Dis. 2020;7(5):ofaa153.
27. Wang L. C-reactive protein levels in the early stage of COVID-19. Med Mal Infect. 2020;50(4):332–4.
28. Cao X. COVID-19: immunopathology and its implications for therapy. Nat Rev Immunol. 2020;20(5):269–70.
29. Parsons PE, Eisner MD, Thompson BT, Matthay MA, Ancukiewicz M, Bernard GR. NHLBI acute respiratory distress syndrome clinical trials network. Lower tidal volume ventilation and plasma cytokine markers of inflammation in patients with acute lung injury. Crit Care Med. 2005;33(1):1–6.
30. Yang AP, Liu JP, Tao WQ, Li HM. The diagnostic and predictive role of NLR, d-NLR and PLR in COVID-19 patients. Int Immunopharmacol. 2020;84:106504.

31. Rahi MS, Jindal V, Reyes SP, Gunasekaran K, Gupta R, Jaiyesimi I. Hematologic disorders associated with COVID-19: a review. Ann Hematol. 2021;100:309–20.
32. Ragnoli B, Da Re B, Galantino A, Kette S, Salotti A, Malerba M. Interrelationship between COVID-19 and coagulopathy: pathophysiological and clinical evidence. Int J Mol Sci. 2023;24(10):8945.
33. Lio KU, Rali P. Coagulopathy in COVID-19. Lung India. 2021;38(Suppl 1):S53–7.
34. Devreese KM, Linskens EA, Benoit D, Peperstraete H. Antiphospholipid antibodies in patients with COVID-19: a relevant observation? JTH. 2020;18(9):2191–201.
35. Han Y, Zhang H, Mu S, Wei W, Jin C, Tong C, Gu G. Lactate dehydrogenase, an independent risk factor of severe COVID-19 patients: a retrospective and observational study. Aging (Albany NY). 2021;12(12):11245.
36. Han H, Ma Q, Li C, Liu R, Zhao L, Wang W, Xia Y. Profiling serum cytokines in COVID-19 patients reveals IL-6 and IL-10 are disease severity predictors. Emerg Microbes Infect. 2020;9(1):1123–30.
37. Sokolski M, Reszka K, Adamik B, Kilis-Pstrusinska K, Lis W, Pomorski M, Protasiewicz M. Antiplatelet therapy prior to COVID-19 infection impacts on patients mortality: a propensity score-matched cohort study. Sci Rep. 2024;14(1):4832.
38. Santoro F, Nuñez-Gil IJ, Vitale E, Viana-Llamas MC, Reche-Martinez B, Romero-Pareja R, Brunetti ND. Antiplatelet therapy and outcome in COVID-19: the health outcome predictive evaluation registry. Heart. 2022;108(2):130–6.
39. Abdel Magid HS, Ferguson JM, Van Cleve R, Purnell AL, Osborne TF. Differences in COVID-19 risk by race and county-level social determinants of health among veterans. Int J Environ Res Public Health. 2021;18(24):13140.
40. Kow CS, Hasan SS. Use of antiplatelet drugs and the risk of mortality in patients with COVID-19: a meta-analysis. J Thromb Thrombolysis. 2021;52:124–9.
41. Zong X, Wang X, Liu Y, Li Z, Wang W, Wei D, Chen Z. Antiplatelet therapy for patients with COVID-19: systematic review and meta-analysis of observational studies and randomized controlled trials. Front Med. 2022;9:965790.
42. Duo H, Jin M, Yang Y, Baheti R, Feng Y, Fu Z, Jiang Y, Zheng L, Wan J, Pan H. Effect of antiplatelet therapy after COVID-19 diagnosis: a systematic review with meta-analysis and trial sequential analysis. PLoS One. 2024;19(2):e0297628.
43. Heestermans M, Poenou G, Hamzeh-Cognasse H, Cognasse F, Bertoletti L. Anticoagulants: a short history, their mechanism of action, pharmacology, and indications. Cells. 2022;11(20):3214.
44. Pirmohamed M. Warfarin: almost 60 years old and still causing problems. Br J Clin Pharmacol. 2006;62(5):509–11.
45. Mangiafico M, Caff A, Costanzo L. The role of heparin in COVID-19: an update after two years of pandemics. J Clin Med. 2022;11(11):3099.
46. Hippensteel JA, LaRiviere WB, Colbert JF, Langouët-Astrié CJ, Schmidt EP. Heparin as a therapy for COVID-19: current evidence and future possibilities. Am J Physiol Lung Cell Mol Physiol. 2020;319(2):L211–7.
47. Godino C, Scotti A, Maugeri N, Mancini N, Fominskiy E, Margonato A, Landoni G. Antithrombotic therapy in patients with COVID-19?-rationale and evidence. Int J Cardiol. 2021;324:261–6.
48. Thachil J, Tang N, Gando S, Falanga A, Cattaneo M, Levi M, Iba T. ISTH interim guidance on recognition and management of coagulopathy in COVID-19. J Thromb Haemost. 2020;18(5):1023–6.
49. Noh H, Lee J, Chow R, Lee J, Simone CB 2nd, Shin HJ, Choi YG. The prognostic role of anticoagulants in COVID-19 patients: national COVID-19 cohort in South Korea. Ann Palliat Med. 2022;11(4):1317–25.
50. Matli K, Chamoun N, Fares A, Zibara V, Al-Osta S, Nasrallah R, Salameh P, Mokhbat J, Ghanem G. Combined anticoagulant and antiplatelet therapy is associated with an improved outcome in hospitalised patients with COVID-19: a propensity matched cohort study. Open Heart. 2021;8(2):e001785.

COVID-19 Target-Based Drug Discovery 16

Ankur Chaudhuri, Izaz Monir Kamal,
and Saikat Chakrabarti

Abstract

The COVID-19 pandemic, initiated by the emergence of SARS-CoV-2 in late 2019, has profoundly impacted global health, with over 704 million infections and seven million deaths. This crisis has not only overwhelmed healthcare systems but also significantly disrupted global economies. The virus's adaptability, particularly in its transmission, has posed ongoing challenges. In response, interdisciplinary efforts have led to significant advancements in vaccine development, drug discovery, and public health strategies. Computational drug discovery has become pivotal, utilizing in silico techniques to screen existing drugs and identify potential inhibitors targeting viral proteins like the spike (S) and other nonstructural proteins. These approaches, including molecular docking and dynamic simulations, have expedited drug discovery by allowing rapid screening of chemical libraries. Additionally, mRNA vaccines have demonstrated high efficacy and rapid production capabilities, marking a significant advancement in vaccine technology. The integration of computational methods with extensive drug libraries and high-performance computing resources has accelerated both drug and vaccine development, highlighting the critical role of global collaboration in managing and mitigating the pandemic's impact.

A. Chaudhuri (✉)
Structural Biology and Bioinformatics Division, CSIR-Indian Institute of Chemical Biology, IICB TRUE Campus, Salt Lake, Kolkata, West Bengal, India
e-mail: ankur.chaudhuri@csiriicb.res.in

I. M. Kamal · S. Chakrabarti (✉)
Structural Biology and Bioinformatics Division, CSIR-Indian Institute of Chemical Biology, IICB TRUE Campus, Salt Lake, Kolkata, West Bengal, India

Academy of Scientific and Innovative Research (AcSIR), Ghaziabad, India
e-mail: saikat@iicb.res.in

M. K. Ghosh, M. Basu (eds.), *Pathogenesis, Management and Socio-Economic Impact of COVID-19 Pandemic*, Medical Virology: From Pathogenesis to Disease Control, https://doi.org/10.1007/978-981-95-1426-7_16

Keywords

High-throughput virtual screening · Molecular docking · Structure-based drug design (SBDD) · Repurposing existing drugs · Non-structural proteins · Spike

Introduction

The COVID-19 pandemic, triggered by the sudden emergence of the SARS-CoV-2 virus in late 2019, has been one of the most significant and far-reaching global health crises of recent times. This virus has affected more than 704 million individuals globally, causing over seven million deaths [1]. While the loss of life is devastating, we are also seeing substantial effects on global economies and healthcare systems. The virus's highly adaptable features, particularly its transmission modes, have spread rapidly across the globe, overwhelming healthcare infrastructures and disrupting daily life. In response to the pandemic, researchers from various fields have come together to address the problem. Efforts have been made to mitigate the viral impact, including developing vaccines, therapies, and public health interventions. The viral adaptability, however, poses ongoing challenges, necessitating continued research and collaboration to create a world where the virus can be managed effectively. Drug discovery has been instrumental in the efforts to combat COVID-19 [2]. Using in silico drug discovery techniques, including computational approaches to screen existing drugs against the virus's structural and non-structural proteins, has been instrumental in understanding the viral mechanisms of action. Large-scale screening of global drug databases has helped identify potential candidates that can prevent or reduce the severity of the disease. Moreover, the development of mRNA-based vaccines has opened new avenues in drug discovery [3], highlighting the importance of rapid and accurate drug development. Although traditional drug development processes often take two to three decades, the urgent need for effective treatments against COVID-19 has led to unprecedented levels of collaboration and resource-sharing worldwide. This collective effort has accelerated the drug discovery process and underscored the importance of a coordinated global response in addressing future pandemics. Target-based drug discovery involves identifying and characterizing specific biological targets, typically proteins or nucleic acids that play a critical role in a disease's pathology. In case of COVID-19, target-based approaches have focused on critical viral proteins essential for the SARS-CoV-2 replication and survival. Researchers aim to disrupt the viral life cycle by designing drugs that specifically inhibit these targets and prevent them from causing disease. Identifying suitable drug targets is a critical first step in this approach.

The SARS-CoV-2 spike (S) protein has been a central focus for virtual screening and computational drug discovery. By simulating interactions between the spike protein and potential small-molecule inhibitors, researchers aim to identify compounds that can block the virus binding to the ACE2 receptor, thereby preventing its entry into the host cells. Docking studies have uncovered several candidate molecules capable of disrupting the function of the S protein [4]. These computational techniques, including molecular docking and molecular dynamic (MD) simulations, have significantly expedited the drug discovery process by enabling the rapid screening of extensive chemical libraries and identifying promising antiviral compounds [5]. In addition, the envelope (E) and membrane (M) proteins have been investigated in docking studies to discover inhibitors that could hinder the viral assembly and release, offering alternative therapeutic options. Beyond structural proteins like the spike (S) protein, non-structural proteins such as the Main Protease (Mpro), RNA-dependent RNA Polymerase (RdRp), Papain-like Protease (PLpro), and Helicase (NSP13) have also emerged as vital targets for SARS-CoV-2 drug discovery [6]. These non-structural proteins are crucial for viral replication and transcription, making them appealing targets for antiviral drug development. Computational approaches, such as structure-based drug design (SBDD) and computer-aided drug design (CADD), have been crucial in expediting the identification of potential inhibitors for these viral proteins. By leveraging virtual screening, molecular docking, and molecular dynamics simulations, researchers can efficiently assess extensive libraries of compounds to predict their binding affinity and stability with these viral targets. Target-based drug discovery combines high-throughput screening, computational modeling, and experimental validation to ensure that the identified compounds are both theoretically promising and effective in inhibiting the virus in laboratory and clinical settings. The availability of comprehensive databases, including those of antiviral and FDA-approved drugs, along with high-performance computing resources, has greatly improved the efficiency of these processes. These tools allow for the rapid screening of millions of molecules, accelerating the identification of viable drug candidates. This integration of advanced computational methods with extensive drug libraries and powerful computing facilities has dramatically enhanced the speed and accuracy of drug discovery during the COVID-19 pandemic.

Recent research in vaccine development has made significant contributions since the pandemic began. Scientists have explored various vaccine platforms, including mRNA, protein subunit, viral vector, and inactivated virus vaccines. Notably, mRNA vaccines, such as those developed by Pfizer-BioNTech and Moderna, have been at the forefront, offering high efficacy and rapid production capabilities. In parallel, computational drug design has significantly aided these efforts by predicting antigenic epitopes, optimizing formulations, and simulating immune responses. This integration has expedited vaccine development, provided critical insights for addressing emerging variants, and improved vaccine durability, contributing to a more effective and equitable global pandemic response.

Structural Organization of SARS-CoV-2

SARS-CoV-2 is an enveloped, positive-sense, single-stranded RNA virus from the Coronaviridae family. The genome contains 14 open reading frames (ORFs), which can be categorized into two parts. ORF1a and ORF1ab, found in the first two-thirds of the viral genome from the 5′ end, are directly translated into two polyproteins, pp1a and pp1ab. These polyproteins are then processed by two viral proteases, papain-like protease (PLpro) and main protease (Mpro), to generate 16 non-structural proteins (Nsps), Nsp1–Nsp16. Collectively, these proteins constitute the replication–translation complex (RTC). The remaining segment near the 3′ end encodes structural proteins, including spike (S), envelope (E), membrane (M), and nucleocapsid (N) proteins. Additionally, this region includes structural genes that code for nine accessory proteins, produced by the ORF3a, ORF3d, ORF6, ORF7a, ORF7b, ORF8, ORF9b, ORF14, and ORF10 genes.

The life cycle of SARS-CoV-2 involves several critical steps for its replication and spread. Initially, the virus binds to the ACE2 receptor on the surface of host cells through its spike protein, facilitating the fusion of the viral envelope with the host cell membrane and allowing the viral RNA to enter the cytoplasm. Inside the host cell, the viral RNA is translated into viral proteins by the host's ribosomes, and the viral RNA-dependent RNA polymerase (RdRp) replicates the viral genome. The viral RNA is also transcribed into sub-genomic mRNAs, translated into structural and non-structural proteins necessary for assembling new virions. These newly synthesized viral proteins and RNA genomes are assembled into new virions in the host cell's cytoplasm, transported to the cell surface, and released into the extracellular environment, enabling them to infect new cells.

Essential Viral Proteins as Drug Targets

The development of drugs or inhibitors aims to block different stages of the SARS-CoV-2 life cycle, including receptor binding (spike inhibitors), proteolytic processing (papain-like protease inhibitors, main protease inhibitors), RNA synthesis (NSP12–NSP16 inhibitors), and assembly (nucleocapsid inhibitors). This section focuses on drug development for these viral targets at various stages of the viral life cycle.

Spike

The spike is a homo-trimeric class I fusion glycoprotein on the virion surface that is responsible for the binding and penetration of SARS-CoV-2 into host cells, making it an attractive antiviral target. In most cases, the spike protein is cleaved by host proteases into S1 subunit (receptor-binding domain) and S2 subunit (membrane-fusion domain), heavily shielded by N-linked and O-linked glycans [7]. The receptor-binding domain (RBD) of the S1 subunit is the only exposed component of the virus, making it a prime target for drug design. Following significant conformational changes, the RBD of the SARS-CoV-2 spike protein binds with high affinity to angiotensin-converting enzyme 2 (ACE2) on the host cell surface [8]. To inhibit this spike–ACE2 interaction or prevent membrane fusion, several types of molecules have been developed, small-molecule inhibitors and peptide inhibitors, including neutralizing antibodies [9]. Several different types of inhibitors have been developed to target this interaction, including small-molecule inhibitors, neutralizing antibodies, and peptide inhibitors. Numerous monoclonal antibodies have been reported for early COVID-19 treatment, though their effectiveness is diminishing with emerging variants. Small-molecule inhibitors and peptide-based therapies are under investigation, with some showing promise in early trials. Engineered proteins like thermostable designed ankyrin repeat proteins (DARPins) have also been explored, though their clinical benefits are still being assessed. Despite advancements, challenges remain in optimizing these treatments and addressing issues like antibody-dependent enhancement and the need for better pharmacokinetic properties in peptide inhibitors [10].

Nsp1

SARS-CoV-2 non-structural protein 1 (Nsp1) blocks mRNA entry into the ribosome, halting host protein synthesis and suppressing immune function, making it a key target for therapies. Montelukast sodium hydrate was identified as a promising candidate by screening FDA-approved drugs. It binds to Nsp1-C-ter with high affinity, forming a stable complex and counteracting Nsp1's inhibition of host protein synthesis. Montelukast also reduces viral replication in infected cells. Thus, it is suggested as a lead compound for developing effective SARS-CoV-2 inhibitors [11].

Nsp3

Papain-like protease (PLpro) is a cysteine protease that plays an important role in both viral protein processing and immune suppression by the virus. It cleaves polyproteins to release viral proteins and removes ubiquitin, suppressing the innate immune system. PLpro inhibitors are classified into three groups based on their mechanisms: covalent inhibitors (VIR250 and VIR251) that form a linkage with the catalytic cysteine of the PLpro [12], non-covalent inhibitors that block substrate access to the catalytic site (GRL0617, F0213, acriflavine, Jun9-72-2, Rac3k and XR8-24) [13], and non-covalent inhibitors targeting allosteric sites (HE9) [13]. Although PLpro inhibition can block viral protein maturation and restore immune responses, designing selective inhibitors is challenging due to structural similarities between PLpro and human deubiquitinating enzymes. Future research is likely to focus on optimizing non-covalent inhibitors to improve selectivity and reduce off-target effects.

Nsp5

The SARS-CoV-2 main protease (Mpro) plays an important role in replication by cleaving the pp1a and pp1ab polyproteins to release viral proteins NSP4 to NSP16. Inhibiting Mpro prevents the maturation of key viral enzymes, blocking viral replication. Various inhibitors of Mpro, including ensitrelvir, nirmatrelvir (Paxlovid), and SIM0417, have been developed to target this crucial enzyme through different mechanisms, such as covalent and non-covalent binding to the catalytic site or allosteric sites [14, 15]. Nirmatrelvir is the most successful inhibitor that has shown significant success in reducing COVID-19-related hospitalizations and mortality, although challenges like drug resistance and interactions remain unclear. It is important to note that Mpro is a stable homodimer and its dimerization interface may be an important binding site for targeting this critical viral enzyme.

Nsp7 and Nsp8

Nsp7 and nsp8 form a primase complex that is involved in the RNA synthesis pathway and enhances the activity of the RNA-dependent RNA polymerase (RdRp). A recently published crystal structure of the nsp7-nsp8 complex [16, 17], specifically focusing on the C-terminus of nsp8, has revealed several potentially druggable pockets at the dimerization interface. These pockets could be targeted to design inhibitors aimed at suppressing SARS-CoV-2 infection.

Nsp12

SARS-CoV-2 RNA-dependent RNA polymerase (RdRp), encoded by NSP12, supports a crucial role in viral RNA replication and transcription process. After processing by Mpro, NSP12 interacts with primase complex (nsp7–nsp8) and helicase (nsp13) to form a replication–transcription complex that performs several biological processes such as RNA synthesis, template unwinding, and RNA capping [18]. Inhibitors targeting these functions include several nucleotide analogues like remdesivir, which has shown effectiveness in early COVID-19 treatment but faces challenges such as lung-specific delivery and low oral bioavailability. Other antiviral agents like favipiravir and molnupiravir have varying degrees of safety and efficacy concerns, with ongoing research to optimize these treatments for SARS-CoV-2 [19].

Nsp13

SARS-CoV-2 helicase, a key enzyme with ATPase activity and RNA unwinding, has a complex structure featuring RNA-binding, zinc-binding, and ATPase domains. RNA-binding site and ATPase active site are the promising drug-binding pockets of the Nsp12. Some small molecules, such as 2-phenylquinoline derivatives and FPA-124, have shown potential in inhibiting helicase activity, but they require further optimization to overcome challenges like limited potency and off-target toxicity [20]. Although no helicase inhibitors have yet progressed to clinical trials for COVID-19, potent inhibitors have been developed for other viral infections, including drugs like amenamevir and pritelivir, which are used to treat herpes infections.

Nsp14

SARS-CoV-2 NSP14 has an exoribonuclease domain and a guanine-*N*7-methyltransferase domain that conduct viral RNA proofreading and capping, respectively. This proofreading mechanism also reduces the efficacy of several nucleos(t)ide inhibitors like ribavirin. Inhibitors targeting NSP14's exoribonuclease domain have been identified, but their potential off-target effects need careful consideration due to structural similarities with human enzymes. The C-terminal guanine-N7-methyltransferase domain of NSP14 is involved in the construction of viral RNA cap structure, which shields the viral RNA from immune detection and degradation. Although several inhibitors targeting this domain have shown moderate anti-SARS-CoV-2 activity in cell cultures, no NSP14 inhibitors have yet progressed to clinical trials for COVID-19.

Nsp15

NSP15 is a uridine-specific endoribonuclease that cleaves 5′-polyuridine tracts in negative-sense viral RNA to evade host immune responses. The Mn^{2+}-dependent active site, responsible for cleaving viral RNA substrates and producing specific termini, presents a potential drug-binding pocket [21]. Although inhibitors like tipiracil, an uracil derivative used in colorectal cancer treatment, have shown weak activity against SARS-CoV-2, further optimization could enhance their efficacy. Given the lack of close human homologues, NSP15 endoribonuclease can be an attractive antiviral target.

Nsp16

SARS-CoV-2 NSP16 and its activator NSP10 form a complex that modifies viral RNA to evade the host's immune response. While inhibitors like SAM derivatives and S-adenosyl-l-homocysteine (SAH) show anti-SARS-CoV-2 activity, their high toxicity and poor membrane permeability limit clinical use. NSP16 shares structural similarity with human CMTR1, a critical enzyme in RNA methylation. A small molecule called 3-deazaneplanocin A can inhibit both CMTR1 and NSP16, reducing viral replication, but it also causes significant side effects, highlighting the risks of targeting the RNA methylation process, as it could impair human mRNA processing and lead to toxicity.

Other Viral Targets

Non-structural proteins like NSP2, NSP4, NSP6, and NSP9 have been studied as antiviral targets, though their inhibitors, primarily derived from drug repurposing, need further refinement. Accessory proteins, such as ORF6, ORF3a, and ORF9b, play an important role in viral replication, immune evasion, and other functions, but are generally non-essential for the viral life cycle. While some molecules have been screened against these proteins, and RNA structures in conserved regions are explored for nucleic acid therapeutics, challenges in manufacturing have prevented any RNA-based inhibitors from progressing to COVID-19 trials [22].

Target Identification for COVID-19

Target identification for COVID-19 has focused on several key viral proteins and host factors critical to the viral life cycle. The interaction between the spike (S) protein and the ACE2 receptor is the initial step in viral infection, making it a prime target for drug development; inhibiting this interaction can prevent viral entry into host cells [23]. The Main Protease (Mpro) is essential for processing viral polyproteins into functional units necessary for replication, with Mpro inhibitors showing promise in halting viral production [24]. The Papain-like Protease (PLpro) processes viral polyproteins and modulates the host immune response, making PLpro inhibitors a valuable target for preventing immune evasion and replication. RNA-dependent RNA Polymerase (RdRp) is crucial for replicating the viral RNA genome, and inhibitors like Remdesivir have demonstrated efficacy in disrupting this process. Additionally, targeting host factors such as TMPRSS2 and Furin, which are involved in the spike protein's activation and entry into cells, offers alternative strategies for drug development. COVID-19 can also induce a cytokine storm, an aggressive immune response that can lead to severe complications; targeting inflammation pathways, such as the interleukin-6 (IL-6) pathway, with drugs like corticosteroids or monoclonal antibodies can help manage this severe immune reaction [25]. Table 16.1, Figs. 16.1 and 16.2 briefly outline the various targets and their respective therapeutic approaches to mitigate COVID-19 infection.

Drug Discovery Approaches

The following sections and Fig. 16.3 briefly describe various computational approaches implemented for drug screening against SARS-CoV-2 target proteins.

High-Throughput Screening (HTS)

High-throughput screening (HTS) is a cutting-edge technique in drug discovery that allows researchers to rapidly evaluate the biological activity of thousands of compounds against specific targets. This method has been particularly impactful in the fight against COVID-19 by identifying potential inhibitors of critical viral proteins, such as the Main Protease (Mpro) and RNA-dependent RNA Polymerase (RdRp). HTS involves automated assays that test large libraries of compounds to find ones that interact effectively with the target proteins. These promising candidates can undergo further refinement and validation in preclinical and clinical studies, significantly accelerating the drug discovery process. For instance, a comprehensive HTS effort screened a library of thousands of compounds to identify inhibitors of the SARS-CoV-2 main protease. This screening resulted in the discovery of several potent inhibitors that demonstrated promise in subsequent studies, highlighting HTS's crucial role in identifying lead compounds for drug development [4]. Similarly, researchers used HTS to evaluate a diverse chemical library for inhibitors of the RNA-dependent RNA polymerase (RdRp) of SARS-CoV-2. This study identified several compounds with strong antiviral activity, demonstrating HTS's ability to uncover effective drug candidates [6].

Virtual Screening

Virtual screening is a computational method used to assess large libraries of compounds to identify those most likely to bind to a specific target protein. This

Table 16.1 Target identification of COVID-19

Target	Function	Drug development focus	Examples of therapeutics
Spike (S) protein	Facilitates viral entry into host cells by binding to the ACE2 receptor.	Inhibitors/blockers to prevent virus entry; target for neutralizing antibodies and vaccines.	Monoclonal antibodies (e.g., Bamlanivimab and vaccines (e.g., Pfizer-BioNTech, Moderna) Casirivimab Sotrovimab
Main Protease (Mpro)	Cleaves polyproteins into functional units essential for viral replication.	Inhibitors to block viral replication; key target for antiviral drug development.	Nirmatrelvir Lopinavir
Papain-like Protease (PLpro)	Processes viral polyproteins and modulates host immune response.	Inhibitors to block immune evasion and viral replication.	GRL-0617
RNA-dependent RNA Polymerase (RdRp)	Replicates the viral RNA genome.	Inhibitors to halt viral replication.	Remdesivir, Favipiravir, Molnupiravir
NSP13	Helicase that unwinds RNA for replication.	Inhibitors to block viral replication.	Puniclagin
Envelop	Involved in viral particle formation and release.	Inhibitors targeting envelope proteins to prevent viral budding.	Sinapic acid, Rutin Doxycycline
Nucleocapsid	Binds viral RNA and assists in virion assembly.	Inhibitors targeting nucleocapsid proteins to disrupt viral assembly.	Caftazidime, Sennoside A Tannic acid
Membrane	Contributes to the structure and function of the viral membrane.	Targeting membrane proteins to interfere with viral replication.	Caffeic acid, Ferulic acid
Viral assembly proteins	Proteins involved in assembling and packaging viral particles.	Inhibitors targeting viral assembly processes to prevent new virion formation.	(No specific drugs are approved so far; research is ongoing)
Viral RNA interference	Methods targeting viral RNA for degradation or inhibition.	RNA-based drugs to interfere with viral RNA processes, reducing viral load.	(Experimental approaches; research is ongoing)
Host factor: TMPRSS2	Serine protease involved in priming the spike protein for entry into host cells.	Inhibitors to block viral entry by targeting host proteases.	Camostat mesylate Nafamostat
Host factor: Furin	Protease is responsible for activating the spike protein.	Inhibitors to prevent activation of the spike protein and viral entry.	(No specific drugs are approved so far; research is ongoing)

(continued)

Table 16.1 (continued)

Target	Function	Drug development focus	Examples of therapeutics
Host factor: ACE2	Receptor for spike protein binding, facilitating viral entry.	Drugs to alter ACE2 interactions or levels; potential for therapeutic modulation.	(No specific drugs are approved so far; research is ongoing)
Host factor: Cathepsin L	Cellular protease involved in spike protein activation and viral entry.	Inhibitors to block the activation of spike protein and viral entry.	(No specific drugs are approved so far; research is ongoing)
Host immune modulators	Modulator to enhance antiviral activity.	Drugs to boost or modulate the host immune response to better fight the virus.	Interferons (e.g., interferon-beta)
Host cytokine storm pathways	Hyper-inflammatory response causing severe complications.	Targeting inflammatory pathways to mitigate severe disease; modulation of the immune response.	Dexamethasone Tocilizumab (IL-6 inhibitor)
Host cell signaling pathways	Pathways involved in viral replication or immune response modulation.	Targeting host cell signaling pathways to disrupt viral life cycle or enhance immune response.	Experimental approaches; research is ongoing

technique is crucial for quickly narrowing down potential drug candidates from extensive chemical libraries, especially in response to emerging infectious diseases like COVID-19. During the COVID-19 pandemic, virtual screening has been vital in identifying potential inhibitors of SARS-CoV-2. It involves simulating interactions between compounds and viral proteins to predict binding affinities and prioritize candidates for further testing [26]. Virtual screening was used to find inhibitors of the SARS-CoV-2 main protease (Mpro), an essential enzyme for viral replication. By screening a library of over 10,000 compounds, they identified several with strong binding affinities to Mpro, leading to further experimental validation and optimization. Similarly, virtual screening was employed to evaluate compounds' ability to inhibit the SARS-CoV-2 RNA-dependent RNA polymerase (RdRp). This process involved analyzing compound libraries against the RdRp structure, identifying several promising inhibitors, and demonstrating how virtual screening accelerates the discovery of effective antiviral agents.

Molecular Docking

Molecular docking is a computational technique to predict how small molecules, such as drugs, interact with proteins at the atomic level. This method simulates the binding of potential drugs to specific viral targets, such as the spike protein or RNA-dependent RNA Polymerase (RdRp) of SARS-CoV-2. By analyzing these interactions, researchers can identify compounds with high binding affinities and specific

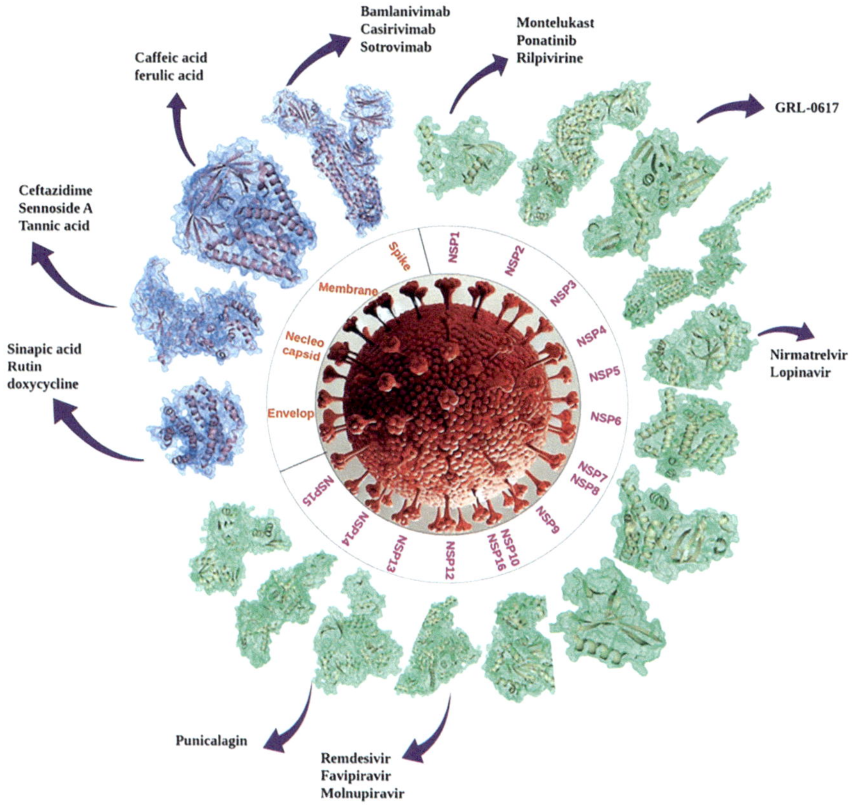

Fig. 16.1 Therapeutic drugs against SARS-CoV-2 proteins

interactions, which are prioritized for further testing. During the COVID-19 pandemic, molecular docking has been pivotal in discovering and evaluating potential antiviral agents. For instance, one study used molecular docking to investigate potential inhibitors of the SARS-CoV-2 main protease (Mpro) via docking a library of small molecules into the Mpro's active site. This research identified several promising inhibitors with strong binding to Mpro, providing valuable insights for drug development [27]. Similarly, another study employed molecular docking to explore how various compounds interact with the SARS-CoV-2 spike protein, identifying several candidates with high binding affinities, which guided further experimental studies [28].

Structure-Based Drug Design (SBDD)

Structure-Based Drug Design (SBDD) leverages the three-dimensional structure of target proteins to develop novel medications. Researchers can design inhibitors that selectively bind and block the protein's function by analyzing the spatial

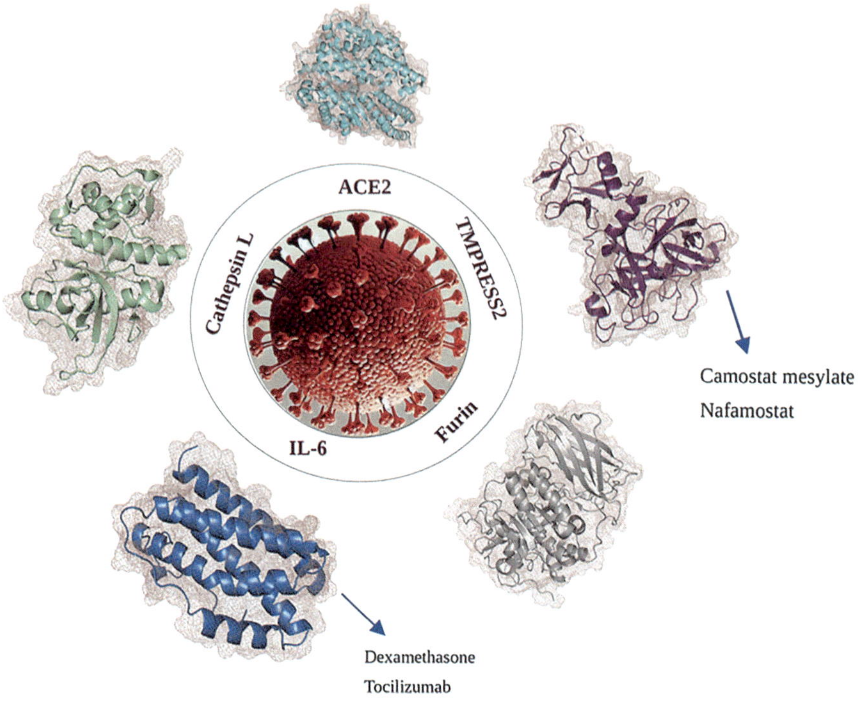

Fig. 16.2 Therapeutic drugs against host factors

arrangement of atoms in a protein's active site. This approach is precious for targeting proteins critical to the viral life cycle. In the context of COVID-19, SBDD has been employed to develop inhibitors against essential SARS-CoV-2 proteins. For example, a study focused on the crystal structure of the SARS-CoV-2 main protease (Mpro) to identify potential binding sites and develop effective inhibitors [6].

Ligand-Based Drug Design (LBDD)

Ligand-Based Drug Design (LBDD) employs known ligands—molecules that bind to a target protein—to guide the development of new drugs, especially when the protein's structure is incomplete or poorly defined. LBDD utilizes existing ligands' structural and chemical properties to design novel, effective compounds. During the COVID-19 pandemic, LBDD has been crucial in identifying potential therapeutics for the SARS-CoV-2 spike protein. Researchers used LBDD to design inhibitors based on ligands that bind to the spike protein, creating novel compounds that can disrupt the interaction between the spike protein and human receptors, facilitating the development of new antiviral agents.

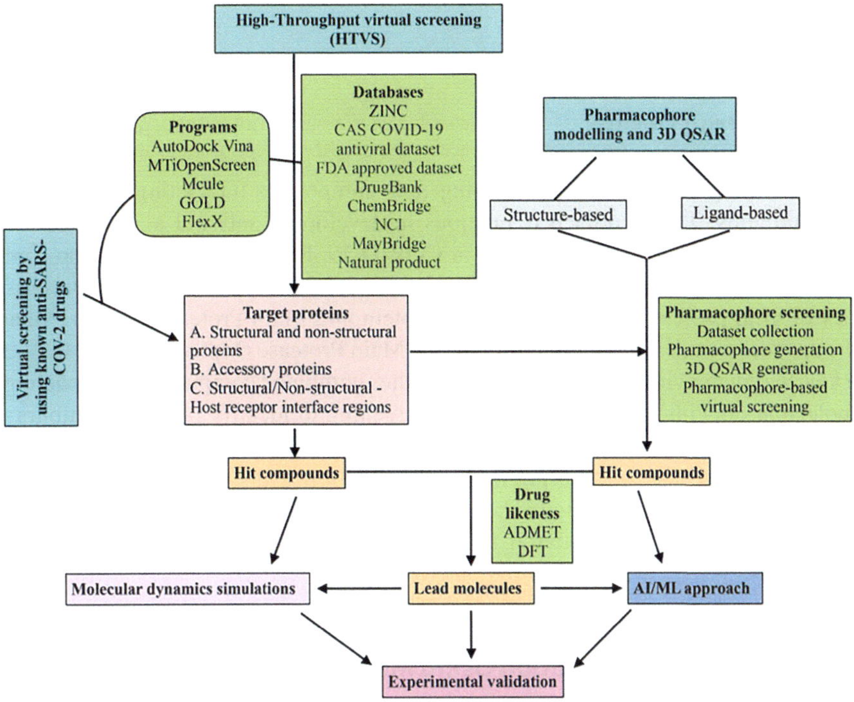

Fig. 16.3 Flow chart of computer-aided drug designing approaches against SARS-CoV-2

Repurposing Existing Drugs

Drug repurposing, or drug repositioning, involves evaluating existing drugs approved for other indications to determine their potential efficacy against new diseases. This approach can considerably cut down both time and expenses. Repurposing existing drugs is beneficial in drug development since these medications have already been thoroughly tested for safety. In the context of COVID-19, drug repurposing has been instrumental in identifying several effective treatments. Remdesivir, originally developed for treating Ebola virus disease, was repurposed for COVID-19 after showing promise in preclinical studies [29]. The study found that Remdesivir significantly shortened the recovery time in hospitalized patients with severe COVID-19 compared to a placebo. This research led to regulatory agencies' emergency use authorization of Remdesivir for treating severe COVID-19 cases. Dexamethasone, a corticosteroid with known anti-inflammatory properties, was repurposed for COVID-19 based on its potential to reduce inflammation in severe cases. The RECOVERY trial demonstrated that Dexamethasone reduced mortality among hospitalized COVID-19 patients requiring oxygen or mechanical ventilation [30]. This landmark study highlighted the drug's effectiveness in managing severe

COVID-19 and led to its widespread use in clinical practice to enhance patient outcomes.

De novo Drug Design

De novo drug design involves creating new compounds interacting with critical viral proteins. This approach is precious in developing antiviral agents aimed at SARS-CoV-2, the virus responsible for COVID-19. Researchers can develop compounds with high specificity and potency by focusing on the structural characteristics of viral proteins, such as the spike protein and Main Protease (Mpro). Many groups focus on designing inhibitors for the Main Protease (Mpro) of SARS-CoV-2, a key enzyme in viral replication. These studies combined computational modeling, molecular docking, and high-throughput screening to identify potential inhibitors. For example, Zhang and colleagues successfully designed novel small molecules that effectively bind to the Mpro active site, inhibiting its enzymatic function and disrupting the viral life cycle [31]. This research identified several promising lead compounds with potent antiviral activity, advancing them to preclinical evaluation stages.

Docking Complex Validation Using MD

Validating docking complexes with molecular dynamic (MD) simulations is crucial in drug discovery to confirm the stability and reliability of predicted protein–ligand interactions. Tools like GROMACS [32] and DESMOND [33] are commonly employed to assess complex's dynamic behaviors over time. Key metrics include Root Mean Square Deviation (RMSD), which evaluates the average deviation of atomic positions to determine stability, and Root Mean Square Fluctuation (RMSF), which measures residue flexibility across the complex. Hydrogen Bond Analysis examines bond formation and persistence, which are critical for stable interactions, while the Radius of Gyration assesses the complex's compactness and structural stability. For instance, GROMACS [32] validated a novel inhibitor of the SARS-CoV-2 spike protein, demonstrating stable binding with low RMSD and consistent hydrogen bonding. GROMACS [32] and DESMOND [33] were employed for molecular dynamics (MD) simulations to validate the inhibitors of the SARS-CoV-2 proteins. Molecular dynamic simulations revealed significant stability of the protein–ligand complexes, suggesting that the inhibitors effectively bind to the protein's active site. Binding free energy calculations were performed using the gmx_MM-PBSA [34] tool to further quantify the binding strength and stability of the complexes. The MM-PBSA/MM-GBSA method [35, 36] estimates the binding free energy by evaluating different energetic components, including molecular mechanics (MM) energy, solvation-free energy, and the entropic contribution [37].

Results from the MM-PBSA/MM-GBSA calculations demonstrated binding affinity between the inhibitors and SARS-CoV-2 proteins, with highly negative free energy values, indicating stable interactions. The Interaction Entropy (IE) module was also utilized to account for the entropic contribution to the binding free energy [37]. This module enhances the accuracy of the thermodynamic calculations by extracting entropy directly from MD simulation ensembles. Overall, the combination of MD simulations and MM-PBSA/MM-GBSA analysis [35, 36] provided detailed insights into the molecular interactions between the inhibitors and the target proteins, supporting their potential as effective antiviral agents.

Pharmacophore Screening Approach

Structure-based and ligand-based pharmacophore modeling are two complementary approaches used to design several different inhibitors against SARS-CoV-2. In structure-based modeling, the 3D structure of viral proteins (such as spike protein or main protease) is used to identify key molecular interactions between the target proteins and potential inhibitors, often by analyzing the binding pocket's shape, hydrogen bonding, and hydrophobic interactions. On the other hand, ligand-based modeling relies on known inhibitors or drugs of SARS-CoV-2 proteins to derive common geometric features responsible for activity, such as specific functional groups or spatial arrangements. Both methods are used to generate essential features or pharmacophores that can guide virtual screening or the design of novel antiviral compounds.

Key Drug Candidates and Clinical Trials

Table 16.2 highlights a range of therapeutic agents developed for COVID-19, categorized by their mechanisms and clinical findings. Remdesivir and Favipiravir are antiviral drugs targeting RNA-dependent RNA polymerase (RdRp), with evidence supporting their efficacy in reducing recovery times and viral loads. Lopinavir/Ritonavir and Molnupiravir are protease inhibitors with mixed results, while Paxlovid combines a protease inhibitor with a CYP3A4 inhibitor to improve effectiveness. Monoclonal antibodies like Bamlanivimab, Casirivimab/Imdevimab, and Sotrovimab target the spike protein, showing promise in reducing hospitalization risk. Immunomodulators such as IL-6 inhibitors and corticosteroids have been beneficial in managing severe cases by reducing inflammation. Peptide inhibitors and interferons are under investigation, offering potential new treatments. Convalescent plasma has shown mixed results, and Tocilizumab and Dexamethasone are established treatments for severe cases. The landscape of COVID-19 therapeutics is dynamic, with ongoing research and development aimed at optimizing treatment strategies and improving patient outcomes.

Table 16.2 Therapeutic agents against target proteins of SARS-CoV-2

Drug candidate	Target	Description	Clinical trial findings	Development status
Remdesivir	RNA-dependent RNA Polymerase (RdRp)	Antiviral drug originally developed for Ebola; reduces recovery time in hospitalized COVID-19 patients.	Shown to reduce recovery time in hospitalized patients with COVID-19.	Emergency use authorization granted.
Favipiravir	RNA-dependent RNA Polymerase (RdRp)	Initially developed for influenza; broad-spectrum activity against RNA viruses.	Some studies suggest reduced viral load and improved patient outcomes in COVID-19.	Approved in some countries; ongoing trials.
Lopinavir/ Ritonavir	Main Protease (Mpro)	Protease inhibitors developed for HIV; repurposed for COVID-19 to inhibit SARS-CoV-2 protease.	Mixed results; some studies show limited efficacy in reducing symptoms or improving survival rates.	Not recommended for routine use.
Bamlanivimab	Spike protein	Monoclonal antibody designed to prevent SARS-CoV-2 from entering cells.	Demonstrated ability to reduce hospitalization risk in high-risk COVID-19 patients.	Emergency use authorization granted.
Casirivimab/ Imdevimab	Spike protein	Combination monoclonal antibodies that bind to distinct sites on the spike protein of SARS-CoV-2, preventing viral entry into cells.	Shown to reduce hospitalization rates and improve outcomes in high-risk patients.	Emergency use authorization granted.
Sotrovimab	Spike protein	Monoclonal antibody targeting the spike protein, designed to neutralize SARS-CoV-2 and prevent viral entry into cells.	Reduced risk of hospitalization and death in high-risk patients.	Emergency use authorization granted.
Molnupiravir	RNA-dependent RNA Polymerase (RdRp)	Oral antiviral medication that introduces errors into the RNA of SARS-CoV-2, inhibiting viral replication.	Reduced risk of hospitalization and death in high-risk patients.	Emergency use authorization granted.

(continued)

Table 16.2 (continued)

Drug candidate	Target	Description	Clinical trial findings	Development status
Paxlovid (Nirmatrelvir/ Ritonavir)	Main Protease (Mpro) and CYP3A4 inhibitor	Oral antiviral combination with Nirmatrelvir (Mpro inhibitor) and Ritonavir (CYP3A4 inhibitor) to enhance Nirmatrelvir levels.	Shown to reduce hospitalization and severe outcomes in high-risk COVID-19 patients.	Emergency use authorization granted.
Small molecules	Various targets	Includes new RdRp inhibitors and protease inhibitors explicitly designed for SARS-CoV-2.	Actively investigated; potential for inhibiting viral replication or modulating immune response.	In various stages of clinical trials.
Immuno-modulators	IL-6, Cytokines	Drugs such as IL-6 inhibitors (e.g., Tocilizumab) and corticosteroids (e.g., Dexamethasone) are used to reduce inflammation and cytokine storm in severe COVID-19 cases.	Effective in reducing inflammation and improving outcomes in severe cases of COVID-19.	Approved for severe cases; ongoing research.
Peptide inhibitors	Viral proteins and host receptors	Peptide-based inhibitors that block interactions between viral proteins and host receptors, preventing viral entry or replication.	Still under exploration; potential to provide new therapeutic options for COVID-19.	Early stages of development.
Interferons	Various viral and host targets	Natural proteins that enhance the immune response against viral infections.	Shown to reduce viral load and improve outcomes in some studies; mixed results in others.	Approved in some regions; ongoing trials.
Convalescent plasma	SARS-CoV-2 antibodies	Plasma from recovered COVID-19 patients contains antibodies against SARS-CoV-2.	Mixed results; some studies suggest potential benefit in reducing severity, others show limited impact.	Used in some regions; research is ongoing.

(continued)

Table 16.2 (continued)

Drug candidate	Target	Description	Clinical trial findings	Development status
Tocilizumab	IL-6 receptor	Monoclonal antibody targeting IL-6 receptor, used to mitigate severe inflammatory response in COVID-19 patients.	Shown to reduce mortality and improve outcomes in severe COVID-19 cases.	Approved for use in severe cases.
Dexamethasone	Glucocorticoid receptor	Corticosteroid that reduces inflammation and modulates the immune response.	Proven to reduce mortality in severely ill COVID-19 patients.	Widely used in severe cases.
Baricitinib	Janus kinase (JAK) 1 and 2	Janus kinase inhibitor that modulates immune response and inflammation.	Improve outcomes in hospitalized COVID-19 patients with supplemental O_2.	Emergency use authorization granted.

Challenges and Future Directions

Drug discovery for COVID-19 faces significant challenges, including the emergence of new viral variants, drug resistance, and the need for scalable and cost-effective therapies. The emergence of new variants, such as those identified in the United Kingdom, South Africa, and India, poses challenges to vaccine efficacy and therapeutic interventions. These variants may carry mutations that alter viral proteins, potentially leading to reduced effectiveness of existing drugs. Addressing drug resistance and escape mechanisms is essential for long-term efficacy. Ongoing research efforts focus on understanding the mechanisms of drug resistance and escape to develop strategies that minimize the risk of resistance and enhance the durability of therapeutic interventions. To overcome these challenges, ongoing research efforts are focused on developing strategies that target multiple viral proteins or host factors. Combination therapies, which involve using multiple drugs with different mechanisms of action, hold promise for enhancing therapeutic efficacy and reducing the likelihood of drug resistance. Collaborative research efforts across academia, industry, and government agencies are essential for addressing the complexities of drug discovery and development in the face of a global pandemic. Continued investment in drug discovery, innovative approaches, and global collaboration will be vital in addressing COVID-19 and future pandemics.

Balancing Efficacy and Safety in Drug Development

Balancing efficacy and safety is a critical challenge in drug development, particularly for COVID-19. Developing drugs that are effective against the virus while ensuring human safety for widespread use is crucial. The rapid development and deployment of COVID-19 therapies have highlighted the need for robust clinical trials to thoroughly assess both efficacy and safety. Concurrently, vaccines have been the cornerstone of efforts to control the pandemic, but therapeutic approaches remain vital for treating those who are infected. Effective antiviral drugs are essential for managing severe cases and providing options for individuals who cannot be vaccinated. The interplay between vaccine and therapeutic approaches will continue to shape the future of COVID-19 management.

The Role of AI and Machine Learning in Accelerating Drug Discovery

Artificial intelligence (AI) and machine learning (ML) are revolutionizing drug discovery, particularly addressing urgent global health challenges like COVID-19. These advanced technologies excel at processing vast datasets to identify potential drug candidates, predict interactions between drugs and their targets, and streamline the drug development process. Beyond COVID-19, AI and ML have been applied across various stages of drug discovery, including virtual screening, lead optimization, and drug efficacy and toxicity predictions. For example, AI-driven tools like ADMET LAB2.0 and ADMET LAB3.0 [38, 39] have been instrumental in optimizing drug properties such as solubility and bioavailability while predicting potential side effects. AI's success in predicting protein structures, especially when traditional methods like X-ray crystallography fall short, has cemented its role as a valuable tool in drug design. AI enhances the accuracy of protein structure predictions [40] and advances protein multimerization and docking complex analysis. By offering a cost-effective and time-saving alternative to labor-intensive and expensive experimental approaches, AI significantly accelerates drug discovery, presenting innovative solutions for developing treatments for current and future health crises.

Future Pandemics and Preparedness in Drug Discovery

The global impact of the COVID-19 pandemic has emphasized the critical importance of being prepared to discover new drugs. Establishing a resilient global infrastructure for rapid drug development is vital for effectively responding to potential future pandemics. This necessitates significant investment in research and development, encouragement of international collaboration, and harnessing innovative technologies to expedite the drug discovery process.

Conclusion

The global response to COVID-19 has led to unprecedented advances in drug discovery aided by scientists worldwide coming together to develop new therapies and vaccines. The target-based drug discovery approach has played a central role in identifying and developing potential drugs that specifically target critical components of the SARS-CoV-2 virus. While significant progress has been made, challenges remain, including new viral variants, the need for safe and effective treatments, and the ongoing threat of future pandemics. The COVID-19 pandemic has demonstrated the importance of innovation, collaboration, and preparedness in addressing global health challenges. As we continue to navigate the evolving landscape of COVID-19, the lessons learned from this pandemic will inform future drug discovery efforts, ensuring that we are better equipped to respond to the next global health crisis.

Author's Contributions Conceived the idea and review structure: AC and SC; Wrote the manuscript: AC, SC, and IMK; Revised and edited the manuscript: AC, SC, and IMK. All the authors have read, agreed, and approved the final draft of the manuscript.

Funding SC acknowledges CSIR-IICB for funding and infrastructure support. AC acknowledges ICMR for post-doctoral Research Associateship [BMI/11(55)2022]. IMK acknowledges DBT, Government of India, for his Ph.D. fellowship (DBT/2019/IICB/1213).

Declaration

Conflicts of Interest There is no competing and conflict of interest.

References

1. Worldometer. COVID-19 coronavirus pandemic. 2023. Retrieved 30 Aug 2024.
2. Arshad U, et al. Prioritization of anti-SARS-Cov-2 drug repurposing opportunities based on plasma and target site concentrations derived from their established human pharmacokinetics. Clin Pharmacol Ther. 2020;108(4):775–90.
3. Sahin U, et al. COVID-19 vaccine BNT162b1 elicits human antibody and TH1 T cell responses. Nature. 2020;586(7830):594–9.
4. Gordon DE, et al. A SARS-CoV-2 protein interaction map reveals targets for drug repurposing. Nature. 2020;583(7816):459–68.
5. Jin Z, et al. Structure of Mpro from SARS-CoV-2 and discovery of its inhibitors. Nature. 2020;582(7811):289–93.
6. Zhang L, Lin D, Sun X, Curth U, Drosten C, Sauerhering L, Becker S, Rox K, Hilgenfeld R. Crystal structure of SARS-CoV-2 main protease provides a basis for design of improved α-ketoamide inhibitors. Science. 2020;368(6489):409–12.
7. Cao Y, et al. Potent neutralizing antibodies against SARS-CoV-2 identified by high-throughput single-cell sequencing of convalescent patients' B cells. Cell. 2020;182(1):73–84.e16.
8. Shang J, Ye G, Shi K, Wan Y, Luo C, Aihara H, Geng Q, Auerbach A, Li F. Structural basis of receptor recognition by SARS-CoV-2. Nature. 2020;581(7807):221–4.
9. Shi R, Shan C, Duan X, Chen Z, Liu P, Song J, Song T, Bi X, Han C, Wu L, Gao G, Hu X, Zhang Y, Tong Z, Huang W, Liu WJ, Wu G, Zhang B, Wang L, Qi J, Feng H, Wang FS, Wang

Q, Gao GF, Yuan Z, Yan J. A human neutralizing antibody targets the receptor-binding site of SARS-CoV-2. Nature. 2020;584(7819):120–4.
10. Hansen J, et al. Studies in humanized mice and convalescent humans yield a SARS-CoV-2 antibody cocktail. Science. 2020;369(6506):1010–4.
11. Afsar M, Narayan R, Akhtar MN, Das D, Rahil H, Nagaraj SK, Eswarappa SM, Tripathi S, Hussain T. Drug targeting Nsp1-ribosomal complex shows antiviral activity against SARS-CoV-2. Elife. 2022;11:e74877.
12. Rut W, Lv Z, Zmudzinski M, Patchett S, Nayak D, Snipas SJ, El Oualid F, Huang TT, Bekes M, Drag M, Olsen SK. Activity profiling and crystal structures of inhibitor-bound SARS-CoV-2 papain-like protease: a framework for anti-COVID-19 drug design. Sci Adv. 2020;6(42):eabd4596.
13. Li G, Hilgenfeld R, Whitley R, De Clercq E. Therapeutic strategies for COVID-19: progress and lessons learned. Nat Rev Drug Discov. 2023;22(6):449–75.
14. Noske GD, de Souza Silva E, de Godoy MO, Dolci I, Fernandes RS, Guido RVC, Sjö P, Oliva G, Godoy AS. Structural basis of nirmatrelvir and ensitrelvir activity against naturally occurring polymorphisms of the SARS-CoV-2 main protease. J Biol Chem. 2023;299(3):103004.
15. Song L, Gao S, Ye B, Yang M, Cheng Y, Kang D, Yi F, Sun JP, Menéndez-Arias L, Neyts J, Liu X, Zhan P. Medicinal chemistry strategies towards the development of non-covalent SARS-CoV-2 Mpro inhibitors. Acta Pharm Sin B. 2024;14(1):87–109.
16. Biswal M, Diggs S, Xu D, Khudaverdyan N, Lu J, Fang J, Blaha G, Hai R, Song J. Two conserved oligomer interfaces of NSP7 and NSP8 underpin the dynamic assembly of SARS-CoV-2 RdRP. Nucleic Acids Res. 2021;49(10):5956–66.
17. Courouble VV, Dey SK, Yadav R, Timm J, Harrison JJEK, Ruiz FX, Arnold E, Griffin PR. Revealing the structural plasticity of SARS-CoV-2 nsp7 and nsp8 using structural proteomics. J Am Soc Mass Spectrom. 2021;32(7):1618–30.
18. Gordon CJ, Tchesnokov EP, Woolner E, Perry JK, Feng JY, Porter DP, Götte M. Remdesivir is a direct-acting antiviral that inhibits RNA-dependent RNA polymerase from severe acute respiratory syndrome coronavirus 2 with high potency. J Biol Chem. 2020;295(20):6785–97.
19. Painter WP, Holman W, Bush JA, Almazedi F, Malik H, Eraut NCJE, Morin MJ, Szewczyk LJ, Painter GR. Human safety, tolerability, and pharmacokinetics of molnupiravir, a novel broad-spectrum oral antiviral agent with activity against SARS-CoV-2. Antimicrob Agents Chemother. 2021;65(5):e02428-20.
20. Otsuka Y, Kim E, Krueger A, Shumate J, Wang C, Bdiri B, Ullah S, Park H, Scampavia L, Bannister TD, Chung D, Spicer TP. High throughput screening for SARS-CoV-2 helicase inhibitors. SLAS Discov. 2024;29:100180.
21. Kim Y, Jedrzejczak R, Maltseva NI, Wilamowski M, Endres M, Godzik A, Michalska K, Joachimiak A. Crystal structure of Nsp15 endoribonuclease NendoU from SARS-CoV-2. Protein Sci. 2020;29(7):1596–605.
22. Finkel Y, et al. The coding capacity of SARS-CoV-2. Nature. 2021;589(7840):125–30.
23. Hoffmann M, Kleine-Weber H, Schroeder S, Krüger N, Herrler T, Erichsen S, Schiergens TS, Herrler G, Wu NH, Nitsche A, Müller MA, Drosten C, Pöhlmann S. SARS-CoV-2 cell entry depends on ACE2 and TMPRSS2 and is blocked by a clinically proven protease inhibitor. Cell. 2020;181(2):271–280.e8.
24. Zhu W, Shyr Z, Lo DC, Zheng W. Viral proteases as targets for coronavirus disease 2019 drug development. J Pharmacol Exp Ther. 2021;378(2):166–72.
25. Stone JH, et al. Efficacy of tocilizumab in patients hospitalized with Covid-19. N Engl J Med. 2020;383(24):2333–44.
26. Wang Q, Zhang Y, Wu L, Niu S, Song C, Zhang Z, Lu G, Qiao C, Hu Y, Yuen KY, Wang Q, Zhou H, Yan J, Qi J. Structural and functional basis of SARS-CoV-2 entry by using human ACE2. Cell. 2020;181(4):894–904.e9.
27. Jin Z, Du X, Xu Y, Deng Y, Liu M, Zhao Y, Zhang B, Li X, Zhang L, Peng C, Duan Y, Yu J, Wang L, Yang K, Liu F, Jiang R, Yang X, You T, Liu X, Yang X, Bai F, Liu H, Liu X, Guddat LW, Xu W, Xiao G, Qin C, Shi Z, Jiang H, Rao Z, Yang H. Structure of Mpro from SARS-CoV-2 and discovery of its inhibitors. Nature. 2020;582(7811):289–93.

28. Liu Z, Xiao X, Wei X, Li J, Yang J, Tan H, Zhu J, Zhang Q, Wu J, Liu L. Composition and divergence of coronavirus spike proteins and host ACE2 receptors predict potential intermediate hosts of SARS-CoV-2. J Med Virol. 2020;92(6):595–601.
29. Beigel JH, et al. ACTT-1 Study Group Members. Remdesivir for the treatment of Covid-19 – final report. N Engl J Med. 2020;383(19):1813–26.
30. Horby P, Lim WS, Emberson JR, Mafham M, Bell JL, Linsell L, Staplin N, Brightling C, Ustianowski A, Elmahi E, Prudon B, Green C, Felton T, Chadwick D, Rege K, Fegan C, Chappell LC, Faust SN, Jaki T, Jeffery K, Montgomery A, Rowan K, Juszczak E, Baillie JK, Haynes R, Landray MJ. Dexamethasone in hospitalized patients with Covid-19. N Engl J Med. 2021;384(8):693–704.
31. Ma C, Sacco MD, Hurst B, Townsend JA, Hu Y, Szeto T, Zhang X, Tarbet B, Marty MT, Chen Y, Wang J. Boceprevir, GC-376, and calpain inhibitors II, XII inhibit SARS-CoV-2 viral replication by targeting the viral main protease. Cell Res. 2020;30(8):678–92.
32. Case DA, Cheatham TE, Darden T, Gohlke H, Luo R, Merz KM, et al. The amber biomolecular simulation programs. J Comput Chem. 2005;26:1668–88.
33. Abraham MJ, Murtola T, Schulz R, Páll S, Smith JC, Hess B, et al. Gromacs: high performance molecular simulations through multi-level parallelism from laptops to supercomputers. SoftwareX. 2015;1–2:19–25.
34. Valdés-Tresanco MS, Valdés-Tresanco ME, Valiente PA, Moreno E. gmx_MMPBSA: a new tool to perform end-state free energy calculations with GROMACS. J Chem Theory Comput. 2021;17(10):6281–91.
35. Genheden S, Ryde U. The MM/PBSA and MM/GBSA methods to estimate ligand-binding affinities. Expert Opin Drug Discov. 2015;10(5):449–61.
36. Wang E, Sun H, Wang J, Wang Z, Liu H, Zhang JZH, Hou T. End-point binding free energy calculation with MM/PBSA and MM/GBSA: strategies and applications in drug design. Chem Rev. 2019;119(16):9478–508.
37. Duan L, Liu X, Zhang JZ. Interaction entropy: a new paradigm for highly efficient and reliable computation of protein-ligand binding free energy. J Am Chem Soc. 2016;138(17):5722.
38. Xiong G, Wu Z, Yi J, Fu L, Yang Z, Hsieh C, Yin M, Zeng X, Wu C, Lu A, Chen X, Hou T, Cao D. ADMETlab 2.0: an integrated online platform for accurate and comprehensive predictions of ADMET properties. Nucleic Acids Res. 2021;49(W1):W5–W14.
39. Fu L, Shi S, Yi J, Wang N, He Y, Wu Z, Peng J, Deng Y, Wang W, Wu C, Lyu A, Zeng X, Zhao W, Hou T, Cao D. ADMETlab 3.0: an updated comprehensive online ADMET prediction platform enhanced with broader coverage, improved performance, API functionality and decision support. Nucleic Acids Res. 2024;52(W1):W422–31.
40. Jumper J, et al. Highly accurate protein structure prediction with AlphaFold. Nature. 2021;596(7873):583–9.

Part V

COVID-19: Vaccine Development and Vaccination

COVID-19 Vaccine: Classification, Development, and Clinical Status

17

Shilpi Sarkar, Thirukumaran Kandasamy, Arisha Arora, and Siddhartha Sankar Ghosh

Abstract

The COVID-19 pandemic, caused by the SARS-CoV-2 virus, has impacted millions of people worldwide, necessitating the urgent need for vaccine development and therapies. Vaccination or immunizations remain the most potent approach against COVID-19 due to the rapid structural alterations in the SARS-CoV-2 structure and the lack of particular medications to combat the virus. The swift development of COVID-19 vaccines and widespread implementation of vaccinations during the pandemic have highlighted their critical role in controlling the spread of pathogens and reducing their impact on global health systems. As of March 2023, 183 different COVID-19 vaccine candidates are in clinical development, while some are still in the preclinical development stage. The categorization, development, and current clinical status of these vaccines are emphasized in this comprehensive chapter. Existing avenues for developing vaccines against COVID-19 include protein subunit, viral vector, inactivated or live

Shilpi Sarkar, Thirukumaran Kandasamy, and Arisha Arora have equally contributed to this book chapter.

S. Sarkar · T. Kandasamy · A. Arora
Department of Biosciences and Bioengineering, Indian Institute of Technology Guwahati, Guwahati, Assam, India

S. S. Ghosh (✉)
Department of Biosciences and Bioengineering, Indian Institute of Technology Guwahati, Guwahati, Assam, India

Centre for Nanotechnology, Indian Institute of Technology Guwahati, Guwahati, Assam, India
e-mail: sghosh@iitg.ac.in

M. K. Ghosh, M. Basu (eds.), *Pathogenesis, Management and Socio-Economic Impact of COVID-19 Pandemic*, Medical Virology: From Pathogenesis to Disease Control, https://doi.org/10.1007/978-981-95-1426-7_17

attenuated, and nucleic acid-based vaccines, each with distinct benefits and limitations. Further, the milestones in COVID-19 vaccine development, from the first identification of cases to mass vaccination and pandemic control, along with the challenges and adaptations encountered, were addressed. Additionally, the current clinical status of the vaccines has been summarized to provide insights into their efficacy and safety profiles from extensive clinical trials. Moreover, the impact of regulatory approvals on public health strategies worldwide is discussed, along with studies focused on long-term immunity and the effectiveness of vaccines against the emergence of Variants of Concern (VOC). Overall, this chapter provides a comprehensive understanding of the classification, rapid development, and current clinical landscape of COVID-19 vaccines by emphasizing significant achievements and challenges in controlling the pandemic through immunization.

Keywords

COVID-19 · SARS-CoV-2 · Vaccine platforms · Clinical status · VOC

Introduction

Global public health has been profoundly affected by the outbreak of SARS-CoV-2 and the subsequent COVID-19 pandemic. This virus was initially identified in Wuhan, China, in late 2019 among patients with pneumonia, and was identified as the cause of COVID-19 [1, 2]. On March 11, 2020, the World Health Organization (WHO) declared the outbreak a global pandemic. SARS-CoV-2 primarily targets the upper respiratory tract, entering through mucosal surfaces. While most infections cause mild to moderate illness confined to the upper respiratory tract, some cases can lead to severe, potentially fatal pneumonia [2].

SARS-CoV-2 is a single-stranded RNA (ssRNA) virus within the coronavirus family [1]. This family also includes other zoonotic viruses, such as the SARS-CoV and the Middle East respiratory syndrome coronavirus (MERS-CoV) [3]. SARS-CoV-2 shares 79% of its genetic sequence with SARS-CoV, which caused an epidemic in China in 2003. MERS-CoV continues to cause sporadic outbreaks in the Middle East [3]. The virus produces structural or building block proteins (like membrane protein, nucleocapsid protein, envelope protein, and spike glycoprotein), along with non-structural proteins essential for genome replication, RNA synthesis, and auxiliary proteins. These building block proteins, combined with a lipid bilayer derived from the host, create an enveloped virion that transports viral RNA into host cells [4]. Although accessory proteins are not essential for replication, they often help the virus evade the immune system [4]. The spike glycoprotein, forming trimers on the virion surface, is crucial for determining the host range of virus and tissue specificity [4].

Efforts to find an effective treatment for SARS-CoV-2 have not resulted in definitive breakthrough drugs. Medications such as Remdesivir and Ritonavir were used as emergency options despite mixed study results, ultimately proving inadequate in controlling the pandemic [5, 6]. Vaccines, however, have been pivotal in reducing the morbidity and mortality associated with COVID-19, as well as mitigating socio-economic disruption. The urgent need for safe and effective COVID-19 vaccines accelerated the development and adoption of new vaccine technologies. Over the past decade, various platforms, namely, viral vector, protein subunit, and nucleic acid vaccines, have been developed to respond to emerging pathogens quickly, playing a crucial role in the fight against COVID-19 [7]. As of March 2023, 183 distinct COVID-19 vaccine candidates are in clinical development, with some still in preclinical stages [7].

This chapter provides an in-depth overview of the classification and development rationale of COVID-19 vaccines. It also systematically examines the global landscape of COVID-19 vaccines and their current clinical status.

COVID-19 Vaccines and Classification

Vaccines are biological preparations that enhance immunity against specific diseases. The term 'vaccine' is derived from the Latin phrase *Variolae vaccinae* (cowpox), a concept Edward Jenner demonstrated in 1798, showing that cowpox could prevent smallpox in humans. Over the last century, vaccination has been considered the most successful medical intervention in lowering the death and morbidity rates from infectious diseases [8]. Conventional vaccine generation involves the use of toxins or surface proteins of the pathogen or the dead or attenuated form of the pathogen. Vaccines trigger an immune response in the body, prompting the immune system to produce antibodies and neutralize the pathogen. This process creates immunological memory, allowing the body to respond more rapidly and effectively if exposed to the disease in the future [8]. Therefore, vaccines serve as the most valuable approach against infectious diseases and represent a promising preventive measure against COVID-19.

The significant rise in virus transmission and the escalating mortality rate worldwide have driven the development of potential vaccine candidates against SARS-CoV-2 [7]. Depending on the platform employed, COVID-19 vaccines are classified into two primary groups: whole virus vaccines and vaccines based on viral components. Whole virus-based vaccines encompass live attenuated and inactivated vaccines. The components utilized in vaccine development comprise protein subunits, nucleic acids, viral vectors, and virus-like particles (Fig. 17.1). Every kind of vaccination has a unique composition, benefits, and drawbacks concerning immunogenicity, protection, usability, and efficacy [9].

Whole Virus Vaccines

Whole virus vaccines, which include both inactivated and live-attenuated forms, are among the oldest and most established vaccine types. These vaccines elicit a broad

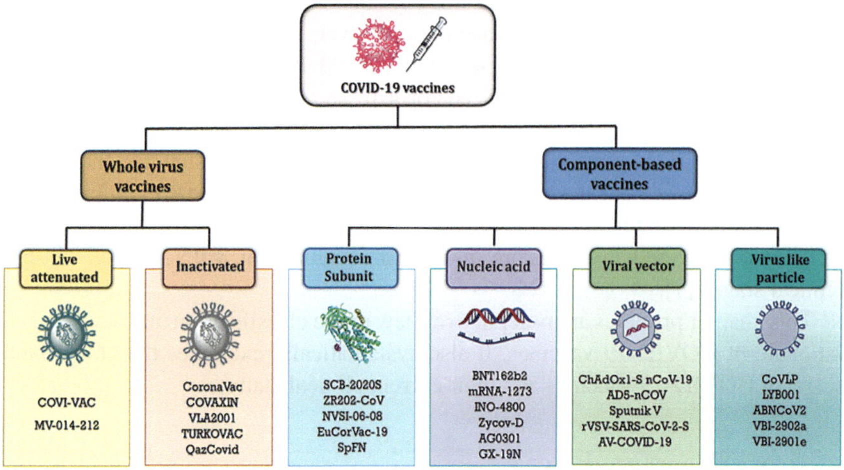

Fig. 17.1 Classification of COVID-19 vaccines based on the platforms used for development

immunologic response because they contain the entire virus with its various surface antigens. However, their production requires handling live virus preparations, necessitating high-level bio-safety installations [8].

Live-Attenuated Vaccines

Vaccines that have been live-attenuated include the entire pathogen in a less virulent form. As of March 30, 2024, only two live-attenuated vaccine candidates, namely, COVI-VAC and MV-014-212, are under development for SARS-CoV-2. Codagenix developed COVI-VAC in cooperation with the Serum Institute of India and is currently under Phase 3 clinical investigation [7]. Attenuated COVI-VAC is accomplished by eliminating the furin cleavage site and introducing 283 silent, deoptimizing mutations. These mutations preserve the viral amino acid sequence but impede viral replication in vivo [10]. Another live-attenuated candidate vaccine, MV-014-212, is in clinical Phase 1 and was developed by Meissa Vaccines, Inc. [7]. It is a modified human respiratory syncytial virus that carries a hybrid SARS-CoV-2 spike protein [10].

The benefits of using live-attenuated SARS-CoV-2 vaccine candidates include their ability to stimulate strong mucosal and cellular immunity, essential for protection without requiring adjuvants. However, this type of vaccine poses safety concerns due to potential fecal excretion, virus transmission, and the risk of generating new variants. Moreover, it requires labor-intensive production with stringent quality control, which decelerates large-scale manufacturing [11].

Inactivated Vaccines

Inactivated vaccines contain preparations of entire pathogens that are inactivated using chemicals like formalin, formaldehyde, heat, or radiation. The immune system can still recognize the targeted pathogen even when it is inactivated while preventing the pathogen from multiplying and causing disease [8]. Currently, there are a total of 23 inactivated SARS-CoV-2 vaccine candidates in clinical trials, such as CoronaVac, COVAXIN, VLA2001, TURKOVAC, and QazCovid-in [7]. CoronaVac, the whole-virion-inactivated SARS-CoV-2 vaccine, was developed by Sinovac Research and Development Ltd., China. Vaccination with CoronaVac was widely implemented in two doses during the pandemic, and it is currently undergoing Phase 4 clinical trials [7, 11]. QazCovid-in vaccine contains the SARS-CoV-2 virus, which is inactivated by formaldehyde and adjuvanted using aluminum hydroxide [12]. This vaccine was developed by the Research Institute for Biological Safety Problems, Republic of Kazakhstan, and it is currently in Phase 3 clinical trial [7, 12]. Another notable inactivated vaccine, COVAXIN, is currently in Phase 3 clinical trials. Covaxin employs the well-established Whole-Virion-Inactivated Vero cell-derived platform technology, utilizing Alhydroxiquim-II as an adjuvant. The Indian Council of Medical Research developed it in collaboration with Bharat Biotech Ltd. [7, 13].

This type of vaccine typically elicits a targeted immune response involving both humoral and cellular mechanisms, with minimal side effects, ensuring a high safety profile for recipients. However, the production of inactivated vaccines faces significant challenges including an intricate manufacturing process and the need to cultivate live infectious SARS-CoV-2 to high titers in biosafety level 3 facilities, which raises safety concerns [11].

Component-Based Vaccines

Subunit vaccines or component-based vaccines are made from components of the pathogen to elicit an immune response. Antigenic components used for such vaccine preparation can be polysaccharides, nucleic acids, or proteins [8].

Protein Subunit Vaccines

The protein subunit contains specific isolated proteins of the virus rather than the entire viral particle. Three major protein components are found in SARS-CoV-2, which include structural, non-structural, and accessory proteins. The primary structural proteins of SARS-CoV-2 are the spike (S), membrane (M), and envelope (E) proteins. In the case of vaccine development, the S protein of SARS-CoV-2 is considered an ideal target because of its high antigenicity and its ability to induce strong immune responses [14]. For example, SCB-2020S entails the trimeric S-protein of SARS-CoV-2 from the B.1.351 variant. It was developed in China by Clover Biopharmaceuticals AUS Pty Ltd. and is undergoing Phase 2 clinical investigation [7]. Another protein subunit vaccine candidate, ZR202-CoV, is designed using the trimeric pre-fusion S protein of the prototype SARS-CoV-2 and is formulated with aluminum hydroxide (Alum) and CpG7909 [15]. This vaccine was created in China by Shanghai Zerun Biotechnology in collaboration with Walvax Biotechnology and Coalition for Epidemic Preparedness Innovations (CEPI) and is currently undergoing Phase 2 clinical trial [7]. Further, a hybrid antigen vaccine, NVSI-06-08, incorporates key mutations from multiple SARS-CoV-2 variants into a single protein. Preclinical studies have shown that it induces a broader immune response against various SARS-CoV-2 variants and is currently being investigated in Phase 3 clinical studies [7, 16]. Additionally, several recombinant protein nanoparticle vaccines were also developed and employed to combat COVID-19, such as EuCorVac-19 and SpFN [7, 17, 18].

The protein subunit vaccine has several advantages, such as the absence of preexisting antivector immunity and no concerns about inadequate virus deactivation, making it very safe for recipients. However, in the case of SARS-CoV-2, this platform also has several drawbacks. The S protein is challenging to express, which further complicates the manufacturing process. Moreover, the preparation procedure is labor-intensive and also necessitates new production protocols and stability assays for each antigen [19].

Nucleic Acid Vaccines

Nucleic acid vaccines, a relatively advanced technology, comprise the genetic material (DNA or mRNA) of the pathogen. Upon delivery to the recipient cells, it is translated into a protein. Further, the antigen-presenting cells (APCs) process and present this protein to immune cells, resulting in an immune response specifically against the desired protein [20].

DNA Vaccines

DNA vaccines use plasmids to deliver genetic materials. It generally involves cloning of the SARS-CoV-2 S gene into bacterial plasmids with a strong mammalian promoter, such as CMV or SV40, and subsequently mass production of the plasmid in a bacterial system. Benefits of such vaccines include the ability to elicit both humoral and cellular immune responses, a simple production process, short preparation periods, and room-temperature storage. However, their low immunogenicity necessitates multiple doses for optimal efficacy, and there is a long-term risk of carcinogenesis due to the potential integration of plasmid DNA into the host cell genome [20]. Currently, several DNA vaccine candidates are in clinical development, such as INO-4800 and Zycov-D are undergoing Phase 3 studies, while AG0301 and GX-19 N are in Phase 2/3 trials. Additionally, some candidates like Covigenix VAX-001, COVIGEN, and GLS-5310 are in Phase 1 or Phase 2 trials. INO-4800, the first DNA vaccine candidate encoding the S protein of SARS-CoV-2, was generated by Inovio Pharmaceuticals in collaboration with the Advaccine (Suzhou, China) Biopharmaceutical Co., Ltd. and International Vaccine Institute [7, 21]. In addition, ZyCoV-D was created by Zydus Cadila Healthcare Ltd. and includes a plasmid DNA with the S gene, along with an additional gene encoding for signal peptide [22].

mRNA Vaccines

The production of mRNA vaccines involves the artificial synthesis of mRNA that encodes the SARS-CoV-2 spike protein. Once inside cells, mRNA vaccines facilitate rapid antigen expression and provide both humoral and cellular immunity. In contrast to other vaccine types, mRNA vaccines represent fewer safety apprehensions and a rapid production time. However, for long-term preservation, such vaccines require extremely cold storage conditions [20].

mRNA vaccines have demonstrated significant efficacy against SARS-COV-2 in numerous preclinical investigations and gained noteworthy success. For example, vaccine candidates such as BNT162b2 (Pfizer) and mRNA-1273 (Moderna) both received emergency use approval from the US FDA and are currently undergoing Phase 4 clinical studies [7]. BNT162b2 is a lipid nanoparticle-formulated, nucleoside-modified RNA vaccine encoding a prefusion-stabilized, membrane-anchored full-length SARS-CoV-2 spike protein [23]. Similarly, the mRNA-1273 vaccine is an mRNA-based vaccine encapsulated in lipid nanoparticles, encoding the prefusion-stabilized full-length spike protein of SARS-CoV-2 [7]. Additionally, there are two mRNA vaccine candidates, namely, ARCoV and PTX-COVID19-B, in Phase 3 clinical studies. Moreover, few mRNA vaccine candidates such as ChulaCov19, CoV2 SAM (LNP) vaccine, mRNA-1273.351., MRT5500, and HDT-301 are currently under Phase 1 or Phase 2 investigation [7].

Viral Vector Vaccines

Viruses that have been genetically modified to carry genes encoding specific antigens from pathogens were utilized in viral vector vaccines. Examples of commonly used viral vectors are adenovirus, measles virus, influenza virus, and poxvirus. These vaccines provide the safety associated with inactivated vaccines while also delivering the immune response benefits and cost-effectiveness of live vaccines. However, it has several disadvantages such as chances of virus recombination and genetic toxicity [24]. As of March 2023, 25 different COVID-19 non-replicating viral vectors and four replicating viral vector vaccine candidates are in clinical development [7].

Non-Replicating Viral Vector

Adenovirus vectors, which are double-stranded DNA viruses without an envelope, are primarily used as replication-defective vectors against COVID-19 [24]. For example, the ChAdOx1-S nCoV-19 vaccine (AZD1222), also known as Covishield, was developed at Oxford University and is currently undergoing Phase 4 clinical trials. It uses a replication-deficient chimpanzee adenoviral vector (ChAdOx1) that includes the gene for the SARS-CoV-2 spike protein (nCoV-19) [7]. AD5-nCOV, also known as Convidecia, was jointly developed by CanSino Biological Inc. in collaboration with Beijing Institute of Biotechnology and is currently in Phase 4 clinical development. Another Phase 4 adenovirus vector-based vaccine candidate, Sputnik V, was developed at Gamaleya Research Institute, Russia. It has two distinct human adenovirus vectors to improve the efficacy and enhance immune response more efficiently that are Ad5 and Ad26 for the initial dose and the booster, respectively [7, 24].

Replicating Viral Vector

Replicating viral vector vaccines generate a large number of virus particles once they enter cells, leading to heightened expression of specific antigens and stimulating a robust and long-lasting immune response against SARS-CoV-2. Unlike non-replicating viral vector vaccines, protein expression from the replicating viral vector can possess toxic effects on the host system [24]. Replicating viral vector vaccine candidates against SARS-CoV-2 includes rVSV-SARS-CoV-2-S and AV-COVID-19. rVSV-SARS-CoV-2-S (IIBR-100), also known as BriLife, uses the recombinant vesicular stomatitis virus (VSV) platform to express the spike protein of SARS-CoV-2. The Israel Institute for Biological Research developed this vaccine and is currently under Phase 2/3 clinical investigation [7]. Aivita Biomedical, Inc., National Institute of Health Research and Development, and Ministry of Health Republic of Indonesia jointly created AV-COVID-19 or Nusantara, which is a dendritic cell-based vaccine. This vaccine is currently being investigated in Phase 2 clinical studies [7].

Virus-Like Particles (VLPs)

Virus-like particles mimic the structure of a virus but are non-infectious. They efficiently present antigens to the host immune system, eliciting a protective and long-lasting immune response. VLPs can be generated using various expression platforms, such as bacteria, yeast, plants, insects, and mammalian cells. Each system has its advantages and disadvantages for VLP production [25]. VLPs designed against SARS-CoV-2 undergoing Phase 3 clinical studies include CoVLP, LYB001, and ABNCoV2. Medicago or CoVLP is a plant-based vaccine for SARS-COV-2, designed by Medicago along with GlaxoSmithKline (GSK). The tobacco plant (*Nicotiana benthamiana*) was utilized as the expression platform for producing this vaccine. LYB001, an RBD-based VLP vaccine, was created by Yantai Patronus Biotech Co., Ltd. Another VLP vaccine, ABNCoV2, was developed by Radboud University utilizing capsid VLP (cVLP) technology [26, 27]. Additionally, Phase 1 or Phase 2 VLP candidates include VBI-2902a, VBI-2901e, and RBD SARS-CoV-2 HBsAg VLP vaccine [7].

Milestones of COVID-19 Vaccine Development

COVID-19 vaccines represent a landmark achievement in the history of medicine and public health. Scientific innovation, expedited vaccine development, and extraordinary global collaboration characterize the immense effort to combat the pandemic. Figure 17.2 portrays the crucial milestone of COVID-19 vaccine development. The first incident of COVID-19 was reported as an instance of pneumonia cases in Wuhan, China, in December 2019. In early 2020, it was named SARS-CoV-2 by Chinese scientists after the successful isolation of the novel coronavirus [1]. The genomic sequence of the virus was quickly shared with the global scientific community, allowing researchers worldwide to begin working on diagnostic tests and vaccines. Further, the outbreak was declared a Public Health Emergency of

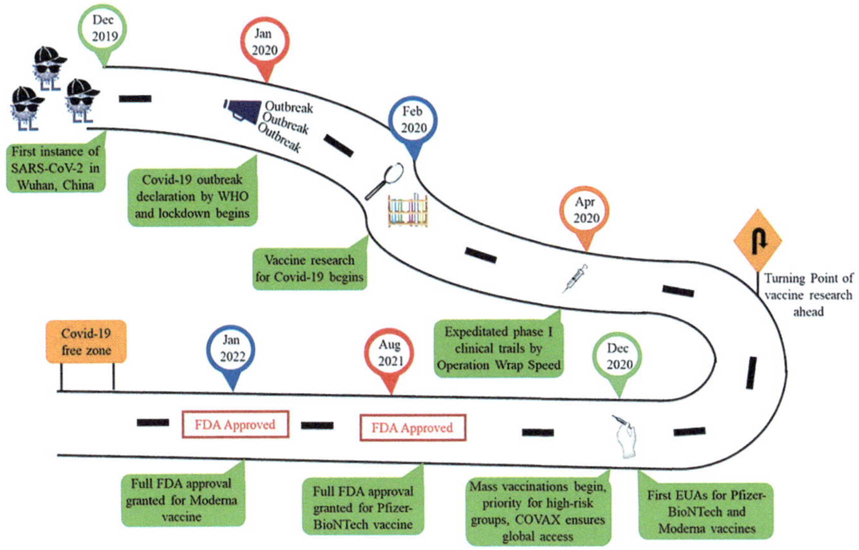

Fig. 17.2 Milestones of COVID-19 vaccine development

International Concern (PHEIC) by the WHO, which underscored the urgent need for coordinated international action, including vaccine development [2].

The development of vaccines for COVID-19 leverages various technological approaches to induce immune responses against the virus. Among these, mRNA vaccines like BNT162b2 from Pfizer-BioNTech and mRNA-1273 from Moderna utilize messenger RNA to produce the SARS-CoV-2 spike protein in the host system, thereby triggering an immune response. These vaccines were among the first to be approved for emergency use, with high efficacy rates and rapid manufacturing capabilities, though they require stringent cold storage conditions [7]. Viral vector vaccines employ a harmless virus to transport genetic material that codes for the spike protein, for example, AZD1222, Ad26.COV2.S, and Sputnik V from AstraZeneca-Oxford, Johnson & Johnson, and Russia, respectively [24]. They are stable at standard refrigeration temperatures, making them easier to distribute, but they face rare blood-clotting issues.

Inactivated virus vaccines, including BBIBP-CorV of Sinopharm and CoronaVac of Sinovac, use killed viruses to elicit an immune response. This well-established technology offers easier production and distribution, especially in low- and middle-income countries, though it often requires multiple doses [24]. Protein subunit vaccines, like NVX-CoV2373 produced by Novavax, use virus fragments to stimulate immunity. They are stable at standard refrigeration temperatures and potentially have fewer side effects, but production can be complex and slower to scale [28]. DNA vaccines, such as ZyCoV-D from Zydus Cadila, use plasmid DNA to code for the spike protein, offering stability at higher temperatures and needle-free delivery, though their efficacy data is less established [22].

Several global initiatives, such as the CEPI, accelerated vaccine development by funding multiple vaccine candidates [29]. Clinical trials of these vaccine candidates were initiated in April 2020, and several vaccine candidates entered Phase 1 clinical trials. These trials aimed to assess the safety, immunogenicity, and appropriate dosage levels of the vaccines. This process of clinical trials was expedited further by the US government with Operation Warp Speed (OWS). OWS provided substantial funding and support to several vaccine developers, including Moderna, Pfizer-BioNTech, and Johnson & Johnson [30]. On December 2, 2020, the Healthcare Products Regulatory Agency (MHRA) and UK Medicines granted the world's first Emergency Use Authorizations (EUA) for the Pfizer-BioNTech vaccine [31]. Subsequently, on December 11, 2020, for Pfizer-BioNTech and on December 18, 2020, for Moderna, the U.S. Food and Drug Administration (FDA) granted EUAs. The full approval for the Pfizer-BioNTech vaccine by the FDA was received in August 2021, followed by January 2022 for Moderna [32]. The mass vaccination campaigns against COVID-19 began in several countries in December 2020. The first doses were prioritized for frontline healthcare workers, elderly individuals, and those at high risk of severe disease [33]. Additional vaccines, such as the AstraZeneca vaccine developed with the University of Oxford and the Johnson & Johnson vaccine, received EUAs and full approvals in early 2021. The COVAX initiative, co-led by Gavi, CEPI, and WHO, aimed to ensure equitable access to COVID-19 vaccines globally. In February 2021, COVAX began distributing vaccines to low- and middle-income countries, marking a significant step towards global immunization efforts [34]. These mass vaccination campaigns required immense logistical planning and coordination involving national governments, international organizations, and private companies. Cold chain infrastructure was established or upgraded to store and transport vaccines that required low temperatures. Public health campaigns were launched to address vaccine hesitancy and misinformation, emphasizing the safety and efficacy of the vaccines. Governments also implemented prioritization strategies to ensure that the most vulnerable populations received the vaccine first, aiming to reduce hospitalizations and deaths.

Challenges and Adaptations

The rollout of vaccines faced logistical challenges, including cold chain requirements, vaccine hesitancy, and disparities in access. Furthermore, the emergence of new variants of concern necessitated ongoing monitoring and potential adaptations of vaccine formulations [35]. The Delta variant, which emerged in late 2020, and the Omicron variant, first identified in November 2021, posed significant challenges due to their higher transmissibility and potential to evade immunity from previous infections or vaccinations. This necessitated continuous surveillance and research to understand the impact of these variants on vaccine effectiveness.

As immunity waned and new variants emerged, booster doses were recommended to enhance and prolong protection. Vaccine manufacturers began developing and testing variant-specific vaccines to address the evolving virus. For example, bivalent vaccines targeting both the original strain and the Omicron variant were developed [36]. Governments and health organizations had to adapt their vaccination strategies to include booster doses and variant-specific vaccines. Public health campaigns were updated to communicate the importance of booster doses in maintaining high levels of immunity and protecting against severe disease. This also involved logistical adjustments to ensure the availability and distribution of booster doses alongside primary vaccination campaigns.

The success of mRNA vaccines spurred interest in mRNA technology for other infectious diseases and even non-infectious conditions such as cancer. Research continued to optimize vaccine platforms for broader and more durable protection. Ongoing surveillance of vaccine efficacy and safety remained crucial. Studies on long-term immunity, vaccine impact on transmission, and the need for future booster doses continued to inform public health strategies. Additionally, research into applications of mRNA vaccines extended to other viruses like influenza and HIV, and even non-infectious diseases such as cancer. These efforts could lead to groundbreaking treatments and preventive measures for a range of health conditions.

The COVID-19 vaccine development process highlighted the potential for rapid vaccine development without compromising safety. Streamlined regulatory processes and robust post-market surveillance were instrumental in ensuring public trust and vaccine safety. The collaboration between governments, international organizations, academic institutions, and pharmaceutical companies was a critical factor in the rapid development and distribution of vaccines. These partnerships set a precedent for addressing future public health crises. The advancements in vaccine technology, particularly mRNA, have broader implications for future pandemic preparedness and vaccine development. These innovations could revolutionize the response to other infectious diseases and health challenges.

Clinical Status of COVID-19 Vaccines

Conventional vaccine development begins with a research or preclinical phase involving laboratory experiments and animal studies, followed by the submission of an investigational new drug (IND) application, after which the vaccine candidate proceeds on to Phases 1, 2, and 3 trials. Preclinical trials are an essential part of the long vaccine development process. This stage aims to assess the safety and protective efficacy of vaccines on animal models before translation to the clinical development phase in human participants. Animal models such as rats, mice, guinea pigs, and rabbits are generally employed to assess the safety, immunogenicity, and tolerance of potential vaccinations. Nevertheless, the species distinctions between these animals and humans may not appropriately mimic the pathology seen in human COVID-19 patients. NHPs (nonhuman primates), on the other hand, have comparable innate and adaptive immune responses to those of humans [37]. Therefore, research on NHPs is crucial for maximizing the scientific value and translation of the vaccine candidates that are under investigation. Following success with preclinical trials (animal experiments), four phases of the clinical trials are the next essential step for developing vaccines.

Phase 1 trials investigate the safety, dose, and tolerance of vaccines; Phase 2 trials focus on evaluating the efficacy and side effects of vaccination; Phase 3 trials confirm the effectiveness of the candidates in comparison to the standard treatments; and Phase 4 continues to study the long-term effects of the vaccines after approval. Once the phase 3 trials conclude and the predefined outcomes have been reached, a biologics license application (BLA) is submitted, which is examined by regulatory authorities, and the vaccine is eventually granted a license. Following this, the large-scale production of the vaccine commenced. The development of SARS-CoV-2 vaccines followed an escalated timeline. By taking clues from the pre-existing SARS-CoV and MERS-CoV vaccine candidates and employing parallel development, the initial phase was reduced to around 6–8 months. Phase 1/2 trials were initiated, and established procedures were followed. Following the interim review of phase 1/2 data, phase 3 studies were started, with many clinical trial stages operating concurrently. The production of various vaccine candidates began at a large scale, albeit at risk. Thereby, these vaccine candidates have only been approved for emergency purposes with strict continuous monitoring (Fig. 17.3).

There are numerous vaccine candidates for COVID-19 in the preclinical phase of research and development. According to Vaccine Tracker, as of March 30, 2023, 199 vaccine candidates are in preclinical trials [7]. These candidates fall under the various established vaccine development platforms such as sprotein subunit vaccines (38.5%), replicating and non-replicating viral vector vaccines (22%), RNA-based vaccines (12.5%), virus-like particles (9.5%), DNA vaccines (8%), inactivated vaccines (5.5%), and live-attenuated vaccines (1.5%) [7]. Apart from these conventional vaccine development platforms, new potential platforms are also being employed, such as cellular-based vaccines, replicating bacterial vectors, and live-attenuated bacterial vector-based vaccines. Interestingly, some of these platforms, as mentioned above, are also being explored for other viruses such as JE (Japanese encephalitis), Zika, MMR (Measles, Mumps, and Rubella), EBOV, MERS, HIV, Influenza, and CHIKV [38].

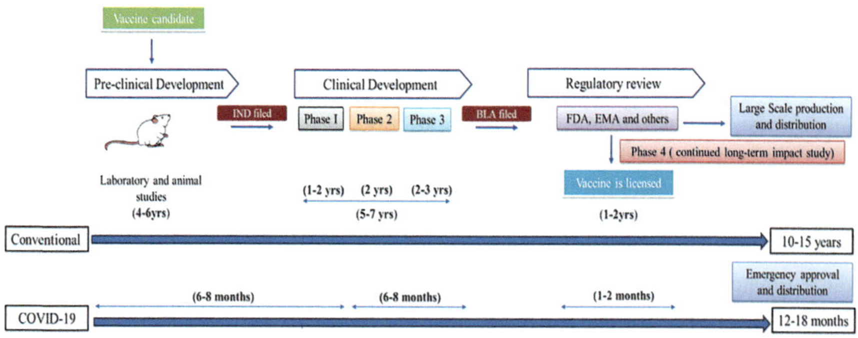

Fig. 17.3 Timeline of conventional vaccine development and accelerated COVID-19 vaccine development process

According to data released by the WHO, as of March 30, 2023, 183 vaccines are approved for clinical trials and are in various phases of the clinical development process. The vaccines approved for clinical trials mainly include a total of 59 protein subunit vaccines (32%), 43 RNA vaccines (24%), 34 replicating and non-replicating viral vector vaccines (16%), 22 inactivated vaccines (12%), 17 DNA vaccines (9%), and 2 live-attenuated vaccines (1%) [7]. The WHO has approved two COVID-19 protein subunit vaccines for emergency use: NVX-CoV2373, cleared for individuals aged 12 years and older, and GBP510, sanctioned for individuals aged 18 years and older. Furthermore, 59 candidates of protein subunit vaccines are permitted for COVID-19 clinical trials, with 23 candidates in Phase 3 and one candidate in Phase 4 (Table 17.1).

The details of COVID-19 vaccine candidates approved for emergency purposes, including their efficacy, mild local and systemic reactions, a few reported adverse reactions, and developers, have been summarized in Table 17.2.

The WHO has approved three inactivated vaccines, the BBIBP-CorV (developed by Sinopharm, China), CoronaVac (developed by Sinovac Biotech, China), and Covaxin (developed by Bharat Biotech, India) [13]. Along with these, 22 inactivated vaccines are in trials, with three candidates in Phase 4 and ten in Phase 3 (Table 17.1).

In the case of live-attenuated vaccine candidates, two vaccines, COVI-VAC and MV-014-212, are in clinical trial Phases 3 and 1, respectively. In this category, no vaccine has been approved by the WHO for emergency purposes. However, five viral vector vaccines have been approved by the WHO to combat COVID-19 (such as Ad26.COV-2-S, Covishield, Convidecia, AZD1222, and Sputnik V). The approval for Sputnik V was terminated due to the manufacturing process of the vaccine had not met the necessary standards [44]. Also, there are 34 viral vector vaccine candidates in clinical trials, with 5 replicating vectors and 26 non-replicating vectors. In addition, three pathogen-specific Artificial Antigen-Presenting Cells (aAPC)-based viral vector vaccines are also approved for clinical trials and are currently in Phase 1/2. Nucleic acid-based vaccines encompass mRNA and DNA platforms. Currently, there are 43 mRNA vaccine candidates in clinical trials. Two mRNA vaccines, namely, BNT162b2 and mRNA-1273, have been approved by the WHO for emergency use. As of March 30, 2023, the WHO has not approved any DNA vaccines for COVID-19. Seventeen DNA vaccines are undergoing clinical trials, with two candidates in Phase 3 (Table 17.1).

Efficacy and Safety of Vaccine Candidates Against VOCs

The effectiveness of currently available vaccinations is seriously challenged by the advent of VOCs. Alpha (B.1.1.7), Beta (B.1.351), Gamma (P.1), Delta (B.1.617.2), and Omicron (B.1.1.529) are the five SARS-CoV-2 VOCs identified by the WHO. Of these variants, Alpha and Delta variants are reported to have a strong contagious rate, while the Beta and Gamma variants were able to evade the immune response evoked by previous immunizations. However, the Omicron variant possessed

Table 17.1 Landscape of COVID-19 candidate vaccines in Phase 3 and Phase 4 clinical development process

Vaccine platform description	Type of vaccine candidate	Current status of clinical evaluation	Respective trial registries and public reports
Protein subunit vaccine	MVC-COV1901	Phase 4	NCT05079633
Protein subunit vaccine	SARS-CoV-2-rs, Recombinant SARS-CoV2-vaccine (CHO cell) (Anhui Biopharmaceutical), VAAT0008, SCB-2019, COVAX-19, FINLAY-FR-2, EpiVacCorona, RBD recombinant vaccine (West China Hospital and Sichuan University), UB-612, CIGB-66, BECOV2, Nanocovax, S-268019, GPB510, Razi Cov Pars, EuCorVac-19, ReCOV, V-01, Recombinant SARS-CoV2-vaccine (CHO cell) (National Vaccine and Serum Institute, China), Noora Vaccine, COVID-19 Vaccine Hipra, SCTV01C, Recombinant Vaccine adjuvanted with Alum + CpG 1018	Phase 3	NCT04611802, ChiCTR2100050849, PACTR2011523101903, NCT05012787, IRCT20150303021315N24, RPCEC00000354, NCT04780035, NCT04887207, NCT05293665, RPCEC00000359, CTRI/2021/08/036074, NCT0492278, NCT05212948, NCT05007951, IRCT20210206050259N3, NCT05572879, NCT05398848, NCT05096832, NCT05599516, IRCT20210620051639N3, NCT0524137, NCT05308576, NCT05433285
Inactivated virus	CoronaVac (Vero cell), Inactivated Vaccine (Vero cell) (Sinopharm), BBIBP-CorV	Phase 4	NCT04756830, NCT05065892, NCT04863638
Inactivated virus	SARS-CoV-2 vaccine (Vero cells) (Institute of Medical Biology and Chinese Academy of Medical Sciences), QazCovid-in, BBV152, Inactivated Vaccine (Vero cell) (Shenzhen Kangtai Biological Products Co. Ltd.), VLA2001, TURKOVAC, Inactivated COVID-19 vaccine (KM Biologics Co. Ltd.), CoviVac, UNAIR, Omicron COVID-19 inactivated vaccine (Vero cell)	Phase 3	NCT04659239, NCT04691908, NCT04641481, NCT04852705, NCT04864561, NCT04942405, jRCT2031210679, NCT05407142, NCT05508477, NCT05374954
Live-attenuated virus	COVI-VAC	Phase 3	ISRCTN15779782

(continued)

Table 17.1 (continued)

Vaccine platform description	Type of vaccine candidate	Current status of clinical evaluation	Respective trial registries and public reports
Viral vector (replicating)	DelNS1-2019-nCoV-RBD-OPT1	Phase 3	ISRCTN15779782
Viral vector (non-replicating)	ChAdOx1-S- (AZD1222), Recombinant novel coronavirus vaccine (Adenovirus type 5 vector) (Beijing Institute of Biotechnology), Ad5-nCoV-IH, Ad26.COV2.S	Phase 4	NCT04760132, NCT04892459, NCT05303584, EUCTR2021-002327-38-NL
Viral vector (non-replicating)	rAd26-S+rAd5-S, BBV154, Convidecia vaccine (Ad5-nCoV)	Phase 3	NCT04530396, CTRI/2022/02/040065, NCT05442684
RNA-based vaccine	mRNA-1273 (Spikevax), BNT162b2, mRNA-1273:351	Phase 4	NCT04760132, NCT04760132, EUCTR2021-000930-32
RNA-based vaccine	CVnCoV, ARCoV, PTX-COVID19-B, SW-BIC-213, ARCT-154, LVRNA009, mRNA-1273.214	Phase 3	NCT04674189, NCT04847102, NCT05534035, NCT05580159, ISRCTN15779782, NCT05428592, NCT05436834
DNA-based vaccine	INO-4800 + electroporation, nCov vaccine (Zydus Candila)	Phase 3	ISRCTN15779782, CTRI/2020/07/026352

various spike protein mutations in the virus, leading to high infectivity [45]. Currently, approved vaccine candidates have shown varying levels of efficacy against these variants of concern. Against the symptomatic infection caused by the Alpha variant, the NVX-CoV2373 protein-based vaccine provides significant protection with 85% efficacy. Also, the approved mRNA vaccines by Pfizer and Moderna suggest an efficacy ranging from 88% to 95%, followed by the viral vector vaccines by AstraZeneca and others, which provide around 70% efficacy [41]. Against the Beta variant (B.1.351) of COVID-19, studies suggest that mRNA vaccines offer significant efficacy of around 75%, whereas the Ad26.COV-2-S and NVX-CoV2373 offer a lower efficacy of about 50–60% [46]. The specific efficacy data against the Gamma variant is limited; however, a few studies highlight that mRNA vaccines have demonstrated good effectiveness of 88% against this variant [45]. Other vaccine platforms, including the inactivated and viral vector, are expected to provide some level of protection (36–50%) against symptomatic infection and severe outcomes of the Gamma variant. In the case of the Delta variant, studies suggest that two doses of the mRNA vaccines showed the best results, with an efficacy of around 87–95% in preventing symptomatic infection. The inactivated and viral vector vaccines showed good results, with an efficacy of 67%, whereas the efficacy studies of the protein subunit vaccines against this variant are under investigation. The performance of the WHO-approved vaccines against the Omicron variant (B.1.1.529) is under investigation with extensive ongoing studies [45].

Table 17.2 List of COVID-19 vaccines approved for emergency use [accessed on 8 August 2023]

Type of Vaccine	Vaccine name	Developer/ manufacturer	Efficacy (%)	Rare reported side effects	Reference
Protein subunit vaccine	NVX-CoV2373	Novavax and the CEPI	82.70	Mild fatigue, muscle pain, headache, rare heart inflammation	[39]
Protein subunit vaccine	GBP510	SK Bioscience	56.70	Pain and swelling at the injection site, joint pain, vomiting/nausea	[40]
Inactivated	Covaxin	Bharat Biotech	81	Skin and musculoskeletal disorders, menstrual abnormalities, and hypothyroidism; stroke and Guillain–Barre Syndrome (1%)	[13]
Inactivated	CoronaVac	Sinovac	50.4–83.5	Pain, swelling, and redness, Encephalopathy	[41]
Inactivated	BBIBP-CorV	Sinopharm	79.3	Thrombocytopenia, abnormal blood pressure, chorioretinopathy	[42]
Viral vector (non-replicating)	ChAdOx1-S (Vaxzevria)/AZD1222	AstraZeneca/Oxford	70.4	Rare side effects: thrombocytopenia and low platelets, Guillain–Barré syndrome	[41]
Viral vector (non-replicating)	Covishield (ChAdOx1_nCoV-19)	Serum Institute of India Pvt. Ltd.	49–95	Fever, fatigue, headache, skin rash, Thrombosis With Thrombocytopenia Syndrome (TTS)	[41]
Viral vector (non-replicating)	Ad26.COV2.S	Johnson & Johnson	66.3	Pain, headache, loss of taste and smell, shortness of breath, deep vein thrombosis, vomiting/nausea	[41]
Viral vector (non-replicating)	Convidecia	CanSino Biologics	58–92	Induration and swelling at the injection site, muscle ache, drowsiness, bleeding, mucosal abnormality	[41]
mRNA	BNT162b2 (Comirnaty)	Pfizer-BioNTech	95	Anaphylaxis, myocarditis, appendicitis	[41, 43]
mRNA	mRNA-1273	Moderna	94.1	Anaphylaxis, pericarditis, dermatologic hypersensitivity reactions	[7, 41]

However, initial reports suggest that these vaccines may offer reduced effectiveness against symptomatic infection caused by the Omicron variant compared to the other VOCs. Further, preliminary data indicate that heterologous booster doses may enhance protection against severe illness and reduce hospitalization cases caused by the Omicron variant [47].

While COVID-19 vaccines have demonstrated promising outcomes in pre-clinical trials and clinical development, with 11 vaccines approved by the WHO for emergency use, there are some concerns regarding vaccine safety. Adverse side effects such as fever, fatigue, and pain, along with other complications like coagulation dysfunction, pericarditis, anaphylaxis, and nervous and lymphatic system diseases, have been reported [48]. However, the incidence of these vaccine-related adverse cases is low, and people with underlying medical conditions are mostly affected. Therefore, the government and related agencies are recommended to expedite the immunization processes and continue monitoring the long-term impact of these vaccines. Further, ongoing research is focused on overcoming the evolving variants and effectively combating the dynamic nature of the COVID-19 pandemic.

Conclusion and Future Prospects

The expedited development of vaccines, emergency approval by regulatory agencies, and mass vaccinations have effectively helped to overcome the COVID-19 pandemic. Currently, many new vaccines based on conventional platforms are being explored along with the evaluation of new platforms, including nanoparticle and VLP vaccines. Primarily, scientific studies highlight that the COVID-19 vaccines have produced a consistent immune response through the combined activity of the Th1 cellular immune response, B-cell immunity-based humoral immune response, and the production of respective memory cells. Nevertheless, the health status of vaccinated individuals and the treatment of side effects of the vaccines should be under continuous monitoring. Also, specific vaccination regimes and follow-up investigations are of utmost importance to particular individuals with pre-existing medical conditions, including senior citizens, organ transplant recipients, cancer patients, and HIV-positive individuals. Also, it has been noted that the efficacy of the vaccines in cancer and organ transplant patients is considerably low, around 50%, which requires considerable improvement.

An important aspect of further management is understanding the nature of evolving VOCs. Despite the success of approved vaccines in tackling the consistently mutated VOCs, the omicron variant posed a serious challenge due to its ability to evade vaccine-induced immunity. This variant possesses mutations in the RBD region of the S protein and also cause upper respiratory tract infection. Conventional vaccines, which are given intramuscularly, primarily target the lower respiratory tract, are therefore less effective against Omicron. This necessitates optimizing and adjusting vaccine strategies, such as using intranasal neutralizing antibodies, inhaled vaccines for the upper respiratory tract, oral antiviral drugs to reduce hospitalization risk, and adjuvants to enhance immune response or reduce side effects. Additionally, nanoparticle vaccines uphold significant potential against the mutant strains, as they can remain in germinal center B cells and produce a higher, prolonged antibody response compared to conventional vaccines. Also, developing vaccines by combining different VOC strains to induce antibodies with multiple epitopes and employing heterologous boosters of different platform vaccines has been observed to provide effective protection against the rapidly evolving strains. In conclusion, COVID-19, though one of the biggest challenges in recent times, brought about a revolution in the medical sciences, microbiology, and the pharmaceutical field with the rapid development of vaccines on the global front. This continued collective action is still imperative in effectively addressing and combating the present scenario of COVID-19 and also any future viral outbreaks.

Author's Contributions Conceived the idea and review structure: SS, TK, AA, and SSG; Wrote the manuscript: SS, TK, and AA; Revised and edited the manuscript: SS, TK, AA, and SSG. All authors have read, agreed, and approved the final draft of the manuscript.

Funding This work was supported by the Department of Biotechnology, Government of India for Translation Program in the DBT Program Support Facility (BT/PR44695/NER/95/1880/2021 and BT/PR41449/NER/95/1687/837 2020), Indian Institute of Technology, Guwahati, India. We also acknowledge the Center for Nanotechnology (ICMR Center for Excellence, Grant no. 5/3/8/20/2019-ITR, INUP, MeitY Grant no. 5(1)/2021-NANO) at IIT Guwahati.

Data Availability The software ChemDraw Professional (version: 16.0) is used to draw some elements of a few figures.

Declarations

Conflicts of Interest There is no competing and conflict of interest.

References

1. Zhu N, Zhang D, Wang W, Li X, Yang B, Song J, et al. A novel coronavirus from patients with pneumonia in China, 2019. N Engl J Med. 2020;382(8):727–33. https://doi.org/10.1056/nejmoa2001017.
2. Wu W, Wang A, Liu M. Clinical features of patients infected with 2019 novel coronavirus in Wuhan, China. Lancet. 2020;395(10223):497–506. https://doi.org/10.1016/S0140-6736(20)30183-5.
3. Zaki AM, Van Boheemen S, Bestebroer TM, Osterhaus AD, Fouchier RA. Isolation of a novel coronavirus from a man with pneumonia in Saudi Arabia. N Engl J Med. 2012;367(19):1814–20. https://doi.org/10.1056/nejmoa1211721.
4. Lamers MM, Haagmans BL. SARS-CoV-2 pathogenesis. Nat Rev Microbiol. 2022;20(5):270–84. https://doi.org/10.1038/s41579-022-00713-0.
5. Spinner CD, Gottlieb RL, Criner GJ, López JRA, Cattelan AM, Viladomiu AS, et al. GS-US-540-5774 Investigators. Effect of remdesivir vs standard care on clinical status at 11 days in patients with moderate COVID-19: a randomized clinical trial. JAMA. 2020;324(11):1048–57. https://doi.org/10.1001/jama.2020.16349.

6. Cao B, Wang Y, Wen D, Liu W, Wang J, Fan G, et al. A trial of lopinavir–ritonavir in adults hospitalized with severe Covid-19. N Engl J Med. 2020;382(19):1787–99. https://doi.org/10.1056/nejmoa2001282.
7. WHO. World Health Organization. COVID-19 vaccine tracker and landscape. 2022. https://www.who.int/publications/m/item/draft-landscape-of-covid-19-candidate-vaccines.
8. Delany I, Rappuoli R, De Gregorio E. Vaccines for the 21st century. EMBO Mol Med. 2014;6(6):708–20. https://doi.org/10.1002/emmm.201403876.
9. Vetter V, Denizer G, Friedland LR, Krishnan J, Shapiro M. Understanding modern-day vaccines: what you need to know. Ann Med. 2018;50(2):110–20. https://doi.org/10.1080/07853890.2017.1407035.
10. Peng F, Yuan H, Wu S, Zhou Y. Recent advances on drugs and vaccines for COVID-19. Inquiry. 2021;58:00469580211055630. https://doi.org/10.1177/00469580211055630.
11. Frederiksen LSF, Zhang Y, Foged C, Thakur A. The long road toward COVID-19 herd immunity: vaccine platform technologies and mass immunization strategies. Front Immunol. 2020;11:560652. https://doi.org/10.3389/fimmu.2020.01817.
12. Zakarya K, Kutumbetov L, Orynbayev M, Abduraimov Y, Sultankulova K, Kassenov M, et al. Safety and immunogenicity of a QazCovid-in® inactivated whole-virion vaccine against COVID-19 in healthy adults: a single-centre, randomised, single-blind, placebo-controlled phase 1 and an open-label phase 2 clinical trials with a 6 months follow-up in Kazakhstan. EClinicalMedicine. 2021;39:101078. https://doi.org/10.1016/j.eclinm.2021.101078.
13. Behera P, Singh AK, Subba SH, Mc A, Sahu DP, Chandanshive PD, et al. Effectiveness of COVID-19 vaccine (Covaxin) against breakthrough SARS-CoV-2 infection in India. Hum Vaccin Immunother. 2022;18(1):2034456. https://doi.org/10.1080/2F21645515.2022.2034456.
14. Martínez-Flores D, Zepeda-Cervantes J, Cruz-Reséndiz A, Aguirre-Sampieri S, Sampieri A, Vaca L. SARS-CoV-2 vaccines based on the spike glycoprotein and implications of new viral variants. Front Immunol. 2021;12:701501. https://doi.org/10.3389/fimmu.2021.701501.
15. Feng GW, Wang ZF, He P, Lan QY, Ni L, Yang YZ, et al. Safety, tolerability, and immunogenicity of a CpG/Alum adjuvanted SARS-CoV-2 recombinant protein vaccine (ZR202-CoV) in healthy adults: preliminary report of a phase 1, randomized, double-blind, placebo-controlled, dose-escalation trial. Hum Vaccin Immunother. 2023;19(2):2262635. https://doi.org/10.1080/21645515.2023.2262635.
16. Kaabi NA, Yang YK, Du LF, Xu K, Shao S, Liang Y, et al. Safety and immunogenicity of a hybrid-type vaccine booster in BBIBP-CorV recipients in a randomized phase 2 trial. Nat Commun. 2022;13(1):3654. https://doi.org/10.1038/2Fs41467-022-31379-0.
17. Lovell JF, Baik YO, Choi SK, Lee C, Lee JY, Miura K, et al. Interim analysis from a phase 2 randomized trial of EuCorVac-19: a recombinant protein SARS-CoV-2 RBD nanoliposome vaccine. BMC Med. 2022;20(1):462. https://doi.org/10.1186/s12916-022-02661-1.
18. Shrivastava S, Carmen JM, Lu Z, Basu S, Sankhala RS, Chen WH, et al. SARS-CoV-2 spike-ferritin-nanoparticle adjuvanted with ALFQ induces long-lived plasma cells and cross-neutralizing antibodies. NPJ Vaccines. 2023;8(1):43. https://doi.org/10.1038/s41541-023-00638-6.
19. Nagy A, Alhatlani B. An overview of current COVID-19 vaccine platforms. Comput Struct Biotechnol J. 2021;19:2508–17. https://doi.org/10.1016/j.csbj.2021.04.061.
20. Ingolotti M, Kawalekar O, Shedlock DJ, Muthumani K, Weiner DB. DNA vaccines for targeting bacterial infections. Expert Rev Vaccines. 2010;9(7):747–63. https://doi.org/10.1586/erv.10.57.
21. Kraynyak KA, Blackwood E, Agnes J, Tebas P, Giffear M, Amante D, et al. SARS-CoV-2 DNA vaccine INO-4800 induces durable immune responses capable of being boosted in a phase 1 open-label trial. J Infect Dis. 2022;225(11):1923–32. https://doi.org/10.1093/infdis/jiac016.
22. Momin T, Kansagra K, Patel H, Sharma S, Sharma B, Patel J, et al. Safety and immunogenicity of a DNA SARS-CoV-2 vaccine (ZyCoV-D): results of an open-label, non-randomized phase I part of phase I/II clinical study by intradermal route in healthy subjects in India. EClinicalMedicine. 2021;38:101020. https://doi.org/10.1016/j.eclinm.2021.101020.

23. Polack FP, Thomas SJ, Kitchin N, Absalon J, Gurtman A, Lockhart S, et al. Safety and efficacy of the BNT162b2 mRNA Covid-19 vaccine. N Engl J Med. 2020;383(27):2603–15. https://doi.org/10.1056/nejmc2036242.
24. Deng S, Liang H, Chen P, Li Y, Li Z, Fan S, et al. Viral vector vaccine development and application during the COVID-19 pandemic. Microorganisms. 2022;10(7):1450. https://doi.org/10.3390/2Fmicroorganisms10071450.
25. Roldão A, Mellado MCM, Castilho LR, Carrondo MJ, Alves PM. Virus-like particles in vaccine development. Expert Rev Vaccines. 2010;9(10):1149–76. https://doi.org/10.1586/erv.10.115.
26. Patil S. A review of virus-like particle-based SARS-CoV-2 vaccines in clinical trial phases. Iran J Pharm Res. 2021;13(2):2901–5. https://doi.org/10.5812/2Fijpr-127042.
27. Tang R, Zeng Y, Zhou Y, Liang Q, Kang W, Yang Z, et al. Safety and immunogenicity of the SARS-CoV-2 LYB001 RBD-based VLP vaccine (CHO cell) phase 1 in Chinese adults: a randomized, double-blind, positive-parallel-controlled study. Expert Rev Vaccines. 2024;23(1):498–509. https://doi.org/10.1080/14760584.2024.2337051.
28. Dunkle LM, Kotloff KL, Gay CL, Áñez G, Adelglass JM, Barrat Hernández AQ, et al. Efficacy and safety of NVX-CoV2373 in adults in the United States and Mexico. N Engl J Med. 2022;386(6):531–43. https://doi.org/10.1056/2FNEJMoa2116185.
29. Halabi S, Gostin LO, Aneja K, Nardi F, Gottschalk K, Monahan J. The Coalition for Epidemic Preparedness Innovations (CEPI) and the partnerships of equitable vaccine access. J Law Med Ethics. 2023;51(2):234–46. https://doi.org/10.1017/2Fjme.2023.85.
30. Kim JH, Hotez P, Batista C, Ergonul O, Figueroa JP, Gilbert S, et al. Operation warp speed: implications for global vaccine security. Lancet Glob Health. 2021;9(7):e1017–21. https://doi.org/10.1016/2FS2214-109X(21)00140-6.
31. Pfizer RH. Pfizer and BioNTech achieve first authorization in the world for a vaccine to combat COVID-19. 2020. https://www.pfizer.com/news/press-release/press-release-detail/pfizer-and-biontech-achieve-first-authorization-world.
32. Lamb YN. BNT162b2 mRNA COVID-19 vaccine: first approval. Drugs. 2021;81:495–501. https://doi.org/10.1007/s40265-021-01480-7.
33. Hasan T, Beardsley J, Marais BJ, Nguyen TA, Fox GJ. The implementation of mass-vaccination against SARS-CoV-2: a systematic review of existing strategies and guidelines. Vaccines. 2021;9(4):326. https://doi.org/10.3390/2Fvaccines9040326.
34. World Health Organization. COVID-19 vaccine doses shipped by the COVAX Facility head to Ghana, marking the beginning of global rollout. World Health Organization [Çevrimiçi]. 2021;24(2). https://www.who.int/news/item/24-02-2021-covid-19-vaccine-doses-shipped-by-the-covax-facility-head-to-ghana-marking-beginning-of-global-rollout.
35. Khan NA, Al-Thani H, El-Menyar A. The emergence of new SARS-CoV-2 variant (Omicron) and increasing calls for COVID-19 vaccine boosters-the debate continues. Travel Med Infect Dis. 2022;45:102246. https://doi.org/10.1016/2Fj.tmaid.2021.102246.
36. Boretti A. mRNA vaccine boosters and impaired immune system response in immune compromised individuals: a narrative review. Clin Exp Med. 2024;24(1):1–7. https://doi.org/10.1007/2Fs10238-023-01264-1.
37. Estes JD, Wong SW, Brenchley JM. Nonhuman primate models of human viral infections. Nat Rev Immunol. 2018;18(6):390–404. https://doi.org/10.1038/s41577-018-0005-7.
38. Batty CJ, Heise MT, Bachelder EM, Ainslie KM. Vaccine formulations in clinical development for the prevention of severe acute respiratory syndrome coronavirus 2 infection. Adv Drug Deliv Rev. 2021;169:168–89. https://doi.org/10.1016/j.addr.2020.12.006.
39. Heath PT, Galiza EP, Baxter DN, Boffito M, Browne D, Burns F, Chadwick DR, Clark R, Cosgrove CA, Galloway J, Goodman AL, Heer A, Higham A, Iyengar S, Jeanes C, Kalra PA, Kyriakidou C, Bradley JM, Munthali C, Minassian AM, et al. Safety and efficacy of the NVX-CoV2373 coronavirus disease 2019 vaccine at completion of the placebo-controlled phase of a randomized controlled trial. Clin Infect Dis. 2023;76(3):398–407. https://doi.org/10.1093/cid/ciac803.

40. Song JY, Choi WS, Heo JY, Kim EJ, Lee JS, Jung DS, Kim SW, Park KH, Eom JS, Jeong SJ, Lee J, Kwon KT, Choi HJ, Sohn JW, Kim YK, Yoo BW, Jang IJ, Capeding MZ, Roman F, Breuer T, et al. GBP510/AS03 Study Group. Immunogenicity and safety of SARS-CoV-2 recombinant protein nanoparticle vaccine GBP510 adjuvanted with AS03: interim results of a randomised, active-controlled, observer-blinded, phase 3 trial. EClinMed. 2023;64:102140. https://doi.org/10.1016/j.eclinm.2023.102140.
41. Cerqueira-Silva T, Andrews JR, Boaventura VS, Ranzani OT, de Araújo Oliveira V, Paixão ES, Júnior JB, Machado TM, Hitchings MDT, Dorion M, Lind ML, Penna GO, Cummings DAT, Dean NE, Werneck GL, Pearce N, Barreto ML, Ko AI, Croda J, Barral-Netto M. Effectiveness of CoronaVac, ChAdOx1 nCoV-19, BNT162b2, and Ad26.COV2.S among individuals with previous SARS-CoV-2 infection in Brazil: a test-negative, case-control study. Lancet Infect Dis. 2022;22(6):791–801. https://doi.org/10.1016/S1473-3099(22)00140-2.
42. Zhang Y, Belayachi J, Yang Y, et al. Real-world study of the effectiveness of BBIBP-CorV (Sinopharm) COVID-19 vaccine in the Kingdom of Morocco. BMC Public Health. 2022;22:1584. https://doi.org/10.1186/s12889-022-14016-9.
43. Fabiani M, Ramigni M, Gobbetto V, Mateo-Urdiales A, Pezzotti P, Piovesan C. Effectiveness of the Comirnaty (BNT162b2, BioNTech/Pfizer) vaccine in preventing SARS-CoV-2 infection among healthcare workers, Treviso province, Veneto region, Italy, 27 December 2020 to 24 March 2021. Euro Surveill. 2021;26(17):2100420. https://doi.org/10.2807/1560-7917.ES.2021.26.17.2100420.
44. Cazzola M, Rogliani P, Mazzeo F, Matera MG. Controversy surrounding the Sputnik V vaccine. Respir Med. 2021;187:106569. https://doi.org/10.1016/j.rmed.2021.106569.
45. Li M, Wang H, Tian L, Pang Z, Yang Q, Huang T, Fan J, Song L, Tong Y, Fan H. COVID-19 vaccine development: milestones, lessons and prospects. Signal Transduct Target Ther. 2022;7(1):146. https://doi.org/10.1038/s41392-022-00996-y.
46. Abu-Raddad LJ, Chemaitelly H, Butt AA, National Study Group for COVID-19 Vaccination. Effectiveness of the BNT162b2 Covid-19 vaccine against the B.1.1.7 and B.1.351 variants. N Engl J Med. 2021;385(2):187–9. https://doi.org/10.1056/NEJMc2104974.
47. Andrews N, Stowe J, Kirsebom F, Toffa S, Rickeard T, Gallagher E, Gower C, Kall M, Groves N, O'Connell AM, Simons D, Blomquist PB, Zaidi A, Nash S, Iwani Binti Abdul Aziz N, Thelwall S, Dabrera G, Myers R, Amirthalingam G, Gharbia S, et al. Covid-19 vaccine effectiveness against the Omicron (B.1.1.529) variant. N Engl J Med. 2022;386(16):1532–46. https://doi.org/10.1056/NEJMoa2119451.
48. Greinacher A, Thiele T, Warkentin TE, Weisser K, Kyrle PA, Eichinger S. Thrombotic thrombocytopenia after ChAdOx1 nCov-19 vaccination. N Engl J Med. 2021;384(22):2092–101. https://doi.org/10.1056/NEJMoa2104840.

SARS-CoV-2: A Comprehensive Insight into DNA, RNA, Subunit, and Viral Vaccines

18

Shaheda Tabassum

Abstract

The global response to the SARS-CoV-2 pandemic has spurred unprecedented advancements in vaccine technology. This chapter explores three critical types of vaccines developed to combat COVID-19: DNA and RNA vaccines, subunit vaccines (protein and virus-like particles), and viral vaccines (replicating and non-replicating). DNA and RNA vaccines represent a revolutionary approach, utilizing genetic materials to instruct cells to produce viral antigens, thereby eliciting an immune response. Subunit vaccines, including protein and virus-like particle vaccines, focus on specific components of the virus to stimulate immunity without the need for live pathogens. Viral vaccines, both replicating and non-replicating, use modified viruses to present antigens and induce protective immunity. Each vaccine type offers unique advantages and challenges, impacting their development, efficacy, and deployment. This chapter provides a comprehensive overview of these vaccine technologies, detailing their mechanisms, clinical progress, and the role they play in controlling the COVID-19 pandemic. Through a comparative analysis, we aim to highlight the strengths and limitations of each vaccine type, offering insights into their future applications in pandemic preparedness and beyond.

Keywords

SARS-CoV-2 · COVID-19 Vaccines · DNA and RNA vaccines · Subunit vaccines · Viral vaccines

---

S. Tabassum (✉)
Department of Zoology, Milli Al-Ameen College (For Girls), Kolkata, West Bengal, India

M. K. Ghosh, M. Basu (eds.), *Pathogenesis, Management and Socio-Economic Impact of COVID-19 Pandemic*, Medical Virology: From Pathogenesis to Disease Control, https://doi.org/10.1007/978-981-95-1426-7_18

Introduction

The unprecedented COVID-19 pandemic, caused by the novel coronavirus SARS-CoV-2, has dramatically reshaped global health priorities, necessitating swift and innovative responses to mitigate its impact. The development and deployment of effective vaccines have been central to controlling the spread of the virus and reducing its associated morbidity and mortality. This chapter provides an in-depth examination of the advanced vaccine technologies that have emerged in response to the pandemic, focusing specifically on DNA, RNA, subunit, and viral vaccines. The onset of the COVID-19 pandemic in late 2019 introduced a global health crisis of unprecedented scale [9]. SARS-CoV-2, a highly transmissible virus with a distinctive spike protein that facilitates entry into host cells, quickly spread worldwide, leading to widespread illness and death. Understanding the structure and function of this spike protein has been crucial for vaccine development, as it serves as the primary target for immune responses [30]. The rapid progression from virus identification to vaccine development was accelerated by unprecedented levels of international collaboration and scientific innovation.

The urgency of the pandemic highlighted the limitations of traditional vaccine development timelines, which typically span several years. In response, researchers and pharmaceutical companies leveraged cutting-edge technologies to develop and test vaccines at an accelerated pace. By mid-2021, multiple vaccines had been authorized for emergency use, demonstrating varying degrees of efficacy and safety. For instance, mRNA vaccines, exemplified by Pfizer-BioNTech's BNT162b2 and Moderna's mRNA-1273, achieved efficacy rates of approximately 95% and 94%, respectively, during clinical trials [1, 29]. These vaccines represented a novel approach, utilizing messenger RNA to instruct cells to produce viral antigens and trigger an immune response. In addition to mRNA vaccines, several other platforms have been explored, including DNA vaccines, subunit vaccines, and viral vector vaccines. DNA vaccines, although less common in practice, offer significant potential due to their ability to induce strong and sustained immune responses with relatively low production costs. RNA vaccines, on the other hand, have demonstrated rapid development capabilities and high efficacy, setting new standards in vaccine technology. Subunit vaccines, such as those developed by Novavax, utilize specific protein components of the virus to elicit immunity and have shown promising results in terms of safety and efficacy [16]. Viral vector vaccines, which use modified viruses to deliver genetic material encoding viral antigens, represent another innovative approach [19].

The deployment of these vaccines has been a monumental achievement, yet it has also revealed several challenges. Ensuring the safety and efficacy of vaccines through rigorous clinical trials has been essential, but the real-world efficacy and safety monitoring continue as vaccines are distributed globally. Moreover, logistical issues, such as the need for cold storage of mRNA vaccines and equitable distribution, have posed significant hurdles [3]. The success of these vaccine technologies underscores the critical role of innovation in addressing global health crises. Moving forward, there are several important areas for future research and development. Advances in vaccine technology to improve stability, reduce production costs, and enhance immunogenicity are needed. Additionally, alternative vaccine delivery methods, such as intranasal or microneedle vaccines, are also being explored to improve accessibility and acceptability. These methods could simplify vaccine administration and expand reach, particularly in underserved regions. Addressing vaccine hesitancy through transparent communication and public education remains crucial for achieving high vaccination coverage and mitigating the impact of the pandemic [20].

In summary, the COVID-19 pandemic has driven rapid advancements in vaccine technology, demonstrating the potential of innovative approaches such as mRNA, DNA, subunit, and viral vector vaccines. The lessons learned during this crisis will shape the future of vaccine development and public health preparedness. Continued research, global collaboration, and investment in these technologies are essential for addressing ongoing and future health challenges, ensuring resilience against emerging infectious diseases, and safeguarding global health (Fig. 18.1).

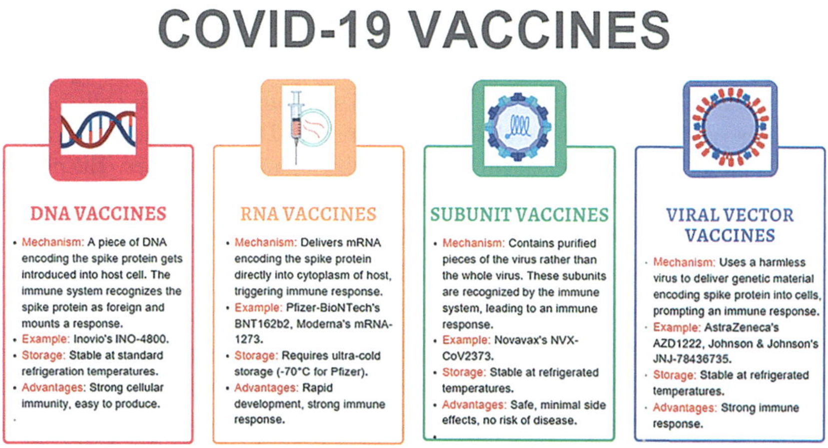

Fig. 18.1 Summary of vaccine types. Overview of DNA, RNA, subunit, and viral vector vaccines, highlighting their mechanisms and examples. (Single vector of DNA, RNA from BioRender)

DNA Vaccines Against SARS-CoV-2: Mechanisms and Applications

DNA vaccines for SARS-CoV-2 are engineered rings of DNA that encode the spike protein antigen, along with promoter and terminator sequences to facilitate expression in mammalian cells. This innovative approach has shown promise in eliciting comprehensive immune responses, including cytolytic T lymphocytes (CTL), T helper cells, and antibodies. The technology offers significant advantages in terms of manufacturing simplicity, scalability for global distribution, and robust safety profiles, making it a viable option for widespread vaccination efforts against COVID-19 [22].

Mechanisms, Design, and Delivery

DNA vaccines work by activating the adaptive immune response. When the plasmid DNA from the vaccine is delivered into the body, it is taken up by cells, which then use their cellular machinery to express the antigen encoded by the DNA. This antigen, often the SARS-CoV-2 spike protein, is presented on the cell surface via two pathways: Major Histocompatibility Complex (MHC) class I (for intracellular antigens) and MHC class II (for extracellular antigens). MHC class I presentation typically occurs in all nucleated cells and activates CD8 T cells, leading to the destruction of infected cells. On the other hand, MHC class II presentation occurs in specialized antigen-presenting cells like dendritic cells and macrophages, activating CD4 T cells, which help further stimulate both CD8 T cells and B cells. B cells produce antibodies, providing a humoral immune response. DNA vaccines induce both strong cellular (CD8 T cells) and humoral (B cells) immune responses. The plasmid DNA can also contain CpG motifs, which act as adjuvants to enhance immune responses. Delivery methods (e.g., intramuscular or intradermal) can influence the type of immune response generated, optimizing the vaccine's effectiveness. Unlike traditional vaccines, DNA vaccines can provide longer-lasting antigen expression, offering better priming of the immune system [34].

DNA vaccines consist of plasmid DNA that includes a bacterial origin of replication and antibiotic resistance genes. However, bacterial backbones can reduce gene expression in mammalian cells due to heterochromatin formation, so removing them enhances antigen expression. To express multiple genes in a single DNA vaccine, strategies such as using multiple expression cassettes with individual promoters, bi-cistronic or multi-cistronic vectors with IRES elements, or virus-derived T2A sequences are employed. Enhancers like the SV40 nuclear localization signal (NLS) and specific transcription factor-binding sequences facilitate nuclear import, while viral promoters like the human cytomegalovirus (CMV) promoter are common, though susceptible to hypermethylation. Eukaryotic or hybrid promoters are preferred for long-term expression. Plasmid DNA is produced in genetically modified *Escherichia coli*, with the production process involving fermentation, cell lysis, clarification, and chromatographic purification. The purified plasmid DNA is then formulated with excipients and adjuvants and filtered through sterilizing filters. Adjuvants enhance immunogenicity by stimulating innate immune responses, with unmethylated CpG motifs naturally stimulating responses via TLR9. Various adjuvants have been explored to enhance the efficacy of these vaccines. Classical adjuvants, such as aluminum salts (alum), are mixed with antigens before administration to slow antigen release and prolong immune cell interaction. However, these adjuvants have shown limited benefits in large animals and humans compared to small animals. Lipid-based adjuvants, including MetaStim™ and ENABL®, are used in veterinary vaccines like West Nile-Innovator® DNA and AgriLabs ExactVac, respectively. While these adjuvants aid in vaccine delivery, their immunogenicity benefits are not well-documented. Nanoparticles, such as liposomes, PLGA, and exosomes, boost immune responses and protect DNA from degradation. Cationic liposomes, magnesium phosphate, and calcium phosphate nanoparticles have demonstrated improved immunogenicity in animal models. Molecular adjuvants encompass a range of substances including TLR agonists, chemokines, cytokines, immune costimulatory molecules, and inhibitors of immune-suppressive pathways. These can be encoded within DNA vaccines or administered separately. Recent clinical trials have tested these adjuvants, including IL-2, IL-12, IL-15, IFN-lambda 3, and GM-CSF, to evaluate their effectiveness in enhancing immune responses [27].

The route of delivery and administration plays a crucial role in determining the immunogenicity of DNA vaccines. Commonly used routes include intramuscular (IM), which was the initial method of administration. However, subsequent studies revealed that intradermal (ID) routes could enhance gene expression and immune responses. The ID route targets Langerhans cells, dendritic cells, and keratinocytes in the skin, which are proficient at presenting antigens to T cells. Subcutaneous (SC) delivery, involving high-pressure injection under the skin, has been shown to evoke stronger immune responses compared to IM. Mucosal routes, such as oral or nasal delivery, are also utilized to generate both mucosal and systemic immune responses [18, 34]. Several physical and chemical methods have been developed to improve DNA delivery. Jet injectors (Biojectors) use compressed CO2 gas to create a high-pressure stream that penetrates the skin, enhancing both cellular immunity and antibody responses. The gene gun employs a biolistic system to deliver DNA-coated microparticles directly into the skin, which requires lower DNA amounts to elicit an immune response. Microneedle arrays consist of small needles that inject DNA directly into skin layers, allowing for controlled delivery. Electroporation uses electrical pulses to create transient pores in cell membranes, thereby increasing DNA uptake, and has shown efficacy and safety in clinical trials. Additionally, nanoparticles and microparticles, made from various synthetic and natural materials such as liposomes, dendrimers, chitosan, and alginate, protect DNA from degradation and enhance cellular uptake. Upon administration, the plasmid DNA translocates to the nucleus of transfected cells, initiating transcription and antigen production. These antigens are then processed and presented on MHC I or II molecules, activating CD8+ cytotoxic T lymphocytes (CTLs) and CD4+ T helper cells, respectively. B lymphocytes capture antigens from transfected cells, thereby triggering humoral immunity [27].

DNA Vaccines Against SARS-CoV-2

As of 2023, several DNA vaccines for COVID-19 are in clinical trials, showcasing various approaches and results (Table 18.1). INO-4800, developed by Inovio Pharmaceuticals, encodes the full-length SARS-CoV-2 S protein and has shown strong immunogenicity in Phase I and II trials, with Phase III trials underway. ZyCoV-D, developed by Cadila Healthcare, is the world's first approved DNA vaccine, showing 67% efficacy against symptomatic COVID-19 and approved for emergency use in India. Other notable DNA vaccines include GX-19 and GX-19N, which target the S protein and additional RBD and N proteins, showing enhanced T-cell responses. VGX-3100, EpiVacCorona, and vaccines by Vaxine Pty Ltd., Entos Pharmaceuticals, and Dynavax Technologies are also in development, employing various DNA platforms to induce strong immune responses. Each vaccine has unique attributes and clinical advancements, contributing to the global effort against COVID-19. These DNA vaccines reflect a global effort to leverage advanced genetic technologies to address the COVID-19 pandemic. Each vaccine demonstrates the potential of DNA-based platforms in providing effective immunization strategies against SARS-CoV-2, contributing to the broader fight against the virus [19, 21, 27].

Challenges and Applications

DNA vaccines offer several significant advantages for combating emerging pathogens like SARS-CoV-2. These vaccines are cost-effective, rapidly manufactured, and scalable, with the entire process—from plasmid construction to purification—completing in just 2–4 weeks. Their development is also highly flexible, allowing for swift design and production for preclinical and clinical studies. DNA vaccines express antigens in their native forms, which helps in proper folding and glycosylation without the typical challenges of protein expression and purification. They are stable at ambient temperatures and can be combined with additional expression cassettes to enhance immune responses. Moreover, pre-existing immunity to the vaccine backbone is not a concern for DNA vaccines, unlike with viral vector vaccines. However, there are challenges associated with DNA vaccines [26]. Naked DNA is rapidly degraded by nucleases, resulting in low uptake by cells and reduced efficiency. Biological barriers, such as cell membranes and endosomes, can hinder DNA vaccines from reaching their target sites. Once inside the nucleus, DNA vaccines may exhibit low antigen expression, potentially leading to reduced immunogenicity. Continuous antigen expression might also risk tolerance development. To address these issues, improved delivery systems are essential. Physical methods like electroporation, jet-injection, and gene guns are employed to enhance DNA delivery and reduce the amount of DNA needed. Natural and synthetic nanoparticles are also explored for their potential in delivering and protecting DNA vaccines. Despite the potential for DNA integration into the host genome, studies show that integration rates are low and generally less than spontaneous mutation rates, minimizing

Table 18.1 DNA vaccines against SARS-CoV-2

Vaccine	Developer	Mechanism	Dose and administration	Efficacy	Information	NCT Trial ID
INO-4800	Inovio Pharmaceuticals	Plasmid encoding full-length S protein	Intradermal injection, followed by electroporation	Phase I/II: Robust humoral and cellular responses	Effective against different variants; 2 mg dose selected for Phase III	NCT04642638
GX-19	Genexine	Plasmid encoding S protein	Intramuscular injection, followed by electroporation	Phase I: Notable T-cell responses	Lower neutralizing antibody levels than convalescent serum	NCT04445389
GX-19 N	Genexine	Plasmid encoding S, RBD, and N proteins	Intramuscular injection, followed by electroporation	Phase I: Enhanced T-cell responses	T-cell responses better than GX-19	NCT04915989
ZyCoV-D	Cadila Healthcare	Plasmid encoding S protein	Intradermal using NFIS device	Phase III: 67% efficacy	First DNA vaccine approved for human use	Momin et al. [26]
pDNA-COVID-19	Entos Pharmaceuticals	Plasmid DNA platform (IDAC technology)	Not specified	Preclinical: Strong immune response	Uses "Intracellular Delivery of Antigen Coding DNA" technology	NCT05171946
SARS-CoV-2 DNA	Dynavax Technologies	Plasmid with CpG 1018 adjuvant	Not specified	Preclinical: Enhanced immunogenicity	Uses CpG 1018 adjuvant technology	NCT04627675
PLX-PAD COVID-19	Pluristem Therapeutics	DNA-based platform	Not specified	Preclinical: Strong immune response	Known for cell therapy products	NCT04614025
ZyCov-D+	Cadila Healthcare	Enhanced DNA vaccine	Not specified	Preclinical: Improved effectiveness	Enhanced version of ZyCov-D	Khobragade et al. [17]

the risk of insertional mutagenesis. Regulatory guidelines and monitoring techniques help ensure safety and efficacy. The FDA recently paused and then resumed the Phase II/III trial of the INO-4800 vaccine, highlighting ongoing scrutiny and refinement in DNA vaccine development [21, 34, 37].

RNA Vaccines: Revolutionary Immunity

Unlike DNA vaccines, mRNA vaccines are a novel type of immunization technology that utilizes synthetic messenger RNA to instruct host cells to produce specific proteins that mimic those of a pathogen (Fig. 18.2). This protein production stimulates the immune system to recognize and mount a defense against the actual pathogen. Unlike traditional vaccines, mRNA vaccines do not contain live virus particles. They offer rapid development and adaptability, proving effective in combating emerging infectious diseases such as COVID-19 [43].

Mechanisms and Delivery

The advancement of mRNA vaccines was made possible by two key innovations: improvements in mRNA itself and its delivery system. Katalin Karikó and Drew Weissman, who were awarded the 2023 Nobel Prize in Physiology or Medicine, showed that incorporating nucleoside modifications into mRNA can reduce its rapid degradation in the body. Normally, mRNA triggers immune responses through toll-like receptors (such as TLR3, TLR7, and TLR8), but by using modified nucleosides

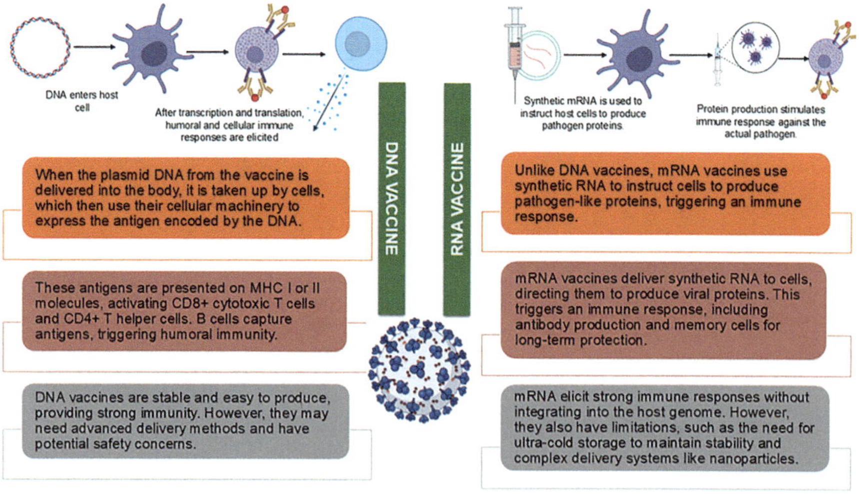

Fig. 18.2 DNA versus RNA vaccines. Comparative illustration of DNA and RNA vaccines, showing their mechanisms, stability, advantages, and differences. (Single vector blank image from BioRender)

like pseudouridine, these responses are diminished. This modification extends the mRNA's lifespan and enhances its expression [2].

mRNA vaccine mechanism involves introducing synthetic mRNA into host cells, where it serves as a blueprint to produce viral antigens, such as the spike protein of SARS-CoV-2. After the mRNA is delivered into cells, it is translated by ribosomes into the target protein. This protein is then processed and presented on the cell surface by major histocompatibility complex (MHC) molecules. This presentation is crucial for initiating an immune response. Helper T cells recognize the MHC–peptide complex [28], leading them to release cytokines such as interferon-gamma (IFN-γ) and interleukin-2 (IL-2), which enhance the activity of other immune cells. Specifically, these cytokines promote the activation and proliferation of B cells, which produce antibodies targeting the viral antigen. Simultaneously, cytotoxic T cells recognize and destroy cells presenting the antigen. This multifaceted immune response generates both antibodies and memory cells, providing long-term protection against future infections by creating a robust defense mechanism that can recognize and neutralize the virus more effectively upon subsequent exposure [11].

The delivery of mRNA vaccines is achieved through the use of lipid nanoparticles (LNPs), which are critical for ensuring the stability and cellular uptake of the mRNA. These LNPs are composed of a phospholipid bilayer, ionizable cationic lipids, cholesterol, and polyethylene glycol (PEG)-containing lipids. The ionizable cationic lipids in the LNPs interact with the negatively charged cell membranes, facilitating the endocytosis of the particles into the cell. Once inside the cell, the LNPs help release the mRNA from the endosomal compartments, ensuring it reaches the cytoplasm where translation occurs. The phospholipid bilayer and cholesterol components of the LNPs contribute to the structural integrity and stability of the nanoparticles, while PEG helps prolong circulation time in the bloodstream and reduce immune system recognition. This sophisticated delivery system not only protects the mRNA from degradation but also ensures efficient uptake and expression, which is essential for eliciting a strong and effective immune response [25, 28, 43].

Notable mRNA Vaccines Against SARS-CoV-2

The rapid development of COVID-19 vaccines highlighted unprecedented scientific and collaborative efforts (Table 18.2). Key vaccines include BNT162b2 (Comirnaty), developed by Pfizer and BioNTech, which was the first COVID-19 vaccine to receive emergency use authorization (EUA) from the US FDA. This mRNA-based vaccine demonstrated 95% efficacy against symptomatic COVID-19 in Phase III trials and was later authorized for use in adolescents and younger children. Real-world studies confirmed its high efficacy, although initial trials did not extensively include pregnant women. Another significant vaccine, mRNA-1273 (Spikevax), created by Moderna in collaboration with NIAID/NIH, also used mRNA technology and was the second to receive EUA. Phase III trials reported 94.1% efficacy in preventing symptomatic SARS-CoV-2 infection, with effective results across

Table 18.2 RNA vaccines against SARS-CoV-2

Vaccine	Developer	Mechanism	Storage requirements	Efficacy	Information	NCT Trial ID
BNT162b2	Pfizer/BioNTech	mRNA encoding S protein	−80 to −60 °C	95% efficacy in Phase III trials	First to receive EUA; effective in adolescents; real-world studies confirm high efficacy	NCT04368728
mRNA-1273	Moderna	mRNA encoding S protein	−20 °C	94.1% efficacy in Phase III trials	Effective in various age groups; ongoing trials for younger populations	NCT04470427
CVnCoV	CureVac	mRNA with lipid nanoparticle carrier	2–8 °C	47% efficacy in Phase III trials	Lower efficacy possibly due to mRNA sequence or emerging variants	NCT04652102
ARCT-021	Arcturus Therapeutics	saRNA with lipid-based delivery system	−20 °C	Early trials: Robust immune response	Lyophilized form simplifies storage	NCT04480957
ARCoV	PLA Academy & Suzhou Abogen	mRNA encoding S protein	Room temperature (1 week)	Early trials: Safe, immunogenic	Notable for room-temperature storage; Phase III trials ongoing	NCT04847102
COVAC1	Imperial College London	saRNA with RNA replicase	Not specified	Preclinical: Promising results	Facilitating vaccine access in low- and middle-income countries	Szubert et al. [39]
HDT-301	HDT Bio & Gennova Biopharmaceuticals	saRNA with lipid inorganic nanoparticle	Not specified	Preclinical: Good immunogenicity	Focus on safety and efficacy; clinical trials in India	NCT04844268

various age groups, including those with chronic conditions like kidney disease. Ongoing trials are assessing its impact on younger populations [6, 23]. CVnCoV, developed by CureVac, used mRNA technology with a lipid nanoparticle carrier and aimed for easier storage but showed only 47% efficacy in Phase III trials, below the WHO's 50% target. This lower performance might be due to the mRNA sequence used or emerging variants. ARCT-021 (LUNAR-COV19), developed by Arcturus Therapeutics, employs saRNA technology with a lipid-based delivery system. It demonstrated robust safety and immune responses in early trials and is unique for its lyophilized form, which simplifies storage. ARCoV, developed by the PLA Academy of Military Sciences and Suzhou Abogen Biosciences, is an mRNA vaccine with the notable advantage of room-temperature storage for up to a week. Phase I and II trials in China have assessed its safety and immunogenicity, with Phase III trials ongoing. COVAC1, from Imperial College London, uses saRNA technology with RNA replicase and has shown promising preclinical results. It is undergoing Phase I trials in the United Kingdom, focusing on facilitating vaccine access in low- and middle-income countries. HDT-301, developed by HDT Bio and Gennova Biopharmaceuticals, combines saRNA with a lipid inorganic nanoparticle. It has shown good immunogenicity in preclinical studies and is advancing through clinical trials in India, emphasizing safety and efficacy assessment [11, 14, 36].

Advantages and Limitations

Vaccine development typically falls into two main categories: gene-based and protein-based. The traditional protein-based approach involves using attenuated or recombinant proteins delivered directly as immunogens to stimulate an adaptive and humoral immune response. In contrast, gene-based vaccines, which use DNA or RNA vectors, involve delivering genetic materials into host cells. This genetic material instructs the cells to produce antigens that induce an immune response. mRNA vaccines, a subset of gene-based vaccines, offer several notable advantages over traditional methods. First, they benefit from a streamlined and rapid manufacturing process. Once the sequence encoding the target antigen is available, mRNA synthesis can be quickly initiated, and the process is scalable and cell-free, simplifying production. Additionally, mRNA vaccines express the target protein rapidly after transfection, with mRNA remaining in the cytoplasm rather than integrating into the genome, which enhances their biosafety compared to DNA vaccines. The transient nature of mRNA and its lack of interaction with the host genome further contribute to its safety profile. mRNA vaccines also leverage the host's translation machinery to produce antigens that closely mimic the native protein, including necessary post-translational modifications [23, 28].

However, mRNA vaccines face challenges, particularly regarding storage and transportation. They require ultralow temperatures to maintain stability, posing logistical and economic difficulties, especially in warmer regions. Despite advancements in lipid nanoparticle technologies that have improved mRNA stability, this remains a significant hurdle. In addition to conventional mRNA vaccines, self-amplifying RNA (saRNA) vaccines represent an innovative approach. saRNA vaccines can replicate after delivery, allowing them to express more target antigens at lower doses compared to conventional mRNA vaccines. These vaccines are derived from self-replicating RNA viruses and have been tested against various infectious diseases, including rabies, influenza, RSV, HIV, and Ebola, demonstrating their potential for efficient and scalable immunization [25].

Subunit Vaccines: Overview of Assembling Protection

Subunit vaccines are a type of vaccine that include only the essential antigens of the virus or bacteria, rather than using the whole pathogen. These antigens are typically proteins or polysaccharides that can stimulate a protective immune response. Because subunit vaccines do not contain live components of the pathogen, they are considered very safe and have fewer side effects compared to whole-cell vaccines. The antigens used in subunit vaccines can be produced through recombinant DNA technology, where the gene encoding the antigen is inserted into bacteria, yeast, or mammalian cells to produce large quantities of the antigen. These antigens are then purified and used to formulate the vaccine. This method allows for precise control over the vaccine's composition and consistency. Subunit vaccines can either be protein subunit vaccines or viral-like particle (VLP) vaccines, each having its unique characteristic [44].

Protein Subunit Vaccines and Viral-Like Particle Vaccines

Protein subunit vaccines, like those targeting SARS-CoV-2, use fragments of viral proteins, particularly the spike protein, to stimulate an immune response without causing disease. These vaccines are notable for their strong safety profile, making them a preferred option for COVID-19 management. Development methods include reverse vaccinology, virus-like particles, and recombinant modifications, supported by advanced manufacturing techniques for large-scale production. Challenges include the need for cold-chain storage and booster doses, but their benefits in widespread immunization are significant [12].

Virus-like particles (VLPs) are nanoscale structures formed from viral proteins, resembling natural viruses in shape and antigenic properties without containing viral genetic material, thus ensuring safety. VLPs effectively present concentrated antigenic epitopes, enhancing phagocytosis by dendritic cells and stimulating both cellular and humoral immune responses. SARS-CoV-2 VLP vaccines offer distinct advantages compared to other platforms. They mimic the virus's natural structure, promoting strong cellular and humoral immune responses without containing viral genetic material, ensuring safety. VLPs exhibit good thermal stability and can be mass-produced and stored effectively. Additionally, their surfaces can be modified, such as by adding adjuvants, enhancing their efficacy. These characteristics make VLPs an ideal platform for COVID-19 vaccine development [31].

Mechanism and Production

Protein subunit vaccines are created by producing antigenic proteins from pathogens either through culturing pathogens or synthesizing recombinant proteins. These vaccines are commonly produced in live microbial systems, such as yeast. Antigenic genes are inserted into yeast cells, which are then grown in large fermentation bioreactors to produce recombinant protein subunits. After purification, these subunits are combined with preservatives to maintain shelf-life and adjuvants like alum to enhance the immune response. Although this method avoids the risks associated with whole pathogen vaccines, it may require booster doses and adjuvants to achieve potent immunological responses. Protein subunit vaccines have been developed for diseases like pertussis, hepatitis B, and HPV. Protein subunit vaccines use antigenic protein parts of pathogens to induce the formation of germinal centers, leading to the production of highly specific antibodies. Due to their limited size, these subunits lack pathogen-associated molecular patterns (PAMPs), which necessitates the use of adjuvants to activate antigen-presenting cells (APCs). These APCs internalize the antigen and present it to cells of the adaptive immune system, thereby inducing a targeted immune response [5, 12].

Virus-like particle (VLP) vaccines work by mimicking the structure of viruses, which allows them to trigger strong immune responses without causing disease. VLPs, composed of viral structural proteins, are recognized by the immune system due to their similarity to actual viruses. They are taken up by dendritic cells, processed, and presented on MHC I and II molecules, stimulating both CD4+ and CD8+ T cells. This leads to a robust activation of B cells, resulting in a strong humoral immune response with the production of neutralizing antibodies. Additionally, VLPs can be engineered to include adjuvants, enhancing their immunogenicity further. Viral-like particles (VLPs) stimulate strong intrinsic immune responses by being phagocytosed by dendritic cells (DCs), leading to DC maturation and upregulation of surface molecules like CD40, CD80, and CD86. This process is accompanied by the release of chemokines and cytokines such as IL-12, IL-6, and IFN-γ. VLPs also engage toll-like receptors (TLRs) on B cells, promoting antibody class switching and enhancing immune signaling. In terms of adaptive immunity, the highly repetitive surface structure of VLPs facilitates their interaction with the complement system, triggering classical pathway activation and aiding antigen presentation by MHC I and II. This leads to robust humoral and cellular immune responses, with the production of neutralizing antibodies and cytokines such as IL-2, IL-4, IL-10, and IFN-γ, and the activation of specific T cell responses against the antigen [7, 31, 40].

Notable Subunit Vaccines

Numerous protein subunit vaccines against SARS-CoV-2 are in various clinical trial phases (Table 18.3). Notable ones include PIKA COVID-19, CoV2-OGEN1, PepGNP-SARS-CoV-2, SpFN COVID-19, CoVepiT, IN-B009, NBP2001,

Table 18.3 Subunit vaccines against SARS-CoV-2

Vaccine	Developer	Mechanism	Dose and administration	Efficacy	Information	NCT Trial ID
NVX-CoV2373	Novavax	Protein subunit (S protein)	Intramuscular	Phase III: 90% efficacy	Stable at 2–8 °C; uses Matrix-M adjuvant	NCT04611802
EpiVacCorona	Vector Institute	Peptide-based	Intramuscular	Limited efficacy data	Authorized in Russia; peptide-based approach	NCT04527575
SCB-2019	Clover Biopharmaceuticals	Protein subunit using Trimer-Tag technology	Intramuscular	Early trials: High immunogenicity	Trimer-Tag presents spike protein effectively	NCT04672395
Adjuvanted recombinant	Sanofi/ GlaxoSmithKline	Protein subunit with AS03 adjuvant	Intramuscular	Phase III: Moderate efficacy	Enhanced immune response with AS03 adjuvant	NCT04742738
Soberana 02	Finlay Institute	Conjugate vaccine (RBD conjugated to tetanus)	Intramuscular	Phase III: Promising results	Conjugated to tetanus toxoid to boost immune response	NCT04839315

SARS-CoV-2 Vax 1, AdimrSC-2f, EuCorVac-19, AKS-452X, QazCoVac-P, and others. Approved vaccines include EpiVacCorona, ZF2001, CIGB-66, Soberana Plus, MVC-COV1901, Razi Cov Pars, COVOVAX, and COVAX-19. These vaccines are in various stages from Phase I to Phase III clinical trials, aiming to provide robust immunity against SARS-CoV-2. Six VLP-based COVID-19 vaccines have been in different phases of clinical trials since 2021. These vaccines are developed by Serum Institute of India with Accelagen Pty and SpyBiotech (RBD SARS-CoV-2 HBsAg VLP Vaccine), Medicago (Coronavirus-like particle COVID-19 (CoVLP), and VBI Vaccines Inc. (VBI-2902a), The Scientific and Technological Research Council of Turkey (SARS CoV-2 VLP vaccine), Radboud University (ABNCoV2 capsid virus-like particle (cVLP) +/− adjuvant MF59), and Yantai Patronus Biotech (SARS-CoV-2 Vaccine LYB001) [5, 7, 13, 35, 38].

Merits and Limitations

Protein subunit vaccines, like those targeting the SARS-CoV-2 spike protein, are highly antigenic and capable of inducing robust immune responses. Their safety profile is enhanced as they do not contain live viral components, reducing the risk of infection. However, their protective efficacy can be limited due to the inclusion of only specific viral proteins, which may not fully represent the virus's antigenic complexity, potentially leading to unbalanced immune responses. Virus-like particle (VLP) vaccines, on the other hand, mimic the natural virus structure without containing viral genetic material, ensuring safety and reducing the risk of infection. VLPs induce strong cellular and humoral immune responses due to their structural similarity to natural viruses and their ability to be easily phagocytosed and presented to the immune system. VLP vaccines are also adaptable for carrying exogenous antigens or adjuvants, enhancing their versatility. However, they can be challenging to produce at a large scale and may require sophisticated manufacturing processes and infrastructure, which can limit their widespread availability and increase production costs. Overall, both vaccine platforms offer unique advantages in terms of safety and immune response induction but face challenges related to production complexity and achieving comprehensive immune protection [5, 7].

Viral Vaccines: Harnessing Viral Vectors for Defense

Viral vaccines are designed to protect against viral infections by stimulating the immune system to recognize and respond to specific viruses. They achieve this by introducing harmless components related to the virus into the body, which prompts the immune system to mount a defense (Fig. 18.3).

These components can include inactivated or killed virus particles, which are non-replicating but capable of eliciting an immune response; live-attenuated viruses, which are weakened versions of the virus that can replicate but are not virulent; subunit or protein-based vaccines, which contain specific viral proteins or antigens

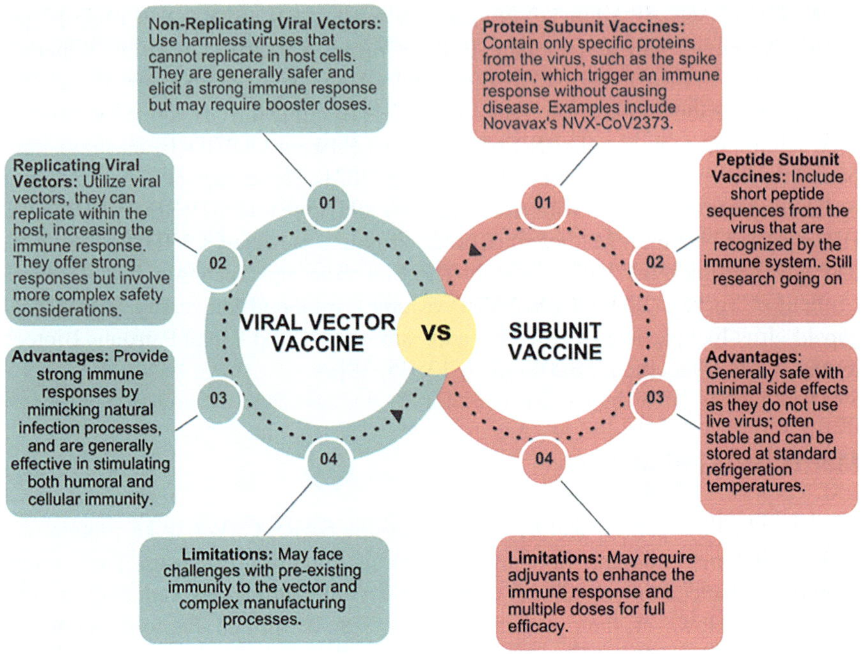

Fig. 18.3 Viral vector and subunit vaccines. Comparison of viral vector (replicating and non-replicating) and subunit vaccines, detailing their mechanisms, limitations, and advantages. (Single vector blank image from BioRender)

without the whole virus; viral vector vaccines, which use a modified virus to deliver genetic material from the target virus to induce an immune response; and mRNA vaccines, which introduce messenger RNA encoding viral proteins into cells, allowing them to produce these proteins and trigger an immune reaction. Each type of viral vaccine trains the immune system to recognize and combat the virus, providing protection against future infections [13].

Types of Viral Vaccines

Viral vaccines can be categorized into replicating and non-replicating types, each serving different functions in immunization strategies. Replicating viral vaccines utilize live-attenuated viruses that have been weakened so they cannot cause disease but can replicate within the host. This replication mimics a natural infection, stimulating a robust and long-lasting immune response. These vaccines induce strong cellular and humoral immunity as the immune system encounters viral antigens over a period of time, providing broad and durable protection. Examples include the measles, mumps, rubella (MMR) vaccine, and the oral polio vaccine. The replication capability of these vaccines enhances their effectiveness but requires careful attenuation to ensure safety. Non-replicating viral vaccines, on the other hand, do not involve live viruses and instead use either inactivated viruses, viral proteins, or viral vectors to elicit an immune response. Inactivated or killed virus vaccines contain virus particles that have been chemically or physically treated to be noninfectious but still capable of stimulating an immune response. These vaccines often require multiple doses to achieve adequate protection due to their inability to replicate and continuously present antigens. Examples include the hepatitis A and inactivated polio vaccines [24, 41]. Subunit or protein-based vaccines consist of specific viral proteins or antigens isolated from the virus and are used to trigger an immune response without the need for the virus to replicate. Additionally, viral vector vaccines utilize a different, harmless virus (the vector) to deliver genetic materials from the target virus into cells, prompting them to produce viral proteins and stimulate an immune response. Examples include the Ebola and some COVID-19 vaccines like the AstraZeneca-Oxford vaccine. These non-replicating vaccines are generally considered safer as they do not involve live virus but often require adjuvants or multiple doses to enhance their efficacy [8, 19].

Mechanism, Delivery, and Insights

Viral vector vaccines operate through various mechanisms depending on the type of vector used. Replicating viral vectors, such as adenoviral vectors, can induce substantial immune responses by replicating within the host. For instance, the Ad5 vector, though effective in generating antibody and CD8+ T cell responses, can be hindered by pre-existing human immunity, potentially reducing its efficacy. To mitigate this issue, nonhuman adenoviral vectors like ChAd63 and rare serotypes like Ad26 are utilized, as they elicit better memory and multifunctionality of CD8+ T cells. Another example includes recombinant measles virus (MV), which directly infects antigen-presenting cells like dendritic cells and macrophages, promoting both cellular and humoral responses. MV vaccines, such as those for chikungunya virus (CHIKV), have demonstrated that pre-existing immunity does not impact their efficacy significantly. Similarly, vesicular stomatitis virus (VSV) vectors can generate both CD4+ and CD8+ T cell responses, showcasing their versatile application [4].

Viral vector vaccines can be administered through several routes, including intramuscular (IM), intranasal, oral, and intradermal (ID) methods. The choice of delivery route affects the nature and strength of the immune response, with intramuscular injection being the most common and practical method, especially in pandemic situations. These vaccines typically induce robust immune responses without requiring additional adjuvants. Although recent trials have explored combining these vaccines with immune-stimulating agents, no significant improvements in immune responses have been observed. The effectiveness of the vaccine often depends on the delivery method and its ability to produce a strong immune response on its own [32].

Replicating viral vector vaccines, which incorporate SARS-CoV-2 genes into a replicating viral vector, are designed to mimic natural infection and elicit strong immune responses. However, challenges such as irregular manufacturing processes and the risk of genomic integration pose significant hurdles. These vaccines aim to activate both T and B cells, generating comprehensive immunity. Clinical trials are ongoing with several candidates, including Covid-19/aAPC, MV-014-212, and DelNS1-nCoV-RBD LAIV in various phases. Although some trials have been discontinued due to inefficacy, these vaccines hold potential for effective SARS-CoV-2 immunization. Non-replicating viral vector vaccines use vectors that do not replicate but carry SARS-CoV-2 genes to provoke an immune response. These vaccines can still produce strong immune responses, but they also face risks like genomic integration. Many candidates are in different stages of clinical trials: Phase I includes MVA-SARS-2-S and VXA-CoV2-1; Phase II includes hAd5-Covid-19 and COVIVAC; and Phase III includes GRAd-COV2 and Covishield. Some non-replicating vaccines, such as Ad5-nCoV and Sputnik V, have already been approved for use, while others are still under development. Despite some trials being halted due to lower effectiveness, the remaining candidates offer promising prospects for effective COVID-19 vaccination [24, 41, 42].

Viral Vaccines Against SARS-CoV-2

Several viral vector vaccines have received approval for use against SARS-CoV-2 (Table 18.4). For instance, the Ad5-nCoV vaccine, which is based on the adenoviral vector Ad5, demonstrated significant effectiveness with 65.7% efficacy against moderate symptoms and 91% against severe symptoms, leading to its approval in ten countries. Currently, Ad5-nCoV is undergoing 11 trials across various phases in six countries. The Sputnik V and Sputnik Light vaccines, developed by Gamaleya, have gained widespread use, with Sputnik V being administered in 74 countries and Sputnik Light in 21. Sputnik V has shown 91.6% overall efficacy, while Sputnik Light has demonstrated 79% efficacy against infection and 88% against hospitalization. Both vaccines are subject to ongoing trials, with Sputnik V undergoing 22 trials and Sputnik Light 4 trials. The Janssen vaccine, Ad26.COV2.S, approved in 85 countries, has shown a 66.9% overall efficacy and significant increases in spike protein-specific and neutralizing antibodies. Its clinical trials span across 18 countries. The Oxford/AstraZeneca vaccines, AZD1222 and Covishield, have overall efficacies of 70.4% and 61%, respectively, with AZD1222 approved in 127 countries and Covishield in 47. AZD1222's efficacy varies with dosing regimens and against different variants, while Covishield has shown 81% effectiveness against the delta variant [32, 41, 42].

Advantages and Disadvantages

Viral vector vaccines offer significant advantages, including versatility in antigen expression and the ability to handle large DNA sequences, making them valuable for rapid vaccine development. The natural source of the antigen ensures effective

Table 18.4 Viral vector vaccines for COVID-19

Vaccine	Developer	Viral vector	Mechanism	Dose and administration	Efficacy	Information	NCT Trial ID
AZD1222 (Vaxzevria)	University of Oxford/ AstraZeneca	ChAdOx1	Adenovirus vector encoding S protein	Intramuscular (2 doses)	Phase III: 70% efficacy	Approved in multiple countries; stable at 2–8 °C	NCT04516746
Covishield	Serum Institute of India/ AstraZeneca	ChAdOx1	Adenovirus vector encoding S protein	Intramuscular (2 doses)	Phase III: 70% efficacy	Licensed version of AZD1222; produced in India	NCT04444674
Ad26.COV2.S (Janssen)	Johnson & Johnson	Ad26	Adenovirus vector encoding S protein	Intramuscular (1 dose)	Phase III: 66% efficacy	Single-dose regimen; effective against severe COVID-19	NCT04505722
Sputnik V	Gamaleya Research Institute	Ad26 and Ad5	Adenovirus vectors encoding S protein	Intramuscular (2 doses)	Phase III: 91.6% efficacy	Two-vector approach; widely used in multiple countries	NCT04530396
Ad5-nCoV	CanSino Biologics	Ad5	Adenovirus vector encoding S protein	Intramuscular (1 dose)	Phase III: 65.7% efficacy	Single-dose regimen; approved for emergency use in several countries	NCT04505722
Vaxzevria (Covishield)	Serum Institute of India/ AstraZeneca	ChAdOx1	Adenovirus vector encoding S protein	Intramuscular (2 doses)	Phase III: 70% efficacy	Produced in India under license from AstraZeneca	NCT04536051
VLA2001	Valneva	Inactivated virus	Traditional inactivated virus	Intramuscular (2 doses)	Phase III: High efficacy	Inactivated virus vaccine; requires standard cold storage	NCT04671017
BBIBP-CorV (CoronaVac)	Sinopharm/ BIBP	Inactivated virus	Traditional inactivated virus	Intramuscular (2 doses)	Phase III: 79% efficacy	Widely used in various countries; approved by WHO	NCT04510207

(continued)

Table 18.4 (continued)

Vaccine	Developer	Viral vector	Mechanism	Dose and administration	Efficacy	Information	NCT Trial ID
BBV152 (Covaxin)	Bharat Biotech	Inactivated virus	Traditional inactivated virus	Intramuscular (2 doses)	Phase III: 81% efficacy	Inactivated virus vaccine; developed with ICMR	NCT04641481
ZyCoV-D	Cadila Healthcare	DNA plasmid	DNA vaccine encoding S protein	Intradermal (3 doses)	Phase III: 66.6% efficacy	World's first approved DNA vaccine; unique intradermal delivery	NCT04939519
COVI-VAC	Codagenix and Serum Institute of India	Codon-optimized DNA plasmid	DNA vaccine with codon-optimized S protein	Intramuscular (2 doses)	Phase I/II: Safety and immunogenicity data	Ongoing trials; aims to induce strong immune response	NCT04619628
IDF-037	International Vaccine Institute	DNA plasmid	DNA vaccine encoding S protein	Intramuscular (2 doses)	Phase I/II: Safety and efficacy data	In collaboration with various international partners	NCT04659486

processing and production, which is crucial during pandemics. However, challenges include the potential for genomic integration and mutagenesis, and the requirement for specialized manufacturing facilities that increase costs. There are also concerns about the safety of genetically modified organisms used in these vaccines. Pre-existing immunity against certain viral vectors, such as Ad5, can limit vaccine effectiveness due to the immune system's pre-established memory cells that recognize and attack the vector, reducing its efficacy. To address this, alternative adenoviral serotypes or vectors from other species are utilized to circumvent pre-existing immunity and enhance vaccine development (2021). Adverse effects associated with viral vector vaccines include a range of systemic and local reactions. For example, in a study involving German healthcare workers who received the AstraZeneca vaccine, 87.2% reported systemic side effects such as fatigue, headache, and fever, and 70.4% experienced local reactions like pain and swelling at the injection site. Other rare side effects include transverse myelitis reported with ChAdOx1 nCoV-19 and Guillain–Barré syndrome (GBS) associated with adenoviral vectors, such as Ad26.COV2.S. The incidence of GBS was noted at 9.8 cases per million doses, with severe cases including one fatality. Additionally, vaccine-induced spike protein interactions with ACE2 receptors have been linked to cardiovascular issues such as thrombosis and thrombotic thrombocytopenia, with cases reported in connection with vaccines like Janssen and ChAdOx1 [4, 32].

Future Perspectives

The rapid development and deployment of vaccines during the COVID-19 pandemic have paved the way for numerous innovations in vaccine technology, setting the stage for future advancements. Several key areas of focus will shape the future of vaccine development and public health preparedness. One promising direction is the continued enhancement of mRNA vaccine platforms. Research is focused on improving the stability and storage requirements of mRNA vaccines to facilitate broader distribution, particularly in low-resource settings. Advances in lipid nanoparticle delivery systems are crucial for enhancing vaccine efficacy and reducing adverse effects [43]. Additionally, self-amplifying RNA vaccines, which require lower doses due to their ability to replicate within the body, are being explored to boost immune responses and reduce manufacturing costs. The emergence of SARS-CoV-2 variants has highlighted the need for vaccines that offer broad protection. Efforts are underway to develop universal coronavirus vaccines targeting conserved regions of the virus, which are less prone to mutation. Such vaccines could provide immunity against multiple coronavirus strains, potentially preventing future pandemics. Innovations in vaccine delivery are essential for improving accessibility and acceptance. Research is ongoing into needle-free delivery methods, such as oral, intranasal, and microneedle patch vaccines. These methods could simplify administration, reduce the need for trained healthcare personnel, and increase vaccine uptake, especially in populations with needle phobia or limited access to healthcare facilities [15]. Public confidence in vaccines is critical for achieving high vaccination coverage. Strategies to combat vaccine misinformation and enhance public trust are vital. Transparent communication, community engagement, and education campaigns are being developed to address vaccine hesitancy and promote informed decision-making. The pandemic has underscored the importance of robust surveillance systems for early detection of emerging infectious diseases. Integrating genomic sequencing and data-sharing platforms can facilitate the rapid identification of new variants and inform vaccine updates [33]. Additionally, establishing global manufacturing networks and stockpiling essential materials will enable a swift response to future outbreaks. Ensuring equitable access to vaccines remains a global priority. International collaborations, such as COVAX, aim to distribute vaccines fairly, particularly to low- and middle-income countries. Strengthening local manufacturing capabilities and improving supply chain logistics are critical for achieving global vaccine equity. In conclusion, the advancements made during the COVID-19 pandemic provide a foundation for future vaccine development. Continued innovation, collaboration, and investment in these areas will enhance our preparedness for future public health challenges and improve global health outcomes [10] (Fig. 18.4).

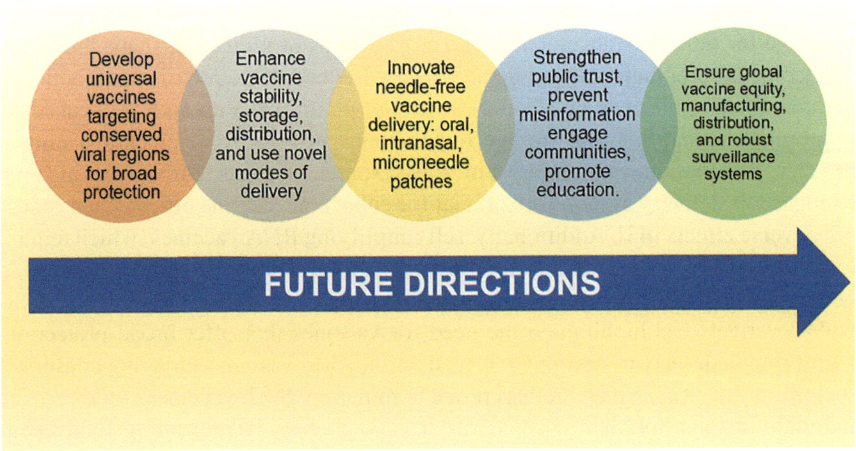

Fig. 18.4 Future directions in vaccine technologies. Key areas of future vaccine innovation including mRNA enhancements, universal vaccines, needle-free delivery, and global equity efforts

Conclusion

The COVID-19 pandemic has highlighted the necessity of rapid and innovative vaccine development to counter emerging infectious diseases. This chapter delves into the remarkable advancements in vaccine technology aimed at combating SARS-CoV-2, focusing on DNA, RNA, subunit, and viral vaccines. Each vaccine platform brings distinct advantages and faces unique challenges, collectively contributing to the global fight against the pandemic. DNA and RNA vaccines signify a revolutionary approach, utilizing genetic instructions to induce an immune response. The swift development and high efficacy rates of mRNA vaccines, such as those developed by Pfizer-BioNTech and Moderna, showcase the potential of genetic vaccines to be rapidly deployed during a public health emergency [19, 20]. Their success marks a milestone in vaccine technology, setting the stage for future advancements. Subunit vaccines, including protein-based and virus-like particle vaccines, provide a targeted approach by using specific viral components to elicit immunity. Vaccines like Novavax's candidate offer strong safety profiles and stability, making them valuable in the global vaccination effort. However, the need for adjuvants and complex manufacturing processes presents challenges. Viral vaccines, both replicating and non-replicating, use modified viruses to deliver antigens. Vaccines such as AstraZeneca/Oxford and Johnson & Johnson demonstrate that viral vector platforms can elicit robust immune responses and offer logistical advantages like easier storage and distribution. Nonetheless, rare but serious side effects necessitate ongoing monitoring and optimization. A comparative analysis reveals that no single vaccine approach is universally superior. The diversity of vaccine types allows for a multifaceted strategy, addressing various logistical, manufacturing, and immunological challenges. This pluralistic approach has been crucial in achieving widespread immunization and controlling the spread of COVID-19 [2].

Looking ahead, the lessons from the rapid development and deployment of these vaccines will shape future pandemic preparedness. Continued innovation, global cooperation, and investment in vaccine research are essential to combat emerging variants and future public health threats. The unprecedented collaboration and scientific breakthroughs during this pandemic have set a new standard for vaccine development, offering hope and resilience against future infectious disease outbreaks [30].

Author's Contributions Conceived the idea and review structure and wrote the manuscript: ST; Revised and edited the manuscript: Dr. Mrinal K. Ghosh, Chief Scientist, CSIR-Indian Institute of Chemical Biology, Jadavpur, Kolkata-700032. The author has read, agreed, and approved the final draft of the manuscript.

Data Availability The software BioRender (https://www.biorender.com/) is used to draw some elements of the few figures.

Declarations

Conflicts of Interest There is no competing and conflict of interest.

References

1. Baden LR, El Sahly HM, Essink B, et al. Efficacy and safety of the mRNA-1273 SARS-CoV-2 vaccine. N Engl J Med. 2021;384:403–16. https://doi.org/10.1056/NEJMoa2035389.
2. Beladiya J, Kumar A, Vasava Y, et al. Safety and efficacy of COVID-19 vaccines: a systematic review and meta-analysis of controlled and randomized clinical trials. Rev Med Virol. 2024;34:e2507. https://doi.org/10.1002/rmv.2507.
3. Cascella M, Rajnik M, Aleem A, et al. Features, evaluation, and treatment of coronavirus (COVID-19). StatPearls; 2024.
4. Chavda VP, Bezbaruah R, Athalye M, et al. Replicating viral vector-based vaccines for COVID-19: potential avenue in vaccination arena. Viruses. 2022;14:759. https://doi.org/10.3390/v14040759.
5. Chavda VP, Ghali ENHK, Balar PC, et al. Protein subunit vaccines: promising frontiers against COVID-19. J Control Release. 2024;366:761–82. https://doi.org/10.1016/j.jconrel.2024.01.017.
6. Fang E, Liu X, Li M, et al. Advances in COVID-19 mRNA vaccine development. Signal Transduct Target Ther. 2022;7:94. https://doi.org/10.1038/s41392-022-00950-y.
7. Gao X, Xia Y, Liu X, et al. A perspective on SARS-CoV-2 virus-like particles vaccines. Int Immunopharmacol. 2023;115:109650. https://doi.org/10.1016/j.intimp.2022.109650.
8. Ghosh MK, Kumar S, Ganguly KK, et al. COVID-19 and cancer: insights into their association and influence on genetic and epigenetic landscape. Epigenomics. 2023;15:227–48. https://doi.org/10.2217/epi-2023-0052.
9. Ghosh MK, Tabassum S, Basu M. COVID-19 and cancer: Dichotomy of the menacing dilemma. MedComm Oncol. 2023;2:e58. https://doi.org/10.1002/mog2.58.
10. Goodson JL, Rota PA. Innovations in vaccine delivery: increasing access, coverage, and equity and lessons learnt from measles and rubella elimination. Drug Deliv Transl Res. 2022;12:959–67. https://doi.org/10.1007/s13346-022-01130-9.
11. Gote V, Bolla PK, Kommineni N, et al. A comprehensive review of mRNA vaccines. Int J Mol Sci. 2023;24:2700. https://doi.org/10.3390/ijms24032700.
12. Heidary M, Kaviar VH, Shirani M, et al. A comprehensive review of the protein subunit vaccines against COVID-19. Front Microbiol. 2022;13:927306. https://doi.org/10.3389/fmicb.2022.927306.
13. Jafari A, Danesh Pouya F, Niknam Z, et al. Current advances and challenges in COVID-19 vaccine development: from conventional vaccines to next-generation vaccine platforms. Mol Biol Rep. 2022;49:4943–57. https://doi.org/10.1007/s11033-022-07132-7.
14. Jamous YF, Alhomoud DA. The safety and effectiveness of mRNA vaccines against SARS-CoV-2. Cureus. 2023;15:e45602. https://doi.org/10.7759/cureus.45602.
15. Kashte S, Gulbake A, El-Amin SF III, Gupta A. COVID-19 vaccines: rapid development, implications, challenges and future prospects. Hum Cell. 2021;34:711–33. https://doi.org/10.1007/s13577-021-00512-4.
16. Keech C, Albert G, Cho I, et al. Phase 1–2 trial of a SARS-CoV-2 recombinant spike protein nanoparticle vaccine. N Engl J Med. 2020;383:2320–32. https://doi.org/10.1056/NEJMoa2026920.
17. Khobragade A, Bhate S, Ramaiah V, et al. Efficacy, safety, and immunogenicity of the DNA SARS-CoV-2 vaccine (ZyCoV-D): the interim efficacy results of a phase 3, randomised, double-blind, placebo-controlled study in India. Lancet. 2022;399:1313–21. https://doi.org/10.1016/S0140-6736(22)00151-9.

18. Kozak M, Hu J. DNA vaccines: their formulations, engineering and delivery. Vaccines (Basel). 2024;12:71. https://doi.org/10.3390/vaccines12010071.
19. Kumar S, Basu M, Ghosh P, et al. COVID-19: clinical status of vaccine development to date. Br J Clin Pharmacol. 2023;89:114–49.
20. Larson HJ, Jarrett C, Schulz WS, et al. Measuring vaccine hesitancy: the development of a survey tool. Vaccine. 2015;33:4165–75. https://doi.org/10.1016/j.vaccine.2015.04.037.
21. Li Y, Tenchov R, Smoot J, et al. A comprehensive review of the global efforts on COVID-19 vaccine development. ACS Cent Sci. 2021;7:512–33. https://doi.org/10.1021/acscentsci.1c00120.
22. Liu MA. DNA vaccines: a review. J Intern Med. 2003;253:402–10. https://doi.org/10.1046/j.1365-2796.2003.01140.x.
23. Machado BAS, Hodel KVS, dos Fonseca LMS, et al. The importance of RNA-based vaccines in the fight against COVID-19: an overview. Vaccines (Basel). 2021;9:1345. https://doi.org/10.3390/vaccines9111345.
24. McCann N, O'Connor D, Lambe T, Pollard AJ. Viral vector vaccines. Curr Opin Immunol. 2022;77:102210. https://doi.org/10.1016/j.coi.2022.102210.
25. Mirtaleb MS, Falak R, Heshmatnia J, et al. An insight overview on COVID-19 mRNA vaccines: advantageous, pharmacology, mechanism of action, and prospective considerations. Int Immunopharmacol. 2023;117:109934. https://doi.org/10.1016/j.intimp.2023.109934.
26. Momin T, Kansagra K, Patel H, et al. Safety and immunogenicity of a DNA SARS-CoV-2 vaccine (ZyCoV-D): results of an open-label, non-randomized phase I part of phase I/II clinical study by intradermal route in healthy subjects in India. EClinicalMedicine. 2021;38:101020. https://doi.org/10.1016/j.eclinm.2021.101020.
27. Pagliari S, Dema B, Sanchez-Martinez A, et al. DNA vaccines: history, molecular mechanisms and future perspectives. J Mol Biol. 2023;435:168297. https://doi.org/10.1016/j.jmb.2023.168297.
28. Park JW, Lagniton PNP, Liu Y, Xu R-H. mRNA vaccines for COVID-19: what, why and how. Int J Biol Sci. 2021;17:1446–60. https://doi.org/10.7150/ijbs.59233.
29. Polack FP, Thomas SJ, Kitchin N, et al. Safety and efficacy of the BNT162b2 mRNA Covid-19 vaccine. N Engl J Med. 2020;383:2603–15. https://doi.org/10.1056/NEJMoa2034577.
30. Pollard CA, Morran MP, Nestor-Kalinoski AL. The COVID-19 pandemic: a global health crisis. Physiol Genomics. 2020;52:549–57. https://doi.org/10.1152/physiolgenomics.00089.2020.
31. Prates-Syed WA, Chaves LCS, Crema KP, et al. VLP-based COVID-19 vaccines: an adaptable technology against the threat of new variants. Vaccines (Basel). 2021;9:1409. https://doi.org/10.3390/vaccines9121409.
32. Robert-Guroff M. Replicating and non-replicating viral vectors for vaccine development. Curr Opin Biotechnol. 2007;18:546–56. https://doi.org/10.1016/j.copbio.2007.10.010.
33. Rodgers AM, Cordeiro AS, Donnelly RF. Technology update: dissolvable microneedle patches for vaccine delivery. Med Dev Evid Res. 2019;12:379–98. https://doi.org/10.2147/MDER.S198220.
34. Shafaati M, Saidijam M, Soleimani M, et al. A brief review on DNA vaccines in the era of COVID-19. Future Virol. 2022;17:49–66. https://doi.org/10.2217/fvl-2021-0170.
35. Sharifzadeh M, Mottaghi-Dastjerdi N, Soltany Rezae Raad M. A review of virus-like particle-based SARS-CoV-2 vaccines in clinical trial phases. Iran J Pharm Res. 2022;21:e127042. https://doi.org/10.5812/ijpr-127042.
36. Silva-Pilipich N, Beloki U, Salaberry L, Smerdou C. Self-amplifying RNA: a second revolution of mRNA vaccines against COVID-19. Vaccines (Basel). 2024;12:318. https://doi.org/10.3390/vaccines12030318.
37. Silveira MM, Moreira GMSG, Mendonça M. DNA vaccines against COVID-19: perspectives and challenges. Life Sci. 2021;267:118919. https://doi.org/10.1016/j.lfs.2020.118919.
38. Suryawanshi YR. An overview of protein-based SARS-CoV-2 vaccines. Vaccine. 2023;41:6174–93. https://doi.org/10.1016/j.vaccine.2023.09.013.

39. Szubert AJ, Pollock KM, Cheeseman HM, et al. COVAC1 phase 2a expanded safety and immunogenicity study of a self-amplifying RNA vaccine against SARS-CoV-2. EClinicalMedicine. 2023;56:101823. https://doi.org/10.1016/j.eclinm.2022.101823.
40. Tariq H, Batool S, Asif S, et al. Virus-like particles: revolutionary platforms for developing vaccines against emerging infectious diseases. Front Microbiol. 2022;12:790121. https://doi.org/10.3389/fmicb.2021.790121.
41. Travieso T, Li J, Mahesh S, et al. The use of viral vectors in vaccine development. NPJ Vaccines. 2022;7:75. https://doi.org/10.1038/s41541-022-00503-y.
42. Wang S, Liang B, Wang W, et al. Viral vectored vaccines: design, development, preventive and therapeutic applications in human diseases. Signal Transduct Target Ther. 2023;8:149. https://doi.org/10.1038/s41392-023-01408-5.
43. Zhao Z, Bashiri S, Ziora ZM, et al. COVID-19 variants and vaccine development. Viruses. 2024;16:757. https://doi.org/10.3390/v16050757.
44. Zhou X, Wang H, Luo Y, et al. Single-injection COVID-19 subunit vaccine elicits potent immune responses. Acta Biomater. 2022;151:491–500. https://doi.org/10.1016/j.actbio.2022.08.006.

COVID-19: Vaccination of Medically Distinct Population

19

Rajni Shaw and Mrinal K. Ghosh

Abstract

The COVID-19 pandemic has resulted in the extensive utilization of genetic vaccines, such as mRNA and viral vector vaccines. Booster vaccinations have also been employed to further combat the ever-evolving coronavirus disease. It is obvious that compared to the general population, patients with pre-existing medical issues are more likely to experience severe coronavirus disease and its related mortality. Patients have been given priority worldwide for vaccination, both for main and booster shots, due to the associated heightened risk. Nonetheless, a great deal of doubt surrounds the effectiveness of the vaccine, the degree of humoral and cellular immune responses in these patients, the likelihood of vaccination-related adverse effects, and the fact that these patients were excluded from crucial clinical studies. A study on the efficacy of COVID-19 vaccination and the gradual decline of immunity was recently published. According to the study, immune function was poorer in those who had received the COVID-19 vaccination 8 months after receiving two doses than in those who had not. These results were particularly noticeable in older people and those with pre-existing medical issues. Hence, we have tried to highlight the importance as well as shortcomings of COVID-19 vaccination in pre-existing patients before the outbreak of the pandemic.

R. Shaw · M. K. Ghosh (✉)
Cancer Biology and Inflammatory Disorder Division, Council of Scientific and Industrial Research-Indian Institute of Chemical Biology (CSIR-IICB), TRUE Campus, CN-6, Sector–V, Salt Lake, Kolkata, India
e-mail: mrinalghosh@iicb.res.in

M. K. Ghosh, M. Basu (eds.), *Pathogenesis, Management and Socio-Economic Impact of COVID-19 Pandemic*, Medical Virology: From Pathogenesis to Disease Control, https://doi.org/10.1007/978-981-95-1426-7_19

Keywords

COVID-19 · Vaccination · Medically distinct population · Cancer · Immune-compromised individuals

Introduction

A class of viruses known as coronaviruses is responsible for mild to moderate upper respiratory tract illnesses in people. A novel strain of SARS-CoV-2 caused COVID-19 towards the end of 2019 [23, 46]. By the end of 2022, COVID-19 had caused more than 600 million infections and over 6.6 million deaths. The infection's most frequently reported symptoms include fever, cough, weariness, dyspnea, and loss of taste or smell. However, septic shock and multiple organ failure may result from pulmonary involvement. On March 11, 2020, WHO declared this outbreak a pandemic. Although there is no currently approved treatment for COVID-19 corticosteroids, tocilizumab and sarilumab are advised for patients who are really sick and in critical condition [24].

The best defense, in any case, is vaccination. Over 11.8 billion doses have been given globally, with at least one dose having been given to 68% of the population. Oxford-AstraZeneca's vaccine is the most commonly utilized, followed by Pfizer-BioNTech and Moderna's mRNA vaccines. The licensed COVID-19 vaccines have all demonstrated outstanding efficacy and safety profiles, except for a few modest adverse effects such as pain and swelling at the site of injection, fever, and exhaustion [43]. Nonetheless, prior research indicated that vaccinations might raise the risk of cardiac events. Most individuals without comorbidities have a mild or moderate course of COVID-19, while patients with pre-existing conditions are far more likely to have severe COVID-19 and related mortality. The existence of other risk variables that are also relevant to the general population, such as advanced age and multiple comorbidities, can contribute to this elevated risk.

Over the past 4 years, COVID-19 vaccination in people with pre-existing conditions has more or less helped patients with better survival rates. Individuals with pre-existing medical illnesses are more susceptible to COVID-19-based hospitalization and death (Fig. 19.1). A person's chance of developing severe COVID-19 rises with the number of underlying medical disorders they have. Vaccination is therefore highly advised for them. It has been discovered that COVID-19 vaccinations are both safe and effective for those who have co-morbidities. Therefore, having one or more underlying medical issues should be an extra justification for choosing to get vaccinated rather than delaying it. Patients with diseases of lung (chronic obstructive pulmonary disease, asthma, bronchiectasis, broncho-pulmonary dysplasia, tuberculosis, pulmonary embolism, cystic fibrosis and respiratory infections other than pneumonia), heart (coronary heart disease, cardiovascular diseases, congestive heart failure, cardiomyopathies, and hypertension), chronic kidney disease, chronic defect in the brain or nerves (cerebral palsy, dementia, mental health conditions such as schizophrenia spectrum disorders, Alzheimer's and Parkinson's disease and other neurological conditions), liver (non-alcoholic fatty liver disease and chronic

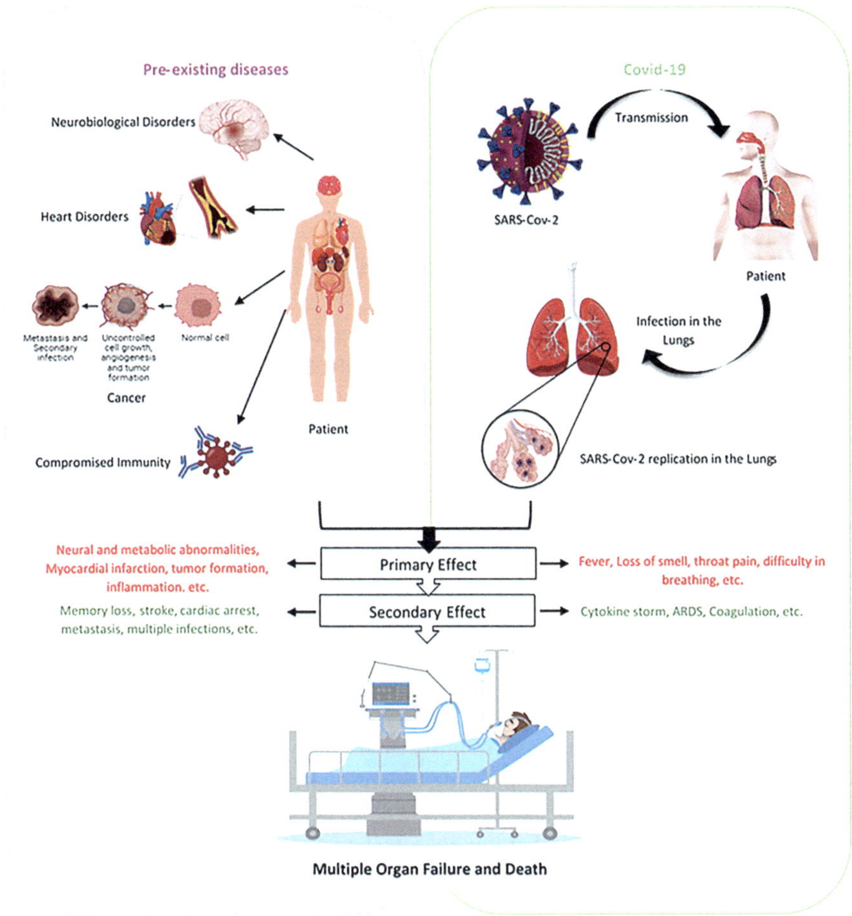

Fig. 19.1 Patients with pre-existing medical illnesses are more prone to COVID-19 infections

liver disease, such as alcohol-related liver disease), immune-compromised individuals (HIV/AIDS, cancer, immunosuppressive medication users undergoing transplantation, and those with genetic illnesses impacting the immune system, such as congenital IgA deficiency, congenital agammaglobulinemia etc.), diabetes, stroke, and obesity are at higher risk of acquiring SARS-CoV-2 infection and therefore can safely vaccinate themselves. More conditions may be added as evidence emerges.

However, reactivation of the shingles-causing virus may be linked to COVID-19 vaccinations, according to certain studies [10]. The term "vaccine-acquired immunodeficiency syndrome" is occasionally used to describe this illness. In addition to COVID-19, the Department of Cardiovascular Surgery at Okamura Memorial Hospital in Shizuoka, Japan, has seen cases of difficult-to-control infections since December 2021. For instance, following open heart surgery, there have been several examples of suspected infections resulting from inflammation, which have proven uncontrollable even after taking multiple medicines for several weeks. There were a few deaths and indications of immunocompromised patients. An increase in infection risk is possible. Hence, future revisions to various medical methods for assessing postoperative prognosis may be necessary [45]. Overall, the effect of COVID-19 vaccination in patients with pre-existing illnesses is debatable and has been discussed below.

COVID-19 Vaccination and Pre-existing Immunity

The European Medicines Agency recommends against giving repeated COVID-19 booster doses since they may not be feasible and negatively impact the immune response. The reduction in immunity stems from multiple sources. First, in the genetic coding, uracil is replaced with N1-methyl pseudouridine. Reduced cellular immunity could arise from the altered protein's ability to activate regulatory T cells [22]. Because of this, the spike proteins do not break down immediately after mRNA vaccinations are administered. For over 4 months, the spike proteins found on exosomes circulate throughout the body. Furthermore, in vivo investigations have demonstrated that mRNA-lipid nanoparticles (LNPs) are extremely inflammatory and concentrate in the liver, spleen, adrenal glands, and ovaries [6, 33]. Spike proteins in the bloodstream harm vascular endothelial cells, which may compromise the immune system organs including the adrenal gland. Newly produced antibodies against the spike protein injure cells and tissues that are ready to produce spike proteins. Furthermore, there is a possibility of antibody-dependent enhancement, in which the protective impact of neutralizing antibodies against infection is diminished by infection-enhancing antibodies. The Wuhan-type vaccine's initial antigenic sin, or the vaccine's lingering immunological memory, may make it ineffective against mutant strains. The COVID-19 flare-up could be caused by these methods [24, 25, 45].

In light of the benefits and drawbacks of mRNA vaccines, vaccination has been widely advised thus far. As the COVID-19 pandemic is better contained, the effects of the vaccine should become more obvious. It has been proposed that the spike proteins in genetic vaccinations may lead to a rise in cardiac problems, particularly acute coronary syndromes [18, 28]. In addition to the risk of infections due to compromised immune systems, there may also be a hidden risk of vaccine-induced organ damage, primarily to the circulatory system, that has not shown any obvious clinical symptoms. As a result, thorough risk assessments are crucial before invasive medical treatments like surgery in critically ill patients. Additionally, randomized controlled trials are required to validate these clinical findings (Fig. 19.2).

Thrombocytopenia Patients

As demonstrated in the United Kingdom and Israel, vaccination is the most effective way to significantly reduce severe illness and death after SARS-CoV-2

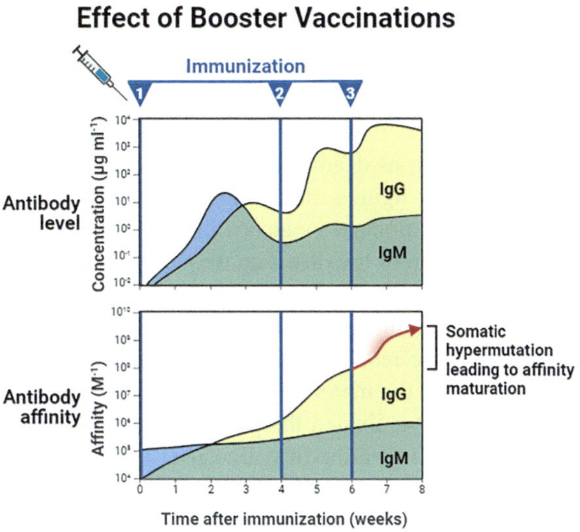

Fig. 19.2 Impact of booster doses on antibody level and affinity

infection. However, there have been some grave concerns raised about an uncommon adverse drug reaction (ADR), such as vaccine-induced immune thrombotic thrombocytopenia (VITT), which is accompanied by low platelets and capillary leak syndrome. In fact, given that 40 million first doses and 29 million second doses had been administered in the United Kingdom by early June, the vaccine's general safety is demonstrated by the low frequency of ADRs. Despite this, 390 thrombotic events—including 71 fatal events—had been documented. It is interesting to note that low platelet numbers and anti-platelet factor-4 (PF4) antibodies were present in the patients, suggesting an aberrant clotting reaction. With postmortem inspection and whole exome sequencing (WES) analysis, a post-vaccine clinical case of fatal thrombosis was studied, whose etiology appeared connected with a pre-existing disease of thrombocytopenia due to myelodysplasia [32].

Cancer Patients

After receiving at least two doses of the vaccine, mass vaccination programs have significantly decreased the incidence of severe COVID-19 in the general population in many parts of the world. Most countries emphasize vaccination of cancer patients due to the increased risk of acquiring severe COVID-19 [17]. These patients, however, were also not included in the crucial clinical studies, raising significant concerns about the safety and effectiveness of the vaccinations that are currently on the market as well as the longevity of vaccine responses in this population. Due to immune system degradation brought on by both disease and therapy, these patients have a higher chance of having a less effective immunological response following vaccination.

The detrimental impact of cancer itself on the risk of severe COVID-19 is especially evident in younger cancer patients (<65 years of age) [34, 42]. Patients with COVID-19 and active cancer had extraordinarily high mortality rates during the first wave; these rates were typically reported to be about 40%. In the subsequent waves that occurred in European countries in 2021, mortality rates decreased to about 25%. However, as death rates are typically provided from cohorts of hospitalized patients, these figures may exaggerate the number of cancer patients. In cancer patients, the estimated incidence of long-term COVID-19 sequelae is 15–30% [3, 38]. In addition to the direct consequences of the pandemic, other factors contributed to an increase in cancer-specific mortality, such as the requirement for frequent treatment adjustments and decreased screening.

While some research has also conducted neutralizing assays, including those against VOCs, the majority of studies examining the COVID-19 vaccine in cancer patients solely evaluated the presence of spike-reactive or RBD (receptor-binding domain)-reactive antibodies. Because T-cell responses have been the subject of fewer studies and frequently smaller subsets, more investigation is required to confirm the effects that have been noted. Generally speaking, cancer patients appear to be more prone than non-cancer patients to experience a weakened immune response after receiving the COVID-19 vaccination. Before the Omicron variant became dominant, the VE (vaccine effectiveness) of mRNA vaccines against COVID-19 hospitalization in cancer patients was estimated to be about 75%, which is lower as compared to 90% in immunocompetent individuals, according to a real-world study involving over 89,000 participants, more than 10,000 of whom had cancer [11]. In this investigation, VE decreased even more as age increased. VE was over 60% in a big US Veterans trial that solely included cancer patients who got mRNA vaccinations [44]. Notably, despite a high incidence of the Delta variation (VE pre-Delta 76%, during Delta 79%), no additional decline in VE was noted [11]. The timing of cancer therapy had an impact on VE, which varied from 54% in patients receiving any conventional therapy to 85% in patients who had not had treatment in the previous 6 months. Sadly, estimates of these effects that were specific to solid and hematological cancers were not given. A prospective cohort study with nearly seven million vaccinated participants in the United Kingdom found that receiving moderate-to-high intensity chemotherapy (HR (hazard ratio) 3.63–4.3), having a stem cell transplant within the previous 6 months (HR 2.5), having hematological cancer (HR 1.86), and having a respiratory tract cancer (HR 1.35), among others, were risk factors for COVID-19-related death even after vaccination with two doses [19]. Data from patients up until late spring of 2021 were included in two of these trials [19, 44]. Since fading immunity and nearly no patients infected with the Delta VOC are factors in the general population's declining VE, these data are not well suited to examine their consequences. In cancer patients, neutralizing reactions to VOCs likewise gradually decline. This finding is consistent with observations made by people without cancer, even though malignancy and VOCs can both significantly lower neutralizing responses, which can lead to significantly lower VE. The patients most likely to be impacted by this consequence are those with hematological malignancies. After two doses of the vaccination, for instance, only 31% of these patients exhibited detectable titers with activity against the Delta variant, but 56% of same patients exhibited detectable titers with neutralizing action against the ancestral Wuhan strain. Importantly, after receiving a booster shot, the proportion of individuals exhibiting measurable neutralizing reactions to VOCs increases. Initial data on Omicron neutralization in cancer patients support the conclusions gained from the general population: after a third vaccination dose, the percentage of patients with solid tumors who had neutralizing responses against Omicron rose from 47.8% to 88.9%. Specifically, following two doses of mRNA vaccine, patients with non-small-cell lung cancer exhibit a 79-fold reduced neutralizing response to Omicron in comparison to healthy individuals. Neutralizing antibodies against Omicron are rarely seen in individuals with hematological malignancies after two vaccination doses; however, following a third treatment, almost 50% of patients have detectable neutralizing antibodies [13].

Risk Factors Affecting Vaccine Response in Patients with Solid Tumors

After receiving two vaccination doses, the vast majority of patients (90–100%) with solid tumors seroconvert (Table 19.1). Immune responses elicited by vaccines are compromised by several cancer therapies. Recent chemotherapy variably defined as receiving chemotherapy from within 28 days to within 6 months of vaccination has been repeatedly found to be a risk factor for lower seroconversion and neutralizing responses, though not in all studies, which is under the reported decrease in VE in this population [15, 16, 26, 27, 44]. Significantly, seroconversion does not appear to be impacted by the vaccination's timing with the schedule of ongoing treatment. To reduce the possibility of overlapping acute side effects, several clinics refrain from

Table 19.1 Seroconversion rates in patients with cancer following two doses of COVID-19 vaccine BNT162b2 (Pfizer BioNtech) [Test used: Anti-S IgG] [7]

			Seroconversion rate after two vaccine doses			
Study	Location	Malignancy	Solid cancer %	Blood cancer %	Healthy control %	*p*-value
Addeo et al	USA and Europe	Solid and hematologic	98%	77%	–	0.002
Monin et al	UK	Solid and hematologic	95%	60%	100%	–
Ehmsen et al	Denmark	Solid and hematologic	93%	66%	–	0.004
Massarweh et al	Israel	Solid malignancies	90%	–	100%	–
Palich et al	France	Solid malignancies	94%	–	100%	–
Pimpinelli et al	Italy	Multiple myeloma and myeloproliferative	–	MM: 78.6% MPM: 88%	100%	MM: 0.003 MPM: 0.038

Abbreviations—*MM* multiple myeloma, *MPM* myeloproliferative malignancies

giving chemotherapy and vaccinations on the same day, but they do not reschedule cancer treatments.

Risk Factors Affecting Vaccine Response in Patients with Hematological Malignancies

In addition to having a lower VE (VE 74% vs. 90% in non-immunocompromised individuals), patients with hematological malignancies are also more likely to acquire diminished immunological responses to the COVID-19 vaccination, according to a large-cohort research conducted in the United States [11]. It is encouraging to note that, despite prior reception of highly immunosuppressive medication, long-term survivors of hematological malignancies, including those who have received stem cell transplants, respond to vaccination similarly to the general population [31, 44]. In terms of vaccine type, mRNA vaccinations appear to produce stronger immunological responses in this population than vaccines vectored by adenoviruses [40]; this appears to be especially true for mRNA-1273 [11, 40]. To date, the majority of research has solely examined the immunological response mediated by antibodies (Table 19.1). A decreased humoral immune response has been linked to several individual risk factors, such as old age, ongoing cancer, and/or lymphoproliferative diseases. There is a continuous link between the use of certain anticancer medicines and a significantly decreased humoral immune response to vaccination.

In conclusion, COVID-19 vaccination-related safety in cancer patients is similar to that of the general population, including the frequency of serious side effects like vaccine-induced immune thrombotic thrombocytopenia. Certain patient populations have reported specific, well-defined toxicity profiles. Overall, for all cancer patients, the advantages of receiving a COVID-19 vaccination exceed the risks, much like they do for the general population.

Neurological Diseases

Patients with specific underlying illnesses, such as neurological problems, are more likely to suffer severe sickness if they contract the COVID-19 virus. More significantly, patients with pre-existing neurological problems have higher mortality rates because they are more vulnerable to potentially fatal consequences such as pneumonia and acute respiratory distress syndrome. Consequently, to reduce the severity of the condition and minimize the risk of infection, COVID-19 immunization is strongly advised for these individuals. It is nonetheless crucial for individuals with neurological problems to confirm their current health status with their neurologists, even though the majority of them qualify for the COVID-19 vaccination. If symptoms are exacerbating or unstable, vaccination should be postponed until the conditions have fully improved. These practical suggestions have been developed based on the most recent data as of 2021. However, as more medical research is to be conducted, standard recommendations may be further obtained.

Cardiac Patients

Lately, there has been a growing concern about the potential correlation between COVID-19 vaccination and adverse cardiac events [20]. The type of vaccine, residual confounding, and demographic characteristics all affect the prevalence of heart illness following vaccination [2, 12, 20]. A total of 349 patients were identified for the current investigation from 150 published case reports and series [21]. Myocarditis was the most frequent adverse event, followed by ischemic heart disease, myopericarditis, and arrhythmia. Furthermore, the majority of instances and adverse cardiac events were documented following the second dosage of the Pfizer and Moderna vaccinations. When vaccination usage was broken down nationwide, Pfizer was the most often used vaccine in the United States, having been given in 57.7% of instances involving adverse cardiovascular events. Pfizer was the most widely used vaccine in Europe, accounting for 58.3% of recorded side effects in Germany. It was found that restoring hemodynamic stability with the help of heart failure and arrhythmia treatment is the major goal of supportive care for myocarditis and other cardiovascular events. NSAIDs, ACEI, ARBs, and PCI were used to treat the majority of patients who needed treatment. Almost all of these patients (96.9%) showed a noticeable improvement following treatment.

One possible cause of myocarditis is molecular mimicry, which is the interaction between vaccine components and certain human proteins that can result in immunological cross-reactivity and autoimmune illness. The greater incidence of myocarditis in the male population compared to the female population may potentially be explained by variations in hormone signaling. Testosterone can suppress immune cells that reduce inflammation while enhancing the activation of T-helper cells. Therefore, even though it is unclear if the cardiovascular side effects that follow the COVID-19 vaccination are related to or caused by the shot, it is certain that they exist and that any treating physician should be aware of them.

In general, sex and age have a significant impact on the incidence of cardiovascular illness following vaccination. The younger population has a higher incidence of post-vaccination cardiovascular events, particularly in cases of myopericarditis and myocarditis, where the mean patient age is 25 and 30, respectively. In a research on the occurrence of myocarditis following COVID-19 vaccination, Ahmed et al. found a trend of male-dominant predominance [1]. Ling et al. and Chou et al. have reported that young males who receive the mRNA COVID-19 vaccine have a higher incidence of myopericarditis than other recipients [8, 29]. Furthermore, the average interval between vaccination and symptom start is 94.6 ± 128 h [21]. This is in line with other research by Pillay et al., Oster et al., and Park et al., which found that myocarditis symptoms started to appear 2–4 days after vaccination [35–37].

A study presented a case series of teenage COVID-19 vaccination-induced myocarditis in which all cases had aberrant repolarization times and ST segment alterations, two hallmarks of acute myocardial damage [39]. It is currently unknown how frequent MI (myocardial infarction) is among younger adults, although they may have additional comorbidities that make them more susceptible to the disease. According to earlier research, the increased incidence may be caused by subclinical coronary disease or arterial spasm that is made worse by post-vaccination arthritis or vasculitis. Young adults may experience arterial spasms for a variety of reasons, including cocaine use, alcoholism, or an adrenergic reaction brought on by stress. They are the source of acute thrombosis, plaque erosion, and spasms [4, 5].

In summary, following the second dosage of the COVID-19 vaccinations, the majority of the adverse cardiovascular events were documented. After vaccinations, myocarditis, myopericarditis, arrhythmia, and ischemic heart disease were the most frequent cardiovascular side effects. Young patients, particularly those who are male, may be more susceptible to myocarditis. Clinical and laboratory results showed that most patients had significantly improved with time. Verification of the potential link between COVID-19 vaccinations and cardiac incidents requires more investigation.

Dialysis Patients

The majority of dialysis patients (DP) were able to elicit an immunological response following their second dose (2D) of the BNT162b2 vaccination, according to the studies. Thirty days after 2D, 94.3% of patients tested positive for SARS-CoV-2 anti-S IgG. Furthermore, they showed that adding a third dose (3D) boosted the seropositivity to 98%. Furthermore, 2 weeks following 2D, 91.8% of DP and 29.6% of kidney transplant recipients developed seropositive results, according to a different study. Additionally, 12 weeks after receiving two doses of the vaccine, it was observed that 95% of patients on peritoneal dialysis, 88% of patients receiving hemodialysis, and 100% of healthy controls exhibited detectable anti-S IgG. They, however, did go into depth about how the anti-S level in hemodialysis patients was much lower than in peritoneal dialysis patients or healthy controls, with median values of 7.0, 21.8, and 134.9 AU/mL, respectively.

In DP, the antibody response to the SARS-CoV-2 vaccination decreased quickly. Ten weeks following the 2D of BNT162b2, it was observed that the percentage of DP with detectable IgG antibodies dropped from 88.9% to 84.37%. Similarly, just 4 months following the BNT162b2 vaccine's second dose, a decrease in anti-S levels was observed. It was noted that seronegativity increased from 9% to >30% and that IgG levels fell even further about 5 months after 2D. Further, it was observed that an undetectable IgG response rose from 6.6% (range 14–30 days) to 20.2% at 5–6 months after vaccination in individuals vaccinated with mRNA-1273, BNT162b2 e Ad26. COV2. Additionally, it was reported that during the 12 weeks following 2D, patients receiving hemodialysis saw a substantial drop in Nabs from 87% to 79%, while those receiving peritoneal dialysis saw a significant decrease from 93% to 85%. Therefore, the quick degradation of antibodies in DP may be the reason why this group has more breakthrough infections than healthy cohorts.

There are known factors that can contribute to a high or poor response in DP to the COVID-19 vaccination. It was shown that in DP vaccinated with BNT162b2, the absence of seroconversion was correlated with the immunocompromised status. They also found that a low response to the vaccine (IgG <500 AU/mL) was associated with low lymphocyte count, age, nutritional condition, and chronic inflammation. Consequently, reduced antibody response was reported following 1D of BTN162b2 in patients <70 years old receiving immunosuppressive treatments as opposed to controls. Furthermore, anti-S IgG concentration to BNT162b2 was found to be negatively impacted by age. However, a significant response (IgG >7000 AU/mL) after immunization was linked to prior SARS-COV-2 infection. When taken as a whole, these findings highlight the significance of carrying out long-term research on this DP following COVID-19 vaccination [14].

Transplant Patients

Vaccination with mRNA COVID-19 vaccines has been associated with a decreased humoral response to SARS-CoV-2 in recipients of solid organs. In their study, Dębska-ślizień et al. [9] examined the antibody response in individuals receiving kidney transplants who were vaccinated with BNT162b2 or mRNA-1273 in comparison to healthy volunteers, both of whom had no history of COVID-19 infection [9]. Following the second dose, seroconversion was observed in 100% of healthy participants and 51.41% of transplant patients. Furthermore, kidney transplant recipients' median antibody titers were eight times lower than those of healthy volunteers. Similarly, IgG anti-S titers were used to assess the immunogenicity of the BNT162b2 vaccination in liver transplant recipients and healthy controls. The seroconversion rate was 75% in transplanted patients and 97.72% in the control group after 3 months of the second dose. Furthermore, compared to the control group, liver transplant recipients had statistically lower titers.

Transplant patients who responded poorly or negatively to the mRNA COVID-19 vaccination had several contributing factors. According to Debska-Ślizien et al. [9], patients who received kidney transplants and were younger (less than 54 years old), had been transplanted for the longest (more than 8 years), had not received corticosteroids or mycophenolate mofetil/Na (MMF/MPS) treatment, and had received the mRNA-1273 vaccine had a higher seroconversion rate. Similarly, non-response to vaccination was substantially correlated with older age (>40 years), a shorter time after liver transplantation (<5 years), and immunosuppressive regimens involving numerous medicines or antimetabolites. Additionally, it was demonstrated that with vaccination within a year of transplantation, low humoral detection was associated with the onset of graft-versus-host disease (GVHD), and lymphopenia (<1 × 109/mL) in recipients of allogeneic hematopoietic cell transplantation (allo-HCT). In a different study, lymphopenia was also noted as a risk factor for a poor response to the BNT162b2 vaccination in doses two and three for the same individuals. Moreover, individuals with non-Hodgkin's lymphoma and those receiving corticosteroid therapy had low antibody detection rates among recipients of autologous hematopoietic stem cells. The BNT162b2 or mRNA-1273-vaccinated patients whose transplants took place 3–6 months and 6–12 months before vaccination, respectively, generated noticeably lower antibody titers than the healthy control group and patients who received the vaccination 12 months after receiving the transplant. In a stratified study, low humoral immune responses to the vaccination were also observed in allogeneic recipients who were older than 65, on immunosuppression for the prevention or treatment of acute GVHD, and in patients who had relapsed. On the other hand, antibody levels were unaffected by the conditioning regimen's strength, the underlying illness (myeloid, lymphoid, or other leukemia), or the existence of ongoing GVHD treatment. All of these studies combined show that the primary determinants of the immunological response to COVID-19 vaccination in transplant recipients are age, the time elapsed between the vaccine and transplantation, and the use of corticoids. As was previously mentioned, innate and adaptive immunity wanes with age. The increased degree of immunosuppression that occurs right after transplantation and gradually lessens to allow for the reconstitution of immune capacity can be used to explain the negative link between the shortest time after transplantation and non-response to vaccination as corticoids weaken the immune system.

Following a spontaneous SARS-CoV-2 infection, COVID-19 mRNA vaccination provided transplant recipients with a more potent immune response stimulant. After administration of BNT162b2 or mRNA-1273, Tylicki et al. [41] reported a seroconversion rate of 100% in vaccinated kidney transplant recipients who had previously been infected with SARS-CoV-2, compared to 45.78% in recipients who had never been infected. They did, however, report that the previously infected group's antibody degradation after second dose was quicker than that of the non-infected group. Compared to recipients of BNT162b2, transplant recipients who received mRNA-1273 had higher antibody titers. When comparing the mRNA-1273 vaccination to BNT162b2, Debska-Ŗlizien et al. [9] found that both the antibody titer's magnitude and seroconversion rates were significantly higher following second dose. Tylicki et al. [41] reported similar results: following the third dose, 82.6% of patients vaccinated with mRNA-1273 and 66% of patients vaccinated with BNT162b2 seroconverted with a median anti-S titer of 1620 BAU/mL and 384.5 BAU/mL, respectively. As previously shown, the threefold increase in mRNA dosage could be a contributing factor to the increased mRNA-1273 vaccination immunogenicity. Because of this, vaccinations containing a larger dose of mRNA may be a better option for immunocompromised people [41].

Transplant recipients' immune responses to SARS-CoV-2 were enhanced upon the administration of the third dosage. The immunogenicity of kidney transplant recipients who got three doses of the BNT162b2 vaccination was assessed by Massa et al. and Tylicki et al. in 2021 [30, 41]. The seroconversion rate rose from 44.3% after second dose to 62.3% after third dose, according to Massa et al. [30]. Additionally, after the third dose, there was an improved serum-neutralizing activity and an approximately four- to fivefold increase in antibody levels. Similarly, after the initial immunization, less than 50% of transplant patients displayed any antibody-mediated response, according to Tylicki et al. [41]. Nonetheless, nearly half of the patients in whom the initial vaccination failed achieved seroconversion with the third dose. Additionally, following the second dose, antibody titers were 30% higher. It is noteworthy to emphasize that even after the third dose, 37.7% of patients in the Massa et al. [30] trial and 21.42% in the Tylicki et al. [41] study remained seronegative. This result demonstrates the necessity for different approaches to enhance this population's COVID-19 vaccination.

Conclusion

The effective containment of the pandemic is seriously threatened by the widespread phenomenon known as vaccine reluctance. The acceptance rate of vaccination varies greatly amongst nations, ranging from more than 90% to less than 50%. Although it appears to affect less than 10% of patients, vaccine reluctance is likely lower in cancer patients. Excessive vaccination hesitancy may also reduce community immunity, which would lessen the amount of protection for cancer and heart patients who are unable to generate a sufficient immune response on their own. Concerns related to vaccination safety, erroneous perceptions of the disease's risks, and/or steadfast confidence in false information are common causes of vaccine hesitation. Crucially, several examples of false information have been explicitly refuted, including the idea that it could raise the rate of miscarriages, although it had no supporting evidence when it was first proposed. There is no proof, nor any kind of reasoning, to support the theory that vaccination against COVID-19 can cause recurrence in cancer patients. Hence, to date, medical professionals have favored vaccination in all populations including medically distinct populations.

Author's Contributions Conceived the idea and review structure and wrote the manuscript: RS & MKG; Revised and edited the manuscript: MKG. The authors have approved the final draft of the manuscript.

Funding This work is jointly supported by the Department of Science and Technology (NanoMission: DST/NM/NT/2018/105(G); SERB: EMR/2017/000992) and Focused Basic Research (FBR), HCT, CSIR, Govt. of India.

Data Availability The software BioRender (https://www.biorender.com/) was used to draw some elements of the figure.

Declarations

Conflicts of Interests There is no competing and conflict of interest.

References

1. Ahmed SK, Mohamed MG, Essa RA, Ahmed Rashad EA, Ibrahim PK, Khdir AA, Wsu ZH. Global reports of myocarditis following COVID-19 vaccination: a systematic review and meta-analysis. Diabetes Metab Syndr Clin Res Rev. 2022;16(6):102513. https://doi.org/10.1016/j.dsx.2022.102513.
2. Al-Ali D, Elshafeey A, Mushannen M, Kawas H, Shafiq A, Mhaimeed N, Mhaimeed O, Mhaimeed N, Zeghlache R, Salameh M, Paul P, Homssi M, Mohammed I, Narangoli A, Yagan L, Khanjar B, Laws S, Elshazly MB, Zakaria D. Cardiovascular and haematological events post COVID-19 vaccination: a systematic review. J Cell Mol Med. 2022;26(3):636–53. https://doi.org/10.1111/jcmm.17137.
3. Barbui T, Iurlo A, Masciulli A, Carobbio A, Ghirardi A, Rossi G, Harrison C, Alvarez-Larran A, Elli EM, Kiladjian J-J, Gasior Kabat M, Marin Sanchez A, Palandri F, Andrade-Campos

MM, Vannucchi AM, Carreno-Tarragona G, Papadopoulos P, Quiroz Cervantes K, Angeles Foncillas M, et al. Long-term follow-up of recovered MPN patients with COVID-19. Blood Cancer J. 2021;11(6):115. https://doi.org/10.1038/s41408-021-00509-0.

4. Baronti A, Gentile F, Manetti AC, Scatena A, Pellegrini S, Pucci A, Franzini M, Castiglione V, Maiese A, Giannoni A, Pistello M, Emdin M, Aquaro GD, Di Paolo M. Myocardial infarction following COVID-19 vaccine administration: post hoc, ergo propter hoc? Viruses. 2022;14(8):1644. https://doi.org/10.3390/v14081644.
5. Bhardwaj R, Kandoria A, Sharma R. Myocardial infarction in young adults-risk factors and pattern of coronary artery involvement. Niger Med J. 2014;55(1):44. https://doi.org/10.4103/0300-1652.128161.
6. BNT162b2 Module 2.4. Nonclinical overview. FDA-CBER-2021-4379-0000681 JW-v-HHS-prod-3-02418.pdf (judicialwatch.org). n.d.
7. Chehade L, Zeitoun J, Lombe D, Irshad S, Van Hemelrijck M, Canfell K, Sullivan R, Mukherji D. COVID-19 vaccination in patients with cancer, a rapid review. Ecancermedicalscience. 2022;16. https://doi.org/10.3332/ecancer.2022.1355.
8. Chou OHI, Mui J, Chung CT, Radford D, Ranjithkumar S, Evbayekha E, Nam R, Pay L, Satti DI, Garcia-Zamora S, Bazoukis G, Çinier G, Lee S, Vassiliou VS, Liu T, Tse G, Wong ICK, Tse G, et al. COVID-19 vaccination and carditis in children and adolescents: a systematic review and meta-analysis. Clin Res Cardiol. 2022;111(10):1161–73. https://doi.org/10.1007/s00392-022-02070-7.
9. Dębska-Ślizień A, Ślizień Z, Muchlado M, Kubanek A, Piotrowska M, Dąbrowska M, Tarasewicz A, Chamienia A, Biedunkiewicz B, Renke M, Tylicki L. Predictors of humoral response to mRNA COVID19 vaccines in kidney transplant recipients: a longitudinal study—the COViNEPH project. Vaccine. 2021;9(10):1165. https://doi.org/10.3390/vaccines9101165.
10. Desai HD, Sharma K, Shah A, Patoliya J, Patil A, Hooshanginezhad Z, Grabbe S, Goldust M. Can SARS-CoV-2 vaccine increase the risk of reactivation of Varicella zoster? A systematic review. J Cosmet Dermatol. 2021;20(11):3350–61. https://doi.org/10.1111/jocd.14521.
11. Embi PJ, Levy ME, Naleway AL, Patel P, Gaglani M, Natarajan K, Dascomb K, Ong TC, Klein NP, Liao I-C, Grannis SJ, Han J, Stenehjem E, Dunne MM, Lewis N, Irving SA, Rao S, McEvoy C, Bozio CH, et al. Effectiveness of 2-dose vaccination with mRNA COVID-19 vaccines against COVID-19–associated hospitalizations among immunocompromised adults—nine states, January–September 2021. MMWR Morb Mortal Wkly Rep. 2021;70(44):1553–9. https://doi.org/10.15585/mmwr.mm7044e3.
12. Fazlollahi A, Zahmatyar M, Noori M, Nejadghaderi SA, Sullman MJM, Shekarriz-Foumani R, Kolahi A, Singh K, Safiri S. Cardiac complications following mRNA COVID-19 vaccines: a systematic review of case reports and case series. Rev Med Virol. 2022;32(4):e2318. https://doi.org/10.1002/rmv.2318.
13. Fendler A, De Vries EGE, GeurtsvanKessel CH, Haanen JB, Wörmann B, Turajlic S, Von Lilienfeld-Toal M. COVID-19 vaccines in patients with cancer: immunogenicity, efficacy and safety. Nat Rev Clin Oncol. 2022;19(6):385–401. https://doi.org/10.1038/s41571-022-00610-8.
14. Fernandes MDCR, Vasconcelos GS, De Melo ACL, Matsui TC, Caetano LF, De Carvalho Araújo FM, Fonseca MHG. Influence of age, gender, previous SARS-CoV-2 infection, and pre-existing diseases in antibody response after COVID-19 vaccination: a review. Mol Immunol. 2023;156:148–55. https://doi.org/10.1016/j.molimm.2023.03.007.
15. Ghosh MK, Kumar S, Ganguly KK, Ghosh P, Tabassum S, Basu B, Basu M. COVID-19 and cancer: insights into their association and influence on genetic and epigenetic landscape. Epigenomics. 2023;15(4):227–48. https://doi.org/10.2217/epi-2023-0052.
16. Ghosh MK, Kumar S, Begam S, Ghosh S, Basu M. GBM immunotherapy: exploring molecular and clinical frontiers. Life Sci. 2024:123018. https://doi.org/10.1016/j.lfs.2024.123018.
17. Giesen N, Sprute R, Rüthrich M, Khodamoradi Y, Mellinghoff SC, Beutel G, Lueck C, Koldehoff M, Hentrich M, Sandherr M, Von Bergwelt-Baildon M, Wolf H-H, Hirsch HH, Wörmann B, Cornely OA, Köhler P, Schalk E, Von Lilienfeld-Toal M. 2021 update of the AGIHO guideline on evidence-based management of COVID-19 in patients with cancer regarding diagnostics, viral shedding, vaccination and therapy. Eur J Cancer. 2021;147:154–60. https://doi.org/10.1016/j.ejca.2021.01.033.

18. Gundry SR. Abstract 10712: observational findings of PULS cardiac test findings for inflammatory markers in patients receiving mRNA vaccines. Circulation. 2021;144(Suppl_1). https://doi.org/10.1161/circ.144.suppl_1.10712.
19. Hippisley-Cox J, Coupland CA, Mehta N, Keogh RH, Diaz-Ordaz K, Khunti K, Lyons RA, Kee F, Sheikh A, Rahman S, Valabhji J, Harrison EM, Sellen P, Haq N, Semple MG, Johnson PWM, Hayward A, Nguyen-Van-Tam JS. Risk prediction of covid-19 related death and hospital admission in adults after covid-19 vaccination: national prospective cohort study. BMJ. 2021;374:n2244. https://doi.org/10.1136/bmj.n2244.
20. Ho JSY, Sia C-H, Ngiam JN, Loh PH, Chew NWS, Kong WK-F, Poh K-K. A review of COVID-19 vaccination and the reported cardiac manifestations. Singapore Med J. 2023;64(9):543–9. https://doi.org/10.11622/smedj.2021125.
21. Khaity A, Rababah AAM, Abdelwahab OA, Albakri K, Diab RA, Al-Dardery NM, Abbassy M, Al-Hanaqtah BM, Awad AK, Mohamad T. Cardiovascular disease and COVID-19 vaccines: a systematic review and analysis of published cases. European Cardiology. 2023;18:e54. https://doi.org/10.15420/ecr.2023.01.
22. Krienke C, Kolb L, Diken E, Streuber M, Kirchhoff S, Bukur T, Akilli-Öztürk Ö, Kranz LM, Berger H, Petschenka J, Diken M, Kreiter S, Yogev N, Waisman A, Karikó K, Türeci Ö, Sahin U. A noninflammatory mRNA vaccine for treatment of experimental autoimmune encephalomyelitis. Science. 2021;371(6525):145–53. https://doi.org/10.1126/science.aay3638.
23. Kumar A, Singh R, Kaur J, Pandey S, Sharma V, Thakur L, Sati S, Mani S, Asthana S, Sharma TK, Chaudhuri S, Bhattacharyya S, Kumar N. Wuhan to world: the COVID-19 pandemic. Front Cell Infect Microbiol. 2021;11:596201. https://doi.org/10.3389/fcimb.2021.596201.
24. Kumar S, Basu M, Ghosh P, Ansari A, Ghosh MK. COVID-19: clinical status of vaccine development to date. Br J Clin Pharmacol. 2023;89(1):114–49. https://doi.org/10.1111/bcp.15552.
25. Kumar S, Basu M, Ghosh P, Pal U, Ghosh MK. COVID-19 therapeutics: clinical application of repurposed drugs and futuristic strategies for target-based drug discovery. Genes Dis. 2023;10(4):1402–28. https://doi.org/10.1016/j.gendis.2022.12.019.
26. Kumar S, Chatterjee M, Ghosh P, Ganguly KK, Basu M, Ghosh MK. Targeting PD-1/PD-L1 in cancer immunotherapy: an effective strategy for treatment of triple-negative breast cancer (TNBC) patients. Genes Dis. 2023;10(4):1318–50.
27. Kumar S, Basu M, Ghosh MK. E3 ubiquitin ligases and deubiquitinases in colorectal cancer: emerging molecular insights and therapeutic opportunities. Biochim Biophys Acta Mol Cell Res. 2024;1871:119827.
28. Lai FTT, Li X, Peng K, Huang L, Ip P, Tong X, Chui CSL, Wan EYF, Wong CKH, Chan EWY, Siu DCW, Wong ICK. Carditis after COVID-19 vaccination with a messenger RNA vaccine and an inactivated virus vaccine: a case–control study. Ann Intern Med. 2022;175(3):362–70. https://doi.org/10.7326/M21-3700.
29. Ling RR, Ramanathan K, Tan FL, Tai BC, Somani J, Fisher D, MacLaren G. Myopericarditis following COVID-19 vaccination and non-COVID-19 vaccination: a systematic review and meta-analysis. Lancet Respir Med. 2022;10(7):679–88. https://doi.org/10.1016/S2213-2600(22)00059-5.
30. Massa F, Cremoni M, Gérard A, Grabsi H, Rogier L, Blois M, Couzin C, Hassen NB, Rouleau M, Barbosa S, Martinuzzi E, Fayada J, Bernard G, Favre G, Hofman P, Esnault VLM, Czerkinsky C, Seitz-Polski B, Glaichenhaus N, Sicard A. Safety and cross-variant immunogenicity of a three-dose COVID-19 mRNA vaccine regimen in kidney transplant recipients. EBioMedicine. 2021;73:103679. https://doi.org/10.1016/j.ebiom.2021.103679.
31. Matkowska-Kocjan A, Owoc-Lempach J, Chruszcz J, Kuźnik E, Szenborn F, Jurczenko L, Wójcik M, Banyś D, Szenborn L, Ussowicz M. The COVID-19 mRNA BNT163b2 vaccine was well tolerated and highly immunogenic in young adults in long follow-up after haematopoietic stem cell transplantation. Vaccine. 2021;9(10):1209. https://doi.org/10.3390/vaccines9101209.
32. Mauriello A, Scimeca M, Amelio I, Massoud R, Novelli A, Di Lorenzo F, Finocchiaro S, Cimino C, Telesca R, Chiocchi M, Sun Q, Wang Y, Shi Y, Novelli G, Melino G. Thromboembolism after COVID-19 vaccine in patients with preexisting thrombocytopenia. Cell Death Dis. 2021;12(8):762. https://doi.org/10.1038/s41419-021-04058-z.

33. Ndeupen S, Qin Z, Jacobsen S, Bouteau A, Estanbouli H, Igyártó BZ. The mRNA-LNP platform's lipid nanoparticle component used in preclinical vaccine studies is highly inflammatory. IScience. 2021;24(12):103479. https://doi.org/10.1016/j.isci.2021.103479.
34. On Behalf of the LEOSS Study Group, Rüthrich MM, Giessen-Jung C, Borgmann S, Classen AY, Dolff S, Grüner B, Hanses F, Isberner N, Köhler P, Lanznaster J, Merle U, Nadalin S, Piepel C, Schneider J, Schons M, Strauss R, Tometten L, Vehreschild JJ, et al. COVID-19 in cancer patients: clinical characteristics and outcome—an analysis of the LEOSS registry. Ann Hematol. 2021;100(2):383–93. https://doi.org/10.1007/s00277-020-04328-4.
35. Oster ME, Shay DK, Su JR, Gee J, Creech CB, Broder KR, Edwards K, Soslow JH, Dendy JM, Schlaudecker E, Lang SM, Barnett ED, Ruberg FL, Smith MJ, Campbell MJ, Lopes RD, Sperling LS, Baumblatt JA, Thompson DL, et al. Myocarditis cases reported after mRNA-based COVID-19 vaccination in the US from December 2020 to August 2021. JAMA. 2022;327(4):331. https://doi.org/10.1001/jama.2021.24110.
36. Park DY, An S, Kaur A, Malhotra S, Vij A. Myocarditis after COVID-19 mRNA vaccination: a systematic review of case reports and case series. Clin Cardiol. 2022;45(7):691–700. https://doi.org/10.1002/clc.23828.
37. Pillay J, Gaudet L, Wingert A, Bialy L, Mackie AS, Paterson DI, Hartling L. Incidence, risk factors, natural history, and hypothesised mechanisms of myocarditis and pericarditis following covid-19 vaccination: living evidence syntheses and review. BMJ. 2022:e069445. https://doi.org/10.1136/bmj-2021-069445.
38. Pinato DJ, Tabernero J, Bower M, Scotti L, Patel M, Colomba E, Dolly S, Loizidou A, Chester J, Mukherjee U, Zambelli A, Dalla Pria A, Aguilar-Company J, Ottaviani D, Chowdhury A, Merry E, Salazar R, Bertuzzi A, Brunet J, et al. Prevalence and impact of COVID-19 sequelae on treatment and survival of patients with cancer who recovered from SARS-CoV-2 infection: evidence from the OnCovid retrospective, multicentre registry study. Lancet Oncol. 2021;22(12):1669–80. https://doi.org/10.1016/S1470-2045(21)00573-8.
39. Puchalski M, Kamińska H, Bartoszek M, Brzewski M, Werner B. COVID-19-vaccination-induced myocarditis in teenagers: case series with further follow-up. Int J Environ Res Public Health. 2022;19(6):3456. https://doi.org/10.3390/ijerph19063456.
40. Thakkar A, Gonzalez-Lugo JD, Goradia N, Gali R, Shapiro LC, Pradhan K, Rahman S, Kim SY, Ko B, Sica RA, Kornblum N, Bachier-Rodriguez L, McCort M, Goel S, Perez-Soler R, Packer S, Sparano J, Gartrell B, Makower D, et al. Seroconversion rates following COVID-19 vaccination among patients with cancer. Cancer Cell. 2021;39(8):1081–1090.e2. https://doi.org/10.1016/j.ccell.2021.06.002.
41. Tylicki L, Dębska-Ślizień A, Muchlado M, Ślizień Z, Gołębiewska J, Dąbrowska M, Biedunkiewicz B. Boosting humoral immunity from mRNA COVID-19 vaccines in kidney transplant recipients. Vaccine. 2021;10(1):56. https://doi.org/10.3390/vaccines10010056.
42. Venkatesulu BP, Chandrasekar VT, Girdhar P, Advani P, Sharma A, Elumalai T, Hsieh CE, Elghazawy HI, Verma V, Krishnan S. A systematic review and meta-analysis of cancer patients affected by a novel coronavirus. JNCI Cancer Spectr. 2021;5(2):pkaa102. https://doi.org/10.1093/jncics/pkaa102.
43. Vitiello A, Ferrara F, Troiano V, La Porta R. COVID-19 vaccines and decreased transmission of SARS-CoV-2. Inflammopharmacology. 2021;29(5):1357–60. https://doi.org/10.1007/s10787-021-00847-2.
44. Wu JT-Y, La J, Branch-Elliman W, Huhmann LB, Han SS, Parmigiani G, Tuck DP, Brophy MT, Do NV, Lin AY, Munshi NC, Fillmore NR. Association of COVID-19 vaccination with SARS-CoV-2 infection in patients with cancer: a US Nationwide Veterans Affairs Study. JAMA Oncol. 2022;8(2):281. https://doi.org/10.1001/jamaoncol.2021.5771.
45. Yamamoto K. Adverse effects of COVID-19 vaccines and measures to prevent them. Virol J. 2022;19(1):100. https://doi.org/10.1186/s12985-022-01831-0.
46. Yesudhas D, Srivastava A, Gromiha MM. COVID-19 outbreak: history, mechanism, transmission, structural studies and therapeutics. Infection. 2021;49(2):199–213. https://doi.org/10.1007/s15010-020-01516-2.

Part VI

COVID-19 Associated Comorbidities, Economic, and Societal Impact

Neurological Damages in COVID-19 Patients

20

Abdolreza Esmaeilzadeh, Armin Jahani Maleki, Amirhosein Moradi, Fereshteh Ebrahimi, Mahdis Nazari, Amir Siahmansouri, Kimia Esmaeilzadeh, Mahsa Bazargan, and Babak khademe Rahgosha

Abstract

The SARS-CoV-2 virus emerged in late 2019 and caused COVID-19 with more than 774 million confirmed cases and over seven million deaths. The main organ infected by the virus is the respiratory system, while studies have reported several involvements of the extrapulmonary organs including the neurological system. The neurological manifestations of COVID-19 range from headache, neuropathy, olfactory disturbances, dizziness, and ataxia to more severe disorders such as ischemic or hemorrhagic stroke, encephalitis, and Guillain-Barre syndrome. The neural system involvement by the SARS-CoV-2 is mediated by

A. Esmaeilzadeh (✉)
Department of Immunology, Zanjan University of Medical Sciences, Zanjan, Iran

Cancer Gene Therapy Research Center, Zanjan University of Medical Sciences, Zanjan, Iran
e-mail: a46reza@zums.ac.ir

A. J. Maleki · A. Moradi · M. Nazari · A. Siahmansouri
Student Research Committee, School of Medicine, Zanjan University of Medical Sciences, Zanjan, Iran

F. Ebrahimi
Department of Genetics and Molecular Medicine, Zanjan University of Medical Sciences, Zanjan, Iran

K. Esmaeilzadeh
Department of Medical Nanotechnology, School of Advanced Medicine Sciences, Tabriz University of Medical Sciences, Tabriz, Iran

M. Bazargan
Department of Immunology, School of Medicine, Shahid Beheshti University of Medical Sciences, Tehran, Iran

B. k. Rahgosha
Departments of Neurology, Zanjan University of Medical Sciences, Zanjan, Iran

M. K. Ghosh, M. Basu (eds.), *Pathogenesis, Management and Socio-Economic Impact of COVID-19 Pandemic*, Medical Virology: From Pathogenesis to Disease Control, https://doi.org/10.1007/978-981-95-1426-7_20

the ACE2 receptor, which studies have shown that neurons and glial cells expressing this receptor. Consistently, postmortem COVID-19 patients' autopsy reports showed the presence of the virus genome sequence in the brain tissue. Moreover, this virus could infect the cerebrovascular system and cause vasculitis and cerebrovascular accident. Cytokine storm secondary to infection with this virus can also be another possible mechanism for neurologic injury. However, despite various efforts, the exact mechanism of neurological involvement of COVID-19 patients has not been found yet, and further experimental studies are needed. In this chapter, we discuss the possible mechanisms of neurological involvement and the neurological manifestations of COVID-19.

Keywords

COVID-19; Neurological damage; Neuropathy; Encephalopathy; Encephalitis

Introduction

The spread of Severe Acute Respiratory Syndrome Coronavirus 2 (SARS-CoV-2) began in late 2019 and rapidly spread worldwide, causing a global pandemic. The SARS-CoV-2 virus, a member of the *Coronaviridae* family with a single-stranded RNA genome, caused Coronavirus disease 2019 (COVID-19) [1, 2]. The World Health Organization (WHO) has reported more than 774 million cases and over seven million deaths thus far [3]. Symptoms of COVID-19 include cough, fever, shortness of breath, fatigue, myalgia, muscle weakness, headache, neuropathy, pneumonia, sepsis, and, in severe cases, cytokine storms, acute respiratory distress syndrome (ARDS), and cerebrovascular accident (CVA) [1, 4]. Additionally, many COVID-19 patients have exhibited neurological manifestations and neurologic disorders, such as headaches, olfactory disturbances, dizziness, ataxia, hypophysis apoplexy, and seizures [4, 5]. While the main target organ of this virus is the lungs, the infection can also spread to other organs [1, 6]. The central nervous system (CNS) and peripheral nervous system (PNS) involvement have been reported in COVID-19 patients. Damage to nerve cells during SARS-CoV-2 infection is mainly attributed to immune system activity and oxidative stress. Lung cell damage and changes in oxygen levels in COVID-19 patients are accompanied by hypoxia, reactive oxygen species (ROS) production, and tissue damage, which can lead to neurological disorders and ischemia. Furthermore, due to immune system responses and inclined levels of inflammatory cytokines such as interleukin (IL)-1 and IL-6 in response to the SARS-CoV-2 virus, there is a possibility of production of self-reactive T cells targeting the brain tissue [5]. The blood-brain barrier (BBB) and cerebrospinal fluid (CSF) protect the CNS from infection and inflammatory response to this infection. Additionally, nerve cells have immune receptors to detect pathogens and prevent their entry into the brain with the help of leukocytes. However, inflammation and immune system responses in COVID-19 patients increase the permeability of the blood-brain barrier, facilitating the entry of the virus into the brain. Virus penetration into neural tissue leads to threatening damage and neurological disorders for the patient. This can manifest as headaches, dizziness, anosmia, impaired consciousness, and even ischemic stroke [7]. Ischemic strokes in COVID-19 patients mainly occur due to endothelial dysfunction and disseminated intravascular coagulation. Infection of neural tissue and viral entry into the brain are not easily treatable because the blood-brain barriers prevent the passage of therapeutic molecules. Therefore, awareness of neurological disorders is essential for selecting the appropriate treatment approach and reducing adverse patient outcomes [8]. In this chapter, we have discussed the mechanisms of SARS-CoV-2 entry into the brain, its spread, and different manifestations of neurological involvement in COVID-19 patients.

Transmission of the SARS-CoV-2 Virus to the Brain

The enclosed SARS-CoV-2 virus particle contains a single-stranded, positive-sense RNA genome. SARS-CoV-2 is a member of the beta-coronavirus family, which encompasses several viruses that are capable of infecting humans. SARS-CoV-1 exhibits a high degree of genetic similarity to SARS-CoV-2, as evidenced by a genome correspondence of roughly 80% and the simultaneous utilization of the human ACE2 primary cell entry receptor. The spike protein of SARS-CoV-2 possesses a receptor-binding region that exhibits a greater affinity for the ACE2 receptor compared to the spike protein of SARS-CoV-1. This observation potentially accounts for the heightened transmissibility of SARS-CoV-2 [9]. The SARS-CoV-2 virus primarily uses the human ACE2 receptor to enter cells, according to current consensus. The union of cell and viral membranes is additionally assisted by the TMPRSS2 protease, which irreversibly cleaves the viral spike protein that is bound to the ACE2 receptor. Nevertheless, cathepsin B or L proteases might work just as well as TMPRSS2 during the mechanism for entering cells [10]. The blood-brain and blood-cerebrospinal fluid (CSF) barriers are designed to hinder the entry of infections and hazardous substances into the brain; nonetheless, they are not completely inaccessible. Neurotropic viruses may penetrate the BBB in several ways, but the most prevalent route is the hematogenous pathway. This approach involves entering the circulation, which provides viremia [11]. Upon entering the bloodstream, viruses can traverse the BBB by transcytosis or by infecting endothelial cells. Upon entering the bloodstream, viruses can traverse the blood-brain barrier by transcytosis, which involves infecting endothelium cells and infected monocytes. This process is known as the "Trojan Horse" mechanism. Additionally, viruses can enter cells through paracellular means, which occurs when inflammation produced by the virus disrupts the endothelial cells' tight connections [12]. An alternative pathway that is not dependent on viremia involves the interplay between dynein and kinesin proteins in facilitating the virus's transportation into the central nervous system (CNS) through infected motor or sensory neurons. The CNS can be infiltrated by viruses via olfactory sensory neurons. Respiratory coronaviruses are often spread by the latter mechanism [12]. Neurological difficulties have been linked to

SARS-CoV-2, and these consequences are now acknowledged as primary symptoms alongside the usual respiratory manifestations. The prevalent neurological symptoms encompass headache, fatigue, unsteady gait, ataxia, and seizures, along with peripheral nervous system (PNS) manifestations such as amnesia, visual impairment, neuropathic pain, and malaise. Guillain-Barre syndrome (GBS), polyneuritis, encephalitis, meningitis, and encephalopathy, as well as cerebral hemorrhage and infarction, are among the most significant emerging neurological disorders [13]. The neurological complications present themselves as central nervous system (CNS)-related complications, including dizziness, headache, impaired consciousness, acute cerebrovascular disease, ataxia, and seizures. They can also manifest as peripheral nervous system (PNS) manifestations, such as loss of smell and taste, neuropathic pain, vision impairment, and skeletal muscular injury. The prevalence of neurological problems was shown to be greater in those diagnosed with severe COVID-19 compared to those diagnosed with moderate COVID-19. The previous manifestations are contingent upon the infection of host target cells by SARS-CoV-2, specifically unciliated bronchial epithelial cells and type II pneumocytes in the lung. This infection occurs after the virus binds to cell surface receptors, including basigin (BSG; CD147), angiotensin-converting enzyme 2 (ACE2), and neuropilin-1 (NRP-1) [14, 15]. The priming of the viral spike (S) protein, a crucial step for viral entry following attachment to host cell receptors, necessitates the presence of cellular proteases such as TMPRSS2, cathepsins, and furin [10]. The analysis of single-nuclear RNA sequencing (RNA-seq) data from the human brain indicates that there is a lack of expression of ACE2 on various cells inside the brain and its microvasculature. Nevertheless, several brain cell types had elevated levels of additional SARS-CoV-2 receptors, including BSG and NRP-1. Furthermore, it is worth noting that host cell proteases exhibit varying degrees of expression in the majority of brain cells [16]. Research has documented the existence of SARS-CoV-2 in the cerebrospinal fluid (CSF) and postmortem brain tissue of individuals who have contracted COVID-19 and developed encephalitis. Nevertheless, there are conflicting results that suggest the neurological consequences may be attributed to significant systemic inflammation rather than the direct infiltration of the brain. SARS-CoV-2 has been proposed to enter the central nervous system (CNS) using similar pathways as other human coronaviruses (HCoVs), such as the hematogenous route or by utilizing retrograde or antegrade transport mechanisms from peripheral neurons to the CNS [17]. One potential mechanism behind the hematogenous pathway involves the attachment of SARS-CoV-2 receptors to BBB endothelial cells, followed by transcytosis across endothelial cells, ultimately leading to the brain. Endothelial cell infection does not entail any viral replication [16]. The higher expression levels of BSG and NRP1 compared to ACE2 in the brain microvasculature suggest a greater likelihood of SARS-CoV-2 using these receptors for entry into the central nervous system (CNS) [18]. Another process that has been suggested includes the infection of immune cells that express ACE2, such as monocytes, granulocytes, and lymphocytes, sometimes referred to as the "Trojan horse" mechanism [19]. Subsequently, the compromised immune cells can transport SARS-CoV-2 to the central nervous system (CNS), where it can invade the brain. The presence of SARS-CoV-2 viral

RNA was identified in the lung macrophages of individuals with COVID-19 [20]. However, it is necessary to verify the replication of the virus in immune cells and the infiltration of the immune system into the brain. Another approach involves traversing the tight connections of damaged endothelial cells, which is known as the paracellular route [21]. SARS-CoV-2 has the potential to enter the central nervous system (CNS) through peripheral nerves, particularly the olfactory sensory neurons. The potential for retrograde or antegrade transport into the central nervous system (CNS) may be facilitated by the elevated levels of ACE2 and the priming protease, transmembrane serine protease 2 (TMPRSS2), in sustentacular cells, olfactory bulb, and stem cells of the olfactory epithelium [22].

Mechanisms of Neurological Damage in COVID-19 Patients

The possible pathway that SARS-CoV-2 can reach the CNS and PNS was discussed. However, the answer to how this virus causes neurological damage is mostly unknown. Different neurological involvements have been reported in COVID-19 patients. Acute cerebrovascular accident (CVA) is one of the most severe and life-threatening complications and has occurred in COVID-19 patients [23]. The hypercoagulable state, embolic events, endotheliopathy, and microangiopathic thrombosis are the main underlying causes of increased risk of CVA in COVID-19 patients [5]. Direct interaction between the ACE2 receptor of the endothelium and the virus results in vasculitis. Moreover, this interaction causes increased luminal pressure in the cerebral vessels [24]. Increased levels of the D-dimer and other coagulation abnormalities, such as thrombocytopenia, have been reported in COVID-19 patients [25]. The evaluation of the severe COVID-19 patients showed that the coagulation parameters, except the platelets, were elevated, and higher D-dimer and fibrinogen levels were associated with the severity of the disease [26]. Hypercoagulable state and endothelial injury promote the production of thrombosis. Consistently, secondary to obstruction of the cerebrovascular system with this thrombosis, ischemic stroke happens in COVID-19 patients [27]. Notably, increased levels of inflammatory cytokines could activate the extrinsic coagulation pathway, leading to fibrin and blood clotting formation and ischemic stroke. Increased levels of angiotensin 2 and declined cerebral renin-angiotensin cause the dysregulation of the vascular system and impaired function of endothelium, which cause hypertension and hemorrhagic stroke [28]. Intracranial hemorrhage (ICH) was also reported in COVID-19 patients, which could happen secondary to disruption of the BBB by SARS-CoV-2 and destabilizing the tight junction protein [27]. However, a study showed that risk factors for ICH in patients were mainly therapeutic interventions including extracorporeal membrane oxygenation (ECMO), anticoagulation, and mechanical ventilation [29]. Hypoxic brain injury is another mechanism that causes neurologic manifestations in COVID-19 patients. Systemic hypoxia in COVID-19 patients is associated with inflammatory response, cytokine release, and reduced brain tissue oxygen levels. Hypoxia-inducible factor-1 (HIF-1α) activation, which promotes cellular adaptation to hypoxia, causes inflammation. Higher cytokine levels, besides low levels

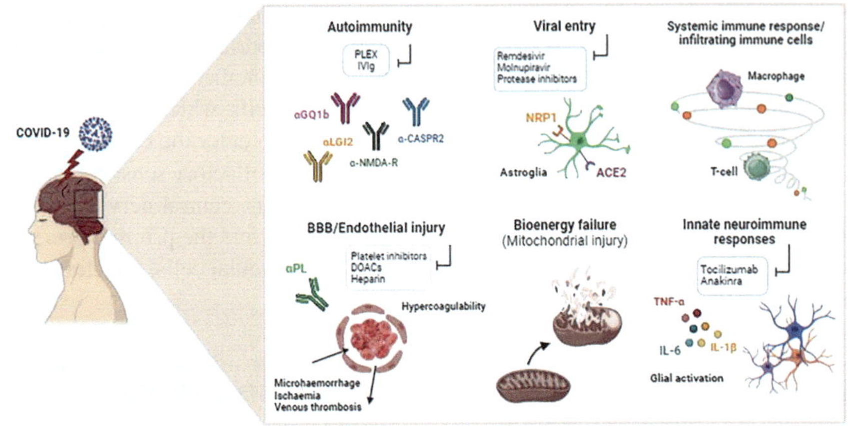

Fig. 20.1 The mechanisms of neurological involvement in COVID-19 patients

of oxygen in brain tissue, might be associated with impaired cognitive function in patients [30]. ACE2 expression was observed in the olfactory bulb, neurons, and oligodendrocytes. The olfactory epithelium served as a possible passage for SARS-CoV-2 to reach the brain. Consistently, loss of smell and taste has been reported in COVID-19 patients and suggestive of the infection of the olfactory epithelium with the virus [31]. Therefore, the intracranial infection might be associated with headache, encephalitis, and other neurologic complications. Consistently, autopsies of COVID-19 patients showed the presence of the genome sequence of the SARS-CoV-2 in brain tissue [32]. Of note, detailed mechanisms of the neurologic manifestation of COVID-19 are discussed in the following part of this chapter (Fig. 20.1).

Manifestations of Nervous System Involvement in COVID-19 Patients

Cerebrovascular Accident (CVA)

SARS-CoV-2 has affected the vascular system through different mechanisms. One of these mechanisms is that the SARS-CoV-2 virus directly infects the vascular system via the ACE2 receptor and causes endothelial dysfunction. Moreover, the virus activates the platelets through platelet angiotensin-converting enzyme 2 (ACE2) receptors, resulting in increased expression of platelet integrin $\alpha IIb\beta 3$ and P-selectin, which cause degranulation and aggregation of the platelets [24]. Also, SARS-CoV-2 indirectly activates factor X, leading to tissue factor expression and activation of the extrinsic pathway. Myocardial ischemia and heart dysfunction were reported in patients with severe COVID-19, which could contribute to the formation of intracardiac thrombosis and cause an embolic stroke [8]. The most common manifestation of COVID-19 patients with CVA was acute ischemic stroke, and intracerebral hemorrhage was less common. Moreover, cerebral venous thrombosis (CVT) is a rare cause of stroke, which was reported in COVID-19 patients, and similarly, CVT has been reported after vaccination with COVID-19 vaccines [23, 33]. An observational study of 219 patients with confirmed COVID-19 reported that ten patients were diagnosed with acute ischemic stroke and one with intracerebral hemorrhage. Patients with cerebrovascular disease (CVD) were older and more likely to have multiple comorbidities such as diabetes, hypertension, and a medical history of CVD. Consistently, the serum analysis of these patients showed higher levels of the C-reactive protein and D-dimer [34]. A systematic review and meta-analysis of 294,249 COVID-19 patients showed that the incidence of acute cerebrovascular accident (CVA) is 2.6%. Risk factors for CVA in COVID-19 patients were coronary artery disease, atrial fibrillation, hypertension, and diabetes [23]. Another study showed that CVA was associated with poor outcomes in patients, and this association was not influenced by age, gender, hypertension, or diabetes [35]. However, another study reported that males had a higher risk of hemorrhagic stroke than females, while there is no difference between males and females in the developing risk of ischemic stroke [36]. Notably, COVID-19 patients with stroke were younger and had higher NIHSS scores and frequency of large vessel occlusion in comparison to patients who experienced stroke without infection [27]. Moreover, ischemic stroke incidence was higher in COVID-19-recovered patients compared to the general population within 9 months of the infection with SARS-CoV-2 [37]. Therefore, early recognition of a hypercoagulable state and related consequences, such as CVA in COVID-19 patients and prophylactic administration of anticoagulants could be crucial and enhance the prognosis of the patients.

Encephalopathy and Encephalitis

Encephalopathy is a clinical syndrome referring to an alteration in mental status and all of the brain functions, and encephalitis is considered one of the etiologies of encephalopathy. Encephalitis is an inflammation of the brain's parenchyma, and the etiologies can be infectious or noninfectious. In the infectious category, the pathogens cause inflammation in the CNS, and they could enter the CNS through two main pathways: hematogenous and neuronal [38]. Moreover, based on the recent evidence, COVID-19 could participate in the encephalitis processes; data also supported that patients who received the COVID-19 vaccine could also develop encephalitis. The mechanism by which SARS-CoV-2 could enter the brain remains unknown and discussed in detail in previous parts. In this part, we are going to discuss briefly the encephalitis associated with SARS-CoV-2 directly and through the autoimmune processes [39].

Infectious Encephalitis Followed by COVID-19

There have been numerous case reports of encephalitis induced by COVID-19 since 2019. Studies revealed that it was an uncommon complication (<1%) of the COVID-19 patients, but it was 6.7% in severely ill COVID-19 patients, and was higher in older patients [40]. The clinical manifestations of COVID-19-associated encephalitis were mostly fever (70.37%), altered mental status (53.7%), decreased consciousness (33.3%), seizure (29.62%), headache (20.37%), weakness (18.51%), and drowsiness (16.66%). Moreover, other symptoms including psychiatric symptoms and neuromuscular symptoms were also reported. It should also be mentioned that the CSF analysis was positive only in 79.41% of the patients. The mortality rate of COVID-19-associated encephalitis was 13.4% to 28.26% [40, 41].

Autoimmune Encephalitis Followed by COVID-19

Researchers showed that some COVID-19 patients developed Autoimmune Encephalitis (AE) as well, since the identification of some autoantibodies in the serum of COVID-19 patients raised the possible relation of SARS-CoV-2-induced AE [42]. COVID-19 patients who developed AE presented at first mostly presented disturbance of consciousness (30.9%), including drowsiness, delirium, confusion, and unconsciousness. Furthermore, most of the patients developed seizures (45%) and psychiatric symptoms (42%). The seizures were mostly generalized tonic-clonic seizures, and the psychiatric manifestations were abnormal behavior, persecutory delusions, and auditory hallucinations. There were also other clinical symptoms such as neuromuscular symptoms, cerebellar syndromes, and extrapyramidal movements. Regarding the types of AE, studies showed the following results. Fifty percent of the cases were an unknown type of COVID-19-associated AE, 23.8% were anti-N-Methyl-D aspartate (NMDA) encephalitis, 9.5% were limbic encephalitis, 7.1% were antimyelin oligodendrocyte glycoprotein (MOG) encephalitis, and other types were also found. Regarding different treatment strategies, almost all patients received intravenous immunoglobulin, methylprednisolone, oral prednisone therapy, and plasmapheresis. Considering the different treatment strategies, most of the patients fully recovered. In this regard, 29 (69.04%) of the patients recovered completely, 8 (19.04%) patients showed partial recovery, 1 (2%) patient showed multiorgan damage, and 2 (4.76%) patients died due to intracranial hemorrhage and sepsis [43].

In conclusion, this part showed the importance of COVID-19-associated encephalitis both through direct viral infection and autoimmune mechanisms. Although it is a rare complication of COVID-19, clinicians should pay attention to this life-threatening complication both in hospitalized and discharged COVID-19 patients, particularly in severely ill and older patients.

Headache and Dizziness

COVID-19 patients could experience specific or nonspecific neurologic signs and symptoms. Numerous studies have pointed to headache and dizziness as common nonspecific neurologic symptoms among COVID-19 patients. Headache and dizziness might be the initiating symptom or be a sign of severe, complicated COVID-19 infection [44]. Several pathophysiologic mechanisms might be responsible for the development of headaches. Vigorous release of cytokines such as IL-1, TNF-α, and

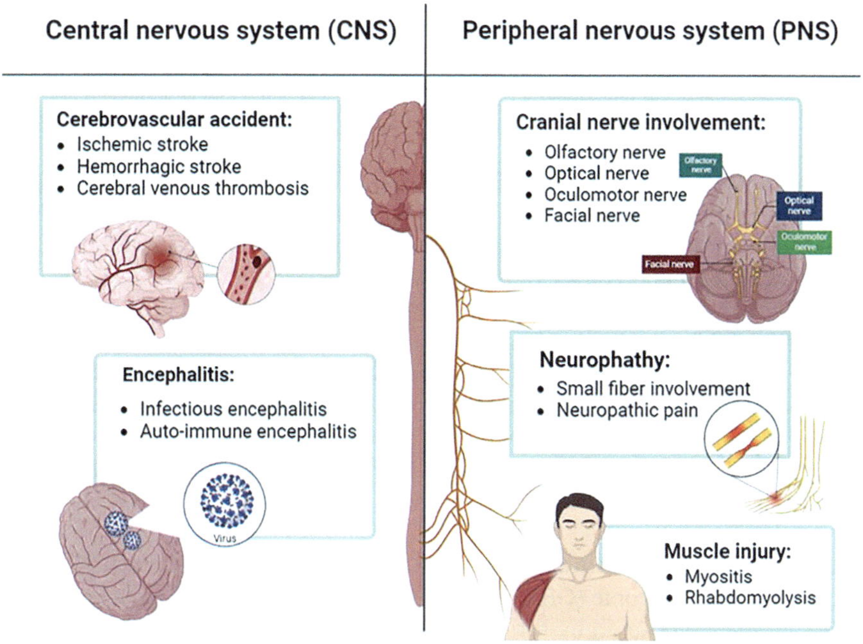

Fig. 20.2 Neurological manifestation of COVID-19

IL-6, direct SARS-CoV-2 invasion to the CNS, and cerebral hypoxia are recommended as reasons for headache [7]. The presence of headache after the seventh day of onset of COVID-19 is strongly associated with cytokine storm. Headache could be concomitant with other neurologic signs such as seizure, alteration in consciousness level, neck stiffness, ataxia, cognitive, and psychiatric as a result of encephalitis, or meningeal irritation. The prevalence of headache and dizziness following COVID-19 varied from about 10.1–29% and 6.4–10% respectively [45]. No significant difference was found between the severe and nonsevere groups in association with the experience of headaches. Furthermore, headache in company with fatigue, attention deficit, and dyspnea is a major complaint in long COVID-19 syndrome. COVID-19 patients may experience various types of headache, including tension and migraine type [46]. Up to now, there is no specific treatment for COVID-19-associated headache, and routine management is recommended to be applied. Noteworthy, SARS-CoV-2 vaccination could reduce the possibility of long-term COVID-19 syndrome to as little as half [47] (Fig. 20.2).

The Peripheral Nervous System Manifestations

While significant focus is on COVID-19's CNS effects, there is also growing evidence of its impact on the PNS, leading to various neurological issues spanning disturbances in cranial nerves to conditions like myasthenia gravis, Guillain-Barre syndrome, and other neuropathies. The exact mechanisms are still debated, but theories suggest the virus may spread through the olfactory nerve or sensory fibers of other nerves like the vagus and glossopharyngeal.

Olfactory Dysfunction

The sudden loss of smell (anosmia) in COVID-19 patients, a widely discussed phenomenon, is thought to be caused by damage to the olfactory nerve. Studies have found high smell and taste dysfunction rates in infected patients, with 85.6% to 88.0% affected [48]. The exact mechanism is unclear, but theories suggest that COVID-19 may temporarily disable olfactory receptor neurons (ORNs), leading to loss of smell. Recent evidence suggests that infected sustentacular support cells (SUS) and changes in mucus and cilia contribute to smell alteration, which is usually followed by rapid recovery as support cells regenerate and cilia regrow from surviving olfactory neurons. Despite some doubts, the sudden onset and recovery of anosmia in COVID-19 is considered unique, prompting ongoing research into the molecular mechanisms involved [49].

Guillain-Barre Syndrome, Its Variants, and Myasthenia Gravis (MG)

Guillain-Barre syndrome (GBS) is an autoimmune neuropathy often triggered by a recent infection caused by various pathogens, including *Campylobacter jejuni, Haemophilus influenzae,* and *cytomegalovirus (CMV).* SARS-CoV-2 has been associated with an increased incidence of GBS, with a significant rise in prevalence from 0.93 to 2.43 per 100,000 cases [50]. Those at higher risk include males over 40 and those with severe respiratory symptoms. The average time between initial symptoms and GBS onset is 14 days [51]. As an important sequela of COVID-19, GBS is generally believed to be a result of a mistaken attack on ganglioside components of axons and myelin sheath, caused by an immune-mediated response against the virus's surface glycoproteins, since the virus has not been detected in the CSF of any SC2-GBS patient. Acute demyelinating inflammatory polyneuropathy (AIDP), a common subtype, is characterized by numbness, tingling, and progressive weakness in the extremities. However, overlapping symptoms may not fall into specific variant categories [52]. GBS variants, including Miller-fisher syndrome (MFS), polyneuritis cranialis (PNC), and pharynx-cervical-brachial (PCB), usually present with specific symptoms such as ophthalmoplegia, areflexia, ataxia, and dysphagia. In GBS patients with COVID-19, laboratory findings may include an elevated protein level (greater than 0.55 g/L) without an increase in white blood cells, known as albuminocytologic dissociation (DCA). Electrophysiological studies showing demyelinating neuropathy and imaging findings may include signal hyperintensity and contrast enhancement of the peripheral nerves, nerve plexus, and cauda equina [52, 53]. Treatment of COVID-19 patients with GBS is similar to that for nonCOVID-19 GBS patients. However, the clinical outcome of SC2-GBS is often worse than that of non-SC2-GBS patients, with a higher need for ICU admission and mechanical ventilation. Early diagnosis and treatment are important to enhance the overall outcome of the infection. It is unclear whether respiratory failure in SC2-GBS results from GBS-induced neuromuscular paralysis, the immune response against the virus, or additional sepsis.

Myasthenia gravis (MG) is a neuromuscular junction immune-mediated disorder, where the immune system mistakenly attacks the postsynaptic membrane. Based on systematic studies, COVID-19, like other viral infections, can worsen symptoms of MG, causing a myasthenic crisis or triggering new-onset symptoms. These symptoms may include general muscle weakness, inability of eyelid closure and bilateral ptosis, dysarthria, and dysphagia, often worsening during the day and improving after rest. This condition can be due to molecular mimicry or breakdown of self-tolerance mechanisms, leading to increased T-cell signaling and inflammation, which can trigger cytokine storm, ARDS, and multiorgan failure [54]. ARDS caused by COVID-19, combined with respiratory muscle fatigue from MG, results in respiratory failure and increases the need for intubation. Results of a study by Camelo-Filho et al. [55] showed that 87% of patients with MG and COVID-19 required ICU admission, and 73% required mechanical ventilation. The management of MG patients with COVID-19 involves continuing immunosuppressive drug therapy, such as corticosteroids, with expert guidance. Research suggests that corticosteroids can improve short-term outcomes, while other immunosuppressive drugs and IVIG/PLEX therapy may also be beneficial.

Cranial Nerve Involvement

Cranial nerve deficits in COVID-19 patients, affecting up to 36% of adults, and limited reports in children [56]. The exact mechanism is unclear, with theories suggesting retrograde virus travel from peripheral neurons to the brain, potential neurotoxic effects of certain medications, or immunological responses similar to Guillain-Barre syndrome. Indeed, the virus's surface similarity to human glycoproteins allows it to infiltrate and harm neural tissue abundant in ACE2 receptors [57, 58]. The virus can cause either polycranial or isolated neuropathy, with symptoms often resolving within a few weeks. Olfactory dysfunction is prevalent, followed by facial (VII) and oculomotor (III) nerve problems. Additional cranial nerves may be affected, including IV (trochlear), VI (abducens), IX (glossopharyngeal), and X (vagus), usually in combination with other PNS symptoms. Glossopharyngeal and vagal nerve involvement is more common in severe cases, leading to impaired food intake and longer hospital stays. Rare cases of acute abducens nerve palsy (ANP) have been reported, characterized by unilateral diplopia and limited external gaze. Imaging tests in these cases are typically normal but may occasionally show mild muscle thinning [59]. Less commonly affected cranial nerves include the vagus (X), vestibulocochlear (VIII), glossopharyngeal (IX), and trigeminal nerves (V). The hypoglossal (XII) nerve is generally the least affected. The trigeminal nerve is often affected in combination rather than in isolation. In contrast, the vestibulocochlear nerve (VIII) is commonly affected in isolation, typically presenting with unilateral symptoms such as vertigo, sensory-neural hearing loss (SSNHL), and tinnitus [60].

Optic (II) and Oculomotor (III) Involvement

Generally, Ocular symptoms in COVID-19 are rare and may occur before the onset of other symptoms. Optic neuritis, a rare phenomenon of SARS-CoV-2 infection, typically presents with headache, gradual visual loss, unilateral eye pain, and decreased visual acuity. In some cases, this condition is associated with multiple sclerosis, but brain lesions are typically absent. Rarely, COVID-19 can trigger myelin oligodendrocyte glycoprotein (MOG)-IgG-related optic neuritis and myelitis, characterized by bilateral lesions in the optic nerve and spinal cord [61]. Despite all, the link between SARS-CoV-2 and demyelinating diseases remains unclear, with further study needed to determine if the virus triggers or exacerbates these conditions [62]. Isolated oculomotor nerve palsy is an unusual complication of SARS-CoV-2, often occurring in patients with mild symptoms. It can be the initial symptom of COVID-19 and typically presents as painless unilateral ophthalmoplegia, blepharoptosis, and diplopia with preserved pupillary light reflex. The prognosis is good, with complete recovery often occurring quickly. Treatment may include early discharge and outpatient follow-up, with short-term usage of oral steroids [63].

Facial Nerve (VII) Impairment

Facial nerve paralysis (FNP) is often linked to infections, including *herpes simplex virus, varicella-zoster, Lyme disease, tuberculosis, and HIV.* A study found a 3.5% incidence of facial nerve palsy (FNP) in 2020, higher than the previous year, suggesting a possible association with COVID-19 [64]. The average age of infected patients with FNP was reported to be 31.9 years, with both genders experiencing unilateral or bilateral symptoms [65]. FNP can arise independently, but can also be a complication of GBS. Symptoms of facial nerve paralysis in COVID-19 cases usually include sudden unilateral weakness in facial muscles, difficulty closing one eye, and mouth angle deviation, often preceded by mastoid pain. Some patients may experience fatigue, chills, and rigors before developing facial paralysis [66]. A study comparing the severity of paralysis before and after the COVID-19 pandemic found no significant difference [58]. CSF findings are not specific, and MRI results can vary from normal to displaying contrast enhancement of the facial nerve in these individuals [67]. Similarly, contrast enhancement of cranial nerves VI, III, VII, and VIII was evident in some patients with Guillain-Barré syndrome (GBS) and cranial polyneuropathies [68]. Early intervention with supportive care and prednisolone (1 mg/kg/day for 5–10 days) can improve outcomes and minimize nerve damage. Most patients experience full recovery or significant improvement within days to weeks, suggesting a positive prognosis for COVID-19-related facial palsy.

Neuropathy

Neuropathy is a condition in which the nervous system is affected. Several conditions, mostly infectious, diabetes, and nutritional (low vitamins), have been demonstrated that could affect the nerves and cause neuropathy. In this regard, the studies also showed that neuropathy is common among COVID-19 patients. In cross-sectional studies done by Oaklander et al., 59% of the patients developed neuropathy, and most of them were reported as small-fiber neuropathy after mild SARS-CoV-2 infection. Moreover, Williams et al. showed that neuropathic pain associated with COVID-19 was common, and the prevalence of COVID-19-related neuropathic pain was 6.7% in hospitalized patients and 34.4% as a long-term consequence of COVID-19. They also reported that COVID-19 severity, azithromycin use, and depression were the risk factors for COVID-19-related neuropathic pain development [69]. Since there were no recently valuable data supporting the neuropathies and neuropathic pain associated with COVID-19, physicians should be familiar with the neuropathies associated with COVID-19, and further research is warranted to evaluate the incidence, prevalence, manifestation, and treatment of it.

Autonomic Nervous Complications

Autonomic dysfunction (AD) has been reported among patients with mild and severe COVID-19. AD could be a manifestation of early-onset SARS-CoV-2 infection or long COVID-19 syndrome [70]. The clear pathophysiology of AD is not well known. Direct invasion of the SARS-CoV-2 through hematogenous or neuronal pathways to the hypothalamus or medulla may explain AD. Also, indirect effects of autoantibodies, rennin-angiotensin-aldosterone axis imbalance, hypoxia, and systemic inflammation could be involved [71]. Postural orthostatic tachycardia syndrome, inappropriate tachycardia syndrome, neurocardiogenic syncope, and hyperhidrosis are reported even more than 3 months after the acute COVID-19 infection. Increased salt and water intake, fludrocortisone, beta-blockers, and alpha-2 agonists may be effective in managing AD following COVID-19 [72].

Muscle Injury

Muscle stiffness or weakness is an important manifestation of patients with COVID-19. Decreased muscle membrane excitability, decreased neural activation, fibrosis, fiber atrophy, and alterations in blood flow and metabolic function are underlying causes of muscle weakness. Notably, malnutrition, hypoxemia, comorbidities, and medication could worsen muscle injury in COVID-19 patients [73]. In seven retrospective studies, an overall 25.1% of COVID-19 patients experienced myalgia. Patients with muscle injury had higher levels of the C-reactive protein, creatine kinase, and D-dimer than patients without muscle injury. Moreover, patients with muscle injury showed multiple organ involvement, such as kidney and liver abnormalities compared to patients with no muscle injury [74]. In the context of muscle injury, rhabdomyolysis has been reported in hospitalized COVID-19 patients as a manifestation of this disease. Therefore, physicians should be aware of an accurate evaluation of the patients for timely intervention to prevent the consequences of muscle injury, including rhabdomyolysis [75]. An observation of patients with COVID-19 showed that some patients had muscle weakness up to 1 year after acute infection. Physical rehabilitation with regular physical exercise, besides neuromuscular electrical stimulation, improves the myopathy of COVID-19 patients, particularly in ICU-admitted COVID-19 patients [73] (Table 20.1).

Table 20.1 The CNS manifestation of COVID-19

Complication	Symptoms	Laboratory findings	Imaging findings/ EEG findings	References
Cerebrovascular accident (CVA)				
Ischemic stroke	Speech impairment Weakness of upper or/and lower extremities Disorientation Swallowing difficulty Paresthesia of limbs Weakness/ paresthesia in the facial muscle	Higher C-reactive protein Higher ferritin levels Elevated D-dimer	Decreased parenchymal attenuation Tissue swelling (mass effect) Hyperattenuated artery sign (arterial occlusion)	Menezes et al. [23]
Hemorrhagic stroke			Hyperdense collection of blood, often with surrounding hypodense edema	
Intracranial hemorrhage			The presence of isodense or hypodense regions within the hyperdense intracerebral hemorrhage represents active bleeding.	
Cerebral venous thrombosis (CVT)			Hyperdense thrombosis Delta signs or pseudo-delta sign	

(continued)

Table 20.1 (continued)

Complication	Symptoms	Laboratory findings	Imaging findings/ EEG findings	References
Encephalitis				
Infectious encephalitis	Fever. Seizures. Headache. Movement disorders. Sensitivity to light. Sensitivity to sound. Neck stiffness. Loss of consciousness.	CSF: Lymphocytic pleocytosis Increased protein Increased opening pressure Neutrophilic pleocytosis	Brain MRI: Medial mesencephalic hyperintensity Bilateral diffuse white matter hyperintensity Brain CT: Diffuse brain edema Normal brain CT and MRI were also reported. EEG: Generalized slowing Bilateral slowing Intermittent delta activity	Koupaei et al. [41]
Auto-immune encephalitis	Flu-like symptoms (in the early phase of the disease) Psychiatric symptoms (Hallucinations, delusions, inappropriate sexual) Altered mental status Seizure			Stoian et al. [42]
Headache and dizziness				
Headache	The low onset of the headache. Head usually hurts on both sides Pain is dull or feels like a band around the head Pain may radiate to the back part of the neck or head	NR	NR	He et al. [44]
Peripheral nervous system (PNS)				
Olfactory dysfunction	Loss of smell	NR	NR	Lechien et al. [48]
Guillain-Barre syndrome (GBS)	Symptoms of AIDP, a common subtype of GBS include numbness, tingling, and progressive weakness in the extremities.	CSF: Elevated protein level (greater than 0.55 g/L) without an increase in white blood cells, known as albuminocytologic dissociation	Signal hyperintensity and contrast enhancement of the peripheral nerves, nerve plexus, and cauda equina EEG: Demyelinating neuropathy	Filosto et al. [50]

(continued)

Table 20.1 (continued)

Complication	Symptoms	Laboratory findings	Imaging findings/ EEG findings	References
Myasthenia gravis (MG)	General muscle weakness Inability of eyelid closure Bilateral ptosis Dysarthria Dysphagia	NR	EEG: Diffuse or focal slow abnormalities Epileptiform pattern	Tugasworo et al. [54]
Cranial nerve involvement				
Optic nerve	Headache Gradual visual loss Unilateral eye pain Decreased visual acuity	NR	Bilateral lesions in the optic nerve and spinal cord	Rodríguez et al. [62]
Oculomotor nerve	Painless unilateral Ophthalmoplegia blepharoptosis Diplopia	NR	NR	Tan et al. [63]
Facial nerve	Sudden unilateral weakness in facial muscles Difficulty closing one eye Mouth angle deviation Mastoid pain	CSF: NS	MRI: Results can vary from normal to displaying contrast enhancement of the facial nerve	Zammit et al. [64]
Neuropathy				
Small fiber involvement	Tingling or prickling sensation Hypersensitivity to touch and temperature changes	NR	NR	Oaklander et al. [76]
Neuropathic pain	Shooting, stabbing, or burning sensation	NR	NR	Williams et al. [69]
Autonomic nerve involvement				
Autonomic involvement	Postural orthostatic Tachycardia syndrome Inappropriate tachycardia syndrome Neurocardiogenic syncope Hyperhidrosis	NR	NR	Jamal et al. [72]

(continued)

Table 20.1 (continued)

Complication	Symptoms	Laboratory findings	Imaging findings/ EEG findings	References
Skeletal muscle injury				
Myalgia	Deep, steady ache, or random sharp pains Pain form may vary between patients	Increase in creatinine kinase (CK) level and lymphocyte count Increased levels of the C-reactive protein and D-dimer	NR	Soares et al. [73]
Rhabdomyolysis	Muscle aching or swelling Muscle stiffness or weakness Feeling exhausted Brown or dark red urine, or very little urine Fever Nausea and vomiting	Increase in creatinine kinase (CK) level Presence of myoglobin in the urine	NR	Byler et al. [75]

Abbreviations: Acute demyelinating inflammatory polyneuropathy (AIDP), not reported (NR), not specific (NS)

Conclusion

SARS-CoV-2 can present with a wide spectrum of neurological symptoms in the early stage of the disease or as a long-term consequence of the disease [13]. The neurological symptoms can vary from headache and myalgia to life-threatening conditions, including CVA and encephalitis. Consistently, some of these manifestations, including stroke, encephalitis, and GBS, can be associated with the severity of the disease, long-term hospitalization, and, in some cases, cause mortality in COVID-19 patients [23, 77]. Therefore, early recognition of severe neurological manifestations can be crucial and life-saving, and give physicians a chance to do timely therapeutic or prophylactic intervention.

Conflicts of Interest The authors declare no conflict of interest.

Funding None.

Data Availability The software Biorender (https://www.biorender.com/) is used to draw some elements of the few figures.

References

1. Mohamadian M, Chiti H, Shoghli A, Biglari S, Parsamanesh N, Esmaeilzadeh A. COVID-19: virology, biology and novel laboratory diagnosis. J Gene Med. 2021;23:e3303.
2. Esmaeilzadeh A, Ebrahimi F, Jahani Maleki A, Siahmansouri A. EG. 5 (Eris) and BA. 2.86 (Pirola) two new subvariants of SARS-CoV-2: a new face of old COVID-19. Infection. 2024:1–7.
3. World Health Organization (WHO). weekly-epidemiological-update-on-covid-19. 2024.
4. Esmaeilzadeh A, Elahi R, Siahmansouri A, Maleki AJ, Moradi A. Endocrine and metabolic complications of COVID-19: lessons learned and future prospects. J Mol Endocrinol. 2022;69:R125–50.
5. Sarkar S, Karmakar S, Basu M, Ghosh P, Ghosh MK. Neurological damages in COVID-19 patients: mechanisms and preventive interventions. MedComm. 2023;4:e247.
6. Elahi R, Hozhabri S, Moradi A, Siahmansouri A, Jahani Maleki A, Esmaeilzadeh A. Targeting the cGAS-STING pathway as an inflammatory crossroad in coronavirus disease 2019 (COVID-19). Immunopharmacol Immunotoxicol. 2023;45:639–49.
7. Dai X, Cao X, Jiang Q, Wu B, Lou T, Shao Y, Hu Y, Lan Q. Neurological complications of COVID-19. QJM Int J Med. 2023;116:161–80.
8. Page EM, Ariëns RA. Mechanisms of thrombosis and cardiovascular complications in COVID-19. Thromb Res. 2021;200:1–8.
9. Zhang H, Penninger JM, Li Y, Zhong N, Slutsky AS. Angiotensin-converting enzyme 2 (ACE2) as a SARS-CoV-2 receptor: molecular mechanisms and potential therapeutic target. Intensive Care Med. 2020;46:586–90.
10. Hoffmann M, Kleine-Weber H, Schroeder S, Krüger N, Herrler T, Erichsen S, Schiergens TS, Herrler G, Wu N-H, Nitsche A. SARS-CoV-2 cell entry depends on ACE2 and TMPRSS2 and is blocked by a clinically proven protease inhibitor. Cell. 2020;181(271–280):e278.
11. Shehata GA, Lord KC, Grudzinski MC, Elsayed M, Abdelnaby R, Elshabrawy HA. Neurological complications of COVID-19: underlying mechanisms and management. Int J Mol Sci. 2021;22:4081.
12. Swanson PA II, McGavern DB. Viral diseases of the central nervous system. Curr Opin Virol. 2015;11:44–54.
13. Wildwing T, Holt N. The neurological symptoms of COVID-19: a systematic overview of systematic reviews, comparison with other neurological conditions and implications for healthcare services. Ther Adv Chronic Dis. 2021;12:2040622320976979.
14. Wang K, Chen W, Zhang Z, Deng Y, Lian J-Q, Du P, Wei D, Zhang Y, Sun X-X, Gong L. CD147-spike protein is a novel route for SARS-CoV-2 infection to host cells. Signal Transduct Target Ther. 2020;5:283.
15. Cantuti-Castelvetri L, Ojha R, Pedro LD, Djannatian M, Franz J, Kuivanen S, van der Meer F, Kallio K, Kaya T, Anastasina M. Neuropilin-1 facilitates SARS-CoV-2 cell entry and infectivity. Science. 2020;370:856–60.
16. Iadecola C, Anrather J, Kamel H. Effects of COVID-19 on the nervous system. Cell. 2020;183(16–27):e11.
17. Solomon IH, Normandin E, Bhattacharyya S, Mukerji SS, Keller K, Ali AS, Adams G, Hornick JL, Padera RF Jr, Sabeti P. Neuropathological features of Covid-19. N Engl J Med. 2020;383:989–92.
18. Baig AM, Khaleeq A, Ali U, Syeda H. Evidence of the COVID-19 virus targeting the CNS: tissue distribution, host–virus interaction, and proposed neurotropic mechanisms. ACS Chem Neurosci. 2020;11:995–8.
19. Trojanowicz B, Ulrich C, Kohler F, Bode V, Seibert E, Fiedler R, Girndt M. Monocytic angiotensin-converting enzyme 2 relates to atherosclerosis in patients with chronic kidney disease. Nephrol Dial Transpl. 2017;32:287–98.

20. Abassi Z, Knaney Y, Karram T, Heyman SN. The lung macrophage in SARS-CoV-2 infection: a friend or a foe? Front Immunol. 2020;11:550815.
21. Bost P, Giladi A, Liu Y, Bendjelal Y, Xu G, David E, Blecher-Gonen R, Cohen M, Medaglia C, Li H. Host-viral infection maps reveal signatures of severe COVID-19 patients. Cell. 2020;181(1475–1488):e1412.
22. Li YC, Bai WZ, Hashikawa T. The neuroinvasive potential of SARS-CoV2 may play a role in the respiratory failure of COVID-19 patients. J Med Virol. 2020;92:552–5.
23. Menezes RG, Alabduladhem TO, Siddiqi AK, Maniya MT, Al Dahlawi AM, Almulhim MWA, Almulhim HW, Saeed YAA, Alotaibi MS, Alarifi SS. Cerebrovascular disease in COVID-19: a systematic review and meta-analysis. Infez Med. 2023;31:140.
24. Zhang S, Liu Y, Wang X, Yang L, Li H, Wang Y, Liu M, Zhao X, Xie Y, Yang Y. SARS-CoV-2 binds platelet ACE2 to enhance thrombosis in COVID-19. J Hematol Oncol. 2020;13:1–22.
25. Zhang X, Yang X, Jiao H, Liu X. Coagulopathy in patients with COVID-19: a systematic review and meta-analysis. Aging (Albany NY). 2020;12:24535.
26. Len P, Iskakova G, Sautbayeva Z, Kussanova A, Tauekelova AT, Sugralimova MM, Dautbaeva AS, Abdieva MM, Ponomarev ED, Tikhonov A. Meta-analysis and systematic review of coagulation disbalances in COVID-19: 41 studies and 17,601 patients. Front Cardiovasc Med. 2022;9:794092.
27. Nannoni S, de Groot R, Bell S, Markus HS. Stroke in COVID-19: a systematic review and meta-analysis. Int J Stroke. 2021;16:137–49.
28. Gorog DA, Storey RF, Gurbel PA, Tantry US, Berger JS, Chan MY, Duerschmied D, Smyth SS, Parker WA, Ajjan RA. Current and novel biomarkers of thrombotic risk in COVID-19: a consensus statement from the international COVID-19 thrombosis biomarkers colloquium. Nat Rev Cardiol. 2022;19:475–95.
29. Daly SR, Nguyen AV, Zhang Y, Feng D, Huang JH. The relationship between COVID-19 infection and intracranial hemorrhage: a systematic review. Brain Hemorrhages. 2021;2:141–50.
30. Boldrini M, Canoll PD, Klein RS. How COVID-19 affects the brain. JAMA Psychiatry. 2021;78:682–3.
31. Ziuzia-Januszewska L, Januszewski M. Pathogenesis of olfactory disorders in COVID-19. Brain Sci. 2022;12:449.
32. Fabbri VP, Foschini MP, Lazzarotto T, Gabrielli L, Cenacchi G, Gallo C, Aspide R, Frascaroli G, Cortelli P, Riefolo M. Brain ischemic injury in COVID-19-infected patients: a series of 10 post-mortem cases. Brain Pathol. 2021;31:205.
33. Esmaeilzadeh A, Maleki AJ, Moradi A, Siahmansouri A, Yavari MJ, Karami P, Elahi R. Major severe acute respiratory coronavirus-2 (SARS-CoV-2) vaccine-associated adverse effects; benefits outweigh the risks. Expert Rev Vaccines. 2022;21:1377–94.
34. Li Y, Li M, Wang M, Zhou Y, Chang J, Xian Y, Wang D, Mao L, Jin H, Hu B. Acute cerebrovascular disease following COVID-19: a single center, retrospective, observational study. Stroke Vasc Neurol. 2020;5:279–84.
35. Pranata R, Huang I, Lim MA, Wahjoepramono EJ, July J. Impact of cerebrovascular and cardiovascular diseases on mortality and severity of COVID-19–systematic review, meta-analysis, and meta-regression. J Stroke Cerebrovasc Dis. 2020;29:104949.
36. Ong B, Conda LE, Uchino K, Mendoza T, Hernandez-Diaz A, Cho S. Sex differences on cerebrovascular complications in hospitalized COVID-19 patients: a meta-analysis (P11-5.030). in neurology. AAN Enterprises; 2024. p. 1280.
37. Zuin M, Mazzitelli M, Rigatelli G, Bilato C, Cattelan AM. Risk of ischemic stroke in patients recovered from COVID-19 infection: a systematic review and meta-analysis. Eur Stroke J. 2023;8:915–22.
38. Jordá-Marcos R. Encephalitis. in critical care infectious diseases textbook. Springer; 2001. p. 515–31.
39. Abdelhady M, Husain MA, Hawas Y, Elazb MA, Mansour LS, Mohamed M, Abdelwahab MM, Aljabali A, Negida A. Encephalitis following COVID-19 vaccination: a systematic review. Vaccine. 2023;11:576.

40. Siow I, Lee KS, Zhang JJ, Saffari SE, Ng A. Encephalitis as a neurological complication of COVID-19: a systematic review and meta-analysis of incidence, outcomes, and predictors. Eur J Neurol. 2021;28:3491–502.
41. Koupaei M, Shadab Mehr N, Mohamadi MH, Asadi A, Abbasimoghaddam S, Shekartabar A, Heidary M, Shokri F. Clinical symptoms, diagnosis, treatment, and outcome of COVID-19-associated encephalitis: a systematic review of case reports and case series. J Clin Lab Anal. 2022;36:e24426.
42. Stoian A, Stoian M, Bajko Z, Maier S, Andone S, Cioflinc RA, Motataianu A, Barcutean L, Balasa R. Autoimmune encephalitis in COVID-19 infection: our experience and systematic review of the literature. Biomedicine. 2022;10:774.
43. Xue H, Zeng L, He H, Xu D, Ren K. Autoimmune encephalitis in COVID-19 patients: a systematic review of case reports and case series. Front Neurol. 2023;14:1207883.
44. He Y, Bai X, Zhu T, Huang J, Zhang H. What can the neurological manifestations of COVID-19 tell us: a meta-analysis. J Transl Med. 2021;19:1–35.
45. Brola W, Wilski M. Neurological consequences of COVID-19. Pharmacol Rep. 2022;74:1208–22.
46. Aghagoli G, Gallo Marin B, Katchur NJ, Chaves-Sell F, Asaad WF, Murphy SA. Neurological involvement in COVID-19 and potential mechanisms: a review. Neurocrit Care. 2021;34:1062–71.
47. Nuzzo D, Cambula G, Bacile I, Rizzo M, Galia M, Mangiapane P, Picone P, Giacomazza D, Scalisi L. Long-term brain disorders in post Covid-19 neurological syndrome (PCNS) patient. Brain Sci. 2021;11:454.
48. Lechien JR, Chiesa-Estomba CM, De Siati DR, Horoi M, Le Bon SD, Rodriguez A, Dequanter D, Blecic S, El Afia F, Distinguin L. Olfactory and gustatory dysfunctions as a clinical presentation of mild-to-moderate forms of the coronavirus disease (COVID-19): a multicenter European study. Eur Arch Oto-Rhino-Laryngol. 2020;277:2251–61.
49. Carrillo-Larco RM, Altez-Fernandez C. Anosmia and dysgeusia in COVID-19: a systematic review. Wellcome Open Res. 2020;5:94.
50. Filosto M, Piccinelli SC, Gazzina S, Foresti C, Frigeni B, Servalli MC, Sessa M, Cosentino G, Marchioni E, Ravaglia S. Guillain-Barré syndrome and COVID-19: an observational multicentre study from two Italian hotspot regions. J Neurol Neurosurg Psychiatry. 2021;92:751–6.
51. Scheidl E, Canseco DD, Hadji-Naumov A, Bereznai B. Guillain-Barré syndrome during SARS-CoV-2 pandemic: a case report and review of recent literature. J Peripher Nerv Syst. 2020;25:204–7.
52. Bentley SA, Ahmad S, Kobeissy FH, Toklu HZ. Concomitant Guillain-Barré syndrome and COVID-19: a meta-analysis of cases. Medicina (Kaunas); 2022. p. 58.
53. Finsterer J, Scorza FA. Guillain-Barre syndrome in 220 patients with COVID-19. Egypt J Neurol Psychiat Neurosurg. 2021;57:55.
54. Tugasworo D, Kurnianto A, Retnaningsih, Andhitara Y, Ardhini R, Budiman J. The relationship between myasthenia gravis and COVID-19: a systematic review. Egypt J Neurol Psychiatr Neurosurg. 2022;58:83.
55. Camelo-Filho AE, Silva AMS, Estephan EP, Zambon AA, Mendonça RH, Souza PVS, Pinto W, Oliveira ASB, Dangoni-Filho I, Pouza AFP, Valerio BCO, Zanoteli E. Myasthenia gravis and COVID-19: clinical characteristics and outcomes. Front Neurol. 2020;11:1053.
56. Knoflach K, Holzapfel E, Roser T, Rudolph L, Paolini M, Muenchhoff M, Osterman A, Griese M, Kappler M, von Both U. Case report: unilateral sixth cranial nerve palsy associated with COVID-19 in a 2-year-old child. Front Pediatr. 2021;9:756014.
57. Heidari ME, Nazemi P, Feizabad E, Beiranvand F, Afzali M. Cranial nerve involvement among COVID-19 survivors. Front Neurol. 2023;14:1182543.
58. Choi GW, Yon DK, Choi YS, Lee J, Park KH, Lee YJ, Park DC, Kim SH, Byun JY, Yeo SG. Comparing the clinical manifestations of Bell's palsy between pre-COVID-19 pandemic and COVID-19 pandemic periods. J Clin Med. 2023;12:1700.

59. Tata G, Isik S, Diktas H, Genc G, Bulut S. Abducens nerve palsy in a patient with COVID-19: a case report. Med Bull Sisli Etfal Hosp. 2024;58:131–4.
60. Doblan A, Kaplama ME, Ak S, Basmacı N, Tarini EZ, Göktaş ŞE, Güler S, Müderris T. Cranial nerve involvement in COVID-19. Am J Otolaryngol. 2021;42:102999.
61. Akaishi T, Sato DK, Nakashima I, Takeshita T, Takahashi T, Doi H, Kurosawa K, Kaneko K, Kuroda H, Nishiyama S, Misu T, Nakazawa T, Fujihara K, Aoki M. MRI and retinal abnormalities in isolated optic neuritis with myelin oligodendrocyte glycoprotein and aquaporin-4 antibodies: a comparative study. J Neurol Neurosurg Psychiatry. 2016;87:446–8.
62. Rodríguez-Rodríguez MS, Romero-Castro RM, Alvarado-de la Barrera C, González-Cannata MG, García-Morales AK, Ávila-Ríos S. Optic neuritis following SARS-CoV-2 infection. J Neurovirol. 2021;27:359–63.
63. Tan Y-J, Ramesh R, Tan Y-H, Tan SML, Setiawan S. COVID-19 and isolated oculomotor nerve palsy: clinical features and outcomes. Clin Neurol Neurosurg. 2023;225:107601.
64. Zammit M, Markey A, Webb C. A rise in facial nerve palsies during the coronavirus disease 2019 pandemic. J Laryngol Otol. 2020;134:905–8.
65. Tonkal A, Alamri AA, AlMaghrabi SJ, Mozahim NF, Mozahim SF, Alsubaie SA, Alsehly AA, Alshuaibi RO, Alotaibi LA, Qashgari FS. Cranial nerve impairment associated with COVID-19 infections: a systematic review. Cureus. 2022;14:e31997.
66. Estakhr M, Tabrizi R, Ghotbi Z, Shahabi S, Habibzadeh A, Bashi A, Borhani-Haghighi A. Is facial nerve palsy an early manifestation of COVID-19? A literature review. Am J Med Sci. 2022;364:264–73.
67. Mahammedi A, Saba L, Vagal A, Leali M, Rossi A, Gaskill M, Sengupta S, Zhang B, Carriero A, Bachir S, Crivelli P, Paschè A, Premi E, Padovani A, Gasparotti R. Imaging of neurologic disease in hospitalized patients with COVID-19: an Italian multicenter retrospective observational study. Radiology. 2020;297:E270–e273.
68. Bigaut K, Mallaret M, Baloglu S, Nemoz B, Morand P, Baicry F, Godon A, Voulleminot P, Kremer L, Chanson JB, de Seze J. Guillain-Barré syndrome related to SARS-CoV-2 infection. Neurol Neuroimmunol Neuroinflamm. 2020;7:335–43.
69. Williams LD, Zis P. COVID-19-related neuropathic pain: a systematic review and meta-analysis. J Clin Med. 2023;12:1672.
70. Milovanovic B, Djajic V, Bajic D, Djokovic A, Krajnovic T, Jovanovic S, Verhaz A, Kovacevic P, Ostojic M. Assessment of autonomic nervous system dysfunction in the early phase of infection with SARS-CoV-2 virus. Front Neurosci. 2021;15:640835.
71. Barizien N, Le Guen M, Russel S, Touche P, Huang F, Vallée A. Clinical characterization of dysautonomia in long COVID-19 patients. Sci Rep. 2021;11:14042.
72. Jamal SM, Landers DB, Hollenberg SM, Turi ZG, Glotzer TV, Tancredi J, Parrillo JE. Prospective evaluation of autonomic dysfunction in post-acute sequela of COVID-19. J Am Coll Cardiol. 2022;79:2325–30.
73. Soares MN, Eggelbusch M, Naddaf E, Gerrits KH, van der Schaaf M, van den Borst B, Wiersinga WJ, van Vugt M, Weijs PJ, Murray AJ. Skeletal muscle alterations in patients with acute Covid-19 and post-acute sequelae of Covid-19. J Cachexia Sarcopenia Muscle. 2022;13:11–22.
74. Nepal G, Rehrig JH, Shrestha GS, Shing YK, Yadav JK, Ojha R, Pokhrel G, Tu ZL, Huang DY. Neurological manifestations of COVID-19: a systematic review. Crit Care. 2020;24:1–11.
75. Byler J, Harrison R, Fell LL. Rhabdomyolysis following recovery from severe COVID-19: a case report. Am J Case Rep. 2021;22:e931616–1.
76. Oaklander AL, Mills AJ, Kelley M, Toran LS, Smith B, Dalakas MC, Nath A. Peripheral neuropathy evaluations of patients with prolonged long COVID. Neurol Neuroimmunol Neuroinflam. 2022;9:e1146.
77. Chen X, Laurent S, Onur OA, Kleineberg NN, Fink GR, Schweitzer F, Warnke C. A systematic review of neurological symptoms and complications of COVID-19. J Neurol. 2021;268:392–402.

COVID-19 Comorbidities in Cancer

21

Kritika Prasad, Saurav K. Saha, Sourav Sanyal,
Arundhaty Pal, and Tapas K. Sengupta

Abstract

The Coronavirus disease 2019, known as COVID-19, caused by the severe acute respiratory syndrome coronavirus 2 or SARS-CoV-2 infection, first came into notice in December 2019 and was announced as a pandemic by the World Health Organization in March 2020. Despite people spanning all age groups being susceptible to COVID-19, individuals with preexisting diseases are more vulnerable. Due to the immunocompromised condition, cancer patients posed an elevated risk of severity, complications, as well as death in the COVID-19 pandemic. This chapter provides insight into the SARS-CoV-2 infections in cancer patients and related outcomes. Here, we discuss the risk factors linked to COVID-19 and cancer, underlying molecular mechanisms, and the impact of COVID-19 treatments on cancer patients.

Keywords

COVID-19 · Cancer · Comorbidity

Abbreviations

ACE2	Angiotensin-converting enzyme 2
ADT	Androgen-deprivation therapy
AST	Aspartate aminotransferase

K. Prasad · S. K. Saha · S. Sanyal · A. Pal · T. K. Sengupta (✉)
Department of Biological Sciences, Indian Institute of Science Education and Research, Kolkata, Kolkata, India
e-mail: senguptk@iiserkol.ac.in

M. K. Ghosh, M. Basu (eds.), *Pathogenesis, Management and Socio-Economic Impact of COVID-19 Pandemic*, Medical Virology: From Pathogenesis to Disease Control, https://doi.org/10.1007/978-981-95-1426-7_21

CD	Clusters of differentiation
COPD	Chronic obstructive pulmonary disease
COVID-19	Coronavirus disease 2019
CRP	C-reactive protein
ERG	Erythroblast transformation-specific (ETS)-related gene
GCSF	Granulocyte colony-stimulating factor
HCQ	Hydroxychloroquine
ICU	Intensive care unit
IFN	Interferon
IL-6	Interleukin 6
IRF	Interferon regulatory factor
ISG	IFN-stimulated genes
JAK/STAT	Janus Kinase/signal transducers and activators of transcription
LDH	Lactate dehydrogenase
NF-κB	Nuclear factor kappa-light-chain-enhancer of activated B cells
NIH	National Institute of Health
NTproBNP	N-terminal pro B-type natriuretic peptide
ORF	Open reading frame
RdRp	RNA-dependent RNA polymerase
RNA	Ribonucleic acid
SARS-CoV-2	Severe acute respiratory syndrome coronavirus 2
TMPRSS2	Transmembrane serine protease 2
TNF	Tumor necrosis factor
WHO	World Health Organization
α-HBDH	α-Hydroxybutyrate dehydrogenase

Introduction

The Coronavirus disease 2019 (COVID-19) first came to notice on 31st December 2019 in Wuhan, China. Later, within a month, the World Health Organization (WHO) designated it as a threat to world public health owing to the alarming rise in cases. To date, the virus is known to have infected approximately 774 million people, out of which nearly seven million have lost their lives [44]. Data provided by the Ministry of Health and Family Welfare (MoHFW), Government of India, indicates that as of May 24, 2024, the severe acute respiratory syndrome coronavirus 2 (SARS-CoV-2) infection count was 45,038,999 in India. Of these, 432 cases are still active, 44,505,387 (98.81%) were cured or discharged, and 533,612 (1.18%) died due to COVID-19 [21]. These statistics are quite an indication of the severity of the disease.

The human-to-human transmission of this virus makes it even deadlier. The major symptoms include respiratory problems, loss of smell, cough, and body aches [39]. Immunosuppressed individuals are more prone to getting affected by COVID-19. Numerous reports have linked the COVID-19 comorbidity with diseases like respiratory illness, cancer, and cardiovascular diseases. This chapter will shed some light on COVID-19 comorbidities in patients having cancer beforehand.

Cancer is the abnormal cell growth in any part of the body with the potential to spread. Various treatments, like chemotherapy, are used to cure cancer, which has multiple side effects. Cancer weakens the immune system, thus escalating the susceptibility of cancer-infected patients to COVID-19. The infection rate across various cancer types with increasing susceptibility is leukemia>lymphoma>myeloma>lung cancer>esophageal cancer [27]. Compared to normal counterparts, cancer patients are at a threefold higher risk of mortality from COVID-19 infections [6]. There are multiple factors that increase this susceptibility. According to sources, cancer patients who died of COVID-19 had low hemoglobin, high cytokine, and high inflammatory markers [19]. In lung cancer patients, age, smoking, and hypertension contributed to COVID-19 severity [27]. In thoracic cancers, tumors, smoking, and chemotherapy treatment cause a higher chance of being infected by COVID-19, whereas in breast cancer, age, and hypertension accounted for the reasons for higher infection chances [27]. Thus, it is comprehensible that cancer patients possess a higher risk in comparison to individuals without existing comorbidity.

Prevalence of SARS-CoV-2 Infection in Cancer Patients

Various parameters like age, hormonal status, weak immune system, and comorbidities make individuals susceptible to SARS-CoV-2 infection [3, 7, 38]. COVID-19 infection has detrimental effects on patients suffering from diabetes, infection caused by Human Immunodeficiency Virus (HIV), hypertension, cardiovascular diseases, cancer, chronic obstructive pulmonary disease, and other comorbidities [7]. Onder and coworkers reviewed the medical records in detail for 355 COVID-19 patients who passed away in Italy [25]. This subsample included people with atrial fibrillation (24.5%), ischemic heart disease (30%), stroke history (9.6%), diabetes (35.5%), dementia (6.8%), and active malignancy (20.3%). Among these patients, 48.5% had three or more diseases, compared to only 0.8% who had none, 25.1% who had one, and 25.6% who had two diseases. It can be postulated from this study by Onder et al. that, besides SARS-CoV-2 infection, these comorbidities raised the likelihood of death.

Cancer is a condition marked by aberrant cell proliferation almost anywhere in the body that has the ability to spread to other organs. In order to evade the antitumor immune response, cancer cells suppress the immune system and rewire immune cells to secrete protumor substances [2]. Moreover, the immune system may become weakened by certain cancer treatments such as chemotherapies or targeted therapies. This is due to the possibility that these therapies may also exert antiproliferative effect on other rapidly proliferating cells like white blood cells, including B lymphocytes and T lymphocytes present in the bone marrow [13]. Other than that, many cancer patients require frequent visits to hospitals or medical centers to monitor their condition or to avail treatment facilities. Immunodeficient cancer patients thus become more prone to SARS-CoV-2 infection [2].

According to a retrospective investigation carried out by Rogado et al., 2.39% of the 1878 patients with COVID-19 who attended Hospital Universitario Infanta

Leonor of Madrid had cancer, with lung cancer accounting for 37.7% of all the cancer cases [31]. Furthermore, they reported that nearly half of the lung cancer patients passed away due to COVID-19 (52.3%), whereas the overall death count was 10.22% [31]. Interestingly, the median age of lung cancer patients who passed away was 72, whereas the median age of those who survived was 64.5 [31]. Therefore, in this case, it can be presumed that both lung cancer and age determined COVID-19 severity. Dai and coworkers conducted a detailed cohort study on the aftermath of SARS-CoV-2 infection on cancer patients in the Hubei Province of Wuhan in China, where the initial COVID-19 outbreak was identified [6]. A total of 641 COVID-19 patients who were similar regarding their gender, age, and hospitalization time were part of the study, among whom 105 patients were suffering from cancer and 536 were noncancer patients. It was found that the death rate, the rate of having at least one critical symptom, and the probability of requiring intensive care and invasive mechanical ventilation were much higher in cases of cancer patients in contrast to noncancer patients [6]. Dai et al. also found that the death rate and chances of having severe conditions were higher in hematological cancers that include myeloma, lymphoma, and leukemia, followed by lung cancer [6]. They also correlated the impact of SARS-CoV-2 infection with cancer stage and discovered that individuals with stage IV metastatic cancer with active COVID-19 infection have elevated death rates and high chances of having severe conditions, among which cancer patients with lung metastasis showed the most vulnerability to dying from COVID-19 and needing ICU admission and invasive mechanical ventilation [6]. When Dai and coworkers compared the effects of different cancer treatments on COVID-19 patients who were also suffering from cancer, they found that compared to patients who underwent chemotherapy or radiotherapy, those who got surgery or immunotherapy typically had higher death rates and a higher likelihood of experiencing critical symptoms [6]. A meta-analysis performed by Venkatesulu and coworkers reported an increased risk of death and higher intubation rates among cancer patients suffering from COVID-19, in contrast to noncancer patients with COVID-19 [41]. Furthermore, they found that hematological cancers accounted for the highest cancer subtype mortality (33.1%), followed by lung cancer (28%), gastrointestinal cancer (19.8%), and breast cancer (10.9%) [41]. Interestingly, no correlation was found between mortality and the use of a certain kind of anticancer treatment [41]. Another study conducted by Trapani and coworkers in Milan, Lombardy region, found that, out of the 909 individuals who passed away from COVID-19, 16.5% had cancer, including those who had already been treated and those undergoing active treatments [40]. Trapani et al. also reported that less than 1% of the sample had no comorbid conditions, whereas approximately 99% of the population had multiple noncommunicable comorbid diseases [40]. Therefore, it can be concluded from the above observations that the death rate and chances of having life-threatening conditions due to SARS-CoV-2 infection are quite high in individuals suffering from comorbid diseases, especially cancer (Fig. 21.1).

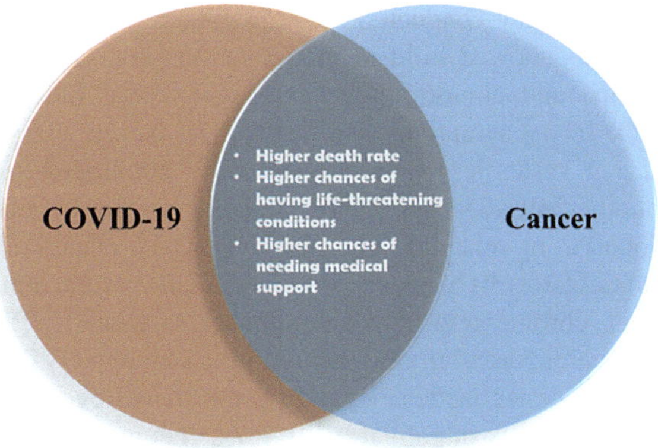

Fig. 21.1 Venn diagram depicting that SARS-CoV-2 infection in cancer patients has more harmful consequences compared to patients suffering from COVID-19 or cancer alone

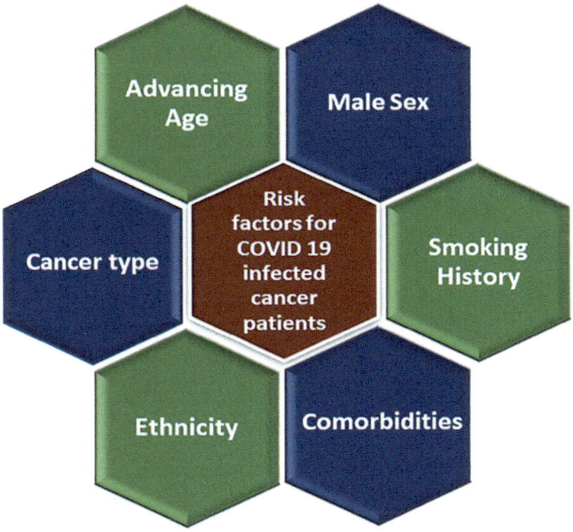

Fig. 21.2 Common risk factors for COVID-19 and cancer

Common Risk Factors Related to COVID-19 and Cancer

Cancer patients are regarded as being at a higher risk of infection with COVID-19 because of their compromised immunity, as already discussed in the previous section. Moreover, regular visits to the hospitals for treatments and, therefore, close contact with COVID-19-infected individuals increase the chance of infection [17]. Initial studies from China reported that individuals suffering from cancer with active COVID-19 infection have a reduced probability of survival than normal individuals, with an approximate mortality of 20–29% [35]. The higher fatalities in cancer patients are related to several factors like old age, smoking status, comorbidities, sex, cancer type, and ethnicity [27] (Fig. 21.2).

In a study conducted on 1253 cancer patients with COVID-19 infection, most patients showed mild to moderate symptoms; however, advancing age, smoking status, and comorbidities were seen to be independent contributors to severe COVID-19 symptoms and death [35]. Nevertheless, the type of cancer, sex, and recent systemic cancer therapy were not linked with severe outcomes [35]. In another investigation, Mehta et al. also found that older age in cancer patients was found to be related to elevated COVID-19-related deaths [19]. In the same report, investigators have also emphasized the involvement of comorbidities like hypertension, chronic obstructive pulmonary disease (COPD), diabetes, cerebrovascular, and cardiovascular ailments with the severe COVID infection. Moreover, anemia, high D-dimer, lactic acid, and Lactate dehydrogenase (LDH) levels were associated with mortality in patients [19]. Russell et al. also reported the contributions of ethnicity, age, sex, smoking history, and comorbidities in severe COVID-19 infection and deaths among patients with active cancer [34]. In a study on 108 cancer patients affected with COVID-19 infection, the main affected cancers were breast cancer, followed by lung cancer and prostate cancer. The symptoms include fever, cough, diarrhea, and vomiting, which were very similar to those of noncancer patients, but in cancer patients, the incidence of severe COVID-19 had high median age, heart disease, smoking, diabetes, and dyspnea. Deeper analysis showed elevated proinflammatory cytokine levels and increased expression of biomarkers involved in infection-related to severe COVID-19 among cancer patients. Some of these include interleukin-6 (IL-6), C-reactive protein (CRP), and procalcitonin. In addition, COVID-19 might also cause injuries to the vital organs like the heart, kidney, and liver. Moreover, LDH, D-dimer, α-hydroxybutyrate dehydrogenase (α-HBDH), N-terminal pro B-type natriuretic peptide (NTproBNP), and aspartate aminotransferase (AST) were also found to be high in severe COVID-19-infected patients with cancer [48]. Kuderer et al. reported more deaths in COVID-19-infected patients with active cancer outside the ICU in comparison with patients in the ICU. ICU admission rates were higher in cancer patients with smoking history, comorbidities, obesity, and recent surgery [14]. In several studies, it was seen that among COVID-19-infected cancer patients, the most predominant was lung cancer, and higher death rates were observed in these patients [27]. Among different comorbidities, it was found that cancer patients with various heart-related ailments like coronary artery disease, hypertension, congestive heart failure, and chronic lung diseases manifest an elevated risk of dying of COVID-19 infection [19]. In a similar study on 260 COVID-19 patients (57 cancer & 203 noncancer), patients with cancer required hospital stay and reintubation compared to the noncancer groups. Older age (≥ 60), smoking status, obesity, diabetes, and other comorbidities were linked with severe COVID-19 infection and high mortality rates. Breast and prostate cancer were found to be prevalent in the group. Mortality rates were higher in patients admitted to the ICU [1]. Though all types of cancer patients have been regarded to be at a higher threat of getting infected by COVID-19, the fact that cancer is a heterogeneous disease with different cancer subtypes, stages, and tumor grades cannot be ignored. Hence, it might not always be justified to assume that most cancer patients are susceptible to COVID-19 infection. In various reports, it has been shown that cancer patients with hematological malignancies are more prone to infection by COVID-19. Mehta et al. in their study reported 28% mortality in patients having solid tumors [19]. Among the most frequently reported cancers in the US population, the annual mortality percentage for lung cancer was around 59.2%, whereas for breast cancer, it was 15.2%, and for colorectal and prostate cancer, it was 36% and 17.4%, respectively. This data reflected an amplified death rate among these populations. Furthermore, this group reported higher mortality (37%) among the patients with hematologic malignancies because these patients are more immunocompromised due to frequent myelosuppressive therapy. Another study by Lee et al. also showed that patients having hematological malignancies (leukemia, lymphoma, and myeloma) had been observed to show more extreme COVID-19 infection in comparison to patients having solid tumors [16]. This group has compared adult cancer patients registered in the UK Coronavirus Cancer Monitoring Project (UKCCMP) during 18th March to eighth May 2020, with the population of cancer patients without COVID infection from the UK Office for National Statistics (2017 data) [16]. To conclude, it is quite evident that advancing age, sex, smoking status, and comorbidities have been independent contributors to severe COVID-19 infection in cancer patients, though the role of race and ethnicity is not very clear.

Signaling Pathways Common to COVID-19 and Cancer

Signaling pathways are central to determining the development, progression, and course of any disease. Various common signaling mechanisms contribute to the progression and complications of COVID-19 as well as cancer, as discussed below.

Role of Cytokine Signaling

The immune system was found to become hyperactive due to COVID-19, resulting in the release of cytokines in an uncontrolled manner [23]. An elevated level of proinflammatory cytokines, which includes Interleukin-6, Interleukin-10, Interferon-γ, and Tumor Necrosis Factor-α, was observed in COVID-19 patients with severe infections [49]. According to a study, Interleukin-6, Interleukin-7, Interleukin-2, Interleukin-10, Interferon-γ, Granulocyte Colony-Stimulating factor (G-CSF), and Tumor Necrosis Factor-α are some key players in the pathogenesis of the COVID-19 disease [23]. The increased death rates in COVID-19 individuals are reported to be caused due to the "cytokine storm" induced by the virus. The cytokine storm is characterized by a heightened immune response with the release of proinflammatory cytokines in huge amounts [28]. Interleukin-6 particularly plays an essential role in contributing to the complexities in COVID-19 individuals (Fig. 21.3).

It was reported that Interleukin-6 induces excessive production of proinflammatory cytokines that result in immune dysregulation in the COVID-19-infected individuals [49]. Various factors contribute to the accumulation of Interleukin-6. The

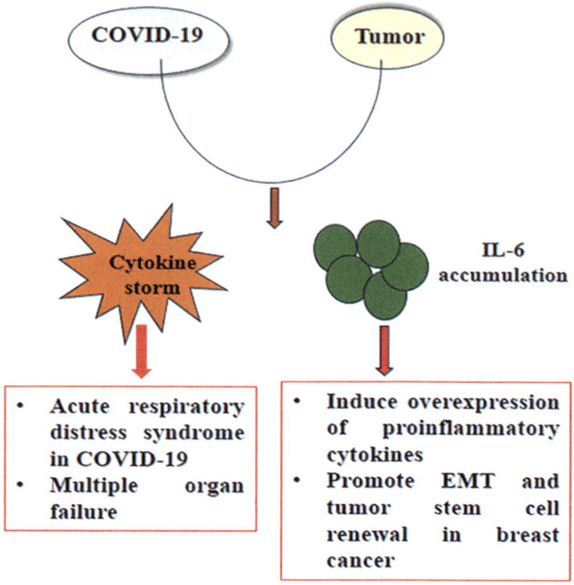

Fig. 21.3 Role of cytokines in COVID-19 and cancer

activation of the NF-κB signaling or other cytokines like tumor necrosis factor (TNF)-α can also lead to interleukin-6 accumulation, which eventually results in the cytokine storm [49].

Interleukin-6 not only has a role in the COVID-19 disease but also contributes to the progression of cancer by inducing the overproduction of proinflammatory cytokines in the tumor microenvironment. Elevated Interleukin-6 levels are seen in various cancer types, for example, lung, pancreatic, and breast cancers. Interleukin-6 stimulates the tumor stem cell renewal in breast cancer. Other than that, it also promotes epithelial-to-mesenchymal transition, thus contributing to metastasis in breast cancer [49]. Moreover, Chimeric antigen receptor T cell therapy also leads to Interleukin-6 production, which can trigger cytokine storm in treating hematological malignancies [9].

Role of TMPRSS2 Signaling

The protease known as the transmembrane serine protease 2 (TMPRSS2) functions to cleave the spike protein of the SARS-CoV-2 into its two components, i.e., S1 and S2. The process is known as S protein priming. The S1 component bears the domain for binding to the receptor [37]. Hence, the S1 component functions in recognition and the initial attachment of the SARS-CoV-2 to its host cell, whereas the S2 component is needed for fusing the virus with the host cell membrane. This process ultimately helps in the viral entry into the host cell. By cleaving the S protein, the TMPRSS2 thus facilitates the virus to interact with the angiotensin-converting enzyme 2 or ACE2 receptor and the ultimate penetration of the virus into its host cell [49] (Fig. 21.4a).

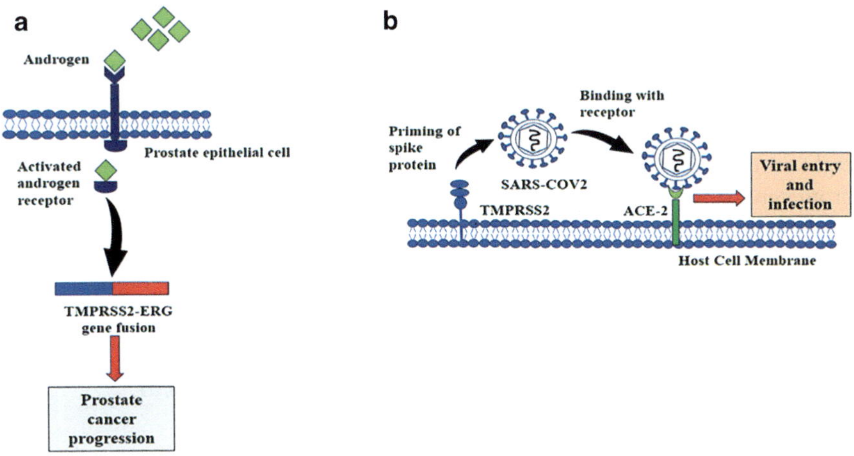

Fig. 21.4 TMPRSS2 in (**a**) prostate cancer and (**b**) COVID-19

The transmembrane serine protease 2 (TMPRSS2) gene is also implicated in prostate cancer development and progression. The TMPRSS2 gene expression is induced when androgen receptors are activated in the presence of their ligand (androgen). The TMPRSS2 can fuse to the ERG (Erythroblast transformation-specific-related gene), which is a transcriptional factor playing a variety of roles in inflammation, apoptosis, and cell proliferation [22]. The TMPRSS2-ERG gene fusion is often observed in prostate cancers, thus it is widely acknowledged as a molecular marker (Fig. 21.4b). Moreover, ERG overexpression is set in an androgen-rich environment because the TMPRSS2 gene promoter contains androgen-sensitive elements [22]. Besides prostate cancer, some types of lung cancer tissues have also been found to bear androgen receptors. This suggests that androgen signaling can also elevate TMPRSS2 expression in lung tissues [20]. Thus, the predominance of male deaths in COVID-19 can be speculated to be a result of androgen levels and TMPRSS2 expression [49].

Role of Interferon Signaling

Interferons (IFNs) belong to the cytokine family and are classified as type I, II, and III IFNs on the basis of their homology and specificity of receptors. The type I interferons are particularly involved in providing innate antiviral immunity. IFN-Is induce the JAK/STAT signaling mechanism, which results in the expression of IFN-stimulated genes (ISGs) and activate a plethora of interferon regulatory factors (IRFs) [49].

Concerning COVID-19, contradictory reports with respect to IFN-1 signaling have surfaced. A study by Zhou et al. demonstrated that COVID-19-affected patients had enhanced expression of IFN-stimulated genes, which suggests a strong IFN-1 response in them [47]. Also, in individuals having severe COVID-19 infection,

IFN-1 response was shown to occur simultaneously with TNF and IL-1-induced inflammatory responses in the peripheral blood mononuclear cells [15]. These reports demonstrate the impact of IFN-1 in the progression of COVID-19. However, SARS-CoV-2 proteins have also demonstrated some inhibitory effects on IFN-1 production or response. The ORF3b of SARS-CoV-2 which encodes one of the eight accessory proteins in the virus is shown to inhibit the activation of the IFN-1 promoter [26].

IFN-I is also involved in inhibiting the proliferation of tumor cells. It also promotes senescence in tumor cells. Thus, defects in the IFN-1 signaling are linked to the progression of tumors [49].

Effect of COVID-19 Treatments on Patients with Cancer

The COVID-19 pandemic exhibited a prominent impact on the diagnosis strategies and treatment regimes of cancer patients in various ways, like slowing down the cancer screening process due to the omnipresent fear of infection [33]. It was observed that cancer survivors were more inclined to develop depressive symptoms than noncancer patients during the COVID-19 period due to factors like loneliness and economic shortcomings [12]. Therefore, cancer patients took both physical and mental tolls as a result of the pandemic (Fig. 21.5). This section focuses on how COVID-19 treatments impacted cancer patients and the various side effects caused by the same.

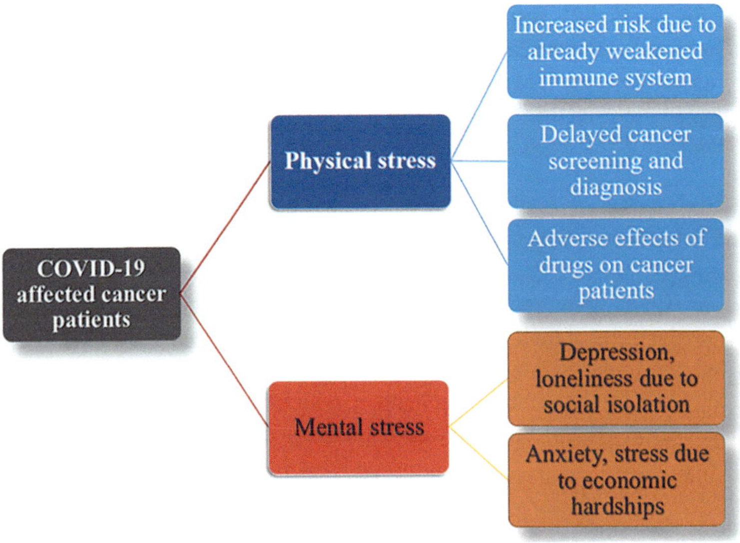

Fig. 21.5 Types of stress experienced by COVID-19-affected patients

Effect of Antiviral Drugs

Various antiviral drugs were used to treat COVID-19, like azvudine, remdesivir, and paxlovid. The potential effect of these drugs on cancer patients becomes significant. The effects of some antiviral drugs are described below.

Azvudine: Azvudine, which is a nucleosidic analogue, was found to hinder the RNA-dependent RNA polymerase (RdRp) activity of SARS-CoV-2 and had a commendable efficacy rate in patients with mild symptoms [46]. In a study conducted by Wang and his coworkers in [42], azvudine formidably prevented cell proliferation in various types of aggressive cancers [42]. Mouse xenograft models also showed that azvudine has tumor inhibition activities. Based on recent studies, patients with aggressive cancer having SARS-CoV-2 infection may get a positive outcome if treated with azvudine.

Hydroxychloroquine (HCQ): HCQ emerged as a life-saving drug with greater efficacy in treating COVID-19 patients. HCQ acts by delaying the termination of the RNA chain and disrupting the stability of the RdRp complex. Luo et al. showed that there were no benefits to using it in lung cancer-affected cancer patients [18]. However, in the clinical trials, when HCQ was administered in combination with azithromycin, the mortality rate of the patients increased [14]. A cohort analysis on 2186 cancer patients with active COVID-19 revealed that HCQ, when administered along with other drugs like azithromycin and tocilizumab, increased the risk of mortality in comparison to HCQ administered alone [29]. Most studies thus indicate that HCQ hardly benefits COVID-19 patients, and thus, later, the National Institute of Health (NIH) stalled its use in clinical trials.

Effect of Interleukin Inhibitors

Inflammation and increased proinflammatory cytokines are one of the important characteristic features of SARS-CoV-2 infection. IL-6 inhibitors like tocilizumab are found to be effective against COVID-19 [45]. However, in COVID-19-affected cancer patients, no benefit of tocilizumab was seen in studies conducted by Guaraldi G et al. and Campochiaro C et al. [4, 10]. The participant pool in both studies is too small to draw any conclusions. Thus, a larger cohort study is needed to assess the drug's effectiveness.

Effect of Corticosteroids, Immune Checkpoint Inhibitors, and ADT

The treatment of individuals with cancer and COVID-19 is more complex than treating a normal individual having COVID-19. Certain medications that are effective against normal COVID patients may turn out to be life-threatening for the cancer-affected patients. For instance, the use of dexamethasone showed a reduction in 28-day mortality compared to the usual patient care group [11]. Cook et al. reported that this drug causes considerable immunological changes by depleting the CD4+ as well as CD8+ cells [5]. Therefore, drugs like dexamethasone, which have immunosuppressive effects, are of grave danger to cancer patients who are undergoing treatment for COVID-19 [27]. Furthermore, cancer patients are also treated with immune checkpoint inhibitors, but some clinicians have speculated that this may lead to a more dangerous COVID-19 phenotype [32]. Thus, a correlation of COVID-19 positive cancer patients and inhibitors of immune checkpoints needs to be further studied, and utmost care should be taken to prescribe such drugs as hospitalization and illness have been seen in patients treated with these kinds of drugs [30].

TMPRSS2 is a gene regulated by androgen whose expression is increased in prostate cancer patients, supporting tumor progression. As discussed in the previous section, TMPRSS2 is a vital and indispensable component for the entry of SARS-CoV-2 into the respiratory epithelial cells. A study done by Montopoli and his coworkers demonstrated that patients with prostate cancer undergoing Androgen-deprivation therapy (ADT) showed partial protection from COVID-19 [24]. However, studies by Welen et al. and Shah et al. showed that ADT has no such beneficial effect in prostate cancer patients [36, 43]. The results of their studies did not suggest any preventive effect of ADT in COVID-19, and thus, these antiandrogens should not be used till further studies on hospitalized patients to treat COVID-19.

Conclusion

The COVID-19 pandemic created havoc in the field of healthcare across the whole world. The condition became even more challenging in the management of people with preexisting comorbidities, including cancer. Due to a lack of healthcare workers and the unavailability of hospital beds during the pandemic, cancer patients underwent delays in accessing treatment and healthcare, which not only deteriorated their health but also increased their risk of infection with SARS-CoV-2 and death. Cancer patients coinfected with COVID-19 worsened the situation of adversity due to their compromised immunity because of treatments like chemotherapy and surgery. Moreover, older age, sex, smoking status, race/ethnicity, cancer type, and comorbidities like heart disease, diabetes, and hypertension made the situation worse. It has been observed that patients having one/more comorbidities have succumbed to severe COVID-19 infection compared to patients without comorbidities. Since signaling mechanisms have a crucial role in the development and progression of diseases, the common signaling pathways in cancer and COVID-19 opened new avenues to consider repurposing drugs used in cancer therapy in the treatment of COVID-19. For instance, ruxolitinib is a drug used in anticancer therapy that inhibits JAK 1 and JAK 2 and has been applied in the SARS-CoV-2 treatment in a quest to suppress cytokine storm [8]. A monoclonal antibody, Bevacizumab, which inhibits the activation of the vascular endothelial growth factor (VEGF) signaling pathways, is introduced in clinical trials for patients showing severe symptoms or acute respiratory distress syndrome [8]. However, it was reported that certain cancer drugs can further intensify the symptoms of COVID-19. Therefore, it becomes much more difficult to treat a COVID-19 patient with cancer, as drugs can have adverse effects on the individual. Antivirals, IL-6 inhibitors, and ADT all have some or other impact on cancer patients as discussed in the previous section. Further cohort studies are an absolute need for validating the effects of various other COVID-19 treatment drugs on cancer patients. The signaling mechanisms that closely knit COVID-19 and cancer need to be studied further to find novel targets and tailor therapeutics effective in managing cancer patients with COVID-19.

Acknowledgments KP, SS, and AP acknowledge the Department of Biotechnology, Govt. of India, and SKS acknowledges the Indian Institute of Science Education and Research Kolkata (IISER Kolkata) for providing fellowships. All authors also acknowledge the Indian Institute of Science Education and Research Kolkata (IISER Kolkata) for providing research facilities.

Authors' Contributions Conceived the idea and review structure: KP, SKS, SS, AP, TKS; Wrote the manuscript: KP, SKS, SS, AP, TKS; Revised and edited the manuscript: KP, SKS, SS, AP, TKS. All authors have read, agreed, and approved the final draft of the manuscript.

Funding This work is supported by a fellowship to KP, SS, and AP from the Department of Biotechnology, Govt. of India. Indian Institute of Science Education and Research Kolkata (IISER Kolkata) provided a fellowship to SKS and funds to TKS.

Data Availability The software Microsoft PowerPoint is used to draw all figures.

Declarations

Conflicts of Interest There is no competing or conflict of interest.

References

1. Aboueshia M, Hussein MH, Attia AS, Swinford A, Miller P, Omar M, Toraih EA, Saba N, Safah H, Duchesne J, Kandil E. Cancer and COVID-19: analysis of patient outcomes. Future Oncol. 2021;17(26):3499–510. https://doi.org/10.2217/fon-2021-0121.
2. Alaeddini M, Etemad-Moghadam S. SARS-Cov-2 infection in cancer patients, susceptibility, outcome and care. Am J Med Sci. 2022;364(5):511–20. https://doi.org/10.1016/j.amjms.2022.05.017.
3. Al-Kuraishy HM, Al-Gareeb AI, Faidah H, Al-Maiahy TJ, Cruz-Martins N, Batiha GE. The looming effects of estrogen in Covid-19: a rocky rollout. Front Nutr. 2021;8:649128. https://doi.org/10.3389/fnut.2021.649128.
4. Campochiaro C, Della-Torre E, Cavalli G, De Luca G, Ripa M, Boffini N, Tomelleri A, Baldissera E, Rovere-Querini P, Ruggeri A, Monti G, De Cobelli F, Zangrillo A, Tresoldi M, Castagna A, Dagna L. Efficacy and safety of tocilizumab in severe COVID-19 patients: a single-Centre retrospective cohort study. Eur J Intern Med. 2020;76:43–9. https://doi.org/10.1016/j.ejim.2020.05.021.
5. Cook AM, McDonnell AM, Lake RA, Nowak AK. Dexamethasone co- medication in cancer patients undergoing chemotherapy causes substantial immunomodulatory effects with implications for chemo-immunotherapy strategies. Onco Targets Ther. 2016;5(3):e1066062. https://doi.org/10.1080/2162402x.2015.1066062.

6. Dai M, Liu D, Liu M, Zhou F, Li G, Chen Z, Zhang Z, You H, Wu M, Zheng Q, Xiong Y, Xiong H, Wang C, Chen C, Xiong F, Zhang Y, Peng Y, Ge S, Zhen B, et al. Patients with cancer appear more vulnerable to SARS-CoV-2: a multicenter study during the COVID-19 outbreak. Cancer Discov. 2020;10(6):783–91. https://doi.org/10.1158/2159-8290.Cd-20-0422.
7. Ejaz H, Alsrhani A, Zafar A, Javed H, Junaid K, Abdalla AE, Abosalif KOA, Ahmed Z, Younas S. COVID-19 and comorbidities: deleterious impact on infected patients. J Infect Public Health. 2020;13(12):1833–9. https://doi.org/10.1016/j.jiph.2020.07.014.
8. El Bairi K, Trapani D, Petrillo A, Le Page C, Zbakh H, Daniele B, Belbaraka R, Curigliano G, Afqir S. Repurposing anticancer drugs for the management of COVID-19. Eur J Cancer. 2020;141:40–61. https://doi.org/10.1016/j.ejca.2020.09.014.
9. Fraietta JA, Lacey SF, Orlando EJ, Pruteanu-Malinici I, Gohil M, Lundh S, Boesteanu AC, Wang Y, O'Connor RS, Hwang WT, Pequignot E, Ambrose DE, Zhang C, Wilcox N, Bedoya F, Dorfmeier C, Chen F, Tian L, Parakandi H, et al. Author correction: determinants of response and resistance to CD19 chimeric antigen receptor (CAR) T cell therapy of chronic lymphocytic leukemia. Nat Med. 2021;27(3):561. https://doi.org/10.1038/s41591-021-01248-2.
10. Guaraldi G, Meschiari M, Cozzi-Lepri A, Milic J, Tonelli R, Menozzi M, Franceschini E, Cuomo G, Orlando G, Borghi V, Santoro A, Di Gaetano M, Puzzolante C, Carli F, Bedini A, Corradi L, Fantini R, Castaniere I, Tabbì L, et al. Tocilizumab in patients with severe COVID-19: a retrospective cohort study. Lancet Rheumatol. 2020;2(8):e474-e484. https://doi.org/10.1016/s2665-9913(20)30173-9.
11. Horby P, Lim WS, Emberson JR, Mafham M, Bell JL, Linsell L, Staplin N, Brightling C, Ustianowski A, Elmahi E, Prudon B, Green C, Felton T, Chadwick D, Rege K, Fegan C, Chappell LC, Faust SN, Jaki T, et al. Dexamethasone in hospitalized patients with Covid-19. N Engl J Med. 2021;384(8):693–704. https://doi.org/10.1056/NEJMoa2021436.
12. Islam JY, Vidot DC, Camacho-Rivera M. Evaluating mental health-related symptoms among cancer survivors during the COVID-19 pandemic: an analysis of the COVID impact survey. JCO Oncol Pract. 2021;17(9):e1258–69. https://doi.org/10.1200/op.20.00752.
13. Kang DH, Weaver MT, Park NJ, Smith B, McArdle T, Carpenter J. Significant impairment in immune recovery after cancer treatment. Nurs Res. 2009;58(2):105–14. https://doi.org/10.1097/NNR.0b013e31818fcecd.
14. Kuderer NM, Choueiri TK, Shah DP, Shyr Y, Rubinstein SM, Rivera DR, Shete S, Hsu CY, Desai A, de Lima Lopes G Jr, Grivas P, Painter CA, Peters S, Thompson MA, Bakouny Z, Batist G, Bekaii-Saab T, Bilen MA, Bouganim N, et al. Clinical impact of COVID-19 on patients with cancer (CCC19): a cohort study. Lancet. 2020;395(10241):1907–18. https://doi.org/10.1016/s0140-6736(20)31187-9.
15. Lee JS, Park S, Jeong HW, Ahn JY, Choi SJ, Lee H, Choi B, Nam SK, Sa M, Kwon JS, Jeong SJ, Lee HK, Park SH, Park SH, Choi JY, Kim SH, Jung I, Shin EC. Immunophenotyping of COVID-19 and influenza highlights the role of type I interferons in development of severe COVID-19. Sci Immunol. 2020a;5(49) https://doi.org/10.1126/sciimmunol.abd1554.
16. Lee LYW, Cazier JB, Starkey T, Briggs SEW, Arnold R, Bisht V, Booth S, Campton NA, Cheng VWT, Collins G, Curley HM, Earwaker P, Fittall MW, Gennatas S, Goel A, Hartley S, Hughes DJ, Kerr D, Lee AJX, et al. COVID-19 prevalence and mortality in patients with cancer and the effect of primary tumour subtype and patient demographics: a prospective cohort study. Lancet Oncol. 2020b;21(10):1309–16. https://doi.org/10.1016/s1470-2045(20)30442-3.
17. Liu C, Zhao Y, Okwan-Duodu D, Basho R, Cui X. COVID-19 in cancer patients: risk, clinical features, and management. Cancer Biol Med. 2020;17(3):519–27. https://doi.org/10.20892/j.issn.2095-3941.2020.0289.
18. Luo J, Rizvi H, Preeshagul IR, Egger JV, Hoyos D, Bandlamudi C, McCarthy CG, Falcon CJ, Schoenfeld AJ, Arbour KC, Chaft JE, Daly RM, Drilon A, Eng J, Iqbal A, Lai WV, Li BT, Lito P, Namakydoust A, et al. COVID-19 in patients with lung cancer. Ann Oncol. 2020;31(10):1386–96. https://doi.org/10.1016/j.annonc.2020.06.007.
19. Mehta V, Goel S, Kabarriti R, Cole D, Goldfinger M, Acuna-Villaorduna A, Pradhan K, Thota R, Reissman S, Sparano JA, Gartrell BA, Smith RV, Ohri N, Garg M, Racine AD, Kalnicki S,

Perez-Soler R, Halmos B, Verma A. Case fatality rate of cancer patients with COVID-19 in a New York hospital system. Cancer Discov. 2020;10(7):935–41. https://doi.org/10.1158/2159-8290.Cd-20-0516.
20. Mikkonen L, Pihlajamaa P, Sahu B, Zhang FP, Jänne OA. Androgen receptor and androgen-dependent gene expression in lung. Mol Cell Endocrinol. 2010;317(1–2):14–24. https://doi.org/10.1016/j.mce.2009.12.022.
21. Ministry of Health and Family Welfare, G. o. I. COVID-19 INDIA as on: 24 May 2024, 08:00 IST (GMT+5:30). 2024. Retrieved May 24, 2024 from https://www.mohfw.gov.in/
22. Mollica V, Rizzo A, Massari F. The pivotal role of TMPRSS2 in coronavirus disease 2019 and prostate cancer. Future Oncol. 2020;16(27):2029–33. https://doi.org/10.2217/fon-2020-0571.
23. Montazersaheb S, Hosseiniyan Khatibi SM, Hejazi MS, Tarhriz V, Farjami A, Ghasemian Sorbeni F, Farahzadi R, Ghasemnejad T. COVID-19 infection: an overview on cytokine storm and related interventions. Virol J. 2022;19(1):92. https://doi.org/10.1186/s12985-022-01814-1.
24. Montopoli M, Zumerle S, Vettor R, Rugge M, Zorzi M, Catapano CV, Carbone GM, Cavalli A, Pagano F, Ragazzi E, Prayer-Galetti T, Alimonti A. Androgen-deprivation therapies for prostate cancer and risk of infection by SARS-CoV-2: a population-based study (N = 4532). Ann Oncol. 2020;31(8):1040–5. https://doi.org/10.1016/j.annonc.2020.04.479.
25. Onder G, Rezza G, Brusaferro S. Case-fatality rate and characteristics of patients dying in relation to COVID-19 in Italy. JAMA. 2020;323(18):1775–6. https://doi.org/10.1001/jama.2020.4683.
26. Palermo E, Di Carlo D, Sgarbanti M, Hiscott J. Type I interferons in COVID-19 pathogenesis. Biology (Basel). 2021;10(9) https://doi.org/10.3390/biology10090829.
27. Pathania AS, Prathipati P, Abdul BA, Chava S, Katta SS, Gupta SC, Gangula PR, Pandey MK, Durden DL, Byrareddy SN, Challagundla KB. COVID-19 and cancer comorbidity: therapeutic opportunities and challenges. Theranostics. 2021;11(2):731–53. https://doi.org/10.7150/thno.51471.
28. Ragab D, Salah Eldin H, Taeimah M, Khattab R, Salem R. The COVID-19 cytokine storm; what we know so far. Front Immunol. 2020;11:1446. https://doi.org/10.3389/fimmu.2020.01446.
29. Rivera DR, Peters S, Panagiotou OA, Shah DP, Kuderer NM, Hsu CY, Rubinstein SM, Lee BJ, Choueiri TK, de Lima Lopes G Jr, Grivas P, Painter CA, Rini BI, Thompson MA, Arcobello J, Bakouny Z, Doroshow DB, Egan PC, Farmakiotis D, et al. Utilization of COVID-19 treatments and clinical outcomes among patients with cancer: a COVID-19 and cancer consortium (CCC19) cohort study. Cancer Discov. 2020;10(10):1514–27. https://doi.org/10.1158/2159-8290.Cd-20-0941.
30. Robilotti EV, Babady NE, Mead PA, Rolling T, Perez-Johnston R, Bernardes M, Bogler Y, Caldararo M, Figueroa CJ, Glickman MS, Joanow A, Kaltsas A, Lee YJ, Lucca A, Mariano A, Morjaria S, Nawar T, Papanicolaou GA, Predmore J, et al. Determinants of COVID-19 disease severity in patients with cancer. Nat Med. 2020;26(8):1218–23. https://doi.org/10.1038/s41591-020-0979-0.
31. Rogado J, Pangua C, Serrano-Montero G, Obispo B, Marino AM, Pérez-Pérez M, López-Alfonso A, Gullón P, Lara M. Covid-19 and lung cancer: a greater fatality rate? Lung Cancer. 2020;146:19–22. https://doi.org/10.1016/j.lungcan.2020.05.034.
32. Rubinstein SM, Steinharter JA, Warner J, Rini BI, Peters S, Choueiri TK. The COVID-19 and cancer consortium: a collaborative effort to understand the effects of COVID-19 on patients with cancer. Cancer Cell. 2020;37(6):738–41. https://doi.org/10.1016/j.ccell.2020.04.018.
33. Rucinska M, Nawrocki S. COVID-19 pandemic: impact on cancer patients. Int J Environ Res Public Health. 2022;19(19) https://doi.org/10.3390/ijerph191912470.
34. Russell B, Moss CL, Shah V, Ko TK, Palmer K, Sylva R, George G, Monroy-Iglesias MJ, Patten P, Ceesay MM, Benjamin R, Potter V, Pagliuca A, Papa S, Irshad S, Ross P, Spicer J, Kordasti S, Crawley D, et al. Risk of COVID-19 death in cancer patients: an analysis from Guy's Cancer Centre and King's College Hospital in London. Br J Cancer. 2021;125(7):939–47. https://doi.org/10.1038/s41416-021-01500-z.
35. Sengar M, Chinnaswamy G, Ranganathan P, Ashok A, Bhosale S, Biswas S, Chaturvedi P, Dhamne C, Divatia J, D'Sa K, Jain H, Laskar S, Moulik NR, Mummudi N, Nair S, Nayak L,

Nayak P, Patkar S, Pawaskar P, et al. Outcomes of COVID-19 and risk factors in patients with cancer. Nat Cancer. 2022;3(5):547–51. https://doi.org/10.1038/s43018-022-00363-4.
36. Shah NJ, Patel VG, Zhong X, Pina L, Hawley JE, Lin E, Gartrell BA, Febles VA, Wise DR, Qin Q, Mellgard G, Joshi H, Nauseef JT, Green DA, Vlachostergios PJ, Kwon DH, Huang F, Liaw B, Tagawa S, et al. The impact of androgen deprivation therapy on COVID-19 illness in men with prostate cancer. JNCI Cancer Spectr. 2022;6(3):pkac035. https://doi.org/10.1093/jncics/pkac035.
37. Shin HJ, Ku KB, Kim HS, Moon HW, Jeong GU, Hwang I, Yoon GY, Lee S, Lee S, Ahn DG, Kim KD, Kwon YC, Kim BT, Kim SJ, Kim C. Receptor-binding domain of SARS-CoV-2 spike protein efficiently inhibits SARS-CoV-2 infection and attachment to mouse lung. Int J Biol Sci. 2021;17(14):3786–94. https://doi.org/10.7150/ijbs.61320.
38. Sica A, Colombo MP, Trama A, Horn L, Garassino MC, Torri V. Immunometabolic status of COVID-19 cancer patients. Physiol Rev. 2020;100(4):1839–50. https://doi.org/10.1152/physrev.00018.2020.
39. Struyf T, Deeks JJ, Dinnes J, Takwoingi Y, Davenport C, Leeflang MM, Spijker R, Hooft L, Emperador D, Domen J, Tans A, Janssens S, Wickramasinghe D, Lannoy V, Horn SRA, Van den Bruel A. Signs and symptoms to determine if a patient presenting in primary care or hospital outpatient settings has COVID-19. Cochrane Database Syst Rev. 2022;5(5):Cd013665. https://doi.org/10.1002/14651858.CD013665.pub3.
40. Trapani D, Marra A, Curigliano G. The experience on coronavirus disease 2019 and cancer from an oncology hub institution in Milan, Lombardy region. Eur J Cancer. 2020;132:199–206. https://doi.org/10.1016/j.ejca.2020.04.017.
41. Venkatesulu BP, Chandrasekar VT, Girdhar P, Advani P, Sharma A, Elumalai T, Hsieh CE, Elghazawy HI, Verma V, Krishnan S. A systematic review and meta-analysis of cancer patients affected by a novel coronavirus. JNCI Cancer Spectr. 2021;5(2):pkaa102. https://doi.org/10.1093/jncics/pkaa102.
42. Wang Q, Liu X, Wang Q, Zhang Y, Jiang J, Guo X, Fan Q, Zheng L, Yu X, Wang N, Pan Z, Song C, Qi W, Chang J. FNC, a novel nucleoside analogue inhibits cell proliferation and tumor growth in a variety of human cancer cells. Biochem Pharmacol. 2011;81(7):848–55. https://doi.org/10.1016/j.bcp.2011.01.001.
43. Welén K, Rosendal E, Gisslén M, Lenman A, Freyhult E, Fonseca-Rodríguez O, Bremell D, Stranne J, Balkhed Å, Niward K, Repo J, Robinsson D, Henningsson AJ, Styrke J, Angelin M, Lindquist E, Allard A, Becker M, Rudolfsson S, et al. A phase 2 trial of the effect of anti-androgen therapy on COVID-19 outcome: no evidence of benefit, supported by epidemiology and in vitro data. Eur Urol. 2022;81(3):285–93. https://doi.org/10.1016/j.eururo.2021.12.013.
44. World Health Organization. COVID-19 epidemiological update–16 February 2024. Retrieved March 10, 2024 from https://www.who.int/publications/m/item/covid-19-epidemiological-update-16-february-2024
45. Xu X, Han M, Li T, Sun W, Wang D, Fu B, Zhou Y, Zheng X, Yang Y, Li X, Zhang X, Pan A, Wei H. Effective treatment of severe COVID-19 patients with tocilizumab. Proc Natl Acad Sci USA. 2020;117(20):10970–5. https://doi.org/10.1073/pnas.2005615117.
46. Zhang JL, Li YH, Wang LL, Liu HQ, Lu SY, Liu Y, Li K, Liu B, Li SY, Shao FM, Wang K, Sheng N, Li R, Cui JJ, Sun PC, Ma CX, Zhu B, Wang Z, Wan YH, et al. Azvudine is a thymus-homing anti-SARS-CoV-2 drug effective in treating COVID-19 patients. Signal Transduct Target Ther. 2021;6(1):414. https://doi.org/10.1038/s41392-021-00835-6.
47. Zhou Z, Ren L, Zhang L, Zhong J, Xiao Y, Jia Z, Guo L, Yang J, Wang C, Jiang S, Yang D, Zhang G, Li H, Chen F, Xu Y, Chen M, Gao Z, Yang J, Dong J, et al. Heightened innate immune responses in the respiratory tract of COVID-19 patients. Cell Host Microbe. 2020;27(6):883–890.e882. https://doi.org/10.1016/j.chom.2020.04.017.
48. Zhou Y, Yang Q, Ye J, Wu X, Hou X, Feng Y, Luo B, Li J, Yang G, Li L, Yang X, Wang B, Zhao S, Li L, Li Q, Xu Z, Wu H, Sun J. Clinical features and death risk factors in COVID-19 patients with cancer: a retrospective study. BMC Infect Dis. 2021;21(1):760. https://doi.org/10.1186/s12879-021-06495-9.
49. Zong Z, Wei Y, Ren J, Zhang L, Zhou F. The intersection of COVID-19 and cancer: signaling pathways and treatment implications. Mol Cancer. 2021;20(1):76. https://doi.org/10.1186/s12943-021-01363-1.

COVID-19, Pneumonia, and Tuberculosis: Challenges and Treatment

22

Yogeshwar Devarakonda, Anindita Ray, Bidisha Ganguly, Sakshi Batra, and Kirtimaan Syal

Abstract

The co-occurrence of COVID-19 with tuberculosis (TB) and pneumonia presents complex challenges in the formulation of treatment regimens, requiring careful consideration of drug interactions in addition to diagnosis. Such coinfections significantly escalate mortality rates, highlighting the critical need for effective treatment guidelines. The COVID-19 pandemic showed a profound effect on the incidence rates of other respiratory infections due to various factors, including lockdowns, quarantine, social distancing, poor diagnostic facilities, and limited access to health care facilities. The biomarkers associated with TB and pneumonia, like cytokines and CRP, may contribute to heightened COVID-19 susceptibility and worse outcomes. The drug interactions among treatments for COVID-19, TB, and pneumonia are not well-characterized and are under investigation. Together, the interplay between COVID-19, TB, and pneumonia necessitates tailored treatment approaches, biomarker monitoring, and collaborative

Yogeshwar Devarakonda and Anindita Ray contributed equally to this chapter.

Y. Devarakonda · B. Ganguly · K. Syal (✉)
Genetics and Molecular Microbiology Laboratory, Department of Biological Sciences, Birla Institute of Technology and Sciences-Pilani, Hyderabad campus, Institute of Eminence, Hyderabad, Telangana, INDIA
e-mail: ksyal@hyderabad.bits-pilani.ac.in

A. Ray
Department of Biotechnology, Birla Institute of Technology and Sciences-Pilani, Dubai, UAE

S. Batra
Department of Respiratory Medicine, Mallareddy Institute of Medical Sciences, Hyderabad, Telangana, India

M. K. Ghosh, M. Basu (eds.), *Pathogenesis, Management and Socio-Economic Impact of COVID-19 Pandemic*, Medical Virology: From Pathogenesis to Disease Control, https://doi.org/10.1007/978-981-95-1426-7_22

research efforts to optimize treatment outcomes in order to address broader public health implications, including the emergence of antibiotic resistance, spread, diagnosis, and disease dynamics.

Keywords

COVID-19 · Tuberculosis · Pneumonia · Respiratory infections

Introduction

Tuberculosis (TB) and pneumonia are ancient diseases known to humankind which continue to afflict millions worldwide, whereas the COVID-19 pandemic became a global health concern and compromised healthcare systems in the last 4 years. TB was declared a global health emergency by the WHO in 1993, and it remains a significant public health concern to date. Around eight million individuals contract active tuberculosis worldwide each year, with nearly two million losing their lives to the disease [21]. Similar to SARS-CoV-2, *Mycobacterium tuberculosis* (MTB) primarily manifests as an infection in the lungs [34, 45]. TB has a long incubation period and a delayed onset of the disease in most patients. Both active TB and severe COVID-19 trigger an uncontrolled release of proinflammatory cytokines, resulting in a cytokine storm [36]. This compromises the integrity of the lung tissues, making them more susceptible to secondary infections and increasing the likelihood of coinfections [36].

However, limited data is available regarding the cases of coinfection involving TB and COVID-19. In high TB-burden countries, the prevalence of TB among hospitalized COVID-19 patients has been estimated to be nearly 2% to 8%[36]. The data from the Western Cape province of South Africa indicated that among 22,308 COVID-19 patients, 10% were previously diagnosed with TB or had active TB [36]. A global study also analyzed data from 767 patients and found that the majority (97.4%) had a history of tuberculosis (TB) before contracting COVID-19 [46]. Also, 9.5% of patients had COVID-19 before being diagnosed with TB, while 16.5% were diagnosed with both diseases simultaneously [46]. These findings suggest that TB is a predisposing factor for COVID-19, but more evidence is required to determine if COVID-19 may reactivate or worsen active TB. Furthermore, according to the WHO, there has been an increase in the new cases of rifampicin-resistant TB (RR-TB) from 2020 to 2021, suggesting the reemergence of drug-resistant TB during the COVID-19 pandemic [16].

On the other hand, an estimated 2.5 million people worldwide succumb to pneumonia each year, making it the single largest infectious cause of death in children under the age of five [18]. In the context of COVID-19, pneumonia results in inflammation and fluid buildup in the alveoli of the lungs. COVID-19 patients have been shown to be more susceptible to secondary bacterial infections, especially in severe cases requiring hospitalization. Adelman et al. reported that 10–33% of the hospitalized patients developed bacterial pneumonia [1].

As per the WHO, the COVID-19 pandemic had a fatality rate of nearly 2.2% in 2020 across the globe, which has been further mitigated by vaccines and novel treatment regimens [19]. Before vaccines became widely available, the lines of treatment for COVID-19 primarily included antiviral therapies, convalescent plasma therapy, corticosteroids, anti-inflammatory drugs, immunomodulatory agents, and antibiotics [12, 31]. Antiviral therapeutics like remdesivir, favipiravir, and lopinavir/ritonavir were used to treat COVID-19 patients [33]. Antiparasitic drug hydroxychloroquine was also considered for its potential antiviral effects [12]. Additionally, convalescent plasma therapy was used as a form of passive immunity, whereas corticosteroids and anti-inflammatory drugs were administered to mitigate harmful inflammation, especially in severe cases [12, 43]. Nevertheless, it is worth mentioning that, in a clinical setting, these regimens demonstrated less efficacy compared to remdesivir against SARS-CoV-2. However, remarkably, the prescription of antibiotics constituted the frontline defense against COVID-19 before the widespread availability of vaccines [12].

This chapter delves into the overlapping markers that predispose COVID-19 patients to TB and pneumonia, thereby reviewing the interplay between COVID-19, TB, and pneumonia. Further, the controversial role of antibiotics, particularly azithromycin, as a COVID-19 therapeutic has been discussed. Additionally, we have discussed other respiratory infections and addressed key factors common to such conditions.

Role of Antibiotics in the Treatment of COVID-19

While COVID-19 is caused by a virus, bacterial coinfections, including pneumonia and TB, often necessitate antibiotic therapy. In fact, antibiotics became the indispensable constituent of the anti-COVID-19 treatment regimen. Most bacterial pneumonias in COVID-19 patients could be effectively treated with broad-spectrum antibiotics [17]. Notably, azithromycin, being a licensed, easily available, and inexpensive safe drug, has been widely used to treat COVID-19 [17]. Interestingly, in vitro studies suggested anti-SARS-CoV-2 activity of azithromycin [8]. Here, azithromycin may alter the pH of cellular compartments such as the Golgi complex and recycling endosomes, potentially interfering with the intracellular activity and replication of SARS-CoV-2 [8]. Additionally, Butler et al. proposed that azithromycin might decrease the levels of furin, an enzyme involved in facilitating the entry of the virus into host cells through cleavage of the spike protein [8]. Azithromycin could impede viral entry by reducing furin levels. Moreover, azithromycin was observed to diminish the levels of proinflammatory cytokines like IL-6, which play a key role in the cytokine storm associated with severe COVID-19 cases [8]. As cytokine release syndrome is implicated in COVID-19 mortality, drugs with immunomodulatory properties, such as azithromycin, have garnered attention as a repurposed drug in the treatment of COVID-19 [42].

However, the clinical evidence supporting the efficacy of azithromycin, either alone or in combination with other drugs like hydroxychloroquine, remains inconclusive [42]. While some studies have reported favorable outcomes, others have raised concerns regarding methodology and the reliability of results. Various small-scale clinical trials have yielded conflicting outcomes, with some patients still testing positive for the virus despite the treatment [29, 42]. Additionally, observational studies have failed to demonstrate significant differences in mortality rates among patients treated with azithromycin-containing regimens compared to the subjects who did not receive such treatments [42]. These discrepancies underscore the need for further rigorous investigation to ascertain the true efficacy and safety profile of azithromycin in the management of COVID-19. According to the combined data of 19 studies (2834 patients), the average rate of antibiotic use in COVID-19 management was revealed to be 74.0%, with only 17.6% of patients developing secondary infections [8]. Another meta-analysis found that merely 7.0% of hospitalized COVID-19 patients had bacterial coinfections [23]. In light of the extensive data indicating low rates of bacterial infections in COVID-19 patients, the unregulated prescription of azithromycin in COVID-19 treatment protocols has raised concerns about its potential impact on antimicrobial resistance and the development of superbugs [22].

COVID-19 and Tuberculosis

TB persists as a predominant cause of mortality among infectious diseases. Earlier, WHO projected that every third individual of the world's population is carrying latent TB infection (without exhibiting active symptoms), which was later revised to one-fourth of the world's population [11, 44]. Each person with latent TB infection (LTBI) carries a roughly 10% risk of progressing to active TB disease over their lifetime, and can increase rapidly to 50% in individuals coinfected with the human immunodeficiency virus (HIV) [37]. The intersection of COVID-19 and TB has increased concerns among health authorities, particularly in regions endemic to TB. The infection by SARS-CoV-2 can have lasting effects on the immune system, irrespective of the severity of the COVID-19 episode. The progression and outcomes of tuberculosis are tuned by host-pathogen interaction and corresponding host immune responses. Interestingly, the interaction between tuberculosis and pathogens like SARS-CoV-2 is multifaceted and can be modulated by several genetic factors [39]. The intricate interplay between these infections and their implications, particularly the long-term effects of tuberculosis treatment in COVID-19 patients, remains an area for further evaluation and monitoring. In parallel, limited data exist on the crosstalk between cytokines produced during coinfection of SARS-CoV-2 and MTB, and their combined impact on human disease outcomes. Both diseases share similar hyperinflammatory signatures characterized by key cytokines like TNF-α, IL-6, and IL-1, which recruit immune cells and enhance inflammatory reactions that result in tissue damage (Fig. 22.1). Like MTB infection, the delayed type I IFN response observed in COVID-19 patients is also associated with hyperinflammation [30]. The involvement of cytokines in such coinfection has been further elaborated in other sections.

Due to limited therapeutic options, managing patients with severe COVID-19 remains a significant challenge. Immunosuppressive agents such as corticosteroids and cytokine blockers have demonstrated efficacy in treating severe cases of COVID-19. However, the use of immunosuppressive therapies carries the risk of opportunistic infections, including bacterial, fungal, and viral infections (Fig. 22.1). The patients coinfected with COVID-19 and TB need careful management of drug regimens due to potential interactions, which require further investigation [25]. COVID-19 progression can vary widely among patients, so it is essential to monitor closely and adjust medications as needed [10].

COVID-19 has severely disrupted TB health services, hindering diagnosis, care, and prevention efforts worldwide. Service providers face challenges due to equipment shortages, movement restrictions, and resource reallocation. TB patients struggle to access services due to COVID-19 fears, stigma, and financial constraints [28]. The survival of a TB patient depends on timely diagnosis, treatment, and prevention of transmission. However, resources are diverted to COVID-19 care, leading to undiagnosed TB cases. With a focus on COVID-19, TB diagnosis and preventive therapy services suffered, potentially leading to increased incidences of TB. Household transmission is also a concern, especially with prolonged family proximity due to social distancing measures [47].

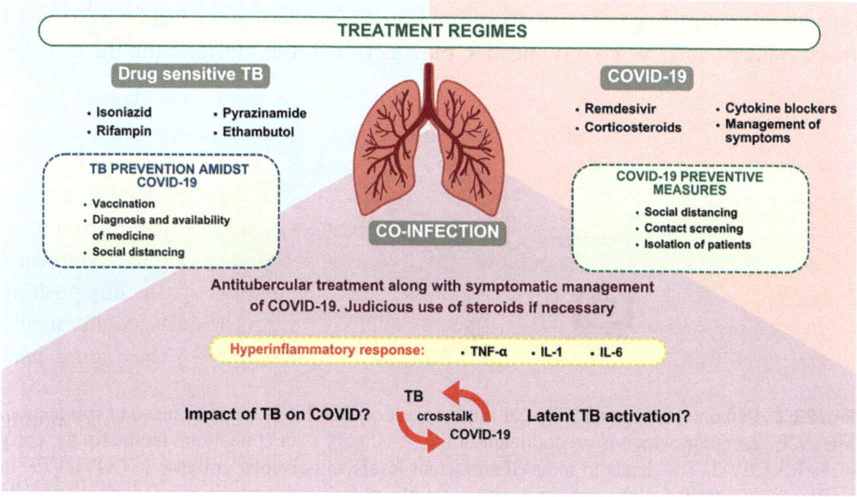

Fig. 22.1 Typical treatment strategies implicated in TB, COVID-19, and coinfection

COVID-19 and Pneumonia

Pneumonia is a lung disease primarily caused by viral and bacterial pathogens, including *Streptococcus pneumoniae*, *Mycoplasma pneumoniae, Klebsiella pneumoniae,* and influenza, involving alveolar inflammation [27]. While viral pneumonia is more common in children, it has become prevalent in adults during the COVID-19 pandemic. The clinical diagnosis depends on identifying symptoms like fever, rigors, malaise, shortness of breath, productive cough, and radiographic evidence of lung consolidation [27]. CT imaging aids in distinguishing COVID-19 pneumonia from bacterial pneumonia, primarily based on the distribution of ground-glass opacities [2, 5]. However, distinguishing between viral and bacterial origin based only on clinical presentation remains challenging due to its significant overlap in symptoms [2].

COVID-19 predominantly affects the lower respiratory tract, inducing hypoxia and ground-glass opacification due to alveolar collapse. High expression of angiotensin-converting enzyme 2 (ACE2) in lung tissue renders type II pneumocytes susceptible to SARS-CoV-2 infection, resulting in reduced surfactant levels and resulting in alveolar collapse [2].

The respiratory tract is exposed to airborne microorganisms and particulate contaminants due to microaspiration events, which is one of the primary routes of entry for microorganisms. While defense mechanisms like cough reflex and immune response normally keep microbial levels low, pneumonia can occur when these mechanisms are compromised or overwhelmed due to virulent bacterial strains invading lung tissue [32]. Alveolar macrophages play a crucial role in responding to lower airway bacteria and recruitment of polymorphonuclear neutrophils (PMN) during virulent pathogenic manifestation (Fig. 22.2). Cytokines released by alveolar macrophages, such as TNF-α and ILs, play a critical role in triggering the immune

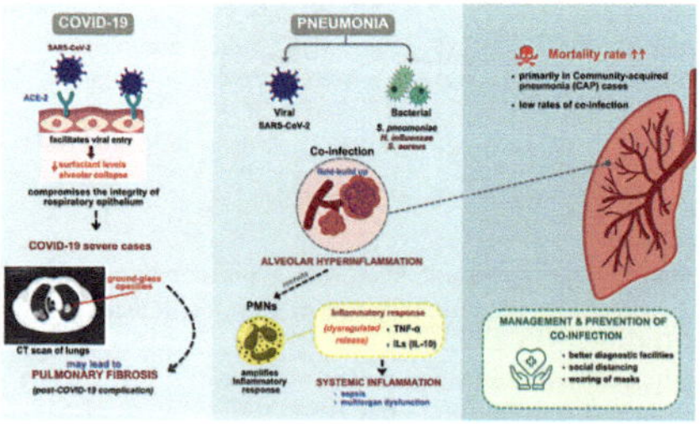

Fig. 22.2 Pathophysiology and clinical implications of COVID-19, pneumonia, and coinfection. The ACE2 receptor, which is predominant in the respiratory tract epithelium, facilitates the entry of SARS-CoV-2. This leads to reduced surfactant levels and alveolar collapse in COVID-19. In severe cases, ground-glass opacities and possible pulmonary fibrosis postinfection may be observed. Pneumonia, whether viral or bacterial in origin, is recognized by fluid accumulation in the alveoli, indicating alveolar inflammation. This inflammatory response recruits polymorphonuclear leukocytes (PMNs) and other immune cells, leading to dysregulated cytokine release and systemic inflammation. Coinfections, particularly community-acquired pneumonia (CAP), increase the risk of mortality

response. Excessive cytokine release can lead to symptoms like fever and contribute to the pathology of bacterial pneumonia, which can also lead to complications like sepsis and multiorgan failure [32].

Bacterial and viral pneumonia can coexist, with bacterial coinfection significantly increasing the mortality rates. Viruses exhaust host defenses potentially resulting in bacterial infections by damaging the respiratory epithelium's barrier makes it easier for bacteria to adhere and invade. However, the prevalence of bacterial coinfection in COVID-19 pneumonia is much lower compared to the 1918 influenza pandemic, as indicated by numerous studies [4, 7, 24]. This may be attributed to variations in public health strategies such as the implementation of nationwide lockdowns, social distancing, and the wearing of masks. Amin-Choudhury and colleagues confirmed that public health policies, such as social distancing practices, can decrease rates of pneumococcal community-acquired pneumonia (CAP) in the context of coinfection. In the case of coinfection, a sevenfold increase in 28-day mortality was observed [4]. Likewise, a systematic review encompassing 31 studies involving 10,762 patients found that 25% of individuals with CAP had viral infections, with the figure rising to 44% in studies where more than half of the patients provided lower respiratory samples [7]. The latter review also suggested that the combination of CAP and viral infection doubled the mortality risk, with a rate of deaths being 2.1 times higher in patients with both bacterial and viral infections [7]. Notably, postmortem examinations of individuals succumbing to severe COVID-19 indicate that at least 32% experienced bacterial superinfections [24]. Ventilator-associated pneumonia (VAP) is identified as the primary cause in over 50% of hospital-acquired infections, with *Pseudomonas aeruginosa* and *Klebsiella pneumoniae* being the most frequently isolated organisms [24].

Viral pneumonia poses therapeutic challenges, predominantly managed with supportive care due to limited antiviral options. Neuraminidase inhibitors like oseltamivir, zanamivir, and peramivir are recommended to reduce morbidity and mortality in Influenza A and B infections [41]. A meta-analysis of ten studies on influenza pneumonia has shown that steroid use was linked to increased risk of death in patients with prolonged ICU stays and higher rates of secondary infections [49]. Overall, the COVID-19 pandemic has challenged healthcare systems worldwide, leading to a scarcity of resources and staff. Hospitals have implemented extensive infection prevention and control (IPC) measures to limit the spread of the virus, which may have contributed to the decline in nosocomial infections. However, the full impact of these preventive measures on healthcare-associated infections (HAIs) remains unclear. The pandemic has resulted in increased awareness of IPC, indirectly benefiting HAI prevention efforts. The closure of health facilities and disruptions in vaccination programs lead to the risk of increased incidences of other diseases, such as child pneumonia, in low- and middle-income countries [38]. It is crucial to focus on addressing these challenges and enhancing healthcare services in the postpandemic period.

Other Respiratory Infections

Coinfections with other microorganisms are common in the COVID-19 subjects. The list of coinfecting pathogens includes influenza, *Staphylococcus aureus, Haemophilus influenzae, Klebsiella pneumoniae, Streptococcus pneumoniae, Pseudomonas aeruginosa, Escherichia coli,* and *Neisseria meningitides [*20*]*. Here, influenza is a highly contagious respiratory illness which annually affects about 9% of the world population, leading to millions of severe cases and deaths. Despite having similar symptoms between influenza and COVID-19, the coinfection initially received little attention during the early period of the pandemic. Further, reports from various countries highlighted the occurrence of coinfection, emphasizing the importance of specific and accurate detection methods. Studies have shown varying rates of influenza coinfection in hospitalized COVID-19 patients. Allou et.al reported that 13.9% of COVID-19 patients admitted to hospitals were coinfected with influenza H1N1, rhinovirus, methicillin-susceptible *S. aureus,* and human coronavirus 229E [3].

Bacterial coinfections in COVID-19 pneumonia are linked to disease severity in COVID-19 patients. A study has been conducted to investigate microbial coinfection in 301 patients with SARS-CoV-2 infection. The analysis revealed that the most prevalent microbes were multidrug-resistant, including *K. pneumoniae* (37%), followed by extremely drug-resistant *A. baumannii* (26%), multidrug-resistant *Escherichia coli* (18.6%), extremely drug-resistant *P. aeruginosa* (8.5%), and other bacteria (9.3%). Additionally, a higher death rate was observed in COVID-19 subjects afflicted with bacterial infection, such as *K. pneumoniae* and *A. baumannii [*35*]*. Post-COVID-19, *P. aeruginosa* infections reveal a notable shift in the antibiotic susceptibility, which signifies the misuse of antibiotics.

The incidence of fungal infection known as Mucormycosis or black fungus has been reported in COVID-19 patients. Mucormycosis (CAM) was primarily observed in COVID19 during the second wave inflicting a mortality rate of nearly 31–50%. Since the pandemic outbreak of COVID-19, fungal coinfections have become increasingly common in critically ill patients, mainly due to the overuse of steroids. These fungal coinfections include COVID-19-associated pulmonary aspergillosis (CAPA) and *Pneumocystis jirovecii* pneumonia (PCP). Severe influenza and SARS-CoV-2-induced pneumonia have similar immunopathological characteristics such as cytokine storm syndrome, airway epithelial damage, and lymphopenia [15]. The diagnosis of CAPA is often missed in COVID-19 patients admitted to ICUs due to less specific radiological signs of infection and concurrent presence of acute respiratory distress syndrome. The lack of understanding of interactions among different respiratory coinfections hampers accurate diagnosis and effective treatment strategies.

Discussion

The co-occurrence of COVID-19 with other respiratory infections such as tuberculosis (TB) and pneumonia presents a complex challenge for clinicians and researchers alike. Several cohort studies demonstrate that cases of coinfection significantly escalate the mortality rate [4, 16, 24, 36, 46]. Broad-spectrum antibiotics like azithromycin have been commonly recommended in such cases, yet their efficacy in addressing COVID-19 specifically remains uncertain [22]. A randomized trial evaluated the effectiveness of azithromycin as part of COVID-19 treatment regimens in comparison to the usual treatment, which revealed insignificant benefit in terms of prognosis or hospitalization rates with azithromycin [8]. While generally considered safe, pre-COVID era studies have linked azithromycin use to a higher risk of cardiovascular complications, especially in COVID-19 patients with preexisting cardiovascular conditions [6]. Together, the role of azithromycin in COVID-19 remains controversial. Moreover, the threefold increase in azithromycin consumption during the pandemic compared to the pre-COVID-19 period may have certainly exacerbated the existing problem of antibiotic resistance (AMR) [22]. The inappropriate use of antibiotics has not only contributed to unnecessary healthcare costs and side effects, but has also negatively impacted other respiratory infections like TB and pneumonia.

The relationship between COVID-19 and TB is particularly intriguing. TB patients seem to be predisposed to COVID-19, but it remains unclear whether COVID-19 itself is a risk factor for TB. Sheerin et al. demonstrated the induction of ACE2 expression in human macrophages during MTB infection [40]. It is well established that ACE2 is crucial to facilitate the entry of viruses like SARS-CoV-2 into host cells; however, the underlying mechanisms behind the elevated levels of the receptor are under investigation. Likewise, their hypothesis suggests that individuals with existing MTB infection may be at higher risk of severe COVID-19, with overlapping mechanisms potentially leading to TB progression [40]. The progression of latent TB to active TB upon COVID-19 coinfection still remains poorly understood.

Similarly, it is uncertain whether active TB increases susceptibility to COVID-19 and vice versa. Hence, it is necessary to explore the potential pathophysiological interface between these conditions to gain a better understanding of their interplay. On the contrary, it is important to note that in most severe COVID-19 cases, fatalities occur mostly due to complications arising from pneumonia and not solely due to the SARS-CoV-2 infection [48]. The predominant causative agent of pneumonia is *Streptococcus pneumoniae*, and in some cases, it can evolve into a multipathogenic infection [27]. It is challenging to manage pneumonia alongside COVID-19, as it complicates the treatment regimens and exacerbates the symptoms.

Moreover, biomarkers differentially expressed in TB and pneumonia may serve as potential risk factors for COVID-19. For example, elevated levels of cytokines (TNF-α, IL-6, GM-CSF) or CRP associated with TB or pneumonia, respectively, indicate a preexisting dysregulated immune state [30]. Such dysregulation may predispose individuals to heightened susceptibility to COVID-19 or worsened disease outcomes upon infection. Through the identification and monitoring of these biomarkers, healthcare professionals may be better equipped to assess the risk profile of individuals with TB or pneumonia and tailor interventions to mitigate the impact of COVID-19.

The potential interaction between drugs used to treat COVID-19, TB, and pneumonia is not clear. The coadministration of medications may lead to drug-drug interactions, which can affect therapeutic outcomes. Additionally, some drugs may share similar mechanisms of action or adverse effects, necessitating careful prescription to prevent complications. Moreover, it is beneficial to optimize the combination therapy to enhance treatment outcomes and minimize the risk of treatment failure. Thus, further research into the compatibility and efficacy of drug combinations is necessary to make informed clinical decisions and improve patient care [26].

During the COVID-19 pandemic, practices such as lockdowns and social distancing measures led to a significant shift in the incidence of other respiratory diseases (Fig. 22.3). This slowed down the spread of the pathogens, including influenza, rhinovirus, and respiratory enterovirus infections, thereby limiting the development of immunity at the population level. A study noted the changes in the number of

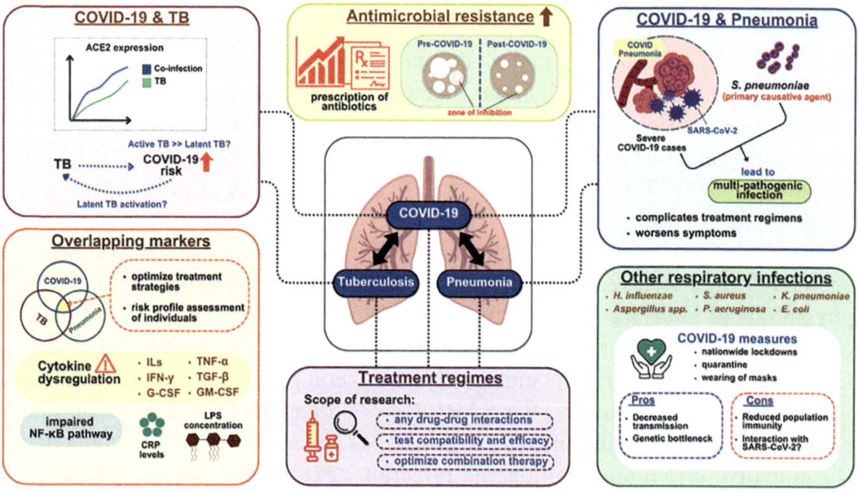

Fig. 22.3 Interplay between COVID-19, tuberculosis, and pneumonia and their impact on coinfection dynamics and treatment strategies. The COVID-19 pandemic has not only heightened concerns about antimicrobial resistance through increased antibiotic use but has also fostered a complex relationship between SARS-CoV-2 and diseases like tuberculosis and pneumonia. Understanding this crosstalk between cytokine dysregulation and shared biomarkers is pivotal for unraveling the underlying mechanisms

cases during and post-COVID-19 pandemic [9]. Likewise, the evolution of respiratory viruses relies on maintaining a diverse genetic pool, but evidence suggests that the reduction in cases during the pandemic has led to a genetic bottleneck, resulting in reduction of genetic diversity in some respiratory viruses, such as influenza [14]. However, the interaction between SARS-CoV-2 and such pathogens is not fully understood and requires further research.

These insights into COVID-19 coinfections are critical not only for developing targeted therapeutic strategies and improving patient outcomes, but also for improving overall preparedness and response measures. Understanding the dynamics of coinfections can help healthcare systems in better utilization of resources, improve diagnostic and treatment protocols, and strengthen surveillance efforts. This can serve as a basis for public health measures aimed at preventing and controlling the spread of infectious diseases, thereby contributing to the provision of more resilient and effective health services to the public.

Acknowledgments KS acknowledges DBT, Govt. of India, and BITS Pilani, Hyderabad campus, for funding. YD acknowledges UGC, Delhi, for Junior Research Fellowship. The authors acknowledge the Canva web tool, which was used for preparing illustrations.

Author's Contributions Conceived the idea and review structure: KS, YD, AR, BG, SB; Wrote the manuscript: KS, YD, AR, BG, SB; Revised and edited the manuscript: KS, YD, AR, BG, SB. The author has read, agreed, and approved the final draft of the manuscript.

Funding This work is jointly supported by DBT, Govt. of India, and BITS Pilani, Hyderabad campus.

Data Availability The software Canva (www.canva.com/en_in/) is used to draw some elements of the figures.

Declarations

Conflicts of Interest There is no competing and conflict of interest.

References

1. Adelman MW, Bhamidipati DR, Hernandez-Romieu AC, Babiker A, Woodworth MH, Robichaux C, et al. Secondary bacterial pneumonias and bloodstream infections in patients hospitalized with COVID-19. Ann Am Thorac Soc. 2021;18(9):1584–7. https://doi.org/10.1513/AnnalsATS.202009-1093RL.
2. Alipoor SD, Jamaati H, Tabarsi P, Mortaz E. Immunopathogenesis of pneumonia in COVID-19. Tanaffos. 2020;19(2):79–82.
3. Allou N, Larsen K, Dubernet A, Traversier N, Masse L, Foch E, et al. Co-infection in patients with hypoxemic pneumonia due to COVID-19 in Reunion Island. Medicine (Baltimore). 2021;100(4):e24524. https://doi.org/10.1097/md.0000000000024524.
4. Amin-Chowdhury Z, Aiano F, Mensah A, Sheppard CL, Litt D, Fry NK, et al. Impact of the coronavirus disease 2019 (COVID-19) pandemic on invasive pneumococcal disease and risk of pneumococcal coinfection with severe acute respiratory syndrome coronavirus 2 (SARS-

CoV-2): prospective National Cohort Study, England. Clin Infect Dis. 2021;72(5):e65-e75. https://doi.org/10.1093/cid/ciaa1728.

5. Batra S, Nair AG, Syal K. An unusual presentation of superior mesenteric venous occlusion in mild COVID-19. Indian J Clin Biochem. 2023;38(2):275–8. https://doi.org/10.1007/s12291-022-01067-3.
6. Bergami M, Manfrini O, Nava S, Caramori G, Yoon J, Badimon L, et al. Relationship between azithromycin and cardiovascular outcomes in unvaccinated patients with COVID-19 and preexisting cardiovascular disease. J Am Heart Assoc. 2023;12(14):e028939. https://doi.org/10.1161/jaha.122.028939.
7. Burk M, El-Kersh K, Saad M, Wiemken T, Ramirez J, Cavallazzi R. Viral infection in community-acquired pneumonia: a systematic review and meta-analysis. Eur Respir Rev. 2016;25(140):178–88. https://doi.org/10.1183/16000617.0076-2015.
8. Chedid M, Waked R, Haddad E, Chetata N, Saliba G, Choucair J. Antibiotics in treatment of COVID-19 complications: a review of frequency, indications, and efficacy. J Infect Public Health. 2021;14(5):570–6. https://doi.org/10.1016/j.jiph.2021.02.001.
9. Chow EJ, Uyeki TM, Chu HY. The effects of the COVID-19 pandemic on community respiratory virus activity. Nat Rev Microbiol. 2023;21(3):195–210. https://doi.org/10.1038/s41579-022-00807-9.
10. Cioboata R, Biciusca V, Olteanu M, Vasile CM. COVID-19 and tuberculosis: unveiling the dual threat and shared solutions perspective. J Clin Med. 2023;12(14) https://doi.org/10.3390/jcm12144784.
11. Cohen A, Mathiasen VD, Schon T, Wejse C. The global prevalence of latent tuberculosis: a systematic review and meta-analysis. Eur Respir J. 2019;54(3) https://doi.org/10.1183/13993003.00655-2019.
12. Conway Morris A, Tong A. Novel treatments and trials in COVID-19: COVID-19 pandemic. 2022:109-20. 2021; https://doi.org/10.1016/B978-0-323-82860-4.00006-9. Epub 2021 Jun 4
13. Copyright © 2024, StatPearls Publishing LLC.
14. Danino D, Ben-Shimol S, van der Beek BA, Givon-Lavi N, Avni YS, Greenberg D, et al. Decline in pneumococcal disease in young children during the coronavirus disease 2019 (COVID-19) pandemic in Israel associated with suppression of seasonal respiratory viruses, despite persistent pneumococcal carriage: a prospective cohort study. Clin Infect Dis. 2022;75(1):e1154–64. https://doi.org/10.1093/cid/ciab1014.
15. Domán M, Bányai K. COVID-19-associated fungal infections: an urgent need for alternative therapeutic approach? Front Microbiol. 2022;13
16. Falzon D, Zignol M, Bastard M, Floyd K, Kasaeva T. The impact of the COVID-19 pandemic on the global tuberculosis epidemic. Front Immunol. 2023;14:1234785. https://doi.org/10.3389/fimmu.2023.1234785.
17. Gbinigie K, Frie K. Should azithromycin be used to treat COVID-19? A rapid review. BJGP Open. 2020;4(2) https://doi.org/10.3399/bjgpopen20X101094.
18. Gothankar J, Doke P, Dhumale G, Pore P, Lalwani S, Quraishi S, et al. Reported incidence and risk factors of childhood pneumonia in India: a community-based cross-sectional study. BMC Public Health. 2018;18(1):1111. https://doi.org/10.1186/s12889-018-5996-2.
19. Hasan MN, Haider N, Stigler FL, Khan RA, McCoy D, Zumla A, et al. The global case-fatality rate of COVID-19 has been declining since May 2020. Am J Trop Med Hyg. 2021;104(6):2176–84. https://doi.org/10.4269/ajtmh.20-1496.
20. Hoque MN, Akter S, Mishu ID, Islam MR, Rahman MS, Akhter M, et al. Microbial co-infections in COVID-19: associated microbiota and underlying mechanisms of pathogenesis. Microb Pathog. 2021;156:104941. https://doi.org/10.1016/j.micpath.2021.104941.
21. Jilani TN, Avula A, Zafar Gondal A, Siddiqui AH. Active tuberculosis. In: StatPearls. Treasure Island: StatPearls Publishing; 2024. ineligible companies. Disclosure: Akshay Avula declares no relevant financial relationships with ineligible companies. Disclosure: Anoosh Zafar Gondal declares no relevant financial relationships with ineligible companies. Disclosure: Abdul Siddiqui declares no relevant financial relationships with ineligible companies.

22. Kournoutou GG, Dinos G. Azithromycin through the lens of the COVID-19 treatment. Antibiotics (Basel). 2022;11(8) https://doi.org/10.3390/antibiotics11081063.
23. Lansbury L, Lim B, Baskaran V, Lim WS. Co-infections in people with COVID-19: a systematic review and meta-analysis. J Infect. 2020;81(2):266–75. https://doi.org/10.1016/j.jinf.2020.05.046.
24. Losier A, Gupta G, Caldararo M, Dela Cruz CS. The impact of coronavirus disease 2019 on viral, bacterial, and fungal respiratory infections. Clin Chest Med. 2023;44(2):407–23. https://doi.org/10.1016/j.ccm.2022.11.018.
25. Maiti K, Syal K, Chatterji D, Jayaraman N. Synthetic Arabinomannan Heptasaccharide glycolipids inhibit biofilm growth and augment isoniazid effects in mycobacterium smegmatis. Chembiochem. 2017;18(19):1959–70. https://doi.org/10.1002/cbic.201700247.
26. Marques L, Costa B, Pereira M, Silva A, Santos J, Saldanha L, et al. Advancing precision medicine: a review of innovative in Silico approaches for drug development. Clin Pharmacol Pers Healthc. 2024;16(3):332.
27. McChlery S, Ramage G, Bagg J. Respiratory tract infections and pneumonia. Periodontol. 2009;49(1):151–65. https://doi.org/10.1111/j.1600-0757.2008.00278.x.
28. McQuaid CF, Vassall A, Cohen T, Fiekert K, White RG. The impact of COVID-19 on TB: a review of the data. Int J Tuberc Lung Dis. 2021;25(6):436–46. https://doi.org/10.5588/ijtld.21.0148.
29. Molina JM, Delaugerre C, Le Goff J, Mela-Lima B, Ponscarme D, Goldwirt L, de Castro N. No evidence of rapid antiviral clearance or clinical benefit with the combination of hydroxychloroquine and azithromycin in patients with severe COVID-19 infection. Med Mal Infect. 2020;50(4):384. https://doi.org/10.1016/j.medmal.2020.03.006.
30. Moreira-Teixeira L, Mayer-Barber K, Sher A, O'Garra A. Type I interferons in tuberculosis: foe and occasionally friend. J Exp Med. 2018;215(5):1273–85. https://doi.org/10.1084/jem.20180325.
31. Naresh K, Avaji PG, Maiti K, Bharati BK, Syal K, Chatterji D, Jayaraman N. Synthesis of β-arabinofuranoside glycolipids, studies of their binding to surfactant protein-A and effect on sliding motilities of M. smegmatis. Glycoconj J. 2012;29(2–3):107–18. https://doi.org/10.1007/s10719-012-9369-2.
32. Pahal P, Rajasurya V, Sharma S. Typical bacterial pneumonia. In: StatPearls. Treasure Island: StatPearls Publishing; 2024. Ineligible companies. Disclosure: Venkat Rajasurya declares no relevant financial relationships with ineligible companies. Disclosure: Sandeep Sharma declares no relevant financial relationships with ineligible companies.
33. Rs N, Reddy M, Batra S, Srivastava SK, Syal K. Vitamin C and its therapeutic potential in the management of COVID19. Clin Nutr ESPEN. 2022;50:8–14. https://doi.org/10.1016/j.clnesp.2022.05.026.
34. Rs N, Sinha SK, Batra S, Regatti PR, Syal K. Promoter characterization of rel Z-bifunctional (pp) pGpp synthetase in mycobacteria. Genes Cells. 2024; https://doi.org/10.1111/gtc.13135.
35. Said KB, Alsolami A, Moussa S, Alfouzan F, Bashir AI, Rashidi M, et al. COVID-19 clinical profiles and fatality rates in hospitalized patients reveal case aggravation and selective co-infection by limited gram-negative bacteria. Int J Environ Res Public Health. 2022;19(9) https://doi.org/10.3390/ijerph19095270.
36. Sanduzzi Zamparelli S, Mormile M, Sanduzzi Zamparelli A, Guarino A, Parrella R, Bocchino M. Clinical impact of COVID-19 on tuberculosis. Infez Med. 2022;30(4):495–500. https://doi.org/10.53854/liim-3004-3.
37. Saunders BM, Britton WJ. Life and death in the granuloma: immunopathology of tuberculosis. Immunol Cell Biol. 2007;85(2):103–11. https://doi.org/10.1038/sj.icb.7100027.
38. Saunders M, Nellums L. The indirect effects of COVID-19 upon global childhood pneumonia. Public Health Pract. 2022;3:100228. https://doi.org/10.1016/j.puhip.2022.100228.
39. Shah T, Shah Z, Yasmeen N, Baloch Z, Xia X. Pathogenesis of SARS-CoV-2 and mycobacterium tuberculosis coinfection. Front Immunol. 2022;13:909011. https://doi.org/10.3389/fimmu.2022.909011.

40. Sheerin D, Abhimanyu, Peton N, Vo W, Allison CC, Wang X, et al. Immunopathogenic overlap between COVID-19 and tuberculosis identified from transcriptomic meta-analysis and human macrophage infection. iScience. 2022;25(6):104464. https://doi.org/10.1016/j.isci.2022.104464.
41. Shobugawa Y, Saito R, Sato I, Kawashima T, Dapat C, Dapat IC, et al. Clinical effectiveness of neuraminidase inhibitors–oseltamivir, zanamivir, laninamivir, and peramivir--for treatment of influenza A(H3N2) and A(H1N1)pdm09 infection: an observational study in the 2010–2011 influenza season in Japan. J Infect Chemother. 2012;18(6):858–64. https://doi.org/10.1007/s10156-012-0428-1.
42. Sultana J, Cutroneo PM, Crisafulli S, Puglisi G, Caramori G, Trifirò G. Azithromycin in COVID-19 patients: pharmacological mechanism, clinical evidence and prescribing guidelines. Drug Saf. 2020;43(8):691–8. https://doi.org/10.1007/s40264-020-00976-7.
43. Syal K. COVID-19: herd immunity and convalescent plasma transfer therapy. J Med Virol. 2020;92(9):1380–2. https://doi.org/10.1002/jmv.25870.
44. Syal K, Chatterji D. Vitamin C: a natural inhibitor of Cell Wall functions and stress response in mycobacteria. Adv Exp Med Biol. 2018;1112:321–32. https://doi.org/10.1007/978-981-13-3065-0_22.
45. Syal K, Maiti K, Naresh K, Avaji PG, Chatterji D, Jayaraman N. Synthetic arabinomannan glycolipids impede mycobacterial growth, sliding motility and biofilm structure. Glycoconj J. 2016;33(5):763–77. https://doi.org/10.1007/s10719-016-9670-6.
46. TB/COVID-19 Global Study Group. Tuberculosis and COVID-19 co-infection: description of the global cohort. Eur Respir J. 2022;59(3) https://doi.org/10.1183/13993003.02538-2021.
47. Togun T, Kampmann B, Stoker NG, Lipman M. Anticipating the impact of the COVID-19 pandemic on TB patients and TB control programmes. Ann Clin Microbiol Antimicrob. 2020;19(1):21. https://doi.org/10.1186/s12941-020-00363-1.
48. Vijay S, Bansal N, Rao BK, Veeraraghavan B, Rodrigues C, Wattal C, et al. Secondary infections in hospitalized COVID-19 patients: Indian experience. Infect Drug Resist. 2021;14:1893–903. https://doi.org/10.2147/idr.S299774.
49. Yang JW, Fan LC, Miao XY, Mao B, Li MH, Lu HW, et al. Corticosteroids for the treatment of human infection with influenza virus: a systematic review and meta-analysis. Clin Microbiol Infect. 2015;21(10):956–63. https://doi.org/10.1016/j.cmi.2015.06.022.

Economic Consequences of the COVID-19 Pandemic

23

Ankita Srivastava and Bhartrihari Pandiya

Abstract

The COVID-19 pandemic had a substantial impact on human lives as well as created seismic economic disruptions, affecting virtually every sector of the world economy. It has disrupted industries from travel and hospitality, creating job losses, and financial instability for millions across the globe. The level of volatility in financial markets has been unprecedented, with investors uncertain and anxious. Though stimulus packages and policies from governments have been put in place, these actions will be question marks when it comes to determining the length of these recoveries by factors like vaccine results and consumer measures. The path to a post-pandemic economy, one that is greener and more inclusive, depends on exercising resilience in innovative ways through coordinated global solidarity during this time of crisis. A coordinated effort and international cooperation can reduce the economic damage and build a more sustainable and inclusive post-pandemic economy. In light of this discussion, this research aims to describe the economic effects of the pandemic, summarizing the literature done in the related area.

Keywords

Economic disruptions · Financial markets · Post-pandemic · Government policy · Economic effects

A. Srivastava (✉) · B. Pandiya
School of Management Studies, National Forensic Sciences University, Gandhinagar, Gujarat, India
e-mail: ankita.srivastava@nfsu.ac.in

M. K. Ghosh, M. Basu (eds.), *Pathogenesis, Management and Socio-Economic Impact of COVID-19 Pandemic*, Medical Virology: From Pathogenesis to Disease Control, https://doi.org/10.1007/978-981-95-1426-7_23

Introduction

The COVID-19 pandemic rapidly spread worldwide and thus was declared a pandemic by the World Health Organization by mid-March 2020. Significant economic disruption resulted from this public health emergency, marking an abrupt end to business operations [41] and showing a market volatility seldom seen since the Great Depression [7]. This health emergency has had significant economic repercussions worldwide, with no country seemingly escaping its impact. The effects extend beyond the economy, influencing the whole society, and leading to a significant change in business practices and consumer behavior. The economic toll of COVID-19 over the past years reflected a rapid and unprecedented decline in production, trade, and employment. It is slowly rebounding, but the historical evidence suggests that some long-term effects might last for decades [5, 10].

Pandemics are generally unbiased in affecting companies irrespective of their size. However, it is believed that small and medium enterprises were less affected in contrast to bigger firms [37]. Entrepreneurial traits, government support, technology utilization, and pro-social leadership were a few reasons for the resilience of small- and medium-sized companies during COVID-19 in India [42]. Another approach that has helped small businesses is crowdfunding, mainly through virtual platforms. Traditional funding sources like banks and venture capitalists became more stringent, and crowdfunding was a boon [33]. Theoretically, crowdfunding is a collective investment model where an assorted set of individuals, generally known as the "crowd," contribute their funds to support a venture. It is considered a substitute to traditional funding sources that offer convenience and inclusivity to project initiators and investors [21].

The pandemic has hit the global economy over the years. By mid-2023, the pandemic had resulted in more than 6.5 million deaths worldwide. The global economic output had declined from its pre-pandemic levels, highlighting a significant adverse effect on global economic growth. Additionally, the pandemic caused substantial interruptions to international trade. In response to the crisis, governments implemented strong policy measures, which led to increased government debt and deficits. Later, following a dip, prices have surged, leading to high rates of inflation [29].

Research on the economic consequences of COVID-19 so far has mainly focused on the immediate impacts of containment and mitigation efforts [13]. However, as governments deploy extensive fiscal measures to combat the pandemic, it is essential to comprehend the potential economic landscape in the coming decades. Past studies typically focused on individual events in a single country or region, tracking their effects over a decade [6, 19]. However, in such huge-scale pandemics, impacts are felt throughout economies and in even broader areas, either due to the widespread nature of the infection or because economic shocks propagate through trade and market integration. In a study by Jordà et al. [25], a broader perspective was taken to examine the macroeconomic effects of pandemics across several European economies, analyzing the aftermath of 15 major pandemics that resulted in at least 100,000 deaths.

Using new data on long-term sovereign debt yields dating back to the fourteenth century [35], researchers estimated the impact of major pandemics on the real (inflation-adjusted) natural rate of interest in Europe. From here on, it is referred to as the "natural rate." Economists refer to the neutral rate of interest as the ideal level where the economy can grow at its full potential while keeping inflation stable. The natural rate is a key economic indicator. For instance, if people start saving more, the supply of available savings increases. Conversely, if economic growth slows down, investment becomes less attractive. In both cases, the natural rate of interest decreases to restore equilibrium.

Jordà et al. [25] proposed that pandemics exert long-lasting effects on interest rates. It was found that the expected rate of interest typically drops by about 1.5 percentage points. Moreover, it takes another two decades for the usual rate to come to its pre-pandemic level. These findings underscore the significant economic impact that pandemics have had throughout history. It is well documented that real safe rates, which closely correlate with the natural rate, can remain depressed for 5–10 years after financial crises [24]. However, the persistence of the effects following pandemics is notably more substantial. After a pandemic, people tend to save more and invest less, which leads to a slower pace of economic growth. Given the current environment, where savings can yield negative returns, it is uncertain whether investors will continue to be as conservative as they have been in the past.

The evidence presented by Jordà et al. [25] aligns with the well-known neoclassical growth model. There is a shift in the relative returns between labor and capital when labor is lost for capital loss. This imbalance can cause interest rates to drop further, amplified by increased savings from survivors, who might either want to rebuild their wealth or become more cautious about spending. If the neoclassical mechanism is correct, there is another aspect where the effects of pandemics should be observed. As the labor-to-capital ratio decreases, the natural rate should drop, but real wages should increase. This pattern is consistent with the predictions of the neoclassical growth model.

If the historical patterns hold after COVID-19, the concept of secular stagnation [39] might be a significant concern for monetary and fiscal policy for at least the next 20 years. Several factors could mitigate the rate of interest fall. First, the death toll from COVID-19, as a proportion of the total population, might be lower than that of some major pandemics in the past. This could be attributed to advances in medical care and public health practices. Second, COVID-19 predominantly affects older adults, who are generally no longer part of the labor force and save comparatively more than younger individuals. This is quite different from pandemics previously, where life expectancy was short and the impact on the working-age population was more significant. Third, the substantial fiscal stimulus is likely to increase public debt, leading to a reduced national savings rate, which might exert upward pressure on real interest rates.

Despite these mitigating factors, a continued period of low real interest rates is likely, albeit with a lesser decline due to the factors mentioned. These lower real interest rates could offer governments the fiscal space needed to implement strong procedures for counteracting the economic outcome.

Literature Review

The pandemic has had profound and widespread economic effects globally. It triggered a tough economic downturn, affecting nearly every sector and demographic. The lockdown measures were implemented worldwide to reduce virus spread, but had significant social and economic repercussions on populations and socio-economic systems. These measures disrupted normal social, commercial, and work activities, altering individuals' and households' daily lives and often increasing their economic vulnerability by reducing their ability to maintain their standard of living. This literature review synthesizes research on the pandemic's economic consequences, focusing on various aspects such as global economic contraction, labor market disruptions, sector-specific impacts, government responses, and long-term implications.

Concerning global economic contractions, several studies have documented the sharp contraction in global economic activity due to the pandemic [3, 16, 17, 38]. According to the International Monetary Fund (IMF), the global economy shrank by 3.5% in 2020, marking the worst peacetime recession since the Great Depression [23]. The World Bank [45] highlighted that the pandemic pushed over 100 million people into extreme poverty, reversing decades of progress. These macroeconomic studies underscore the unprecedented scale and scope of the economic shock. In the domain of labor market disruptions, the pandemic caused significant disruptions in labor markets worldwide. The pandemic led to significant job losses and increased underemployment. Ridhwan et al. [34] and Choi et al. [12] state that unemployment rates spiked globally, with many workers facing reduced hours or temporary layoffs, and also gender inequality [14]. The International Labour Organization (ILO) reported a loss of 8.8% of global working hours in 2020, equivalent to 255 million full-time jobs [22]. As per Adams-Prassl et al. [2], workers who fall in the lower-income bracket and those in precarious employment were affected disproportionately. The shift to remote work, while beneficial for some, created new challenges and widened the digital divide. Different sectors experienced varying degrees of impact from the pandemic. The hospitality, travel, and tourism industries were among the hardest hit, with many businesses closing permanently [31]. Manufacturing and supply chains also faced significant disruptions, leading to shortages and inflationary pressures [20]. In contrast, the technology and e-commerce sectors saw substantial growth, driven by increased demand for digital services and online shopping [43]. Studies by Chetty et al. [11] and Baldwin & Di Mauro [8] highlighted the effectiveness of various stimulus packages, unemployment benefits, and business support measures. However, the scale and success of these interventions varied widely across countries, often reflecting pre-existing economic conditions and institutional capacities.

The economic impact of the pandemic in the long term is yet to come and is gradually unfolding. Initial research suggests the potential for lasting changes in labor markets, with increased adoption of remote work and automation [9]. The pandemic has also accelerated trends toward digitalization and could lead to structural shifts in global trade patterns [15]. Additionally, the increased public and private debt incurred during the crisis may pose challenges to future economic stability and growth [23].

Thus, the objectives of the research work are as follows:

1. To identify and analyze the thematic clusters within the research domain of economic consequences of the COVID-19 pandemic.
2. To highlight the emerging challenges and other variables related to the economic consequences of the COVID-19 pandemic and thus provide theoretical implications along with the scope for future research.

Methodology

This research aims to provide a summary of research to date. Bibliometric analysis can be of help in such cases where the present literature can be analyzed [28, 40] because it provides a systematic and transparent understanding of the research [4]. This concept was initially introduced as a scientific method to observe the evolution of a research area over time using a multi-disciplinary approach [32]. Such analyses facilitate the identification of impact through author data, research domain data, journal data, and emerging themes [36]. VOSviewer is a valuable software tool for conducting bibliometric analysis [44]. Consequently, bibliometric analysis was employed for the identification of the various research themes. Additionally, recent trends in this research area will be explored using bibliometric analysis, highlighting various emerging research areas and leading to new research directions.

The SCOPUS database was considered for the bibliometric analysis. The literature search period was from 2020 to 2024. The last search query was as follows: ((Abstract-Title-Keyword (Economic Consequences) AND Abstract-Title-Keyword (COVID-19) OR Abstract-Title-Keyword (Pandemic)). The total number of research papers finalized was 201. A comprehensive record containing the bibliographical information, citation details, abstract, and related data was downloaded in CSV format for further analysis.

Discussion

Co-authorship analysis acts as a valuable tool to understand research collaboration networks and map collaborations, as shown by the mapping of how researchers collaborate within and across institutions, disciplines, and geographic regions. By visualizing these networks, it becomes easier to identify key researchers and influential groups within a particular field. Co-authorship networks can highlight central figures who have numerous connections, indicating their influential role. In Fig. 23.1, the highly connected authors, like Lisa [27] and Mvogo [30], appear to have numerous connections, indicating the authors have collaborated with many other authors in the network. There are several visible clusters where groups of authors are more closely connected to each other, suggesting strong collaborative

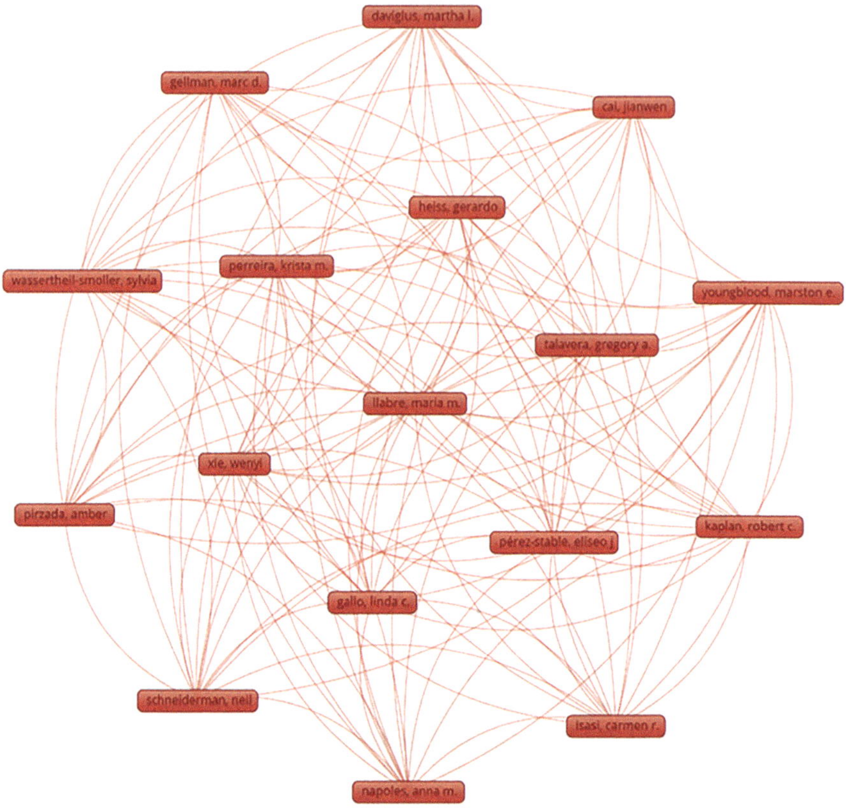

Fig. 23.1 Co-citation analysis

groups. This network can help identify key researchers, potential collaborators, and the structure of research collaboration within a particular field or institution. Understanding these relationships can aid in strategic planning for research projects, funding allocations, and academic networking.

The co-occurrence network analysis helped in the identification of clusters. The clusters show the research themes being studied by previous researchers. The themes identified in the clusters can serve as a future scope for upcoming researchers. Various researchers have worked in different dimensions of this field and discussed their findings. Upon comprehensive screening and evaluation of the research works, some variables related to the research topic were understood. The aforementioned emerging research themes, identified through clusters in the area of the economic consequences of the COVID-19 pandemic, can be further investigated using empirical datasets. Figure 23.2 illustrates the keywords co-occurrence network analysis, where macro clusters have been identified following the bibliometric analysis. The different colors in the co-occurrence diagram represent the connectivity between related keywords within each cluster. The keyword "COVID-19" appears

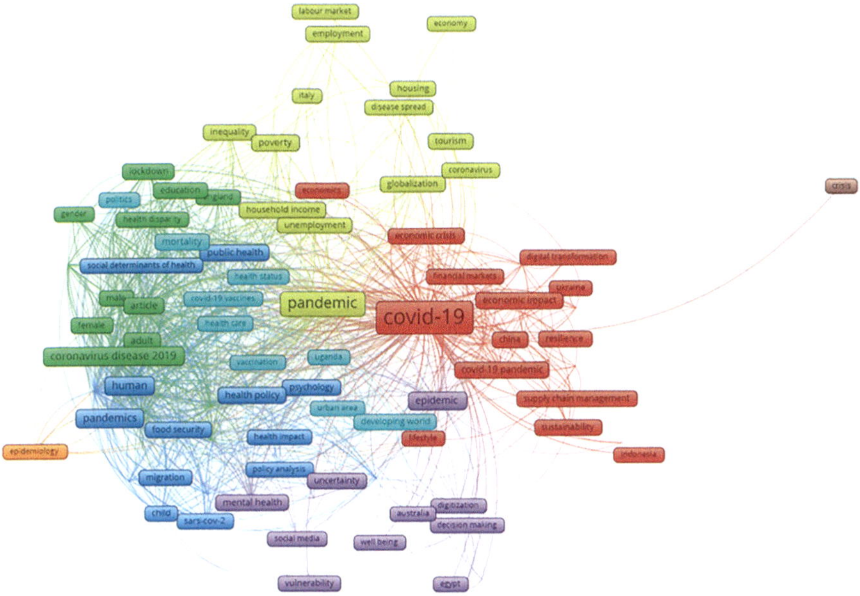

Fig. 23.2 Co-occurrence keyword analysis

most prominently, indicating that it is the central theme, with other keywords related to it. The optimization algorithm detects communities within large networks by combining nodes using heuristics [26]. The keywords co-occurrence analysis provides a unique insight into various areas and their applications across multiple research domains.

A thorough reading of the research papers gave a few insights and variables into the research area, paving the way for emerging research areas. These findings have been discussed in a summarized form below.

Major Macroeconomic Variables

In an interconnected world, the broader impacts of the pandemic have been obvious since its onset. Beyond mortality and morbidity, the pandemic has significantly disrupted various economic activities, particularly global supply chains. Companies worldwide, regardless of their size, that rely on inputs from China have faced production contractions. Restrictions and limitations on transportation between countries have further decelerated economic activity on a global level. Additionally, consumers and firms are in panic, and this has altered typical patterns of consumption. Global financial markets have reacted to these changes, resulting in a sharp decline in global stock indices. The government responded to the pandemic swiftly and comprehensively, introducing a range of policy measures that were unprecedented in scale for emerging economies. These initiatives included large-scale direct income support programs and asset purchase schemes. However, the pandemic has also presented several challenges.

Movement of the Stock Market/Interest Rates

The central banks in numerous countries have reduced interest rates. This measure aims to make borrowing cheaper and stimulate spending to support economic recovery. Important macroeconomic variables like output, investment, lending, interest rates, and inflation are thought to be impacted by past credit accumulation. Abbas et al. [1] findings provide credence to the notion that financial variables are significant drivers of the contemporary business cycle. To reduce past credit accumulated and control inflation post-pandemic, central banks increased interest rates, increasing the cost of borrowing money overall [1].

Rising Inflation

At first, COVID-19's effect on prices around the world was remarkably consistent. The developed world often has the lowest inflation rates, whereas frontier markets (developing economies that have not yet advanced enough to be classified as emerging) usually have the highest rates. 2020 saw a sharp decline in consumer expenditure, global inflation, and a desperate central bank money supply glut. Prices increased almost everywhere in 2021 and 2022 as the world reopened despite supply constraints. That was when the rift opened. This is evident when we examine the average inflation rates for the developed, emerging, and frontier worlds from 2014 onward.

Turkey and Argentina are considered exceptional cases. Turkey's insane interest rate reduction from 2021 to 2023 has left the country with inflation that is currently almost 70%. Furthermore, the causes of Argentina's current 276% inflation rate go back much beyond the pandemic. When the South American nation is taken into account, the mean for frontier markets approaches 25%. It is believed that emerging economies showed more resilience than developed ones. The so-called "frontier" countries, which are still too impoverished or lack adequately accessible and liquid markets to qualify as developing, were unable to contain price increases to the extent that emerging economies demonstrated that they were worthy of the label.

Upon examining the raw data, it is evident that there is still a significant distance to cover in the decline; globally, rates are above inflation, indicating tight policies that are constraining economic activity. Most central banks, particularly those in developed countries, refrained from making policy changes until they were certain that the downside risks had diminished. Consequently, interest rates had to be raised much more quickly and in larger increments than in previous decades due to the delayed response. The abrupt and intricate tightening raised additional concerns regarding the sustainability of budget propositions.

Rising Global Debt

The level of public debt persisted despite the economic recovery that began in 2020 and the significantly higher-than-expected inflation rate. While many governments halted pandemic-related fiscal support, they continued to spend more to spur the economy and respond to surges in the prices of food and energy, which resulted in fiscal deficits that maintained the levels of public debt high.

According to the most recent Global Debt Monitor, the nation's debt decreased by barely 8 percentage points of GDP during the previous 2 years, countering just around half of the rise caused by the pandemic. Private debt, which comprises debt held by households and nonbanking corporations, decreased by 12 percentage points of GDP at a quicker rate. Even still, the decrease was insufficient to eliminate the pandemic spike. Between the middle of the 1970s and the end of 2022, the amount of public debt worldwide tripled, accounting for 92% of GDP, or a little over $91 trillion. Between 1960 and 2022, private debt also tripled, reaching 146 percent of GDP, or about $144 trillion [46] of GDP, or about $144 trillion when the labor is lost there is capital loss.

Unemployment

The global unemployment rate rose by 1.1% to reach 6.5% in 2020. This increase resulted in a global total of 220 million unemployed individuals, marking a rise of 33 million. Additionally, 81 million people withdrew entirely from the labor force. Across Europe, the Americas, Latin America, and the Caribbean, the unemployment rate increased by at least two percentage points. Even before the pandemic, the youth unemployment rate was already three times higher than the adult rate. As per the International Monetary Fund (IMF), the rate of unemployment in the United States touched 8.9%. Numerous workers have been enrolled in job retention schemes supported by the government, especially in industries like tourism and hospitality, which have come to a near standstill. While job vacancies in Australia have rebounded to 2019 levels, they continue to lag when compared to European countries.

Fiscal Deficit

According to data from the Asian Development Bank (ADB), several developing Asian countries have experienced significant negative fiscal balances. In 2020, the pandemic had a dual impact on the region's economies: a sharp decline in government revenue due to business and industrial slowdowns, along with substantial increases in government spending to address the pandemic's economic, societal, and health impacts.

Countries heavily reliant on a single export or service sector were severely affected due to the virus outbreak. Governments in such nations implemented various unprecedented policy measures like asset purchase schemes and direct income support programs, which varied widely in terms of their extent and effectiveness (refer to Fig. 23.1). This disparity can be attributed, in part, to the challenges faced by many low-income countries in accessing credit markets. As a result, high-income countries generally implemented substantial fiscal responses, and countries with low incomes had limited fiscal responses.

Numerous countries experience fiscal imbalances, often referred to as current account deficits, where government spending exceeds income. This leads to an increase in national debt as the government borrows to bridge the spending-income gap. While increasing the fiscal deficit can inject more money into the economy, potentially revitalizing it, long-term deficits can pose risks to stability and economic growth.

Decline in GDP

The implementation of localized lockdowns and restrictions in the past has had a significant impact on recovery. India's fragile economic recovery has been set back by the back-to-back waves of the pandemic. Issues such as rising inequality and limited household balance have further hampered the process of recovery. India, which was experiencing a growth rate of only 4% in 2019–20, saw a contraction of 7–8% in 2020–21 and is now facing another period of low economic growth recovery in 2021. Fiscal policy must play a supportive role in guiding vulnerable businesses and households toward economic recovery. The global economy faced unprecedented challenges in 2020 due to the COVID-19 pandemic, with significant contractions seen across major economies. The G20 countries, representing the world's largest economies, experienced a sharp decline in growth rates, indicating the severity of the economic impact. Countries like the United States, the United Kingdom, and France saw substantial contractions in their GDP, reflecting the widespread disruption caused by the pandemic. In Asia, India and China experienced contrasting economic trends, with India facing a significant decline in growth, while China managed to achieve positive growth. This divergence highlights the varying degrees of resilience and response strategies adopted by different countries in the region. Looking ahead, the IMF's forecast of global growth of 5.2% in 2021 offers a glimmer of hope for economic recovery. However, the recovery process is expected to be uneven, with countries heavily reliant on services facing a slower path to recovery. The pandemic has underscored the importance of coordinated global efforts to address economic challenges and build a more resilient and inclusive economy for the future.

Theoretical Implications

The pandemic has triggered an unprecedented surge of academic work across all disciplines, particularly in economics. This extends a comprehensive set of insights into the themes and trends, as well as the theoretical implications of this expanding body of literature, through the lens of bibliometrics. This is an insightful examination of how the pandemic has altered theoretical orientations and research paradigms in economics through the analysis of citations, co-authorship networks, and thematic clusters. Refocusing our narrative toward the power of crisis economics, looking at how nations overcome profound disturbances to their economic systems, is a much-needed reprieve that this pandemic has forced us once more to remember. Historically, economic resilience has extended well beyond financial and structural robustness, leading to social and health aspects as well.

Theoretical frameworks currently stress that resilience to economic shocks relies upon adaptability, flexibility, and coordination across multiple sectors. These labor market disturbances caused by the pandemic have triggered fresh debates on labor flexibility, remote working, and the gig economy. The sudden move to remote work has set a new dimension of labor market theories, challenging older paradigms that focused on the requirement for physical presence and centralized workplaces. The role of the gig economy has been re-assessed in a way that underscores its weaknesses and the importance of social protection measures. Meanwhile, the disruptions and interruptions of global supply chains were quite visible as well, leading to a reconsideration of theories on globalization. New theoretical models introduce the dangers of over-connected supply chains and the need for backup and diversification. The narrative around de-globalization has been gaining credibility over the past year, suggesting an increasing suitability for more localized or decentralized (and therefore hopefully more resilient) trade and supply chains. Indeed, the pandemic has seen an expansion in the use of big data and real-time analytics in economic research. Instrumented by machine learning, with an increasing volume of highly frequented data sources like mobile phone location data or online transaction records to back up traditional econometric models. This change captures an improvement in both the accuracy and timeliness of economic forecasting as well as policy evaluation. The complexity of the economic impact of the pandemic has required an interdisciplinary approach to research blending economics with public health, sociology, and environmental science. Today, these fields are bridged in economic models that are still a new way of making theory the most convergence part. This represents an enrichment of economic theory through interdisciplinary perspectives and methodologies. The widespread implementation of fiscal and monetary policy responses in the context of the COVID-19 pandemic has led to a rethinking of what is possible regarding policy effectiveness and constraints. Recently, theoretical models have begun to be built that trace the implications of unprecedented government spending and monetary easing over time, including inflationary risks [18], debt sustainability, and putting pressure on crowding out private investment. Central banks are now assigned roles far beyond those traditionally associated with inflation targeting and more toward economic stabilization. The pandemic has revealed limitations in current social safety offerings, prompting theorizations in welfare economics. Following this, models refocus on strengthened and more integrated social protection systems that can flexibly respond to shocks. Debates on the merits of these and related changes include universal basic income, expanded unemployment benefits, and targeted support for traditionally vulnerable populations.

Future Scope of Research

The long-term economic fallouts of the pandemic are uncertain, making it a necessity to have longitudinal studies that can follow recovery and transformation going forward. For instance, future researchers will have to grapple with the possibility of substantial economic transformation: how consumers spend their money; who participates in the labor force, and more broadly, what industries dominate a country's economy. Much more work is needed to explore behavioral economics applications to such issues arising under uncertainty. Further studies might examine how fear, risk perception, and social norms affect economic behavior during the crisis as well as psychological factors will be integrated into theoretical frameworks to make predictions and reconcile economic outcomes. Lockdowns, however, have given environmental economists the closest thing to natural experiments one usually gets in real-world conditions: a temporary relief from pollution and greenhouse gas emissions.

The latest theoretical models will try to understand how economic activity is related to or traded off with both environmental sustainability and public health. Policies of green recovery and sustainable growth could be guided by this line of research. A critical need exists to examine the skills of the workforce offered concerning the impact of these crises. Future work could investigate the impact of these changes in schooling and online learning on long-term educational outcomes and skill formation. The toll of furloughs and layoffs upon reskilling and upskilling initiatives should also be another area for further insight. Global supply chains, however, have proven more susceptible to the pandemic. The usefulness of these analyses may be determined in future research that examines reconfiguring supply chains to make them more resilient, including diversification, regionalization, and introduction of new technologies (blockchain) to bring transparency to the supply chain. Knowing these dynamics can assist businesses and policymakers in managing risk concerning future global disruptions.

Conclusions

The research focusing on the economic aftermath of the pandemic has implications far beyond economics. In certain areas, the crisis has accelerated transformations in economic theories and methodological improvements and prompted new policy dialogues to address the complex impacts of COVID-19. It has exposed weaknesses in established economic models, accelerated existing trends, and sparked fundamental changes in both the theory and practice of economic policies. Research will continue to guide these theoretical frameworks, evolving our comprehension of economic resilience and adaptation in the global economy. Lessons also must be drawn from this crisis to build more resilient, inclusive, and sustainable economies as part of our recovery. Continued research and interdisciplinary efforts will be vital in tackling the multifaceted, evolving challenges. The enduring impact of the pandemic on the economy will affect economic policy and theory well into the future, highlighting the value of adaptability and foresight.

Author's Contributions Conceived the idea and review structure: AS; Wrote the manuscript: AS and BP; Revised and edited the manuscript: AS and BP. The author has read, agreed, and approved the final draft of the manuscript.

Funding This work is not funded by any organization.

Data Availability The software VOS viewer (https://www.vosviewer.com//) was used to draw the figures.

Declarations

Conflicts of Interest There is no competing and conflict of interest.

References

1. Abbas S, Saha T, Sinha A. Price response of top-five renewable energy firms to Russia-Ukraine conflict: an advanced quantile analysis to achieve net-zero in The United States of America. J Clean Prod. 2024;442:141153.
2. Adams-Prassl A, Boneva T, Golin M, Rauh C. Inequality in the impact of the coronavirus shock: evidence from real-time surveys. J Public Econ. 2020;189:104245.
3. Ajmal MM, Khan M, Shad MK. The global economic cost of coronavirus pandemic: current and future implications. Public Adm Policy. 2021;24(3):290–305.
4. Aria M, Cuccurullo C. bibliometrix: an R-tool for comprehensive science mapping analysis. J Informet. 2017;11(4):959–75.
5. Arthi V, Parman J. Disease, downturns, and wellbeing: economic history and the long-run impacts of COVID-19. Explor Econ Hist. 2021;79:101381.
6. Asghar MS, Yasmin F, Alvi H, Shah SMI, Malhotra K, Farhan SA, Rasheed U. Assessing the mental impact and burnout among physicians during the COVID-19 pandemic: a developing country single-center experience. Am J Trop Med Hyg. 2021;104(6):2185.
7. Baker SR, Davis SJ, Levy JA. State-level economic policy uncertainty. J Monet Econ. 2022;132:81–99.
8. Baldwin R, Di Mauro BW. Economics in the time of COVID-19: a new eBook. Vox CEPR Policy Portal. 2020;2(3):105–9.
9. Barrero JM, Bloom N, Davis SJ. Why working from home will stick (No. w28731). National Bureau of Economic Research; 2021.
10. Callegari B, Feder C. A literature review of pandemics and development: the long-term perspective. Econ Disaster Clim Chang. 2022;6(1):183–212.

11. Chetty R, Friedman JN, Hendren N, Stepner M, Team TOI. The economic impacts of COVID-19: evidence from a new public database built using private sector data (Working paper# 27431) 2020.
12. Choi YJ, Kühner S, Shi SJ. From "new social risks" to "COVID social risks": the challenges for inclusive society in South Korea, Hong Kong, and Taiwan amid the pandemic. Polic Soc. 2022;41(2):260–74.
13. Correia S, Luck S, Verner E. Pandemics depress the economy, public health interventions do not: evidence from the 1918 flu. J Econ Hist. 2022;82(4):917–57.
14. Dang HAH, Nguyen CV. Gender inequality during the COVID-19 pandemic: income, expenditure, savings, and job loss. World Dev. 2021;140:105296.
15. Evenett SJ. Sicken thy neighbour: the initial trade policy response to COVID-19. World Econ. 2020;43(4):828–39.
16. Fernandes N. Economic effects of coronavirus outbreak (COVID-19) on the world economy. 2020.
17. Gopinath G. Limiting the economic fallout of the coronavirus with large targeted policies. In: Mitigating the COVID economic crisis: act fast and do whatever it takes; 2020. p. 41–8.
18. Galbács P. Lucas's way to his monetary theory of large-scale fluctuations. J Econ Methodol. 2022;29(1):4–16.
19. Gray NS, O'Connor C, Knowles J, Pink J, Simkiss NJ, Williams SD, Snowden RJ. The influence of the COVID-19 pandemic on mental well-being and psychological distress: impact upon a single country. Front Psych. 2020;11:594115.
20. Guan D, Wang D, Hallegatte S, Davis SJ, Huo J, Li S, Gong P. Global supply-chain effects of COVID-19 control measures. Nat Hum Behav. 2020;4(6):577–87.
21. Hecht K. 'It's the value that we bring': performance pay and top income earners' perceptions of inequality. Soc Econ Rev. 2022;20(4):1741–66.
22. ILO. ILO monitor: COVID-19 and the world of work. 7th ed. International Labour Organization; 2021.
23. IMF. World economic outlook: managing divergent recoveries. International Monetary Fund; 2021.
24. Jorda A, Schularick M, Taylor AM. When credit bites Back. J Money Credit Bank. 2013;45(2):3–28.
25. Jordà Ò, Singh SR, Taylor AM. Longer-run economic consequences of pandemics. Rev Econ Stat. 2022;104(1):166–75. https://doi.org/10.1162/rest_a_01042.
26. Kumar T, Vaidyanathan S, Ananthapadmanabhan H, Parthasarathy S, Ravindran B. Hypergraph clustering by iteratively reweighted modularity maximization. Appl Netw Sci. 2020;5(1):52.
27. Lisa O, Nugroho L, Orban I, Utami W, Nugraha E. The performance of Islamic microfinance institutions in the COVID-19 pandemic: is asset quality important? Sosyoekonomi. 2023;31(58):145–60.
28. Liu Z, Yin Y, Liu W, Dunford M. Visualizing the intellectual structure and evolution of innovation systems research: a bibliometric analysis. Scientometrics. 2015;103:135–58.
29. Martin FM, Sánchez JM, Wilkinson O. The economic impact of COVID-19 around the world. Federal Reserve Bank of St Louis. Review, Second Quarter; 2023. p. 74–88. https://doi.org/10.20955/r.105.74-88.
30. Mvogo G, Djouaka CGA. Effet du mobile money sur la résilience des ménages exerçant des activités génératrices de revenus au Cameroun. Afr Dev Rev. 2022;34(4):459–71.
31. Nicola M, Alsafi Z, Sohrabi C, Kerwan A, Al-Jabir A, Iosifidis C, Agha R. The socio-economic implications of the coronavirus pandemic (COVID-19): a review. Int J Surg. 2020;78:185–93.
32. Pritchard A. Statistical bibliography or bibliometrics. J Doc. 1969;25:348.
33. Reardon, Robert S., COVID-19 and Crowdfunding. 2024. The Palgrave encyclopedia of venture capital., Forthcoming, Available at SSRN: https://ssrn.com/abstract=4715572 or https://doi.org/10.2139/ssrn.4715572.
34. Ridhwan MM, Suryahadi A, Rezki JF, Andariesta DT. The impact of COVID-19 on the labour market and the role of E-commerce development in developing countries: evidence from Indonesia. J Asia Pac Econ. 2023;30:1–44.

35. Rogoff KS, Rossi B, Schmelzing P. Long-run trends in long-maturity real rates 1311–2021. (No. w30475). National Bureau of Economic Research; 2022.
36. Sarin S, Haon C, Belkhouja M, Mas-Tur A, Roig-Tierno N, Sego T, Carley S. Uncovering the knowledge flows and intellectual structures of research in technological forecasting and social change: a journey through history. Technol Forecast Soc Chang. 2020;160:120210.
37. Shafi M, Liu J, Ren W. Impact of COVID-19 pandemic on micro, small, and medium-sized enterprises operating in Pakistan. Res Glob. 2020;2:100018.
38. Sherif M. The impact of coronavirus (COVID-19) outbreak on faith-based investments: an original analysis. J Behav Exp Financ. 2020;28:100403.
39. Summers LH. Reflections on the ‘new secular stagnation hypothesis. In: Secular stagnation: facts, causes and cures; 2014. p. 27–40.
40. Sun Y, Grimes S. The emerging dynamic structure of national innovation studies: a bibliometric analysis. Scientometrics. 2016;106:17–40.
41. Takes I (2020) Mitigating the COVID economic crisis: act fast and do whatever.
42. Thekkoote R. Factors influencing small and medium-sized enterprise (SME) resilience during the COVID-19 outbreak. TQM J. 2024;36(2):523–45.
43. UNCTAD. COVID-19 and e-commerce: a global review. United Nations Conference on Trade and Development; 2021.
44. Van Eck NJ, Waltman L. Text mining and visualization using VOSviewer. 2011. arXiv preprint arXiv:1109.2058.
45. World Bank. Poverty and shared prosperity 2020: reversals of fortune. World Bank Publications; 2021.
46. Zafar A. Emerging markets: anatomy, characteristics, and history. In: Emerging markets in a world of chaos: pathways for economic growth and development. Cham: Springer; 2023. p. 23–110.

COVID-19 Pandemic and Its Societal Impact: A Philosophical/Theoretical Perspective in Human Quantum Mechanics

24

Christian Aspalter

Abstract

This chapter uses a philosophical/theoretical approach of *human entanglement theory* (cf. Aspalter, Soc Dev Issues 46:5, 2024; Aspalter, Human entanglement theory: a quantum approach to the study of all-encompassing human communication, Springer Nature, Singapore, 2024) and applies it to the societal impact of the pandemic, particularly via public policy and media, their forces, and developments. Foucauldian power relations are good examples of human entanglements that control and not only influence people's behavior, especially their decision making and communication choices, especially also within people's minds, through the use of language and communication. The least common denominators that are used by the human entanglement theory are words, which are the quanta/atoms of societies and humanity. The author is following a post-Luhmannian approach that is, besides all social and cultural communications, also integrating all intra-personal communications, as well as human-to-machine, machine-to-machine, and machine-to-human communications, plus biological communications. The resulting perspective of *human quantum mechanics*, as applied in this chapter, is inspired by the latest developments of physical quantum mechanics (as well as quantum computing). The essence of human quantum mechanics and human entanglement theory is outlined and explained in this chapter, and conclusions are drawn as regard their impact on society, particularly major public policy developments, media developments, and mass psychology during the recent COVID-19 pandemic, with a special focus on modern liberal democracies all over the world.

C. Aspalter (✉)
Department of Social Sciences, BNU-HKBU United International College, Zhuhai, China
e-mail: christian@uic.edu.cn

M. K. Ghosh, M. Basu (eds.), *Pathogenesis, Management and Socio-Economic Impact of COVID-19 Pandemic*, Medical Virology: From Pathogenesis to Disease Control, https://doi.org/10.1007/978-981-95-1426-7_24

Keywords

Human quantum mechanics · Human entanglement theory · COVID-19 pandemic · Societal impact · Unconstitutional public policies · Democracies

Introduction

While applying a true transdisciplinary perspective and nonstatistical meta-analytical method, this chapter is going to analyze in a true *explorative and philosophical* manner global public policy during the times of the COVID-19 pandemic as a global mass event, a global mass psychological-cum-political event (in the tradition of Gustave Le Bon, 1895), while building on the new quantum-inspired *theory of human entanglement*, which covers (explains and serves) all disciplines from the microscopic psychological and lingual level all the way to the macroscopic cultural and mass psychological public policy level, and that on a global scale—as humanity cannot be put into boxes, not into containers of nation states, nor into containers of disciplines or this or that theory containers/traditions.

Hence, a transdisciplinary, philosophical, meta-theoretical, and global humanity-focused and all-encompassing communication-focused approach is needed. For this, we will turn to, and take advantage of, the recently set up and fleshed out *human entanglement theory* [5, 6], which builds on new insights derived from quantum physics, and that today are not only derived from and applied to the quantum level, but also in the macroscopic and cosmic level of the physical universe [16, 31, 45, 46, 48].

Pandemic's Impact on Society

What has been going on over the last couple of years needs to be processed, yet to be fully understood, and many still need to be held accountable for the things they did and/or did not do. Unquestionably, these times were utterly tragic on a multitude of levels and in a myriad of situations.

Damages often cannot be paid back to victims of denunciation (cf, e.g., [43]), as damage was exerted on personal and as well as mass (social) psychological levels. A revisiting and full reappraisal of the damages and victims' suffering—for example, in the form of restorative justice (cf [17])—can/could start to heal wounds, knowing that many wounds will be not identified and not healed, not even close enough to be able to speak of a true healing process, or a justice process, that has been started to get going.

People will still run around scared for the rest (or much of the rest) of their lives, covering up their faces, to scratch out some sense and feeling of security from a now precarious, threatening, and viciously looming lifeworld of their own. This is particularly rather easily visible with young people putting on masks all the time, even 1 year after the pandemic has been declared over.

Of course, there are always flu viruses out there, always and everywhere, and there are also young people with heart problems and lack of immune system capabilities. But, today, people are detached from reality, detached from their former friends, speaking through cloth and plastic fabrics for hours on time to the ones that are the closest to them—while they still very much, and very often per day, touch the very same door handles/knobs and elevator buttons with their bare hands, and shake hands, still like all other people.

Mental health has declined not just for many individuals, and that rather permanently for the rest of their lives (in most cases), but also for societies as a whole. What was that? That thing of mass behavior (in terms of Gustave Le Bon [30], and Sigmund [15]), including mass government behavior, mass self-policing behavior, mass-shunning behavior, mass-physical-exclusion practices, which has denied human rights of, for example, visiting one's own families, walking one's own dog, taking out one's own family for a walk, or a hike up a lonely mountain. Why was fishing and biking only possible while being deprived of a full sniff of air, a full breath of air, every time one tried and struggled to get a new mouthful of air?

Also, why are we scared (and perhaps should be) to also pose these very questions (knowing it may cause denial of publication and such)?

On a massive scale, our societal psychology has been altered, and permanently. We have been trained to hunt certain opinions and views, to punish, taint, and exclude them, knowingly and not-knowingly (subconsciously). This essay here is a philosophical essay in the (former, old) tradition of free speech and free critical (deep) thinking, that every democracy needs, every science needs, as its daily bread. Thus, we may dare to ask and demand *from us* in the here and now: "Give us today our daily bread [wit] ... And lead us not into temptation [*again*] but deliver us from the evil one [*'our' hate of and harm to others*]" (own particular interpretation/application, Matthew 6, The Bible [8]).

In Search of Philosophy

We are in search of *philosophy*, and there are many wise thinkers and books out there, and back there in the history of humankind, and that around the world (of course). We are in the dark, as we sit here in Plato's cave and *Weber's iron cage. We think we are free, and yet we are not*, as Johann Wolfgang von Goethe noted so rightfully (cf. Box 24.1).

Box 24.1 On the myth of freedom in thought

"[n]o one is more hopelessly enslaved than those who falsely believe to be free" [7].

This is so because our thoughts have been and always are pre-engineered by our past thinking and action, which again were engineered by outside communication and (especially government and media) communication and action (cf. [5, 6]).

Kong Fu-Zi, alias Confucius [28], has told us, as his one and main message, and last message of the Analects (Book 20), that *if we do not know the power of words, we will never understand human action*. The worst misery of humankind is the practice, and the thousand-year-long inheritance of the practice, of giving wrong names, of naming things the wrong way (cf also Confucius [28], The Analects, Book 13). Concepts are compound systems of words or names. Concepts and deep-going and far-reaching notions of good and evil (cf [34, 35]) are chiseled into our bones and our human mind; they are engineered into the wheels and mechanics of our common senses [19].

We are merely the *product* of human history, as are the workings of our individual and collective (societal) human minds, as they have been washed out and formed by thousands of years of language usage, thousands of years of cultural, social, and political *moralizing norms,* and the therefrom resulting *moralizing and moralized practices*. Friedrich Nietzsche [34, 35] knew this, as did Michel Foucault [14] (cf Box 24.2).

Box 24.2 On the myth of freedom in thought and feeling

Friedrich Nietzsche on the *continuous flow of the causality of thoughts and feeling*; that is, as he called them, "sensations":

"Immediate self-observation is not enough, by a long way, to enable us to learn to know ourselves. We need history, for the past continues to flow through us in a hundred channels. We ourselves are, after all, nothing but our own sensation at every moment of this continued flow." ([35]: 343).

The COVID-19 pandemic has left its mark on humanity, on the human mind, the mind of billions. Our psychology, our norms of, and felt/lived degrees of freedom have been altered, like in a magnetic pole shift.

At the Dawn of Human Quantum Mechanics: Human Entanglement Theory and Its Basics

Quantum Mechanics as an Orientational Foundation for Human Quantum Mechanics

The first wave of publications in human sciences, that is, social sciences, life sciences, humanities, and arts, that draw theoretical and methodological inspiration from quantum physics and quantum technology in general (such as quantum computing) has already swept across all major corners of these human-related science disciplines (cf, e.g., [27]).

In retrospect, the amount of pioneering work is amazing that especially in the last three decades has engendered new paths, new methodologies, and new ways of creating, developing, and testing theories. Quantum physics has revolutionized not only mathematics (with the rise of superstring theory and quantum mathematics, cf [44]) and technical sciences, but also it has started to exert its creative influence on finance, psychology, literature, and, of course, the social sciences as a whole.

This is nothing short of a cross-disciplinary theoretical and methodological revolution that nowadays is sweeping across non-STEM sciences. A great many mainstream representatives have not picked up on this new trend and the rise of new revolutionizing paradigms in terms of creativity and potential, as well as first actual breakthroughs in theory creation and development, plus empirical exploration and testing. But the second wave, which will hinge upon the change of the mainstream theories and practices themselves, will change this.

Building from the works of Niklas Luhmann [32], Friedrich Nietzsche [34, 35], Pierre Bourdieu [10], Michel Foucault [14] and the Buddhist psychology teachings of Tich Nhat Hanh [21] and Aspalter [2] has set up and then developed over the last one and a half decades a post-Luhmannian theory of social-cum-private communication.

Recently, this led to the establishment of a meta-theory for all of human sciences, that is, every science and discipline that uses or is based on human communication, and hence human action. With this new meta theory, all connections contingencies/dependencies/causal forces that exist in between and within people are now the subject of a single general theory that combines formerly segregated fields of theory-making with regard to human communication and human action, by the way of focusing and theorizing upon the *least common denominator* of all human communication and hence action, that is, the quanta (or atoms) of society, which are *words*, or anything that is or can be expressed by *words*. This is the *human entanglement theory* [5, 6]. Luhmann [32] already saw words in between people as the atoms of society. For some time now, Aspalter [2] has seen words both on an interpersonal and intrapersonal level as the atoms, and we shall rather say now quanta, of society.

From the human body and mind to the macroscopic world of world poverty and world hunger, the ways of human decision-making are contingent on a number of factors (cf Table 24.1).

Table 24.1 Analyzing universal factors of human decision making, thoughts, feelings, and actions

Human decision-making is processing, constantly altering/reinterpreting, and responding to:	Description
1. Communicational sensations/forces	(a) *Sensory information*, i.e., received/understood and expected outside and inside communication per se.
2. Contextual sensations/forces	(b) The conditions and actions that are and have been *created by* outside (interpersonal, "lifeworld contextual") and inside (intrapersonal, "lifeworld situational") *communication in the past.*
3. Relational sensations/forces	(c) The *invisible and visible forces of—relations in—people's lifeworlds* (cf [20]), societies, cultures, political/administrative/juridical systems, welfare state systems, financial/economic, and educational systems (all across the world).

As with the new developments regarding the development toward a unified field theory in physics (cf. [37]), social science and other human sciences are finding themselves at the brink of a new era. This new beginning is marked by entanglements, that is, invisible forces that guide and interfere with human decision-making and behavior, in the words of Michel Foucault [14], these are *power relations*.

Human entanglement theory, and hence the new world of *human quantum mechanics*, has extended (and has to extend) the concept and theory of Foucauldian power relations to include all kinds of invisible, intangible forces that push and prevent human decisions, human communications, and hence, human behavior. These are now simply called *human entanglements*, of which there are *different kinds*. There are encouraging and directing/enabling ("positive") entanglements, as well as dis-encouraging and preventing/forbidding ("negative") entanglements. In total, they are created, located, and operated on different intrapersonal, interpersonal, contextual, environmental, financial, cultural, etc. dimensions.

They interact with each other in the form of Schrödingerian waves of cumulative (aggregate) sensations that trigger/exert (cause) either positive or amplifying/enabling effects, or negative, diminishing/obstructing effects.

Box 24.3 On the multidimensionality of human entanglement sensations/forces
These human entanglement forces operate and spread out on different levels and different dimensions of human interaction and human intra-action at the same time. The parallel workings of these multilevel and multidimensional human entanglement nettings determine cumulatively at any given time and over time human decision-making, human thoughts and speech, as well as therefore human action and nonaction.

The cumulative effects of human entanglements are the ground forces that decide intrapersonal and interpersonal human communication and decision-making on a probability basis. Hence, the probabilities/probability distributions of each individual entanglement—of in total an infinite number of human entanglements—are what drives us, what blocks us, and shocks us, what motivates, entices, and induces us.

The degrees of human freedom (real self-determination, and deviation from expected social, legal/administrative, cultural, etc. norms) are smaller than most people on a personal level realize, as they operate in a world of societal, cultural, lingual, and economic forces that act as stealthily hidden puppeteers that force our thoughts, interpretations, and common sense, even logic *in a probabilistic manner* (cf also [19]).

The forces of media, and social media of course, in all these are super important, as they amplify human decisions and human communications a million times over (cf, e.g., [23]).

We are programmed by our own language, by our own history, and by the history of our own language, by our culture and politics, society, and community: who used to say, how, when, why, and what, in whichever context or in general.

Culture and language are subject to hidden forces of a grand magnitude, which we can call *meta-entanglement forces*. What is right and wrong, what is good or evil (cf [34]), is decided by what at any given moment in time, at any given place, at any given context is seen to be the only right choice, in terms of acting, speaking, behaving, including thinking and feeling.

In modern societies, modern people are much more suppressed and subdued than they ever get to realize—in form of norms and regulations, language and cultures, laws and practices of higher-power institutions with executive, legislative and juridical powers, as well as with their religious, cultural, social, and intellectual powers, and particularly their stealthily hidden power relations.

Quantum mechanics is spearheading the notion and theory that everything is connected to everything [45], and that in the form of quantum entanglements—of which there are different degrees/strengths of entanglements (cf, e.g., [50]).

Building on the inspirations and creativity derived from quantum mechanics, human quantum mechanics now can investigate human action, human communication, in short, the whole range of human behavior, including all the underlying—cumulative probabilistic (*not entirely, but largely deterministic*)—forces that drive and determine human thoughts and feelings, and hence decision-making.

New Developments in Human Quantum Mechanics

Brian Swingle [100], among others, has set the stage for a full-blown revolution yet to come, where not only black holes and wormholes are being understood by quantum mechanics, but also all things between them at the macroscopic scale and quantum particles on the quantum scale. Anton Zeilinger [12] already mentioned that for more than 90 years, physicists were asking the wrong questions about the issue of a possible quantum gravity theory, and hence foretold the breakthrough by Jonathan Oppenheim [37]. Oppenheim [37], (second last paragraph) still very much leaves open and full theoretical and empirically test world where especially quantum entanglements play the number one role. Kanata Goto stated that "spacetime itself does not exist fundamentally, but emerges from some quantum nature" (cited by [39]).

What Oppenheim [37] did was revolutionary; he connected both theory worlds, the quantum and the gravity theory world, but he also changed the macroscopic world, turning it into a more or lesser degree stochastic world, ruled by probabilities and randomness.

In *human* quantum mechanics, given the title of the theory alone, we do not care about the gravity theory or the spacetime theory *in the narrow sense*. Humans are all part of the macroscopic world. Nonetheless, Oppenheim's new work supports actually the idea of fuzziness in *human* quantum mechanics, especially, for example, fuzzy randomness, accumulated fuzzy causality, and fuzzy reality (cf, e.g., also [24]).

By introducing probability and randomness (which is unknown probability, as Zeilinger mentioned, that is, randomness is due to information that is not known, cf [49]), Jonathan Oppenheim [37] has brought the quantum blur into the macroscopic world. This is the author's take on it.

Hence, probability rules all of the macroscopic physical world, to a greater and lesser degree. Spacetime is also exposed to it. In addition, diffusion within the macroscopic world, as well, is "fuzzing" things, both measurements and thus predictions.

For *human quantum mechanics*, which tries to be inspired and learn from quantum mechanics and all of physics, in terms of looking *at relationship between people* (instead of matter), by focusing on *human entanglements* (instead of quantum entanglements), every advance and breakthrough in the physics world may or may not also lead to new breakthroughs in the understanding of the human world.

Human entanglements are still key to the understanding of humanity and of human action, communication, thought, and feelings. For these, human entanglements of all sorts are latent and stealthily hidden—and *over time* strengthening and accumulating—causal forces (legal, financial, psychological, communicative, social, cultural, religious, all of these forces) that are based on probabilities, that is, accumulating compound probabilities/probability distributions, of which there are an infinite number of human entanglements (in the furthest sense).

These infinite human entanglements in the lifeworld of any single human, or any group of humans, are de facto exhibiting *a fuzzy nature*, as nothing ever is "fully" determined from the beginning of the universe or from the beginning of a lifetime onwards (cf [37, 49]).

Hence, in the realm of *human* quantum mechanics, we deal with *a fuzzy reality*, a reality we still can calculate, but based on accumulated probabilities/probability distributions, just like in quantum physics, as well as the macroscopic world (due to both the *blur* created by quantum probabilities and *noise* created by diffusion processes).

Human entanglement theory is a theory of human communications, and therefore, resulting actions and conditions. The essential findings of human entanglement theory are already set up in greater detail (see Table 24.2).

Table 24.2 Basic findings of human entanglement theory

1.	An infinite number of human entanglements *create* human activity (thoughts, feelings, decisions, communication, and actions included) and the entirety of humanity (cf also [45]).
2.	*Human communications create, maintain, amplify, dampen, and undo these human entanglements.* Human forces of all kinds are both, they *are and they are not* (cf Parmenides of Elea [42]), i.e., nonpermanence of human entanglement forces is sufficient to create and sustain human entanglements.
3.	Words are the smallest units (quanta) of human action and humanity, as they create *sensations* (like waves, cf [40]; as well as [29]), and these sensations have the potential to create follow-on contingencies (by way of follow-on sensations). These contingencies can be understood as dependencies, one-sided or double-sided dependencies.
4.	Human entanglements are always more or less probabilistic (cf [9, 11, 37, 41]), and these probabilities/probability distributions are of *a compound nature*, they are composed themselves potentially (in the widest sense) a myriad of probabilities/probability distributions (of a myriad of entanglements) themselves.
5.	In human quantum mechanics, we can measure human entanglements exactly (or as much as possible) and more than one aspect of it. However, there is a catch, as there are *an infinite number* of human entanglements, all entangled (connected) to one another, in direct and/or the most indirect ways.
6.	There is an *intrinsic* "blur" *relationship* to human entanglements, as in general in quantum mechanics (cf [22, 40]). The accumulation of uncertainties, in terms of aggregate probabilities/probability distributions of an (in total) infinite number of human entanglements, strong and weak entanglements, all kinds of entanglements, close and far, will always entail different degrees of blur and noise at the same time. This is needed and natural. *Blurs and noises* are created by human self-determination and lack of complete or any near-complete information in general, and pertaining to specific information as well.
7.	*Everything* in the world of human entanglements *is connected to everything*; human entanglements are connected to all other entanglements by entanglements in their (causal) vicinity; every further entanglement leads necessarily to a myriad of other (stochastically) causal entanglements.
8.	Human entanglements' effects *accumulate*, their positive and/or negative causations—i.e., probabilities/probability distributions—accumulate (cf also, e.g., [25]). This accumulation is *multidimensional*, and progresses (accumulates) across time.

COVID-19 Pandemic and Its Societal Impact Around the World: A Human Quantum Mechanics Perspective

In special situations, an observer can make the most of an observation, as systems and practices of cover-up may not work or not right away, or not in a fully disguising way. Also, emergency situations or special situations in general allow a look into the inner workings of the most complex, high-dimensional systems, such as societies and humanity itself, or governance systems of entire countries and the whole of the world as well.

For sure, the COVID-19 pndemic qualifies for a special emergency situation, where governments were taken by surprise, and societies then again, in second instance, were taken by surprise due to the freedom-breaking unconstitutional laws and regulations implemented—against better constitutional knowledge, that most of these were unconstitutional laws and regulations (cf, e.g., [1, 13, 18, 33, 38]). The game that was played was simple: do whatever you like, and the constitutional courts and high courts need a long time to undo whatever you have done, several months, maybe longer, years even, and then you do new things that again are against the constitutions and await proper constitutional court and high court rulings, until the next unconstitutional law and regulation are set up to replace its predecessor. By then, new unconstitutional laws, regulations, orders, practices, and systems will be in place again.

Box 24.4 Reflection No. 1 on the societal effects of the COVID-19 pandemic
Everything that usually meant something—democratic governance, primacy of constitutional law, common ethics, etc.—did not mean anything, not freedom, not morals, not constitutional laws, all were wiped off the face of governing tables, with the free will of the elites, the expert elites and political/economic elites, with the help of the fear-mongering media elites (which also work for, i.e., belong to, the richest of the rich, of course).

Not a day passed by, with not threatening statistics of death and horror, of the end of the world, telling nature. The media used scores of pictures—as from a conveyor belt—that showed (attractive young) women holding both hands up in the air in despair, or both hands pressed against their heads, and so forth. These were actors or willing participants who got paid for these pictures, or simply willful participants in the game of mental mass suppression (cf, e.g., [36]).

It was a frenzy, a media frenzy. The old saying there is a lot of money to be made when blood is on the streets does in fact reveal not only a lot of truth about capitalism, but also about governance in capitalism and how people are being controlled in their everyday lives, through everyday systems, laws and regulations, but also through "special"-power-based systems, laws and regulations, even and especially also unconstitutional ones (depending on the constitution that is in place in each country at each time). This media frenzy, hence, also not only allowed, but amplified this governance frenzy.

Box 24.5 Reflection No. 2 on the societal effects of the COVID-19 pandemic
The gloves came off, morals and better common and scientific knowledge were ignored, pushed aside with a simple vote of a three- or ten-member—nonconstitutional, and often only "advisory"—"expert" team. Sometimes these experts were putting on military uniforms to make a better impression, i.e., to make a better scare, a better suppression of logic, of common needs and common freedoms of the people (take the case of Austria, cf. [36]).

Into the mix, public policy came caught up in recent past developments and could not catch up with current developments. The outcome of which is a succession of zombie policy strategies, and their strings of zombie policies and their regulations and decisions made (cf [3]).

At first, there was no reaction (when there must have been strong and quick reaction), borders were kept completely open and unchecked. Then, there was window dressing reaction—like a health minister going to the airport with the press waiting there to look at the first passengers coming from Asia being temperature checked (cf [36]).

While the virus kept changing, wave after wave, the policies almost entirely always (by and large) kept being pinned to and aligned with the past wave (the outdated wave) and the past variant—the previous, not the current variant—of the coronavirus. This is what is called zombie politics of the COVID-19 pandemic [3].

Box 24.6 On the wave-like distribution, build-up and chain-reactions, and their follow-on reactions of societal impacts of the COVID-19 pandemic

Human communications caused human actions, and both of these spread like *Schrödingerian waves throughout societies at different dimensions/levels of society*, at different parts of the world, and the entire world

That is, as a demonstrative example here, we can see that the media frenzy turned also politics into a frenzy; the political and media frenzy turned into a recurrent/wave-like psychological frenzies (fear mongering that created feelings of hopelessness, desolation, and down-beat feelings), plus a business frenzy which was created and sustained by political, media and psychology frenzies.

Members of the business elites liked to earn extra profits from masks. Those companies that were also often strongly involved with mask production and vaccine development and production were also highly entangled with the political elite, in terms of ownership and friendship economies, which provides plenty of room for corruption at the highest levels of public governance (cf, e.g., [47]).

The media had a heck of a time, with the *extra fear* (i.e., maximally amplified and extended fear) it *produced*, as it managed to get an all-time high attention from viewers and readers, and hence earn from all-time high company and government advertisement contracts. The media was in a position to create both supply and demand, a perfect position to milk *the capitalist COVID-19 media industrialist machine*.

Box 24.7 What we can learn from this last pandemic: a general take away
From the COVID-19 pandemic we can take away a great deal of learning, for the next pandemic: how to behave, how to make policies in a forward-looking and not a backward-fixed manner, how to identify the political and economic deals and forces that create much of the societal storm, in form of hyped-up fear, pumped-up unconstitutional/human rights violating restrictions, and overblown demeaning punishments.

We need to go back and read the classics, such as Gustave Le Bon [30]. He influenced a great deal of Freud's works (cf esp., e.g., [15]). One of the sharpest observers of *societal behavior* in the history of humankind, Le Bon, already a long time ago, without today having lost its significance in any way to any degree, mentioned that: "The true historical upheavals are not those which astonish us by their grandeur and violence. The only important changes whence the renewal of civilizations results, affect ideas, conceptions, and beliefs. The memorable events of history are the visible effects of the invisible changes of human thought. The reason these great events are so rare is that there is nothing so stable … as the inherited groundwork of its thoughts" ([30]: xiv).

Far from having fundamentally transformed the groundwork of societal, political, and cultural thought worlds, the COVID-19 pandemic has yet been used as an opportunity to worsen the human condition, and all its underlying dimensions. The COVID-19 pandemic did not invent inequality, suppression of the weak and the poorer, powerless segments of society. No these systems were in place before. But, the pandemic managed to make things worse for most people on many levels, in many aspects, all of which accumulated in their effects on human life conditions and changed life realities and experiences.

The COVID-19 pandemic unleashed new legal and illegal control mechanisms, most of them of technical nature, but also in terms of thought systems, negative name-calling (with new negative name creations), black-and-white lobbying and politicizing, everyday behavior as political correct behavior, human-rights subjected to political correct behavior of the day, insider and outsider problems have been worsened and used for political, economic, and social gain by those in control, that is, those that control those that are in control.

The paradigm of culture that we were in, that we are in, has not changed (cf [30]). The key question, however is, have we learned from the COVID-19 pandemic? Has public policy learned from the COVID-19 pandemic? Did it learn also positive things, or only freedom-limiting techniques—like ignoring the constitution, and waiting for its slow corrective mechanism to play out, and then simply implement new unconstitutional laws, regulations, and mechanisms?

Also equally important is, have we learned how the politics of the elites have played out during the pandemic, how they have profited billions from tragedy, fired-up fear and confusion, usage of media for creating and amplifying blaming and shaming, lobbying, and pushing for overblown punitive regimes that give those in power the pleasure of feeling the power to be able—and to unleash their powers—to punish and brand any opposition to any new thought and policy stand regime, even the majority of the population?

Democracy has been defeated by its own means, by the control of power by the elites within *the great complex machine that is called democracy*, where however in reality a minority of to-one-another-hooked-up actors—the super-super rich, the powerful and the media-tycoons (and their owners and chief editors)—are playing the game of fear and control, making money out of despair, cashing in billions and millions while people die in nursing homes, in their homes by the millions (cf also [26]).

Box 24.8 What we can learn from this last pandemic: the loss of human self-determination and the loss of individual freedoms

People lost their lives, their livelihoods, their happiness, their decency, their self-esteem, and hope, alongside their human rights for assembly (cf. [4])

People have lost their right to go out to work and earn their daily money to buy their daily food (as they survive on a hand-to-mouth basis), i.e., to be able to find and/or buy food/water/medicine/etc., to fish or hunt, to exercise, taking a walk outside the house, taking their family for a walk

Also, people lost their rights about the integrity of their bodies by being punished, excluded, and shunned for not following the instructions of the pharmaceutical–industrial–political–medial complex, in terms of freedom-robbing/criminalizing vaccination mandates.

The *dark side of democracy* (not only that of the moon) was—once again—revealed during the COVID-19 pandemic. Democracy is great, as long as things do not end up in terrible disaster or as long as societal psychology is not used as a football by the super–super rich and the super-powerful. The game that was played was *the game of power domination*, fully taking advantage and amplifying Foucauldian power relations that unleashed the monsters that usually hide inside democracy, its hidden caves and cellars (cf [14, 19, 23, 26]).

Conclusions

In terms of human quantum mechanics, among the interesting lessons and take-aways from the COVID-19 pandemic, from a global perspective, we find, for example, that indeed everything during the COVID-19 pandemic was connected to everything, as it is always in general. More interesting still is the fact that there was, on a global and historical scale, an *entanglement "bomb"* that went off. It was facilitated (enabled and amplified) by media frenzies, political, and social frenzies that traveled all across the world and bounced back and forth across the world, from one corona wave to another.

This entanglement "bomb" is not one coherent bomb but *a succession of sets of (Schrödingerian) waves of fear, domination, power abuse, human rights violations, and freedom thefts* on formerly entirely unimaginable grounds, terms, as well as extent.

One hence can refer to the COVID-19 pandemic entanglement "bomb" as a succession of waves of cluster bombings all colluding to fight *the war of the elites against all others*—in the Foucauldian, Gramscian, Chomskyan, and Holcombian sense. That is, the war of the super-super rich and the super-powerful against the majorities in democracies (the common people). As a consequence, the majority of all law-abiding citizens has been—in successive sets of waves—name-called and/or branded, excluded, tortured, and scorned during the pandemic by a great deal of completely unethical, as well as quite often to unimaginable extent also unconstitutional ways, with the great help of the media, elite-dominated governance and systematic disregard of constitutional laws (cf also Foucauldian "civil war" *theory* [14]).

From the theoretical perspective, human entanglement theory, as well as human quantum mechanics in general, is building not only on the inspirations and progress derived from quantum theory in physics (cf [5, 6]) but also on post-Luhmannian theory of communication systems (social-cum-psychological systems), Foucauldian theory of power relations, Gramscian theory of societal/political hegemony, and Chomskyan theory of social, political and cultural domination via mass media control, Holcombian political capitalism theory, among others.

Finally, it has become utterly clear that large historical and global events, and developments are in dire need of theoretical elaboration, analytical dissection, and scientific synthesis, in a recurrent manner and from a multidimensional—and *human-entanglement-theory enriched*—perspective.

References

1. Aaron D. Covid-19 policies and constitutional violations. 2020. https://blog.petrieflom.law.harvard.edu.
2. Aspalter C. Towards a human capital welfare state? J Soc Policy. 2007;6(1):1–46.
3. Aspalter C. Introduction: the new age of zombie politics. In: Aspalter C, editor. Covid-19 pandemic: problems arising in health and social policy. Singapore: Springer Nature; 2023.
4. Aspalter C. The human face of the pandemic. In: Aspalter C, editor. Covid-19 pandemic: problems arising in health and social policy. Singapore: Springer Nature; 2023.
5. Aspalter C. Human quantum mechanics and human entanglement theory: a new paradigm for social sciences and beyond. Soc Dev Issues. 2024;46(2):5.
6. Aspalter C. Human entanglement theory: a quantum approach to the study of all-encompassing human communication. Singapore: Springer Nature; 2024. forthcoming.
7. AZQ, AZ Quotes. 2024. website. https://www.azquotes.com/quote/946167.
8. BH, Bible Hub. 2023. website. https://biblehub.com/bsb/matthew/6.htm.
9. Born M, Jordan P. Zur Quantenmechanik. Zeitschrift für Physik. 1925;34:858–88.

10. Bourdieu P. Cultural reproduction and social reproduction. In: Brown R, editor. Knowledge, education, and social change. London: Tavistock; 1973.
11. Capellmann H. Space time in quantum theory. Found Phys. 2021;51:44. https://doi.org/10.1007/s10701-021-00441-0.
12. Dacey J. Anton Zeilinger: a quantum pioneer. Physicsworld, October 14. 2010. https://physicsworld.com.
13. Dewey S. Covid-19 lockdowns violate the US constitution. 2021. https://www.aier.org.
14. Foucault M. Surveiller et punir: naissance de la Prison. Paris: Èditions Gallimard; 1975.
15. Freud S. Massenpsychologie und Ich-Analyse. Leipzig: Psychoanalytischer Verlag; 1921.
16. Fröwis F, Dür W. Stable macroscopic quantum superpositions. Phys Rev Lett. 2011;106:110402.
17. Gavrielides T, editor. Routledge international handbook of restorative justice. Oxon: Routledge; 2021.
18. GFE, Global Freedom of Expression. Jurisprudence relating to Covid-19; 2021. https://globalfreedomofexpression.columbia.edu.
19. Gramsci A. Prison notebooks, vol. 1–3. New York: Columbia University Press; 2011.
20. Habermas J. Theorie des kommunikativen Handelns: zur Kritik der funktionalistischen Vernunft, vol. 2. Suhrkamp: Frankfurt a.M; 1981.
21. Hanh TN. Dharma talks. given in July/August 1993, in Plum Village, France; 1993.
22. Heisenberg W. Über den anschaulichen Inhalt der quantentheoretischen Kinematik and Mechanik. Z Phys. 1927;43:172–98.
23. Herman ES, Chomsky N. Manufacturing consent: the political economy of the mass media. New York: Pantheon; 2002.
24. Hoffman DD. The case against reality: why evolution hid the truth from our eyes. New York: Penguin; 2020.
25. Hofmann HF. Quantum causality relations and the emergence of reality from coherent superpositions. Found Phys. 2020;50:1809–23.
26. Holcombe RG. Political capitalism. Cambridge: Cambridge University Press; 2018.
27. Holtford T, Horsch A. Social science goes quantum. J Bioecon. 2023;25:99–116.
28. Kong FZ [Confucius]. Confucian analects. In: J Legge (transl.) The Chinese Classics, vol. 1; 1893. https://en.wikisource.org.
29. Laughlin RB. A different universe: reinventing physics from the bottom down. New York: Basic; 2006.
30. Le Bon G. The crowd: a study of the popular mind [Psychologie des foules]. New York: Macmillan; 1896 [1895].
31. Lu H, et al. Experimental realization of a concatenated Greenberger–Horne–Zeilinger state for macroscopic quantum superpositions. Nat Photonics. 2014;8:364–8.
32. Luhmann N. Soziale Systeme. Suhrkamp: Frankfurt a.M; 1984.
33. Mkhize M, Naidoo S. South African high court rules Covid-19 lockdown regulations unconstitutional. 2020. https://www.covafrica.com.
34. Nietzsche FW. Zur Genealogie der Moral: eine Streitschrift, C.G. Leipzig: Naumann; 1887.
35. Nietzsche FW. Human, all too human: a book for free spirits. Wordsworth Editions: Hertfordshire; 2014 [1878].
36. OE24 (2020–2022) websites. https://www.oe24.at. https://epaper-oesterreich.at.
37. Oppenheim J. A postquantum theory of classical gravity? Phys Rev X. 2023;13:041040.
38. Reuters. Spanish court rules Covid-19 home confinement was unconstitutional. 2021. https://www.reuters.com.
39. RIKEN. Quantum 'magic' could help explain the origin of spacetime. 2023. https://phys.org.
40. Swingle, brian, spacetime from entanglement, annual review of condensed matter physics. 20189:345–58.
41. Schrödinger E. Die gegenwärtige Situation in der Quantenmechanik. Naturwissenschaften. 1935;23(48, 49):807–12. 823–828, 844–849.
42. Schrödinger E. Discussion of probability relations between separated systems. Proc Camb Philos Soc. 1935;31:555–63.
43. SEP, Stanford Encyclopedia of Philosophy. Parmenides. 2020. https://plato.stanford.edu.

44. Stensvold A. Blasphemies compared: transgressive speech in a globalised world. Oxon: Routledge; 2020.
45. Sułkowski P. Quantum fields, strings, and physical mathematics. 2021. https://arxiv.org/pdf/2104.03350.pdf.
46. Swingle B. Spacetime from entanglement. Annu Rev Condens Matter Phys. 2018;9:345–58.
47. Vedral V. Quantum entanglement. Nat Phys. 2014;10:256–8.
48. Welt. Spahn kontaktierte für Beschaffung von Schutzmasken Freunde und Bekannte. March 26; 2021. https://www.welt.de.
49. Zeilinger A. Einsteins Schleier: die neue Welt der Quantenphysik. Munich: C.H. Beck; 2003.
50. Zeilinger A. The message of the quantum. Nature. 2005;438:743. https://doi.org/10.1038/438743a.
51. Zhao Y, et al. Creation of Greenberger-Horne-Zeilinger states with thousands of atoms by entanglement amplification. NPJ Quantum Inf. 2021;7:24.

Unraveling the Global Economic Impact: A Region and Sector-Wise Post-COVID Analysis

25

Mrinal K. Ghosh, Sunny Kumar, Gouranga Saha, Shaheda Tabassum, Pratyasha Ghosh, Suresh Bajoria, and Malini Basu

Abstract

The COVID-19 pandemic has profound and far-reaching economic repercussions, disrupting global markets, trade, and labor systems. Lockdowns and decreased spending caused GDP contractions in advanced nations, while supply chain problems, capital flight, and decreased remittances affected emerging economies. Some East Asian countries demonstrated resilience through successful containment and robust digital industries, while emerging nations grappled with limited fiscal capacity and instability in commodity prices. The pandemic had varying effects on different industries: the tourism, hotel, and aviation sectors experienced sharp drops, while the technology and healthcare sectors pros-

Mrinal K. Ghosh, Sunny Kumar, Gouranga Saha and Shaheda Tabassum contributed equally (Equal First Authors).

M. K. Ghosh (✉) · S. Kumar · G. Saha · S. Tabassum
Cancer Biology and Inflammatory Disorder Division, Council of Scientific and Industrial Research-Indian Institute of Chemical Biology (CSIR-IICB), TRUE Campus, CN-6, Sector–V, Salt Lake, Kolkata, India
e-mail: mrinalghosh@iicb.res.in

P. Ghosh
Department of Economics, Betune College, University of Calcutta, Kolkata, India

AIESEC United States, New York, NY, USA

S. Bajoria
Department of Urology, NH-RTIICS, Mukundapur, Kolkata, India

M. Basu
Department of Microbiology, Dhruba Chand Halder (DCH) College - University of Calcutta, Jaynagar Mazilpur, West Bengal, India

M. K. Ghosh, M. Basu (eds.), *Pathogenesis, Management and Socio-Economic Impact of COVID-19 Pandemic*, Medical Virology: From Pathogenesis to Disease Control, https://doi.org/10.1007/978-981-95-1426-7_25

pered. Agriculture had localized labor and logistics challenges but preserved the stability of the food supply overall, while manufacturing saw supply chain interruptions with recovery linked to digital adaptation. In this report, we examine the differential economic impacts of the pandemic across regions and sectors, highlighting the interplay of pre-existing vulnerabilities, policy responses, and recovery trajectories.

Keywords

Post-COVID · Economics · Health · Aviation · Tourism · Oil · GDP loss · and Job loss

Introduction

One of the biggest global problems of the twenty-first century, the COVID-19 epidemic, has had a huge effect on economies all across the world. It has caused previously unheard-of economic harm, taking money away from research and COVID-19 care worldwide. A manpower shortage brought on by the strain on healthcare has disrupted the supply chain and reduced economic activity. Global GDP declined and stock markets fell as a result of massive losses in the manufacturing and travel sectors. A careful study is necessary as new viral varieties appear, requiring alternate approaches to amelioration. Several industries and geographical areas experienced economic disruptions as a result of the virus's quick spread and the imposition of public health measures like travel bans, lockdowns, and social isolation. This extraordinary incident revealed weaknesses in international labor markets, banking institutions, and supply chains, which had a domino effect on the economy [1–3].

The impact of the pandemic differed from developed to developing nations depending on pre-existing structural elements, governmental reactions, and the robustness of healthcare systems. Depending on fiscal incentives and vaccination programs, regions like Europe and North America saw varied degrees of recovery after first experiencing severe economic contractions. On the other hand, the lack of adequate healthcare facilities, constrained financial resources, and dependence on international markets presented emerging countries with additional difficulties [4].

The impact of the pandemic varied by sector, with some industries—including e-commerce, technology, and pharmaceuticals—surging while others, like travel, hospitality, and retail, saw steep drops. The crisis also sped up structural changes, such as the adoption of remote labor, digital transformation, and brings changes in consumer behavior, bringing to light both opportunities and hazards for future economic growth [5]. This study analyzes the differences in regional impacts and sectoral disruptions as it looks at the worldwide economic impact of COVID-19. It seeks to provide a thorough grasp of the economic effects of the pandemic and insights for enhancing resilience against future global shocks by combining the data and research that has already been done.

Economic Hurdle: Region-Specific Insights

Drawing from in-depth analysis reports provided by the World Health Organization (WHO) and the International Energy Agency (IEA), the ensuing paragraphs will illuminate regional perspectives on COVID-19 across the continents (Fig. 25.1). It is imperative to undertake a thorough examination of the current circumstances, the socioeconomic challenges, the repercussions, and prospective pathways. This scrutiny aims to guide countries in effectively navigating the adverse effects of COVID-19 and moving toward a future marked by coexistence.

African Region

Compelled by widespread poverty and a significant portion of the population residing below the socioeconomic threshold for growth, progress toward achieving universal health coverage in the African Region has been sluggish. The impact of COVID-19 in the WHO African Region has been characterized by disruptions to vital health services and a limited capacity for emergency responses to the pandemic. Outpatient services, the quality of care, and the availability of staff, equipment, supplies, and medications have been particularly adversely affected.

The path forward involves the implementation of effective mitigation measures, including improved community communication, the procurement of personal protective equipment and medications from more developed nations, and the promotion of societal and home-based care to fortify the resilience of health systems. Consequently, it is imperative for the African region to focus on developing specific and targeted resilience, coupled with enhanced emergency preparedness capabilities, to confront unforeseen threats [6–8].

American Region

According to reports from the American State Bureau, the total revenue of selected services in the United States experienced an 8.7% decline during the pandemic years. The retail and food services also witnessed a downturn in estimated shipments of manufactured goods in 2020. Notably, 39.2% of the health sector's estimated revenue was attributed to telemedicine services. Similarly, due to the pandemic, there was an increase in wholesale estimated e-commerce from 33.6% in 2019 to 35.6% in 2020. Despite widespread lockdown measures and successful vaccine coverage, there was a notable rise in capital expenditure and public sector spending during the pandemic. However, the estimated collections of state and local government tax revenue displayed declining trends. Consequently, the impact of the COVID-19 predicament resonated across robust economies. Nevertheless, these regions are swiftly embarking on a trajectory of self-recovery to bolster their socioeconomic prospects [6, 7].

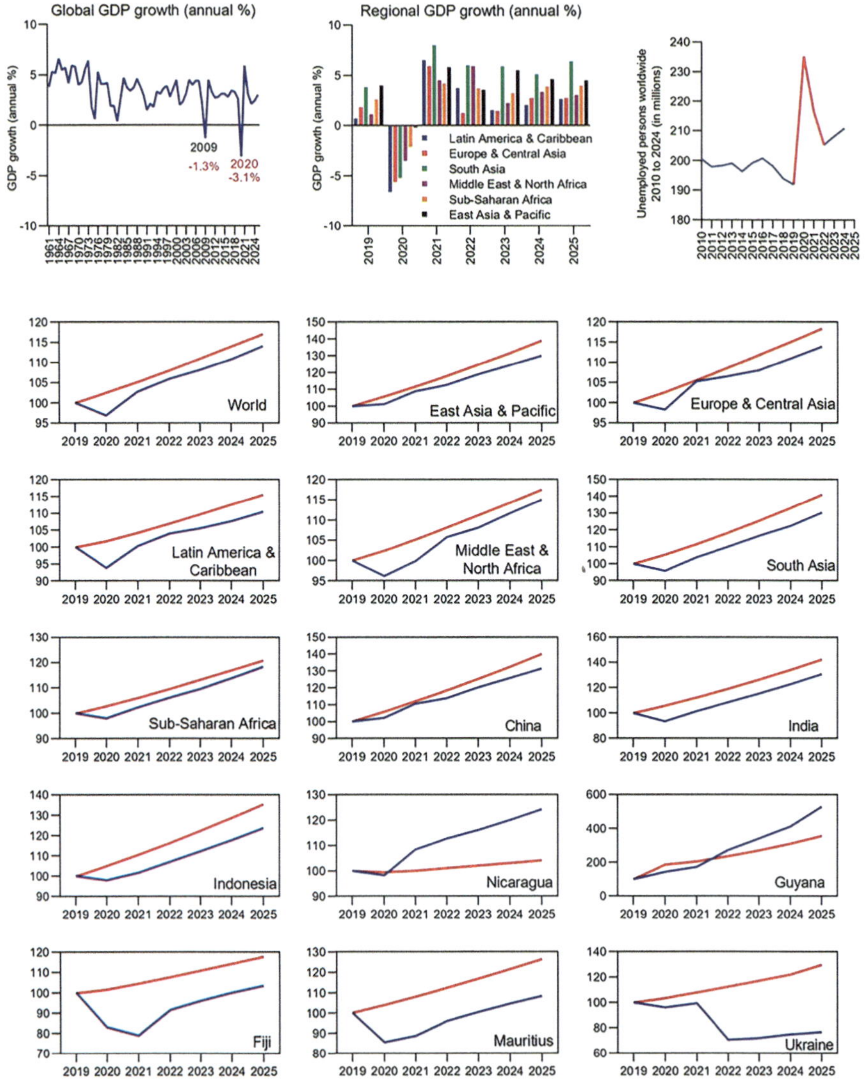

Fig. 25.1 Global, regional GDP change and unemployment in the COVID-19 pandemic. The figure depicts the GDP change and rate of unemployment in the most affected countries

In light of this, it becomes crucial to adopt an equity-focused approach to attain health-related Sustainable Development Goals (SDGs). These regions must enhance their institutional capacity, implement equity-sensitive policy decision-making, and fortify accountability mechanisms to effectively realize their developmental objectives.

European Region

The impact of COVID-19 on access to health services, quality of care, and financial protection was evident even in developed economies, such as those in the European region. A significant 40% decline in essential health services within the WHO European region was attributed to the challenges posed by the pandemic. However, as the latter half of 2021 unfolded, there was a discernible reduction in the extent and severity of these disruptions. This positive trend was driven by a robust vaccination drive and the implementation of more resilient measures [9, 10].

Healthcare services faced considerable impediments due to travel restrictions and financial barriers. Societal inequalities were exacerbated, with older individuals in disadvantaged areas, people with disabilities or from minority populations, and migrants experiencing heightened difficulties in accessing healthcare. Despite these challenges, concerted efforts to accelerate health coverage goals prompted several Member States to exempt COVID-19-related services from co-payments [10].

In response to the ongoing crisis, countries in the WHO European Region are actively working toward mitigating financial barriers to COVID-19 treatment. Their endeavors aim to alleviate the burden of exacerbated societal and financial hardships, signaling a collective commitment to addressing the multifaceted challenges posed by the pandemic [11].

Asian Region

In the wake of the pandemic, there has been a noticeable surge in Chinese activity within the Central Asian region. China is strategically investing and expanding its influence by spearheading infrastructure development and bolstering trade volumes to stimulate economic recovery. Across South Asian economies, including Pakistan, India, Bangladesh, Sri Lanka, Bhutan, Nepal, and the Maldives, substantial strides toward recovery have been observed. These nations have rebounded from the pandemic's impact through infrastructural advancements and a heightened focus on energy trade and commerce.

Conversely, island nations in Asia, like Indonesia, faced significant setbacks due to travel bans and tourism restrictions imposed as a result of COVID-19. The economic ramifications of the pandemic have necessitated more effective strategies for combating and alleviating its impact. To navigate these challenges, countries should prioritize the implementation of routine, high-quality, and reliable health management systems. Additionally, there is a critical need to enhance overall state health through regulatory measures and vigilant monitoring, ensuring a more resilient and robust response to future health crises [8, 12, 13].

Sector-Wise: COVID-19 Economic Impact

The COVID-19 pandemic has caused widespread economic disruption, impacting multiple sectors globally (Fig. 25.2), those issues are discussed below:

Aviation Sector

With travel restrictions and aircraft cancellations leading to a precipitous drop in air traffic and income, the COVID-19 pandemic had a serious effect on the airline sector. According to the Airports Council International (ACI), there will be a 2 billion passenger decline worldwide in Q2 2020 and a 4.6 billion passenger drop for the entire year. This would result in a revenue loss of $39.2 billion in Q2 and $97 billion in 2020. The travel sector in the United States had to cut spending by $519 billion

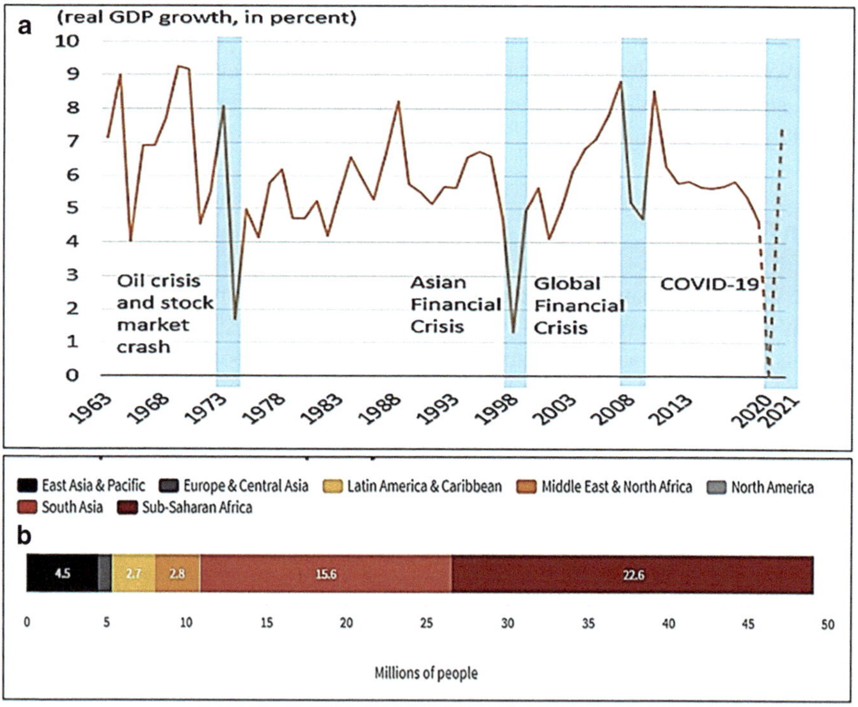

Fig. 25.2 The economic impact of COVID-19 on country-wise real GDP annual growth rate from 2019 to 2023. (**a**) Consensus forecasts the global GDP (%) and the COVID-19 recession that highlighted the fastest and steepest downgrades among all global recessions since 1990 (*This information was obtained from IMF Blog:* https://www.imf.org/ja/Blogs/Articles/2020/04/15/blog-covid-19-pandemic-and-the-asia-pacific-lowest-growth-since-the-1960s). (**b**) The steep decline in real GDP growth percentage due to COVID-19 crisis and the pattern of the dip in GDP across the years in Asian economy. (Source link: https://www.weforum.org/stories/2020/06/great-reset-covid-19-global-leadership-help-most-vulnerable/)

in 2020, which resulted in a $1.2 trillion drop in economic production and the possible loss of 6.9 million jobs in the second quarter. Roughly 57% of the aircraft were kept in storage. The anticipated impact on GDP was $651 million. Airlines adopted flexible policies in order to adjust. For tickets bought before March 1, 2020, American Airlines permitted fee-free flight modifications for travel until December 31, 2021. In April and May 2020, they cut the number of domestic flights by 60–80% and the number of foreign flights by 80–90%. For tickets bought between March and May 2020, Delta Airlines offered unlimited fee-free modifications. Travel was permitted through March 12, 2021. Both domestic and international flight capacity were cut by 80% and 70%, respectively, and tickets that were about to expire were extended [14–19].

Airlines took a number of steps to lessen losses in 2020, such as lowering CEO pay, retiring outdated aircraft, limiting flying capacity, and providing early retirement or voluntary leave. They cut back on employment, onboard services, and nonessential spending (including staff travel, events, and marketing). Airport lounges were shut down, real estate projects were put on hold, aircraft deliveries were postponed, and capital outflows were decreased. Securing loans, selling assets and shares, and halting dividend payments and share buybacks were among the financial tactics [15].

Worldwide flight disruptions and travel restrictions to combat COVID-19 are the root cause of the airline industry's financial difficulties. A dramatic decline resulted from governments locking down impacted areas and stopping the granting of visas. Aircraft manufacturers see order cancellations, while passenger airlines and catering services are most severely impacted. "Stay-at-home" directives have significantly decreased domestic travel in the United States. Major airlines have been severely impacted, including Lufthansa, Emirates, and Qatar Airways. Qatar Airways halted operations of ten A380s, grounded expensive aircraft, and suspended flights to Italy. In order to stop the virus from spreading, Emirates also reduced most of its activities [3, 20].

Significant unemployment has resulted from the COVID-19 outbreak in the aviation sector. There were 0.9 million employment losses in the Middle East, 11.2 million in Asia-Pacific, and 5.6 million in Europe. According to the International Air Transport Association (IATA), the decline in travel demand could jeopardize 25 million aviation employment worldwide.

The economic impact is significant because 65.5 million people, including tourism workers, depend on the aircraft business. Pilots, flight attendants, baggage handlers, and other employees are facing job insecurity as a result of airline cost-cutting. Since aviation is a trading sector, it boosts local economies. In urban areas, extensive unemployment may result in income losses, particularly if unemployment insurance is inadequate [4, 21, 22]. In every location, the COVID-19 epidemic resulted in large financial losses for the aviation sector. Europe suffered a loss of $89 billion, while Asia-Pacific suffered the biggest loss of $113 billion. Africa lost the least amount, at $6 billion. In comparison to 2019, passenger revenue is predicted to drop by $252 billion (44%) globally in 2020. The most hit was the second quarter of 2020, when demand fell by over 70% and airlines spent about $61 billion in cash [4].

Oil Sector

Oil prices fell sharply as a result of the COVID-19 epidemic and OPEC negotiations going awry. OPEC said on March 5, 2020 that it would reduce output by 1.5 million barrels per day (mb/d) in Q2 2020, with OPEC members contributing 1 mb/d and non-OPEC members, such as Russia, contributing 0.5 mb/d. Saudi Arabia increased output to 12.3 mb/d after Russia rejected the idea, which caused oil prices to drop by 30%. Reduced demand, fueled by "stay-at-home" orders and a global economic downturn, is the cause of the oil price decline. According to Rystad Energy, reduced air and road transportation will result in a 10% decrease in oil usage, or roughly 10 mb/d. International financial institutions are providing low-interest loans to emerging economies that are struggling financially. The supply shock will probably keep oil prices low for a long time, even though the demand shock brought on by the pandemic may recover after the virus spreads. Economists distinguish between oil shocks that are caused by demand and those that are driven by supply. Supply shocks, such as the current state of affairs, can have independent economic effects, impacting commodities, bonds, and equity markets, whereas demand shocks are linked to global demand. Falling oil prices in MENA nations have an adverse effect on foreign investment, remittances, and tourism, directly harming exporters and indirectly harming importers. While importers like Lebanon, Jordan, and Egypt are dealing with economic slowdowns that are making their debt loads worse, oil-exporting countries like Algeria and Iran have exhausted their reserves [23–25].

Oil-dependent nations were severely affected by the COVID-19 pandemic. Countries like Venezuela and Nigeria experienced income losses, growing current account deficits, and adverse consequences on their balance of payments as a result of the decline in oil prices and the decline in demand for oil products. The value of their currencies declined in relation to the US dollar as a result of the depletion of their foreign exchange reserves. National budgets of oil-dependent nations were also affected by this drop in oil prices because they were predicated on higher oil prices in 2019. These nations had to either reorganize their budgets or borrow money from the World Bank or IMF to cover the shortages in order to handle the consequent budget deficits and account for the decline in oil prices [23, 25].

Travel restrictions imposed by governments around the world as a result of the COVID-19 outbreak caused a global drop in oil consumption. With a daily consumption of six million barrels, the aviation sector is a significant oil user, contributing 11% of the transportation sector's oil demand. The prohibitions on overseas travel had a major effect on the need for oil, upsetting both supply and demand. The International Energy Agency (IEA) estimated that the world's oil consumption in 2020 was 99.9 million barrels per day, which was about 90,000 barrels less than in 2019. The IEA had initially predicted a daily increase in demand of 825,000 barrels.

Offshore oil production, which is essential to the world's supply of natural gas and oil, was also adversely impacted by the pandemic. Government-backed policies by oil corporations to cut offshore labor to stop the spread of COVID-19 led to further drops in oil production [26–28].

Many market participants closed their holdings prior to the actual settlement of contracts due to the scarcity and growing expenses of storing crude oil. In order to effectively pay a counterparty to sell their contracts, some future investors were even prepared to settle at negative prices. Few sellers in the physical market paid to dump oil, and negative WTI prices were limited to financial markets. Reduced travel and a global economic slowdown brought on by the measures to stop the spread of COVID-19 were the main causes of the drop in oil and petroleum consumption [26].

Tourism Sector

Worldwide travel has been severely disrupted by the COVID-19 epidemic, since lockdowns and "stay-at-home" orders continue to exist in many countries. The sustainability of regional tourism sectors is called into question by the uncertainties surrounding the reopening of popular tourist locations, especially in North America and Europe. Airlines are grounded, and international borders are closed. Social distancing measures will probably be in place even when tourism picks back up, and if limitations are removed too soon, there could be a spike in COVID-19 instances. Prior to the pandemic, nations that depend significantly on tourism, such as Spain and Italy, were already experiencing economic slowdowns as a result of high debt and unemployment. Including earnings from hotels, airlines, and dining establishments, tourism accounted for 14.3% of Spain's GDP and 13% of Italy's GDP in 2019. According to figures from the US Travel Association and Oxford Economics, tourism in the United States accounted for 8.6% of GDP, lost a considerable number of jobs (about eight million), and had a predicted $500 billion drop in travel spending for the year [29–31].

Due to the huge disruptions caused by previous disasters such as the Black Death, Spanish Flu, SARS, Swine Flu, and Ebola, the tourism industry is extremely susceptible to pandemics and macroeconomic shocks. For example, the Spanish flu killed 21 million people and caused travel restrictions for 4 months. The 2009 Swine Flu caused a decline in tourism in Mexico, costing the country $2.8 billion in lost revenue and one million foreign visitors. This trend has been exacerbated by the COVID-19 epidemic, since lockdowns, travel restrictions, and border closures have drastically decreased tourism. People have avoided famous sites out of fear of viral transmission. Travel to the badly impacted areas is further deterred by media coverage, which results in a large number of flight and hotel reservations being cancelled. Tourist spot occupancy rates in Italy, one of the most affected nations, fell to as low as 6% in cities like Rome, Milan, and Venice. According to the World Travel and Tourism Council, there could be a loss of about 50 million jobs in the tourism industry worldwide [17, 31].

Worldwide travel may decline by 45–70% as a result of the COVID-19 epidemic and worldwide efforts to stop its spread. Because so many people around the world are still limited, domestic travel has also suffered. Important industries have drastically cut back on operations, including restaurants, resorts, tour companies, airlines, and cruise lines. Approximately 51 million companies in the hotel and restaurant sectors are having financial difficulties, and many of their employees could lose their jobs as a result. Employees are facing difficulties at work, employment uncertainty, and fewer hours worked. The hardest hit are small and microfirms, especially in developing nations. Without outside assistance, these companies frequently have few assets, are unable to take advantage of financial stimulus measures, and lack access to finance [17].

Approximately 80% of hotels, parks, and tourist attractions in Brazil closed as a result of the severe slump in the country's tourism industry. Brazil's tourism industry is estimated to have lost $6.2 billion as a result of the lack of public transit, and visitor numbers were predicted to decline by 50% by 2020. Around 75,000 French establishments had shuttered by the end of March as a result of "stay-at-home" directives. In addition, some 40,000 eateries and 3000 clubs closed, technically leaving one million workers without jobs. About 80% of employees in the UK's lodging and food services industries were put on leave, and a third would lose their jobs as a result. Hotel occupancy rates fell precipitously in the United States, leading to the layoff of 3.9 million support staff and the loss of 1.6 million hotel employees. Tourism-related industries like aviation, handicrafts, agriculture, lodging, and food services were all negatively affected by the COVID-19 pandemic. Approximately ten million workers in the aviation industry are at risk of losing their jobs as a result of airport closures and airline cancellations [17].

For many Least Developed Countries (LDCs), tourism is a major source of income, accounting for 9.5% of both GDP and employment. This industry has been badly affected by the COVID-19 pandemic because of travel restrictions meant to stop the virus's spread. Global tourism spending is expected to decline by $30–50 billion as a result of a 20–30% decline in foreign visitors, according to the World Tourism Organization (UN-WTO). LDCs are experiencing severe economic losses, particularly those that depend significantly on tourism, including Bangladesh, Malawi, Rwanda, and Tanzania. Due to their reliance on tourism, several of these nations—including 14 Commonwealth LDCs—are at risk. Growth may be hampered in nations like Samoa, the Maldives, and Cabo Verde that have been moving up the economic ladder. With over a million jobs reliant on the tourism industry, African LDCs like Nigeria, Ethiopia, and Kenya have been particularly hard hit by the pandemic [26, 32, 33].

In 2020, Asia is predicted to have the biggest drop in tourism-related income, with China suffering the most. The tourism sector in Europe, which employs roughly 13 million people, loses nearly 1 billion euros in revenue each year, especially in nations like Spain and Italy that are off-limits to foreign visitors. The Catalonia region is expected to lose 11 billion euros, while Italy alone is expected to lose over 55 billion euros. Due to a lack of alternative revenue streams, the pandemic has had a particularly negative impact on Small Island Developing States (SIDS), whose economies rely heavily on tourism (30%). One out of every twenty workers in Sub-Saharan Africa is employed in the tourism industry, which may lose $50 billion, putting two million people at risk of unemployment [34, 35].

Financial Sector

Financial markets were characterized by uncertainty and volatility by the end of February 2020, with equity markets falling over 30% in a matter of weeks, surpassing the pace of the Global Financial Crisis (GFC) of 2008–2009. By March, the US dollar and short-term fund markets showed the symptoms of stress, and the normally liquid US Treasury market also saw a shortage of liquidity. Both businesses and governments found it challenging to borrow money as a result of this disruption, which also affected the credit markets. In response, central banks—including the US Federal Reserve—immediately put into place market-stabilizing strategies that were employed during the Great Financial Crisis, such as asset purchases to guarantee liquidity. Many are taken aback by this forceful engagement since they had anticipated a more measured strategy. Financial institutions and market players, however, are still unsure of the virus's effects, how long mitigation efforts would last, and how well recovery plans would work. The economic downturn sparked worries about fresh threats to the financial system, including the possibility of a financial crisis if things became worse. Although there is some recovery in the equity markets, the volatility reflects the varying macroeconomic outlooks, from pessimism to optimism. Forced asset sell-offs may result from persistent market liquidity problems, particularly as companies' credit ratings declined and their debt profiles deteriorated. Due to issues like declining commodity prices, supply chain interruptions, declining exports as a result of rich nations' recessions, and the expense of local COVID-19 measures, emerging and developing economies were especially vulnerable. Many poor countries lack the financial resources to handle the crisis, in contrast to rich countries that could rely on fiscal packages. In order to obtain short-term dollar liquidity and keep their financial markets stable in the face of uncertainty, some have resorted to the FIMA Repo Facility offered by the US Federal Reserve [36–38].

Banks have suffered greatly as a result of the COVID-19 pandemic, with both businesses and borrowers experiencing financial hardship. In response, banks have improved their digital services to facilitate online banking and remote work for employees. In order to prevent the virus from spreading through cash, the WHO has advised contactless payments. In order to satisfy the growing demand for online banking and other distant services, the Federal Financial Institutions Examinations Council (FFIEC) of the United States has encouraged banks to test digital systems. Travel limitations, staff interruptions, higher phone volumes, a lack of social distancing, insufficient digitization, and an increase in cybercrime are some of the issues that banks are facing as a result of COVID-19. To enhance operations and customer service, solutions include mobile app development, e-learning, AI forecasting, video banking, and predictive AI and machine learning methods [36, 39, 40].

Significant stock market instability is also faced due to the COVID-19 pandemic, which in March 2020 caused four market-wide circuit breakers. If the S&P 500 falls by more than 7%, these breakers, which are put in place to stop panic trading, halt trading for 15 min. They have been activated multiple times since their inception following the 1987 crash, especially during the market volatility of 1997 and 2020. As a result, the Chicago Board of Exchange Volatility Index increased significantly from mid-February, while indices such as the Dow and S&P 500 had strong drops [23, 26].

The COVID-19 pandemic, which began in China's Hubei region, triggered extreme market fluctuations globally. In the United States, market volatility in March 2020 surpassed the levels seen in October 1987 and December 2008. While fluctuations show the signs of easing by the end of April, they remain higher than before the outbreak. The pandemic led to declines in the Shanghai Stock Exchange Composite Index and the Nikkei Stock Average Index. Japan also faced a further economic slowdown with the postponement of the Tokyo 2020 Olympics. In response to the uncertainty, the US Senate passed a two trillion-dollar stimulus package to support the economy [26].

Investor sentiment is also influenced by the WHO and public health experts' pronouncements during the COVID-19 outbreak. While declining markets cause caution and a reluctance to invest, rising markets boost optimism and lessen investor concern. Global market correlation brought out by globalization has an impact on economic policy and asset allocation. Due to personal life disturbances, illness, and widespread anxiety, investors made illogical decisions during the epidemic. Stock market behavior is mainly influenced by the investors' mood in nations with a culture of group consensus and low institutional engagement [41].

Between February 19 and March 23, 2020, stock prices fell by more than 30% as a result of the COVID-19 pandemic's severe impact on financial activities. According to the VIX index, stock market volatility increases even though the reduction is not as sharp as the 57% drop that occurred during the 2007–2009 financial crisis. By March 27, 2020, the market had regained 14% as a result of stimulus packages and Federal Reserve involvement. The Federal Reserve introduced the lending facilities and lowered the interest rates to zero in reaction to the crisis in order to stabilize markets. With output contraction estimates between 10% and 30% and unemployment expected to increase from 3.5% to 6–13%, the US economy entered a severe recession. If the epidemic persists, there may be more disruptions to the long-term effects on labor mobility, small companies, and international trade [23, 26].

Health Sector

Because individuals still need medical treatment regardless of the state of the economy, healthcare has historically held up well during recessions. This custom has been upset by the COVID-19 pandemic, though, since many individuals have postponed or avoided the necessary medical procedures, including elective operations and routine checkups, due to lockdown protocols and infection fears. Patients in primary care offices fell by 70%, while healthcare workers' pay was frozen or cut, and some were put on leave. Healthcare-related unemployment claims increased. People will continue to shun public areas as long as the virus is around, which will impede the sector's recovery. With company closures, layoffs, and pay cuts lowering general spending and worsening the economic cycle, the prolonged recession may make matters worse [42–44].

The global expansion of the coronavirus has put a great deal of strain on hospitals and medical facilities, exposing gaps in access to treatment, especially for impoverished and rural communities. Caravan Health CEO Lynn Bar emphasized that the epidemic revealed the flaws in the fee-for-service healthcare system. Hospital visits have drastically fallen in rural areas, which has resulted in lower remuneration for healthcare providers. La Clínica de la Raza CEO Jane Garcia reported a 25% reduction in staff as a result of fewer patient visits. She underlines the significance of universal coverage and guaranteeing access to medical care, including in situations where people are unable to get to clinics. In this situation, telemedicine has become an essential answer [45–47].

The COVID-19 situation was quickly handled by the federal government. The Federal Reserve put money into the market and lowered interest rates. In order to directly assist American families, provide loans to small businesses, and provide $100 billion to hospitals overburdened with COVID-19 patients, the $2 trillion coronavirus relief law was passed. However, the pandemic-induced recession has resulted in healthcare disruptions. Due to extensive layoffs, many people no longer have employer-sponsored insurance and may turn to Medicaid, which has lower co-payments. Governments may lower Medicaid fees and insurance payments for public sector workers due to tight state and local budgets, which could have long-term effects on healthcare [48].

In the wake of the COVID-19 pandemic, advancements in technology, cost savings, and increased accessibility will be the main focuses of healthcare development. Spending more on healthcare makes a nation more capable of managing the situation. Although efforts are being made to boost the manufacture of diagnostic and therapeutic products, a commercially viable vaccine is still 12–18 months away. The government is funding pharmaceutical companies to test vaccines. While local companies want to make money from the production of medical items, export restrictions on medical equipment are upsetting global supply chains. Payment problems for healthcare professionals and delays in surgeries are common. The rise in popularity of telehealth and digital health has relieved the strain on overburdened hospitals [49].

Prenatal care has been greatly influenced by the COVID-19 pandemic, which has resulted in fewer in-person consultations for expectant mothers. Gynecologists in the United States report seeing fewer patients, with many choosing to use telemedicine. During the epidemic, telemedicine is thought to be a safer alternative. If a mother tests positive for COVID-19, the CDC advises isolating her from her child.

Depending on safety regulations, several hospitals in the United States, the United Kingdom, and Germany restrict or forbid partner access during labor and delivery, raising concerns about partner exclusion from delivery rooms [50–52].

Coordinating COVID-19 treatment with regular services is a challenge for healthcare facilities around the world. Hospitals assigned to COVID-19 care in Japan are overcrowded, and communication with local authorities is challenging. Information delivery to the private sector has been disrupted in Singapore. Although this has not yet affected patient treatment in Japan and Hong Kong, many nations suffer from inadequate intensive care unit capacity and a lack of protective gear. Other issues include shortages of healthcare workers and specialists. Particularly in Japan and Hong Kong, there is a need for better collaboration between public health facilities and private hospitals [21].

Because of the financial difficulties brought on by COVID-19, Moody's Investor Service reduced the credit rating of public and nonprofit hospitals in the United States from stable to negative in March 2020. In addition to increased expenses for healthcare personnel and intensive care unit utilization, the downgrading reflects decreased cash flows as hospitals delayed revenue-generating activities in order to concentrate on COVID-19 cases. Many patients have also lost their health insurance as a result of the unemployment increase. Global shortages of necessary medications are also caused by the pandemic's disruption of the pharmaceutical supply chain, namely the availability of Active Pharmaceutical Ingredients (APIs) coming from China [53, 54].

US healthcare organizations will continue to experience medical, economic, and recruitment challenges even after the COVID-19 curve flattens. At least 150 hospitals were forced to cut benefits, lower salaries, or lay off or furlough employees. Fear of infection may prevent many patients from rescheduling sessions, and hospitals are having difficulty hiring medical staff and obtaining funding for COVID-19 therapy, particularly for those without insurance. Staffing expenses and cancelled elective surgeries add to the revenue loss. $100 billion has been set aside by the US government to help hospitals with their infrastructural, diagnostic, and treatment requirements [53–55].

COVID-19 Pandemic: The Economic Consequences

The COVID-19 pandemic reshaped the global economy, causing sectors like aviation and tourism to plummet while propelling the media and food retail industries forward. Global consumption took a hit as incomes dwindled, prompting a shift toward local economies. The crisis accelerated digital transformations, making remote work and e-commerce indispensable. In the aftermath, businesses are navigating a transformed landscape, balancing continuity and adaptation in the pursuit of stability and innovation. It has led to job losses, business disruptions, and supply chain disruptions. It has also triggered a chain reaction that has unbalanced the socioeconomic homeostasis of society and affected the gross economy of many nations. The lack of money flow may adversely affect the strengthening of a country's health infrastructure and impede the steps toward coping with the pandemic. The pandemic has also led to increased economic, income, and lifestyle uncertainty, unemployment and underemployment pressure, and economic difficulties. These have, in turn, encouraged severe mental health problems, insecurities, worry, fear, and depression, which have overall negatively impacted the socioeconomic progress of any country [14, 47, 56, 57] (Fig. 25.3).

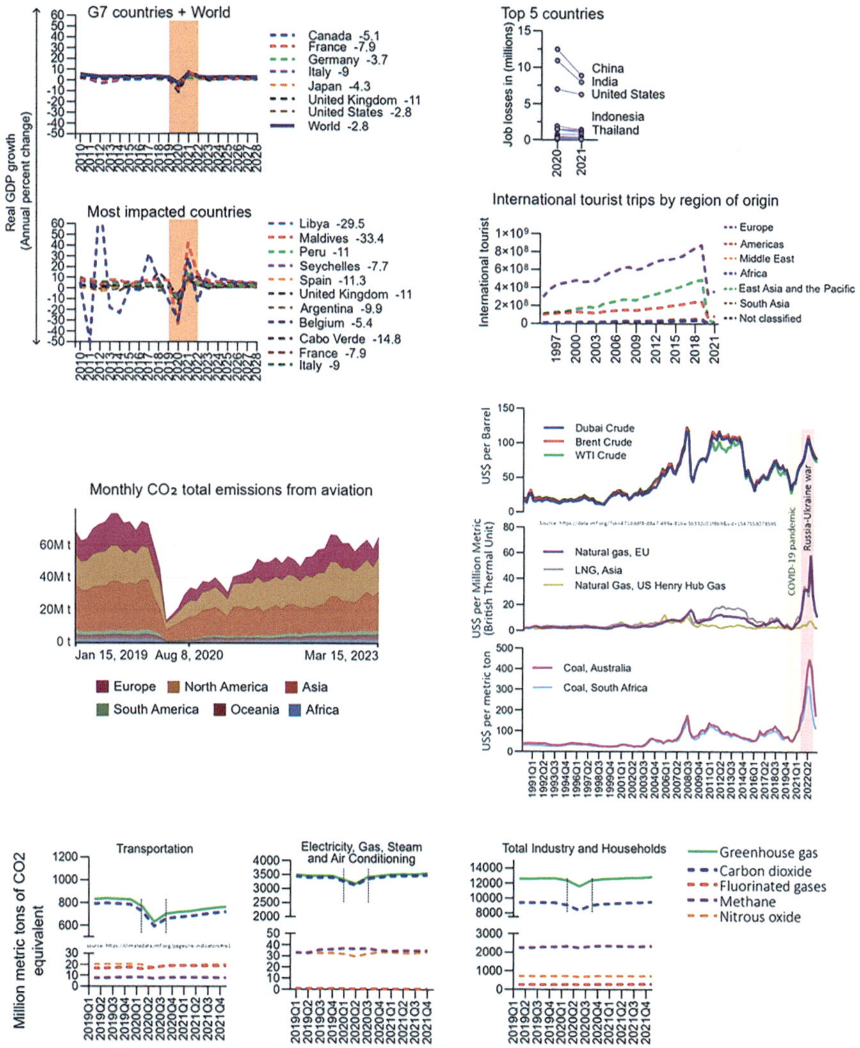

Fig. 25.3 Impact of COVID-19 on the global economy, job losses, transportation, tourism, and household consumption

Trade Disruptions, Job Loss, and Challenges for Vulnerable Populations

Travel bans have gravely impacted the business of trade and tourism at domestic and international levels. This is not a good sign for countries whose economies largely depend on tourism revenue. Also, growth curves and financial deficit rates have shown their worst performance during the past 2 years across many countries, but majorly affected the middle- and low-income countries the most, and pushed them further into economic crisis [58, 59]. According to the Asian Development Bank (ADB), >40% of the gross domestic product (GDP) in the Maldives and Palau is generated from tourism revenue receipts. In many developing member countries (DMCs) of the ADB, travel and tourism receipts make up over 10% of the GDP, underlining the sector's substantial economic contribution in these nations. Therefore, these countries face shattering economic levels and a weakening of the inflow of money due to this pandemic [60]. This pandemic's impact on business entry and exit was studied using official data from the Bureau of Labor Statistics and the US Census Bureau. The data showed that in the early stages of the pandemic, there was a surge in business closures, but many of these were temporary and reopened later. However, there was also a significant increase in permanent business closures, resulting in the loss of 1.2 million jobs in the second quarter of 2020. In the second half of 2020, there was a surge in new business formations, which was consistent with the increase in new business applications. Overall, the number of new businesses surpassed the number of closures, leading to a shift toward smaller firms in terms of employment. The article also discusses the challenges of obtaining timely data on business entry and exit during the pandemic and the importance of the Census Bureau's Business Formation Statistics in providing more up-to-date information on new business applications. Furthermore, dispensing chemists and nonspecialized food stores saw increased sales during and after the first lockdown. Other retailers that relied on customers being able to visit their stores saw a big drop in sales after they were ordered to close. However, some retailers, such as furniture retailers, saw sales return to pre-pandemic levels and grow further. This is likely due to people spending more time at home. This, in turn, adversely affects the healthcare infrastructure and limits grants of money for research related to other diseases like cancer [60–63].

Industrialization in urban cities serves as a pull factor for the immigration of people from rural areas to big cities for job opportunities and the upliftment of conditions of life. But the pandemic has led to the widespread shutdown of factories and industries around the world, thereby impacting the social security status of these people. Larger economic problems in terms of the current and potential future demands for oil and fluctuations in oil prices due to reduced economic activities associated with the pandemic are serious cases of concerns [64–67]. The coronavirus pandemic caused a sharp decline in oil demand in 2020, leading to a major slump in crude oil prices. Russia and Saudi Arabia, two of the largest oil producers, initially disagreed on how to respond to the pandemic. However, they eventually agreed to cut petroleum output in an attempt to stabilize the oil price.

However, this was not enough to prevent oil prices from falling into negative territory in April 2020 due to a lack of storage capacity. This pandemic had a negative impact on the coal and natural gas markets. The demand for both fuels has declined, and prices have fallen. This has had a significant impact on coal and natural gas producers, as well as on the workers who work in these industries [68, 69]. The shift to cleaner energy sources is accelerating, and this is likely to further reduce the demand for coal and natural gas in the coming years. The International Energy Agency (IEA) has published a report on the impact of this pandemic on the growth of renewable energy. The report says that while the amount of new renewable energy capacity added to the grid grew by more than 10% in 2019, the pandemic is expected to cause delays and cancellations of projects in 2020. Lockdowns have disrupted supply chains and construction sites, leading to a reduction in the amount of new renewable energy capacity added in the short term. In addition, some projects may miss deadlines for government incentives and face financial risks. However, some countries have introduced policy changes to address these concerns. The report also highlights the impact of the pandemic on distributed solar PV installations and the decline in demand for transport biofuels. While renewable electricity assets have been largely unaffected, the production of transport biofuels and the consumption of renewable heat in industries have seen significant declines. Global greenhouse gas emissions rebounded by 6.4% in 2021, surpassing pre-pandemic levels. The lion's share of the increase in emissions came from the manufacturing and energy sectors, while transportation and household emissions were relatively lower due to the pandemic's impact on mobility. The emergence of the omicron variant further reduced household and electricity sector emissions. Monitoring these sectors will be crucial as economies fully reopen amid high fossil fuel prices. The Intergovernmental Panel on Climate Change stresses the imperative to peak emissions by 2025, a crucial step to curbing global warming to 1.5 degrees Celsius (as highlighted by the IMF's Climate Change Indicators Dashboard). The global migration rates and labor market fragility can invite severe economic impacts for all nations. Globally, migrant workers are an important part of the labor markets, and the growing incidences of cases may lead to limited migration flows, hindering global economic growth and development. ADB reports that 80,000 people in Bangkok traveled back to their homes during the lockdown. India witnessed a surge in the migrant crisis as thousands of workers mass-moved back to their villages during the lockdown, driven by unemployment and the specter of starvation [70]. A similar sorrowful state was also seen by garment factory workers in Cambodia, Bangladesh, Indonesia, and Myanmar. Among such neglected sections of society, anyone suffering from cancer will get pushed into the danger of aggravated health conditions or even death due to a lack of adequate care and economically strained personal income levels. Addressing challenges in cancer care during strict lockdowns is paramount, focusing on maintaining accessibility and affordability of services. This becomes especially critical for vulnerable groups and economically disadvantaged patients, necessitating innovative solutions despite the suspension of transportation facilities [64–66, 71–75].

Economic Consequences in the Health Sector

According to the estimates from The American Hospital Association, $202.6 billion was lost in revenue in America's healthcare systems. Furthermore, it cost the low- and middle-income countries ~ US $52 billion for an effective healthcare response to combat COVID-19. This burden was further escalated by the regular economic costs associated with procedural patient care, surgeries, and surgical outcomes. The costs of procuring or manufacturing vaccines and distributing them globally have also caused tremendous economic costs and shifting of global budgets toward extreme levels. Without the vaccines developed since 2020, the death toll would have soared even higher. These vaccines stand as formidable defenses, effectively reducing severe illness and death, making them the primary safeguard against the virus. From an overall global economic perspective, the World Bank predicts an approximate 8% shrinkage in global growth, with poorer countries feeling most of the impact. Initially, a lack of preparedness led to great economic hurdles faced by healthcare facilities around the world, such as personal protective equipment (PPE), hospital equipment, sanitizers, and oxygen tanks were in short supply. These deficiencies improved over the years through alternative planning and new pandemic preparedness [66].

Comorbidities, or underlying medical conditions, are closely associated with increased complications and mortality in patients with COVID-19. This is especially true in developing countries, where resources are scarce and socioeconomic status is lower. Moreover, indigenous clans, tribal groups, and numerous ethnic communities exhibit an elevated susceptibility to contracting COVID-19, coupled with an increased risk of mortality. One study encompassing developing countries found that comorbidities were associated with a poor patient outcome due to the economics of less resources, lower socioeconomic background, lack of vaccination drives, and paucity of affordable testing programs to trace, test, and identify cases [76]. Altogether, in the United States, the added total direct medical cost burden due to COVID-19 was estimated to be between $160 billion and $650 billion during the entire course of the pandemic. Additionally, the pandemic has accelerated the adoption of digital technology in healthcare delivery, such as telemedicine and online consultancies, due to the perils of physical contact during the viral outbreak [2, 77].

Global Economic Impact of the Pandemic: Road to Recovery

The global economy grew steadily from 2010 to 2019, but the COVID-19 pandemic caused a sharp decline in 2020. This was the worst economic downturn in decades. In 2020, the GDP plummeted by 3.1%, surpassing the downturn witnessed in 2009 following the global financial crisis, marking a more than twofold decline. Latin America and the Caribbean experienced the largest GDP decline in 2020, with a contraction of 6.6%. South Asia's GDP fell 5.2%, and East Asia and the Pacific saw their first GDP decline since at least the 1960s. The global economy began to recover in 2021, but growth is still below pre-pandemic levels. This stems from various factors, including the slow recovery from COVID-19, inflation, and the war in Ukraine. The World Bank expects global GDP to reach $87 trillion in 2023, which is $4.7 trillion below its pre-pandemic forecast. The economic shortfall is pervasive across all regions. Projections indicate that by 2025, East Asia and the Pacific may experience a roughly 6% dip from its pre-pandemic trajectory, while Latin America and the Caribbean could see a 4% decrease, and South Asia might lag by 8% compared to earlier expectations [1].

The conflict in Ukraine has dealt a severe blow to both the Russian and Ukrainian economies. Ukraine's GDP is estimated to have plunged by 29% in 2022, while Russia saw a 2% contraction. Unfortunately, the region's overall GDP growth is not anticipated to return to its pre-COVID trajectory in the coming years. In addition, the pandemic's impact on India and Indonesia, two of the world's most populous countries, has been profound. India's GDP is projected to be approximately $296 billion lower in 2023 than initially forecast before COVID-19, and Indonesia's GDP is expected to be around $107 billion lower. Regrettably, neither country is anticipated to fully recover the lost ground at least through 2025. China, initially a standout performer during the pandemic, has seen a slowdown in growth. The projected 2025 GDP for China is estimated to be about 6% lower than anticipated before the onset of COVID-19. These economic shifts underscore the enduring challenges faced by diverse regions in the post-pandemic era. The largest percentage change in imports is seen in small economies that rely heavily on imports and import a large number of goods to boost their GDP. In terms of exports, Libya, New Caledonia, the Northern Mariana Islands, and Cape Verde experienced the largest relative losses due to the global demand shock and their trade dependencies on larger economies such as Europe, China, and the United States. In numerous emerging markets and developing economies, the path to economic recovery has been hindered by restricted access to vaccine doses and elevated oil prices. The changing dynamics of imports and exports have been largely influenced by the changing economics of the world due to the pandemic [1, 78–80]. In less developed economies, lower vaccination rates and tighter fiscal and monetary policies have led to the downfall. Furthermore, simultaneous Omicron-driven economic disruptions, climate-related disasters, financial stress, and de-anchoring of inflation expectations have weakened long-term growth drivers. Meeting the challenges ahead requires bolstering global cooperation. This involves ensuring fair vaccine distribution, aligning health and economic policies, and reinforcing debt sustainability. Given the climate shifts accompanying the pandemic, there's an imperative to amplify climate change adaptation globally. This involves ramping up green investments and facilitating energy transition, especially in emerging markets and developing economies (EMDEs). At the national level, tailored policy actions should prioritize investments in renewable energy and technological advancements, fostering targeted and growth-enhancing reforms [80, 81] (Fig. 25.3).

The employment landscape bore the brunt of the COVID-19 impact. In Latin America and the Caribbean, the unemployment rate surged from 7.8% in 2019 to 10% in 2020, remaining persistently high in 2021. North America witnessed a notable increase, with the unemployment rate jumping from 3.9% in 2019 to 8.2% in

2020. South Asia experienced a spike from a 6.3% jobless rate to 9.3% in 2020. Sub-Saharan Africa faced rising rates, climbing from around 6.3% to over 7% in 2021 and 2022. The ripple effects on jobs underscore the profound and lasting impact of the pandemic on global employment dynamics. The increased inflation rates further worsened the situation, especially in Southeast Asia. For instance, in Laos and Thailand, unemployment rates have spiked to unprecedented levels; around 25%; in Vietnam, unemployment rates were at a 10-year high, and five million people lost their jobs in 2020; in the Philippines, 7.3 million lost their jobs and unemployment rate spiked to 18%; in Malaysia, the highest unemployment rate was seen since 1989. In Indonesia, diminishing average household income and increasing poverty were seen. The economic impact hits the people working in the informal sector the most, and the economic crisis works in a feedback loop, thereby pushing them into the dungeons of poverty and escalating the precariously economically ruined conditions of the cancer patients amongst them due to the raging pandemic [82].

A select few nations, Ecuador and the Democratic Republic of Congo among them, are poised to meet or exceed GDP projections from January 2020. Ecuador's remarkable rebound is attributed to a robust COVID vaccination campaign, favorable oil prices, and an expanding domestic credit landscape. Meanwhile, the Democratic Republic of Congo has capitalized on the heightened production and prices of copper and cobalt. Countries like India, Myanmar, Vietnam, and the Philippines have strategically expanded their manufacturing sectors, ensuring uninterrupted goods production even during disruptions such as factory closures in China. This adaptability reflects a proactive approach to economic resilience in the face of global challenges [82, 83]. India imposed some of the world's strictest lockdown measures and limited stimulus spending in 2020, which led to a major GDP drop of 7.2%. However, the economy improved significantly in 2021, driven by recovering domestic demand and rising industrial production. India's vaccine competency and efficient, holistic coverage have largely slowed the decelerating pace of economic growth. Countries are poised to exceed pre-COVID projections significantly, with many reaping substantial benefits from higher commodity prices and recent natural resource discoveries. Nicaragua, buoyed by elevated gold and coffee prices—major exports—has experienced notable gains. Similarly, Guyana is capitalizing on increased production and exports of oil and gas. In contrast, the tourism sector, profoundly impacted by COVID-19, faced severe setbacks. In around 34 countries where tourism constituted over 10% of GDP, economic output was more than 9% lower in 2021 compared to pre-pandemic forecasts. For the 25 countries where tourism makes up 5–10% of GDP, the average shortfall was approximately 6%. This divergence in economic outcomes underscores the varied impact of the pandemic on different sectors and economies [80].

The delayed treatment regime and undue restrictions during the time of the pandemic have not only affected the societal and psychological milieu but have also exposed the grave economic problems of countries around the world. Faced with the fiscal deficit, the burden of the increased cost of living as well as treatment, during the times of the pandemic, has economically strained citizens around the globe.

Travel and movement restrictions that were enforced led to high rates of unemployment, and job and wage loss [84].

The delayed diagnosis of patients with ongoing severe health conditions, such as cancer, during COVID-19 restrictions also has economic consequences for society. For example, in a recent study, it was found that halted or pending colorectal cancer (CRC) screening will transition cancer to a further advanced stage which would in turn influence the CRC mortality rates and higher costs of management of the disease at its advanced stage. This adversely impacts the public health economy, which will escalate with time, making treatments unaffordable and pressurizing the patients. Under the severe COVID-19 conditions, rising unemployment, and catastrophic economic strain, Government systems have collapsed in providing adequate and robust social security schemes, thereby forcing many cancer patients to choose between treatment costs and feeding their families. There has been an unprecedented rise in the expenditure toward traditional, alternative, and/or complementary medicines among cancer patients in Southeast Asia (around 85%), which requires a shift in the budgetary allocations toward healthcare systems, depending upon the situations across the countries [80, 84–87].

Conclusion and Future Perspectives

The global economic environment has been drastically altered by the COVID-19 pandemic, which has highlighted areas of resilience across sectors and countries while also exposing weaknesses. While developing and emerging countries grappled with systemic issues such as supply chain disruptions, budgetary constraints, and diminishing remittances, advanced nations experienced a severe economic downturn. On the other hand, East Asian economies proved resilient thanks to strong digital infrastructure and sensible regulatory initiatives. Sectorally, the crisis fueled growth in technology and healthcare while hastening the collapse in tourism, hospitality, and aviation. Different recovery trends were seen in manufacturing and agriculture, which were impacted by logistical efficiency and digital adaptation.

The pandemic provides important lessons for creating economies that are more robust in the future. To lessen future shocks, policymakers must give priority to supply chain diversification, digital infrastructure investment, and healthcare system strengthening. To stabilize trade and alleviate economic disparities made worse by the crisis, regions should strengthen their international collaboration. Furthermore, supporting digital transformation and green recovery tactics will be essential for long-term success. Through gaining insight from the disparate effects of COVID-19, the world economy can become more robust and better equipped to withstand future shocks.

Acknowledgments The authors are deeply indebted to the researchers and publishers, whose scientific literature helped in this report. We are also thankful to the site BioRender.com, as a few pictorial motifs in the figures were partially created with BioRender.com tools.

Authors' Contributions Conceived the idea and wrote the manuscript: SK, GS, ST, MB, and MKG. All authors have read, agreed, and approved the final draft of the manuscript.

Funding This work is supported by CSIR, Govt. of India.

Conflicts of Interest There is no competing and conflict of interest.

Ethics Approval This article does not contain any studies with human participants or animals performed by any of the authors.

Consent to Participate This article does not contain any studies with human participants and therefore did not require any consent to participate.

References

1. Ahmad T, Haroon, Baig M, Hui J. Coronavirus disease 2019 (COVID-19) pandemic and economic impact. Pak J Med Sci. 2020;36(COVID19-S4):S73–8. https://doi.org/10.12669/pjms.36.COVID19-S4.2638.
2. Garg S, Norman GJ. Impact of COVID-19 on health economics and technology of diabetes care: use cases of real-time continuous glucose monitoring to transform health care during a global pandemic. Diabetes Technol Ther. 2021;23(S1):S15–20. https://doi.org/10.1089/dia.2020.0656.
3. Ghosh MK, Kumar S, Ganguly KK, et al. COVID-19 and cancer: insights into their association and influence on genetic and epigenetic landscape. Epigenomics. 2023;15(4):227–48. https://doi.org/10.2217/epi-2023-0052.
4. Sobieralski JB. Global airline employment and the COVID-19 pandemic: impacts, comparisons, and implications for the future. In: Airlines and the COVID-19 pandemic. Emerald Publishing Limited; 2024. p. 341–68. Accessed 16 Nov 2024. https://www.emerald.com/insight/content/doi/10.1108/S2212-160920240000011013/full/html.
5. Roy S. Economic impact of Covid-19 pandemic. Prepr. 2020;1:29.
6. Ihekweazu C, Agogo E. Africa's response to COVID-19. BMC Med. 2020;18(1):151. https://doi.org/10.1186/s12916-020-01622-w.
7. Obande GA, Bagudo AI, Mohamad S, et al. Current state of COVID-19 pandemic in Africa: lessons for today and the future. Int J Environ Res Public Health. 2021;18(19). https://doi.org/10.3390/ijerph18199968.
8. Goals SD. World health statistics. World Health Statistics; 2022.
9. Marmot M. Review of social determinants and the health divide in the WHO European region. World Health Organization. Regional Office for Europe; 2014. Accessed 27 Nov 2023. https://apps.who.int/iris/handle/10665/108636.
10. Marmot M, Allen J, Bell R, Bloomer E, Goldblatt P. WHO European review of social determinants of health and the health divide. Lancet. 2012;380(9846):1011–29.
11. Organization WH. World mental health report: transforming mental health for all. Published online 2022. Accessed 27 Nov 2023. https://archive.hshsl.umaryland.edu/handle/10713/20295.
12. Chakraborty C, Sharma AR, Bhattacharya M, Agoramoorthy G, Lee SS. Asian-origin approved COVID-19 vaccines and current status of COVID-19 vaccination program in Asia: a critical analysis. Vaccine. 2021;9(6). https://doi.org/10.3390/vaccines9060600.
13. Fitzgerald DA, Wong GWK. COVID-19: a tale of two pandemics across the Asia Pacific region. Paediatr Respir Rev. 2020;35:75–80. https://doi.org/10.1016/j.prrv.2020.06.018.

14. Sidhu GS, Rai JS, Khaira KS, Kaur S. The impact of COVID-19 pandemic on different sectors of the Indian economy: a descriptive study. Int J Econ Financ Issues. 2020;10(5):113–20.
15. Li Y, Wang J. The impact of COVID-19 pandemic on air passenger travel: a focus on empirical findings. Transp Rev. 2024;44(2):461–83. https://doi.org/10.1080/01441647.2023.2244165.
16. Lee JW, Yoon SY. Growing stakeholderism in the airline industry after the COVID-19 pandemic. J Air Transp Manag. 2024;116:102574.
17. Filep S, King B, McKercher B. Reflecting on tourism and COVID-19 research. Tour Recreat Res. 2024;49(2):426–30. https://doi.org/10.1080/02508281.2021.2023839.
18. Vulturius S, Budd L, Ison S, Quddus M. Commercial airline pilots' job satisfaction before and during the COVID-19 pandemic: a comparative study. Res Transp Bus Manag. 2024;53:101108.
19. Unraveling COVID-19 diagnostics: a roadmap for future pandemic – nature cell and science. Accessed 15 Nov 2024. https://cellnatsci.com/2958-695X/article/10-61474-ncs-2024-00026.
20. Bazak YL, Sander B, Werker E, Zhumatova S, Worsnop CZ, Lee K. The economic impact of international travel measures used during the COVID-19 pandemic: a scoping review. BMJ Glob Health. 2024;9(2):e013900.
21. Babarinde TP, Adepoju OO. Desk review of impact of coronavirus on the aviation sector in the United States. Eur J Med Health Res. 2024;2(1):28–35.
22. Ghaleb BDS. The effect of the COVID-19 pandemic on unemployment in the world. J Res Dev. 2024;12:247.
23. Yiming W, Xun L, Umair M, Aizhan A. COVID-19 and the transformation of emerging economies: financialization, green bonds, and stock market volatility. Resour Policy. 2024;92:104963.
24. Pradhan S, Ghose D, Shabbiruddin. Present and future impact of COVID-19 in the renewable energy sector: a case study on India. Energy Sources Part Recovery Util Environ Eff. 2024;46(1):13226–36. https://doi.org/10.1080/15567036.2020.1801902.
25. Rabbani MR. COVID-19 and its impact on supply chain financing and the role of Islamic fintech: evidence from GCC countries. Int J Agile Syst Manag. 2024;17(3):269–91. https://doi.org/10.1504/IJASM.2024.140396.
26. Salman A, Ali Q. Covid-19 and its impact on the stock market in GCC. J Sustain Finance Invest. 2024;14(1):220–36. https://doi.org/10.1080/20430795.2021.1944036.
27. Zahoor Z, Shahzad K, Mustafa AU. Economic impact of Covid 19 on Bangladesh, India, and Pakistan. Int J Manag Res Emerg Sci. 2024;14(1). Accessed 16 Nov 2024. https://www.ijmres.pk/index.php/IJMRES/article/view/479.
28. Al-Mhdawi MKS, Brito M, Onggo BS, Qazi A, O'Connor A. COVID-19 emerging risk assessment for the construction industry of developing countries: evidence from Iraq. Int J Constr Manag. 2024;24(7):693–706. https://doi.org/10.1080/15623599.2023.2169301.
29. Lee LC, Wang Y, Zhang L, Ping L, Zuo J, Zhang H. Impact of COVID-19 on the economic loss and resource conservation of China's tourism industry from the supply chain perspective. Cities. 2024;144:104633.
30. Kumar P, Ekka P. Statistical analysis of the impacts of COVID-19 pandemic on the small and large-scale tourism sectors in developing countries. Environ Dev Sustain. 2023;26(4):9625–59. https://doi.org/10.1007/s10668-023-03112-4.
31. Vayá E, Garcia JR, Suriñach J, Pons E. Effects of the COVID-19 tourism crisis on the Spanish economy. Tour Econ. 2024;30(4):1021–38. https://doi.org/10.1177/13548166231185899.
32. Gunter U, Smeral E, Zekan B. Forecasting tourism in the EU after the COVID-19 crisis. J Hosp Tour Res. 2024;48(5):909–19. https://doi.org/10.1177/10963480221125130.
33. Mane VH. Tourism as a catalyst: employment generation in India's economy. Accessed 16 Nov 2024. https://www.researchgate.net/profile/Vinod-Mane-3/publication/385012908_THE_JOURNAL_OF_ORIENTAL_RESEARCH_MADRAS/links/6711fba1d796f96b8ebd8282/THE-JOURNAL-OF-ORIENTAL-RESEARCH-MADRAS.pdf.
34. King VT, Chin WL. COVID-19 and post-COVID-19 transitions: case study on the Lao PDR. Southeast Asia: Multidiscip J. 2024;24(2):150–63.

35. Drammeh FY. Managing tourism during the COVID-19 pandemic. A systematic review of crisis management in the tourism industry. Cogent Soc Sci. 2024;10(1):2311955. https://doi.org/10.1080/23311886.2024.2311955.
36. Marcu MR. The impact of the COVID-19 pandemic on the banking sector. Manag Dyn Knowl Econ. 2021;9(2):205–23.
37. Kumar S, Maheshwari V, Prabhu J, et al. Social economic impact of COVID-19 outbreak in India. Int J Pervasive Comput Commun. 2020;16(4):309–19.
38. Liu WP, Chu YC. FinTech, economic growth, and COVID-19: international evidence. Asia Pac Manag Rev. Published online 2024. Accessed 17 Nov 2024. https://www.sciencedirect.com/science/article/pii/S1029313223001069.
39. Xie X, Mirza N, Umar M, Ji X. Covid-19 and market discipline: evidence from the banking sector in emerging markets. Int Rev Econ Finance. 2024;89:612–21.
40. Yıldırım AC, Erdil E. The effect of Covid-19 on digital banking explored under business model approach. Qual Res Financ Mark. 2024;16(1):87–107.
41. Miroshnychenko I, Vocalelli G, De Massis A, Grassi S, Ravazzolo F. The COVID-19 pandemic and family business performance. Small Bus Econ. 2024;62(1):213–41. https://doi.org/10.1007/s11187-023-00766-2.
42. Singhal V, Jinger D, Rathore AC, et al. COVID-19, deforestation, and green economy. Front For Glob Change. 2024;6:1305779.
43. Usman SO, Esomchi OS, Nasiru IM, Daniel AV. An assessment of COVID-19 and its impact on Nigeria's socio-economic development. Cogent Soc Sci. 2024;10(1):2306700. https://doi.org/10.1080/23311886.2024.2306700.
44. Chetty R, Friedman JN, Stepner M. The economic impacts of COVID-19: evidence from a new public database built using private sector data. Q J Econ. 2024;139(2):829–89.
45. Blajer-Gołębiewska A, Honecker L, Nowak S. Investor sentiment response to COVID-19 outbreak-related news: a sectoral analysis of US firms. N Am J Econ Financ. 2024;71:102121.
46. Lappegård T, Kornstad T, Dommermuth L, Kristensen AP. Understanding the positive effects of the COVID-19 pandemic on women's fertility in Norway. Popul Dev Rev. 2024;50(S1):129–52. https://doi.org/10.1111/padr.12539.
47. Wang Y, Shi L, Que J, et al. The impact of quarantine on mental health status among general population in China during the COVID-19 pandemic. Mol Psychiatry. 2021;26(9):4813–22.
48. He J, Harkins C, O'Dell K, et al. COVID-19 perturbation on US air quality and human health impact assessment. PNAS Nexus. 2024;3(1):pgad483.
49. Barai MK, Dhar S. COVID-19 pandemic: inflicted costs and some emerging global issues. Glob Bus Rev. 2024;25(3):812–31. https://doi.org/10.1177/0972150921991499.
50. Bahari NI, Sutan R, Abdullah Mahdy Z. The determinants of maternal perception of antenatal care services during the COVID-19 pandemic critical phase: a systematic review. PLoS One. 2024;19(2):e0297563.
51. Gómez-Chávez P, Soriano-Avelar VM, Aguilar-Rodríguez A, Rojas-Russell M, Castro-Porras LV. Risk perception increase due to COVID-19 impacted antenatal care utilization among women in an indigenous community. BMC Pregnancy Childbirth. 2024;24(1):578. https://doi.org/10.1186/s12884-024-06748-w.
52. Zhang R, Byrd T, Qiao S, Torres ME, Li X, Liu J. Maternal care utilization and provision during the COVID-19 pandemic: voices from minoritized pregnant and postpartum women and maternal care providers in deep south. PLoS One. 2024;19(4):e0300424.
53. Maleka NH, Matli W. A review of telehealth during the COVID-19 emergency situation in the public health sector: challenges and opportunities. J Sci Technol Policy Manag. 2024;15(4):707–24.
54. Signorelli C, De Ponti E, Mastrangelo M, et al. The contribution of the private healthcare sector during the COVID-19 pandemic: the experience of the Lombardy region in northern Italy. Ann Ig Med Prev E Comunità. 2024;36(2). Accessed 17 Nov 2024. https://annali-igiene.it/articoli/2024/2/Signorelli%202.pdf.
55. Miller ED. The COVID-19 pandemic crisis: the loss and trauma event of our time. In: Loss and trauma in the COVID-19 era. Routledge; 2024. p. 60–72. Accessed 17

Nov 2024. https://www.taylorfrancis.com/chapters/edit/10.4324/9781032647876-7/covid-19-pandemic-crisis-loss-trauma-event-time-eric-miller.
56. Diana-Elena VC, Lucian B, Alexandra H. Considerations upon the effects of Covid-19 pandemic on the Romanian economic environment. Stud Bus Econ. 2022;17(1):272–89. https://doi.org/10.2478/sbe-2022-0018.
57. Tiwari CK, Pal A, Khandelwal T. Pandemics and their business impacts: a global perspective. Cardiometry. 2022;25:764–72.
58. von Tigerstrom B, Wilson K. COVID-19 travel restrictions and the international health regulations (2005). BMJ Glob Health. 2020;5(5). https://doi.org/10.1136/bmjgh-2020-002629.
59. Costantino V, Heslop DJ, MacIntyre CR. The effectiveness of full and partial travel bans against COVID-19 spread in Australia for travellers from China during and after the epidemic peak in China. J Travel Med. 2020;27(5). https://doi.org/10.1093/jtm/taaa081.
60. De Guzman R, Malik M. Dual challenge of cancer and COVID-19: impact on health care and socioeconomic systems in Asia Pacific. JCO Glob Oncol. 2020;6:906–12. https://doi.org/10.1200/GO.20.00227.
61. Kumar S, Basu M, Ghosh P, Pal U, Ghosh MK. COVID-19 therapeutics: clinical application of repurposed drugs and futuristic strategies for target-based drug discovery. Genes Dis. 2023;10(4):1402–28. https://doi.org/10.1016/j.gendis.2022.12.019.
62. Kumar S, Basu M, Ghosh P, Ansari A, Ghosh MK. COVID-19: clinical status of vaccine development to date. Br J Clin Pharmacol. 2023;89(1):114–49. https://doi.org/10.1111/bcp.15552.
63. Kumar S, Ansari A, Basu M, Ghosh S, Begam S, Ghosh MK. Carbon nanotubes in cancer diagnosis and treatment: current trends and future perspectives. Adv Ther. 2025;8:2400283. https://doi.org/10.1002/adtp.202400283.
64. Prinja S, Pandav CS. Economics of COVID-19: challenges and the way forward for health policy during and after the pandemic. Indian J Public Health. 2020;64(Supplement):S231–3. https://doi.org/10.4103/ijph.IJPH_524_20.
65. Cypress BS. COVID-19: the economic impact of a pandemic on the healthcare delivery system in the United States. Nurs Forum (Auckl). 2022;57(2):323–7. https://doi.org/10.1111/nuf.12677.
66. Kaye AD, Okeagu CN, Pham AD, et al. Economic impact of COVID-19 pandemic on healthcare facilities and systems: international perspectives. Best Pract Res Clin Anaesthesiol. 2021;35(3):293–306. https://doi.org/10.1016/j.bpa.2020.11.009.
67. Forchette L, Sebastian W, Liu T. A comprehensive review of COVID-19 virology, vaccines, variants, and therapeutics. Curr Med Sci. 2021;41(6):1037–51. https://doi.org/10.1007/s11596-021-2395-1.
68. Ceylan Z. The impact of COVID-19 on the electricity demand: a case study for Turkey. Int J Energy Res. 2021;45(9):13022–39. https://doi.org/10.1002/er.6631.
69. Khan K, Su CW, Zhu MN. Examining the behaviour of energy prices to COVID-19 uncertainty: a quantile on quantile approach. Energy. 2022;239:122430.
70. Chakraborty I, Maity P. COVID-19 outbreak: migration, effects on society, global environment and prevention. Sci Total Environ. 2020;728:138882. https://doi.org/10.1016/j.scitotenv.2020.138882.
71. Kolahchi Z, De Domenico M, Uddin LQ, et al. COVID-19 and its global economic impact. Adv Exp Med Biol. 2021;1318:825–37. https://doi.org/10.1007/978-3-030-63761-3_46.
72. Chams N, Chams S, Badran R, et al. COVID-19: a multidisciplinary review. Front Public Health. 2020;8:383. https://doi.org/10.3389/fpubh.2020.00383.
73. Lu X, Lin Z. COVID-19, economic impact, mental health, and coping behaviors: a conceptual framework and future research directions. Front Psychol. 2021;12:759974. https://doi.org/10.3389/fpsyg.2021.759974.
74. Sabetkish N, Rahmani A. The overall impact of COVID-19 on healthcare during the pandemic: a multidisciplinary point of view. Health Sci Rep. 2021;4(4):e386. https://doi.org/10.1002/hsr2.386.

75. Singh K, Kondal D, Mohan S, et al. Health, psychosocial, and economic impacts of the COVID-19 pandemic on people with chronic conditions in India: a mixed methods study. BMC Public Health. 2021;21(1):685. https://doi.org/10.1186/s12889-021-10708-w.
76. Singh AK, Misra A. Impact of COVID-19 and comorbidities on health and economics: focus on developing countries and India. Diabetes Metab Syndr. 2020;14(6):1625–30. https://doi.org/10.1016/j.dsx.2020.08.032.
77. Bel G, Gasulla Ó, Mazaira-Font FA. The effect of health and economic costs on governments' policy responses to COVID-19 crisis under incomplete information. Public Adm Rev. 2021;81(6):1131–46. https://doi.org/10.1111/puar.13394.
78. Basak P, Abir T, Al Mamun A, et al. A global study on the correlates of gross domestic product (GDP) and COVID-19 vaccine distribution. Vaccine. 2022;10(2). https://doi.org/10.3390/vaccines10020266.
79. König M, Winkler A. COVID-19: lockdowns, fatality rates and GDP growth: evidence for the first three quarters of 2020. Inter Econ. 2021;56(1):32–9. https://doi.org/10.1007/s10272-021-0948-y.
80. Global energy review 2021 I. Global energy review 2019. Glob Energy Rev 2019. 2020; (License: CC BY 4.0). https://doi.org/10.1787/90c8c125-en.
81. Goldberg PK, Reed T. The effects of the coronavirus pandemic in emerging market and developing economies: an optimistic preliminary account. Brook Pap Econ Act. 2020;2020(Special Edition):161–235. https://doi.org/10.1353/ECA.2020.0009.
82. Verschuur J, Koks EE, Hall JW. Global economic impacts of COVID-19 lockdown measures stand out in high-frequency shipping data. PLoS One. 2021;16(4):e0248818. https://doi.org/10.1371/journal.pone.0248818.
83. Ashraf BN. Economic impact of government interventions during the COVID-19 pandemic: international evidence from financial markets. J Behav Exp Financ. 2020;27:100371. https://doi.org/10.1016/j.jbef.2020.100371.
84. Kawohl W, Nordt C. COVID-19, unemployment, and suicide. Lancet Psychiatry. 2020;7(5):389–90. https://doi.org/10.1016/S2215-0366(20)30141-3.
85. Drake RE, Sederer LI, Becker DR, Bond GR. COVID-19, unemployment, and behavioral health conditions: the need for supported employment. Admin Pol Ment Health. 2021;48(3):388–92. https://doi.org/10.1007/s10488-021-01130-w.
86. Gezici A, Ozay O. An intersectional analysis of COVID-19 unemployment. J Econ Race Policy. 2020;3(4):270–81. https://doi.org/10.1007/s41996-020-00075-w.
87. Hensher M. Covid-19, unemployment, and health: time for deeper solutions? BMJ. 2020;371:m3687. https://doi.org/10.1136/bmj.m3687.

Navigating Long COVID: Comorbidities and the Global Societal Impact

26

Mrinal K. Ghosh, Shaheda Tabassum, Gouranga Saha, Sunny Kumar, Pratyay Ghosh, Suresh Bajoria, and Malini Basu

Abstract

This review examines the complex interplay between the COVID-19 pandemic, its associated comorbidities, and the broad socio-economic repercussions, with a particular emphasis on the psychological impacts on individuals and society. COVID-19 comorbidities are highlighted as a central theme, exploring the interwoven health challenges and their profound socio-psychological effects. This review seeks to shed light on the far-reaching consequences of the pandemic, encompassing mental health strains, economic disruptions, and societal shifts. Recognizing that many social impacts cannot be alleviated by therapeutic interventions alone, this work offers a comprehensive analysis of the multi-layered

Mrinal K. Ghosh, Shaheda Tabassum and Gouranga Saha contributed equally.

M. K. Ghosh (✉) · S. Tabassum · G. Saha · S. Kumar
Cancer Biology and Inflammatory Disorder Division, Council of Scientific and Industrial Research-Indian Institute of Chemical Biology (CSIR-IICB), TRUE Campus, CN-6, Sector–V, Salt Lake, Kolkata, India
e-mail: mrinalghosh@iicb.res.in

P. Ghosh
Questrom School of Business, Boston, MA, USA

S. Bajoria
Department of Urology, NH-RTIICS, Mukundapur, Kolkata, India

M. Basu
Department of Microbiology, Dhruba Chand Halder (DCH) College - University of Calcutta, Jaynagar Mazilpur, West Bengal, India

M. K. Ghosh, M. Basu (eds.), *Pathogenesis, Management and Socio-Economic Impact of COVID-19 Pandemic*, Medical Virology: From Pathogenesis to Disease Control, https://doi.org/10.1007/978-981-95-1426-7_26

crises brought about by COVID-19, providing valuable insights into the broader implications for both mental health and societal resilience.

Keywords

Long-COVID · COVID-19 · Comorbidity · Societal impact · Economic impact

Introduction

COVID-19 was declared a global pandemic by the World Health Organization (WHO) due to its unprecedented impact on morbidity and mortality. As of December 23, 2022, there have been 651,918,402 confirmed cases and 6,656,601 deaths globally. Vaccination efforts have reached 13,008,560,983 doses administered by December 12, 2022 (Fig. 26.1). It typically incubates for around 5.1 days, with symptoms appearing within approximately 11.5 days. Clinical presentations range from asymptomatic to severe, with common symptoms including fever, cough, shortness of breath, nausea, malaise, stomach problems, and diarrhea. Severe cases can lead to acute respiratory failure, septic shock, and multiple organ failure [1, 2]. COVID-19 heightens risks for those with conditions like Chronic Obstructive Pulmonary Disease (COPD), cardiovascular disease, diabetes, hypertension, and cancer, creating an immune-compromised state. Shared inflammatory responses amplify complications. About 3–4% of COVID-19 cases involve cancer patients, indicating mutual susceptibility. The virus increases respiratory risks, potentially leading to lung cancer. Pre-existing cardiovascular conditions elevate infection and mortality risks, associated with myocarditis and thrombotic events. Direct viral effects and cytokine storms compromise immunity, fostering complications. Ongoing research aims to grasp the intricate pathophysiology of COVID-19-related comorbidities in a succinct manner [3, 4].

COVID-19 has etched a lasting impact on society. The disease has inflicted lasting psychological scars on patients and healthcare providers, with increased stress, behavioral changes, social isolation, substance abuse, and elevated crime rates. This societal upheaval has woven a lasting tapestry of change. This review provides a comprehensive analysis of COVID-19 comorbidities and their extensive societal impact, emphasizing that without dedicated and coordinated societal responses, global hardship will continue. The social ramifications of SARS-CoV-2 prevalence are poised to jeopardize lives and livelihoods for years [5–10].

Urgent development responses are crucial in facing this crisis. These responses should align with long-term aspirations, addressing the psychological fallout of COVID-19 on society. The world must unite to overcome this menace, ensuring both psychological homeostasis. Immediate action is required to steer development trajectories that lead to sustainable recovery and resilience in the face of future challenges. By joining forces globally, we can navigate these uncertain times, safeguarding the well-being of individuals.

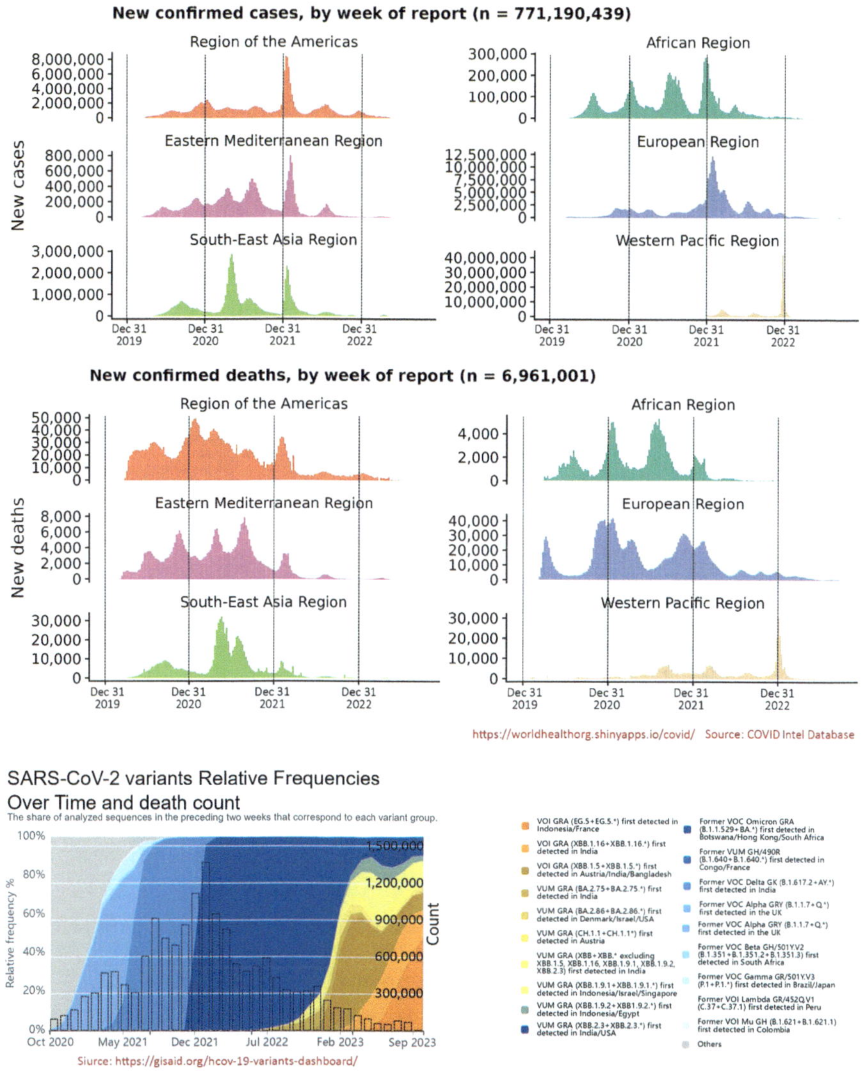

Fig. 26.1 Stats of reported positive cases, deaths, and variant frequencies. The figure illustrates the confirmed cases, deaths, and variants of emergence found throughout the pandemic

Social Impact: A Global Perspective

The COVID-19 pandemic has had a profound and wide-ranging impact on societies around the world. The extent and nature of these impacts can vary by region and population. Here are some of the key social impacts observed during the pandemic (Fig. 26.2):

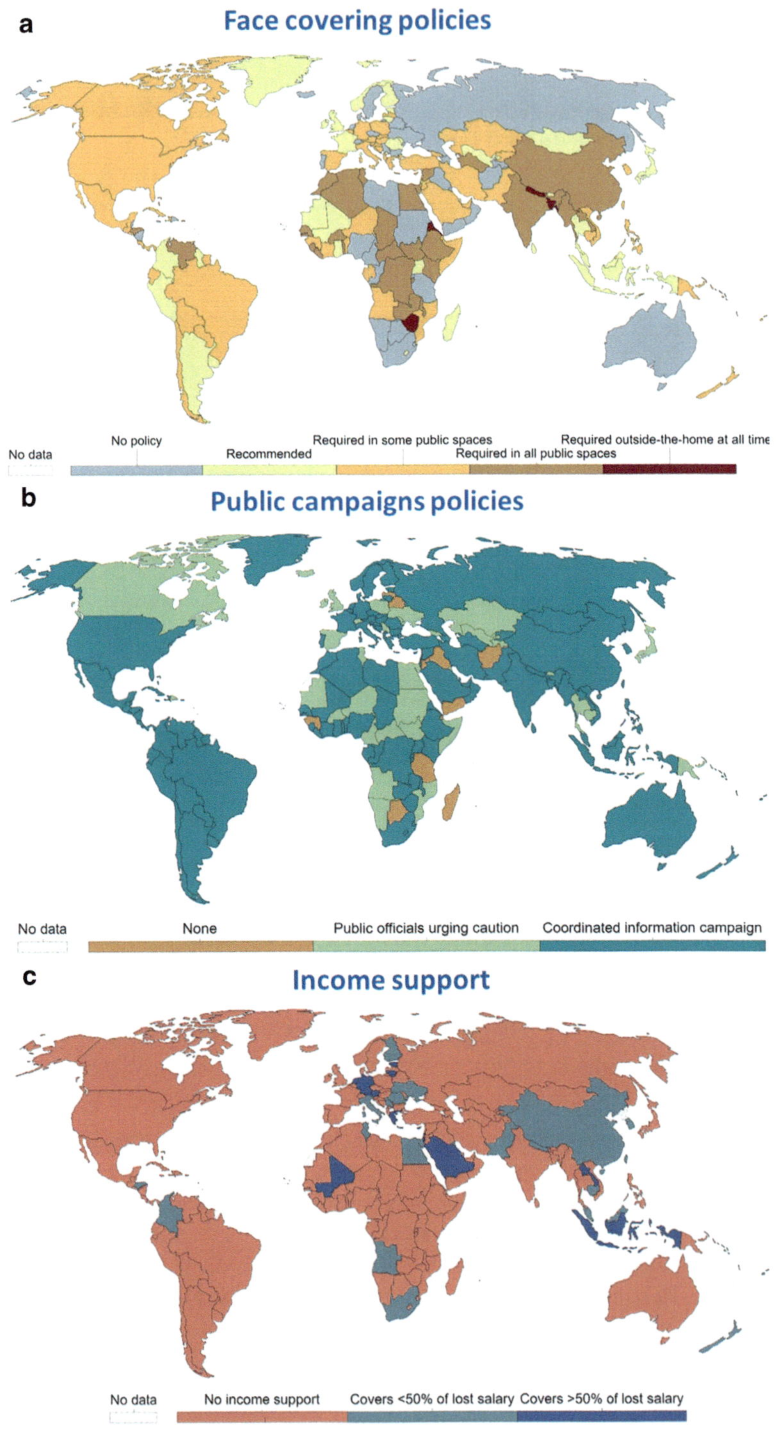

Fig. 26.2 Worldwide follow-up of social policies during the COVID-19 pandemic. The figure depicts the social policies followed during the COVID-19 pandemic, such as (**a**) face covering, (**b**) public campaigns, and (**c**) income support provided to employees worldwide

Healthcare Systems

The COVID-19 pandemic placed immense strain on healthcare systems across the globe, with hospitals struggling to manage the rapid surge in cases. Medical supplies, including PPE, ventilators, and ICU capacity, became severely limited, and healthcare professionals faced burnout as they worked long hours under intense pressure. Non-COVID treatments were often delayed or canceled, compounding health issues for patients with chronic or acute conditions and highlighting the need for a more resilient healthcare infrastructure.

Economic Disparities

Lockdowns, business closures, and economic instability created widespread financial hardship, with millions losing jobs and income. Low-income individuals, gig economy workers, and informal sector employees, who often lack job security or savings, were disproportionately impacted. Economic uncertainty heightened the financial vulnerability of families worldwide, slowing down the global economy and deepening economic inequalities that will likely persist long after the pandemic.

Education

To curb the virus's spread, many countries closed schools, disrupting the education of millions of students and causing significant challenges for parents and educators alike. As education shifted online, disparities in technology and internet access became glaringly evident, with underserved students left behind. The lack of face-to-face interaction also impacted learning outcomes, exacerbating long-standing educational inequities and raising concerns about the long-term effects on student development.

Mental Health

Mental health emerged as a critical issue globally, driven by pandemic-related uncertainties, social isolation, and grief from losing loved ones. Traditional mourning practices were often restricted, adding to the emotional toll on families and communities. Anxiety, depression, and stress-related disorders surged, impacting not only individuals but also families and communities, underscoring the need for accessible mental health support and resources.

Remote Work and Social Isolation

The pandemic's shift to remote work has fundamentally changed work dynamics, pushing many to adopt new technologies and adapt to home-based setups. While remote work offered flexibility, it also blurred work-life boundaries and presented challenges for those in unsuitable or crowded living conditions. Social isolation from prolonged lockdowns amplified feelings of loneliness and mental health strain, particularly for the elderly and those without strong support networks.

Community Solidarity

In response to the crisis, communities around the world demonstrated resilience and solidarity through grassroots initiatives like mutual aid groups, food drives, and volunteer networks. Individuals stepped up to support one another, highlighting the power of community spirit. This surge in volunteerism revealed the essential role of social support networks in times of crisis, fostering hope and unity amid adversity.

Digital Transformation

The rapid pivot to digital solutions in work, education, healthcare, and daily interactions accelerated the adoption of technology but also widened existing digital divides. While some adapted quickly, others faced barriers due to limited access, lack of digital skills, or affordability issues. This digital transformation underscored the need for equitable access to technology and digital literacy to prevent marginalized communities from being further excluded.

Government Response and Trust

Government responses to the pandemic varied widely, from rapid lockdowns to vaccine rollouts, significantly shaping public trust in institutions. Countries that communicated effectively and acted decisively generally saw greater compliance, while misinformation and disinformation posed serious challenges. The rise in misinformation not only undermined public health efforts but also highlighted the importance of transparent, consistent communication to maintain public trust and social cohesion.

Racial and Ethnic Disparities

COVID-19 exposed and intensified racial and ethnic disparities, with marginalized groups experiencing higher infection rates and worse health outcomes. These disparities are often rooted in pre-existing socio-economic and health inequities, such as limited access to healthcare, crowded housing, and employment in high-exposure jobs. The pandemic's unequal impact across racial and ethnic lines underscores the importance of addressing systemic inequalities to improve health equity.

Global Solidarity and Cooperation

Lastly, the pandemic emphasized the interconnectedness of the global community, sparking unprecedented collaboration in areas like vaccine development, research, and resource-sharing. Nations shared data, resources, and expertise, fostering global solidarity. However, challenges in equitable vaccine distribution and nationalism also highlighted the need for coordinated global health responses to ensure that all nations, regardless of wealth, have the resources to combat future pandemics.

The social impact of the COVID-19 pandemic is complex and multifaceted, affecting various aspects of daily life and revealing both vulnerabilities and strengths within societies. The long-term consequences continue to unfold as communities strive to recover and build resilience in the face of ongoing challenges [11–17].

COVID-19: Comorbidity and Socio-Psychological Impact

Extensive research has revealed that the elderly, especially those in long-term care facilities or suffering from serious underlying medical conditions, like respiratory diseases, cardiovascular diseases, cancer patients, smokers, and transplant recipients are at increased risk of COVID-19 infection and subsequent higher mortality rates [18] (Fig. 26.3). However, COVID-19 pandemic has influenced the societal framework of the world in respect to their mental and psychosocial health. Thus, the authors are interested to shed light on major COVID-19 comorbidities along with their socio-psychological impact.

Respiratory Diseases

COVID-19 comorbidity of patients with respiratory diseases: There has been mounting evidence that the phenotype of asthma and associated comorbidities are important factors in assessing the SARS-CoV-2 infection and disease severity. Asthmatic patients are highly vulnerable to this infection because of their inherent antiviral immune response deficiencies and a proclivity for exacerbation. Furthermore, patients with asthma have a chance of aggravated COVID-19 symptoms due to their underlying pathophysiology. The mortality rate in asthma patients suffering from severe COVID-19 is 32.5%. Asthma exacerbations are mostly caused by respiratory infections like the rhinovirus, coronavirus, and influenza, and are frequently seen in the pediatric population. Such patients are at higher risk of contracting COVID-19 infection due to their impaired upper and lower respiratory tracts and slower immunological response to viral infections [19, 20]. Reports have

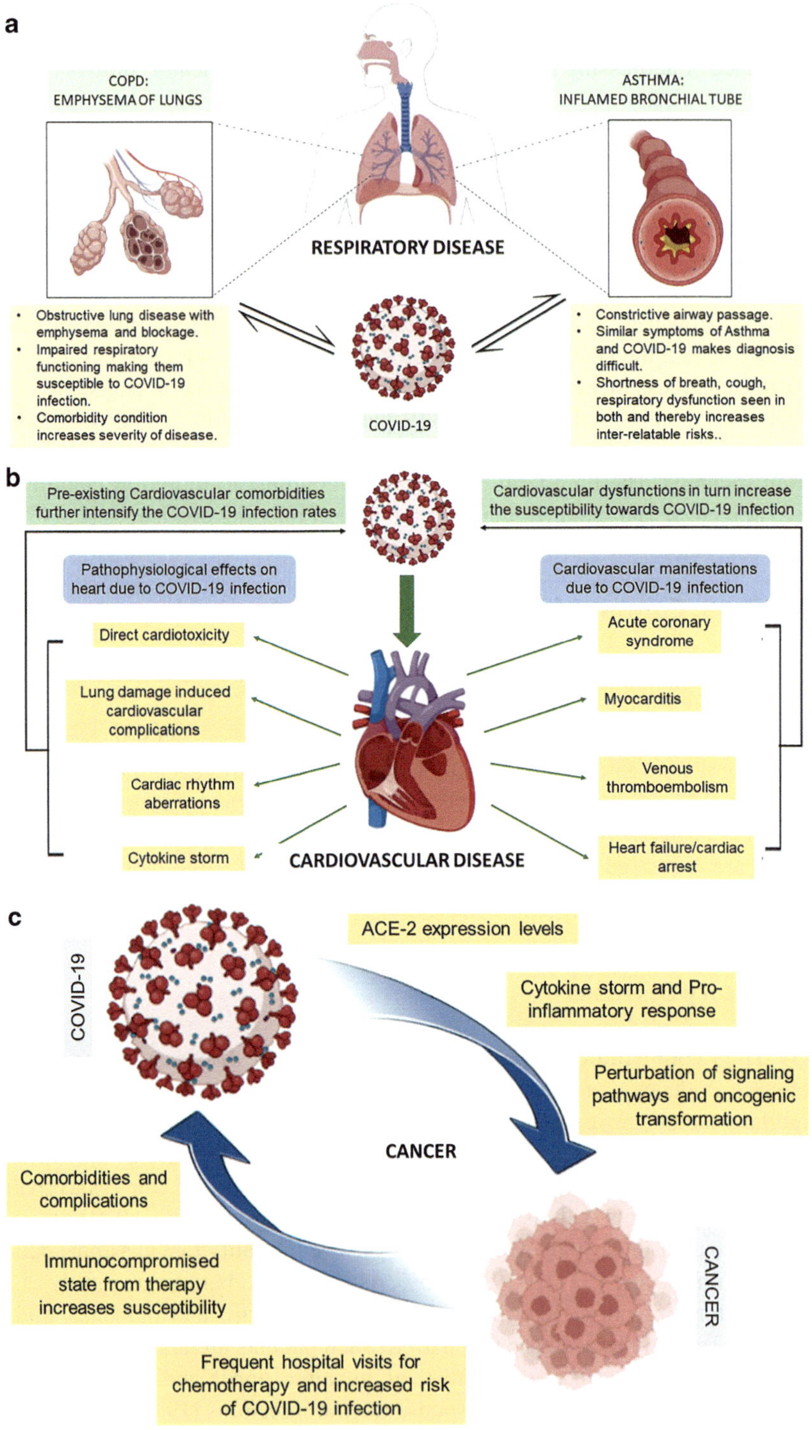
a
COPD:
EMPHYSEMA OF LUNGS
ASTHMA:
INFLAMED BRONCHIAL TUBE
RESPIRATORY DISEASE
• Obstructive lung disease with emphysema and blockage.
• Impaired respiratory functioning making them susceptible to COVID-19 infection.
• Comorbidity condition increases severity of disease.
COVID-19
• Constrictive airway passage.
• Similar symptoms of Asthma and COVID-19 makes diagnosis difficult.
• Shortness of breath, cough, respiratory dysfunction seen in both and thereby increases inter-relatable risks..
b
Pre-existing Cardiovascular comorbidities further intensify the COVID-19 infection rates
Cardiovascular dysfunctions in turn increase the susceptibility towards COVID-19 infection
Pathophysiological effects on heart due to COVID-19 infection
Cardiovascular manifestations due to COVID-19 infection
Direct cardiotoxicity
Lung damage induced cardiovascular complications
Cardiac rhythm aberrations
Cytokine storm
Acute coronary syndrome
Myocarditis
Venous thromboembolism
Heart failure/cardiac arrest
CARDIOVASCULAR DISEASE
c
COVID-19
ACE-2 expression levels
Cytokine storm and Pro-inflammatory response
Perturbation of signaling pathways and oncogenic transformation
CANCER
Comorbidities and complications
Immunocompromised state from therapy increases susceptibility
Frequent hospital visits for chemotherapy and increased risk of COVID-19 infection
CANCER

Fig. 26.3 COVID-19 and its comorbidity with respiratory, cardiovascular, and cancer. The diagram provides a comprehensive insight into the crosstalk between COVID-19 infection with respiratory, cardiovascular, and cancer patients. (**a**) First of all, the interlinkage between COVID-19 and respiratory diseases is illustrated through the mechanism in which they influence each other's susceptibility. In COPD and asthma patients, there is constrictive blockage of respiratory passageways along with lung dysfunction. These lead to higher vulnerability to contracting COVID-19 infections, along with increasing the chances of exposure to the virus. (**b**) Secondly, the diagram also schematically illustrates the possible crosstalk between COVID-19 infection and pre-existing cardiovascular comorbidities, indicating inter-linkage between them and highlighting the fact that they mutually increase susceptibility to each other, leading to deteriorating health conditions. (**c**) The diagram illustrates the interlinkage between COVID-19 and cancer. COVID-19-infected patients show increased ACE-2 expression, cytokine storm due to inflammatory pathways, and perturbations of key signaling pathways that can increase the risk of cancer progression and aggressiveness. Similarly, in cancer, the compromised immune system due to chemotherapy makes the patient susceptible to viral attack. Furthermore, complexities, comorbidities, and frequent hospital visits increase the chances of exposure to the virus

indicated that Chronic Obstructive Pulmonary Disease (COPD) is a notable risk factor for severe COVID-19 disease. A study spanning 1590 COVID-19 patients in China revealed that COPD conferred an odds ratio of 2.681 for ICU admission, mechanical ventilation, or death. In New York, USA, among 13,442 diagnosed COVID-19 patients, COPD showed a significant prevalence, underscoring its association with severe outcomes. Moreover, patients with COPD have worse outcomes after COVID-19 infection. Meta-analysis reported an 88% increased ICU admission or death risk among COVID-19 patients with COPD.

Furthermore, increased hospitalization rates due to COVID-19 infection in a New York cohort were positively correlated with COPD. The ISARIC (International Severe Acute Respiratory and Emerging Infection Consortium) cohort, consisting of 20,000 hospitalized COVID-19 patients, identified nonasthma chronic pulmonary diseases as a significant risk factor for accelerated mortality in these individuals [21–23]. By employing fast-track diagnostics for the respiratory pathogen, Sreenath et al. reportedly found 33 other respiratory pathogens upon testing 191 patients with SARS-CoV-2, and 46.6% of patients had co-infection with one or more additional pathogens. The intricate association of COVID-19 with respiratory ailments can be attributed to the increased pulmonary expression of ACE2, the viral receptor for binding SARS-CoV-2, in COPD and most other viral diseases, making the patients more susceptible to COVID-19 [24]. The timely identification of co-infections can yield improved health for COVID-19 patients by preventing worsening disease outcomes.

Socio-psychological consequences: The social impact of asthma has a deep influence on socioeconomic homeostasis. With the complications arising due to COVID-19 infections, things have further worsened from a sociological aspect. Childhood asthmatic conditions often negatively affect the lifestyle and normal growth of the child. There are significant limitations in leading a normal life for such children. Childhood asthma patients are subjected to some restrictions on their physical and social activities, leading to deleterious effects on a child's daily activities, schooling, family time, and finances for treatment. Currently, with the pandemic decelerating the pace of lives globally, the complicated scenario of asthma patients susceptible to COVID-19 further complicates the case. Since children are a vulnerable group of the population for COVID-19-associated risks, therefore, asthma patients suffer from higher rates of anxiety and stress due to the danger of SARS-CoV-2 infection [25, 26]. Furthermore, frequent hospital visits and requiring treatment modalities, asthma patients face a severe risk of contracting the disease due to their exposure to the virus. The societal impact of increasing cases of asthma-associated COVID-19 can adversely affect the psychology of the population and trigger widespread psychological pressure and impaired mental health. With both diseases showing similar symptoms, there is a greater societal risk of infection spread without understanding the root cause, because there may be chances of errors in diagnosis, differentiating the two diseases. Current asthma guidelines suggest the use of oral corticosteroids in acute exacerbations ubiquitously for hospitalized episodes for immunosuppression and clinical recovery. Conversely, the WHO advises against corticosteroids use for suspected cases of COVID-19, thereby bringing a conflicting approach between the two. Therefore, the chances of COVID-19 infections increase in asthma patients. This pushes them toward social isolation, discriminatory behavior, and increased stress [27, 28].

Amidst the scenario of COPD-associated COVID-19 complications, self-management intervention is crucial for COPD disease management. There is a need for optimized physical health; reduced impairments in daily life; increased emotional and social well-being; and better quality of life with family, friends, and community [29, 30]. However, the recent rise in COVID-19-associated stress, including reduced mobility and fear of symptoms, has significantly impacted the societal fabric. There has been increased pressure on the existing healthcare systems, leading to widespread mental health issues. It has also affected and disrupted social relationships, increased isolation, and higher psychological symptoms in cases of comorbidities. Anxiety and depression in COPD patients are inversely associated with positive social support. On the other hand, negative social interactions such as social criticism, isolation, and disagreements intensify the existing levels of depression and anxiety in the patients (Fig. 26.4).

Meta-analysis studies have revealed that there is a tremendous psychological impact of COVID-19 on healthcare workers, with rates of prevalence of 23.2% in anxiety and 22.8% in depression. These values were found to be significantly higher among patients with comorbidities compared to healthy controls. Furthermore, medical workers suffered from greater psychological distress compared to administrative staff. Within the healthcare staff, psychological distress was more prevalent among women and nurses compared to their male counterparts and doctors. This is further aggravated by increasing pressure on the existing medical systems due to rising cases. Additionally, social isolation and financial

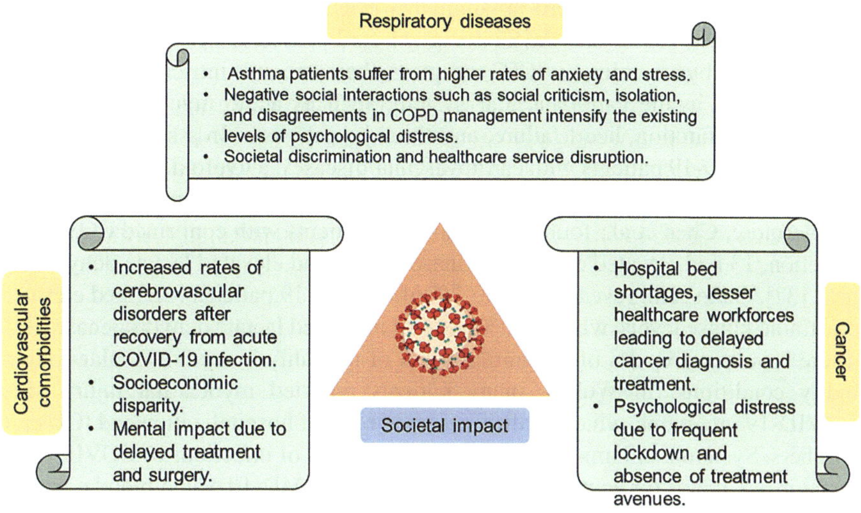

Fig. 26.4 The socio-psychological consequences of COVID-19 comorbidities. The figure schematically provides an insight into the societal impact of respiratory ailments, cardiovascular comorbidities, and cancer with COVID-19. It highlights the psychological pressure on patients associated with each type of comorbidity and its impact on their mental health due to vulnerability and infection of COVID-19

security correlated with COVID-19 infections. The symptoms of social isolation and loneliness are closely tied to anxiety and depression, exacerbating societal inequality and distress, further complicated by the challenges posed by COVID-19 [28, 31–34]. Optimizing social support interventions and the identification of factors that influence positive societal impact and self-management are needed to properly overcome the sociological ramifications of respiratory disorders associated with the COVID-19 pandemic.

Cardiovascular Diseases

COVID-19 comorbidity of patients with cardiovascular diseases: The pathophysiology of SARS-CoV-2 is marked by heightened production of inflammatory cytokines (IL-6 and TNF-α), triggering systemic inflammation and contributing to multi-organ dysfunction, particularly within the cardiovascular system. Prevalent comorbidities among COVID-19 patients include hypertension (56.6%) and diabetes (33.8%). Additionally, cardiac injury correlates with inflammation biomarkers like IL-6 and C-reactive protein (CRP), establishing a connection between viral infection-induced inflammatory hyperactivity and myocardial injury [35]. It has been found that around 2.5–16% of COVID-19 patients suffer from cardiovascular disease. Pondering upon the mechanism of action, it is known that RAS is involved in blood pressure regulation and cardiovascular activities. While angiotensin-2 regulates vasoconstriction and pro-inflammatory effects. There is a need for a proper balance between ACE1 and ACE2 for good cardiovascular health. Since SARS-CoV-2 infects by attaching to ACE2 receptors, therefore, it is intricately linked with complications in the cardiovascular system, such as acute myocardial damage, myocardial infarction, heart failure, and thromboembolism. In a study reported in 44,000 COVID-19 patients with cardiovascular diseases, a fivefold increase in mortality was found compared to patients without cardiovascular problems [36]. Furthermore, Chen et al., found that among 99 patients with confirmed COVID-19 infection, 13 had elevated creatinine kinase and 75 had elevated lactate dehydrogenase [37]. Guan et al. revealed that 13.7% of COVID-19 patients exhibited elevated creatinine kinase levels, while 37.2% showed increased lactate dehydrogenase [38]. There have been reports of a significant risk of mortality in cardiovascular comorbidity conditions. In Wuhan, many patients reported myocardial injury after COVID-19 infections, which further increased rates of hospitalization and ICU care numbers. Systemic inflammation that is characteristic of infections of COVID-19 is a major risk factor for acute myocardial infarction (AMI). Elevated blood coagulation processes play a role in the emergence of coronary complications in individuals with COVID-19. Additionally, 7–17% of COVID-19 patients were observed to experience heart rhythm disturbances. Conversely, patients with underlying cardiovascular conditions are more prone to getting infected with SARS-CoV-2, with higher chances of aggravated pneumonia and severe symptoms [19]. Therefore, cardiovascular implications cause poor prognosis in COVID-19 patients, necessitating precocious detection and optimal therapeutic strategies.

Socio-psychological consequences: From the point of view of cardiovascular diseases and COVID-19, the increased risk of myocardial infarction (MI), stress cardiomyopathy, and disseminated intravascular coagulopathy in critically ill patients leads to higher case severity and mortality rate. Long-term follow-up has demonstrated increased rates of cerebrovascular disorders, ischemic heart disease, and heart failure after recovery from acute COVID-19 infection. Due to the global lockdown measures, a significant decline was observed in the degree of primary percutaneous coronary intervention procedures [39]. This led to a subsequent rise in the time from symptom onset to hospital admission and the interval between arriving at the hospital and receiving balloon angioplasty (door-to-balloon time) for ST-elevation myocardial infarction (STEMI) patients. The fear of catching COVID-19 has hugely impacted the usual STEMI volumes at hospitals. This is a matter of concern for medical societies, such that there are high chances of unnecessary deaths or poor outcomes for patients who survive [40]. According to the American College of Emergency Physicians Hospitals, hospitals have seen around a 60% reduction in admissions for heart attacks. Other emergency room volumes have also decreased by 50% raising suggestive fears in the medical community of the potentially fatal effects of COVID-19. There has been a sharp 38% drop in patients being treated for ST-elevated myocardial infarction (STEMI) heart attacks. Multiple factors related to reduced and avoidance of medical care have been seen. Due to impaired finances, people found it difficult to approach hospitals for treatment modalities. Also, the hospitals and health support staff were overburdened with rising COVID-19 cases, which led to lesser attention to the therapeutic avenues needed for other diseases and cardiovascular abnormalities.

In society, there were rising incidences of social discrimination and phobias of the disease. A constant sense of fear of COVID-19, misdiagnosis, or the pharmacological management of disease led to less frequent therapeutic support and gravely impacted the social determinants of health. Nevertheless, there were frequent reports of out-of-hospital cardiac arrest that were found to be strongly associated with COVID-19 incidences (Fig. 26.4). During the pandemic, there was an increased number of cardiac arrest and stroke cases due to the mounting mental and psychological pressure of the looming threat of the virus. It was also reported that there was large-scale community mortality within elderly care homes, thereby highlighting the high susceptibility of the elderly population to the cardiovascular repercussions of COVID-19 [36, 41–43]. Furthermore, in terms of socioeconomic disparity, it had been reported in Brazil that high rates of cardiovascular mortality were notable in less developed cities, suggesting the impact of societal status and developmental status on disease-associated mortalities. People living in the most deprived areas of the United Kingdom were found to be twice as likely to die from COVID-19 as their well-off counterparts. The mortality rate in the most economically disadvantaged areas stood at 128.3 per 100,000 people, a stark contrast to the rate of 58.8 per 100,000 in the least deprived areas. Also, the detrimental impact of poor socioeconomic status leads to impaired mental health and higher risks of heart ailments during the pandemic times. The additional pressure of social isolation and compulsory quarantine restrictions drastically reduced the mental health of global populations. Evaluative reports suggest that the rate of post-traumatic stress symptoms is 28–34%, low mood reach is 73%, and irritability reach is 57%. These stresses are positively correlated with cardiovascular complications. There have been ample reports highlighting the susceptibility of minority ethnic groups to COVID-19. Therefore, the societal hierarchy leaves a long-lasting impact on the social and mental health of individuals. A recent analysis indicates a social gradient in COVID-19 mortality, revealing a higher confirmed case rate per 100,000 in Black British individuals compared to White ethnic groups. The incidence of COVID-19 was 0.61% in Black/Black British participants, 0.32% in "other" ethnicities and in Asian/Asian British, 0.30% in Chinese, and 0.14% in White. Pakistani and Bangladeshi groups are disproportionately residing in less privileged neighborhoods compared to White British individuals, potentially leading to inadequate access to health benefits and societal support, particularly during a pandemic [41, 44]. Despite the cases dropping with rising vaccination rates, there has been very little vaccine access in the socially deprived regions. Not only has the population not been suitably vaccinated, but there have been high rates of vaccine hesitancy that have complicated matters further. Crowded or poor-quality housing and occupational exposure positively correlated with COVID-19 susceptibility due to the health issues associated with overcrowding and limited supplies of medicine and preventive measures in those deprived regions of low- and middle-income settings. Fear of COVID-19 and delaying care result in more serious heart damage and even death. There is a need to look into the socioeconomic barriers and reason out methods to overcome the disabilities associated with healthcare and social support avenues [45–47].

Cancer

COVID-19 comorbidity of cancer patients: In the backdrop of the COVID-19 pandemic, cancer patients emerge as a highly vulnerable population, their weakened immune systems stemming from oncogenic pathophysiology and anticancer treatments. Additionally, their susceptibility to infection is heightened during hospital visits for chemotherapy or routine check-ups, exposing them to COVID-19 patients or virus-contaminated areas. Noteworthy findings from cohort studies shed light on the intricacies of this scenario. In a study of 105 COVID-19-infected cancer patients, lung cancer emerged as the most prevalent (20.95%). Another cohort study highlighted that Stage IV cancer patients experienced a higher incidence of SARS-CoV-2 infections. Chinese research by Liang et al. underscored that cancer patients faced an elevated risk of severe COVID-19 events, such as intensive care unit admission, invasive ventilation, or death, compared to their noncancer counterparts. Further insights from a multi-center study of 105 cancer patients with COVID-19 indicated increased infection rates among those with hematological malignancies and metastatic solid tumors. The COVID-19 and Cancer Consortium (CCC19) cohort study corroborated these concerns, reporting higher mortality and severity of infection in COVID-19-infected cancer patients compared to the general population. Alarmingly, patients undergoing chemotherapy or surgery faced a significantly elevated risk (75%) of infections [48–50]. COVID-19 cancer patients who received immunotherapy showed 66.67% of symptom severity and 33.33% of death. It was also found that cancer patients with COVID-19 had prolonged hospital stays and a poorer prognosis. Clinical outcomes of cancer patients, such as age, severity, ACE2 expression, and comorbidities, are important factors contributing to their susceptibility toward COVID-19 infection. The broad impact of COVID-19 on cancer patients reportedly causes mortality to increase by 20% in a study in the United Kingdom. In a large cohort study of 10,404 participants, cancer survivors showed a 60% increased risk of a positive COVID-19 test compared to normal individuals. In the United States, there was a 3% rise in cancer-related mortality during the pandemic era due to delayed diagnoses, inefficient patient care, and increased exposure to the SARS-CoV-2 virus. Moreover, the patients with cancer face unprecedented complications in getting adequate care due to the upsurge in COVID-19 cases, owing to the pressure on hospitals and the increased risk of contracting infections easily through such sources. In the realm of lung cancer surgical treatment, a concerning finding emerged: 5 out of 41 patients became infected with SARS-CoV-2 post-surgery. The infected cohort experienced significantly higher morbidity and mortality compared to their noninfected counterparts. Notably, the diagnosis of cancer in the United Kingdom was intricately linked to elevated postoperative 90-day mortality. This underscores a critical imperative for stringent surgical protocols to avert infections and mitigate associated risks, emphasizing the pressing need to reassess the intricate relationship between COVID-19 and cancer [48, 51, 52]. Formulating effective treatment strategies is crucial for navigating complications and proactively preventing further risks in this complex intersection [53].

Socio-psychological consequences: The mental well-being of cancer patients has taken a toll amid the COVID-19 pandemic, manifesting in heightened stress, severe anxiety fuelled by infection fears, and a reduction in societal social support. The global healthcare landscape has witnessed substantial transformations, with countries fortifying their infrastructure, judiciously managing essential resources like ventilators, personal protective equipment, vaccine doses, and medicinal stocks. These measures aim to prevent overwhelming hospital systems, ensuring they can effectively handle the rising influx of COVID-19 patients (Fig. 26.4). However, this strategic focus on pandemic response has inadvertently marginalized the treatment of other diseases, with cancer care suffering significantly. The apprehension of contracting COVID-19 further complicates matters for patients visiting hospitals, even for routine check-ups. In this challenging scenario, a delicate balance between pandemic response and maintaining essential healthcare services, especially for cancer patients, becomes imperative [54, 55]. The dynamics of healthcare are undergoing a significant shift in response to the COVID-19 pandemic, necessitating the reorganization of doctors and healthcare workers with revised priorities. Despite critical efforts in infrastructure strengthening and economic resource allocation, shortages in hospital beds and the healthcare workforce are surfacing. This has, unfortunately, led to the postponement of cancer diagnosis and treatment. The current landscape of the COVID-19 outbreak is profoundly impacting cancer care. Preventive measures, such as restricting in-hospital visits to mitigate contagious contact infections, are causing delays in primary diagnostic care and impeding curative interventions like surgery, radiotherapy, and chemotherapy/immunotherapy. The challenge lies in finding a delicate balance between addressing the pandemic's demands and ensuring continued and timely care for cancer patients [56]. The impact of delays in diagnosis and treatment, in turn, has caused great patient distress and psychological impact. As the world is trying to cope with the grave situation and trying to get back to the "pre-pandemic state," governing bodies have recommended the continuation of most oncology-related operations. However, there arises the dilemma of judgment and the decision of choosing between imperative operations and the ones that can be safely delayed. The governmental resource system, along with balancing the risk of cancer progression with the risk of an ever-increasing COVID-19 infection trail, has made the entire situation very complex and has greatly impacted the progress of society with a deteriorating healthcare system plagued with such difficult scenarios. There is an increased public health burden caused by disruptions to cancer screening and diagnosis during the pandemic, leading to global chaos. An estimated 39–90% decrease has been seen in the screening of major cancer types such as breast, lung, prostate, colon, and cervical cancers according to the projections of the American Cancer Society. They have also found that approximately 5000 cases were identifiable in a short period during pandemic [57, 58]. The unnecessary hindrance in the path of diagnosis and treatment can escalate the disease conditions and later on will require more intensive therapeutic strategies to halt rates of COVID-19-influenced cancer mortality. Several modeled evaluations have begun in various parts of the world to estimate the short- and long-term effects of disruptions in cancer screening. One such modeling study has been done by Working Group 2 of the COVID and Cancer Modelling Consortium (CCGMC). They are dealing with analyzing the impact of disruptions in breast screening, along with a comparative study of treatment and recovery options under diversified international settings [59, 60].

The norms of social distancing during the pandemic have significantly impacted the psychological well-being of cancer patients, resulting in mental disorders, loss of self-confidence, and diminished inner motivation across various societal strata. A recent meta-analysis revealed that post-traumatic stress symptoms were prevalent in 23.9% of the general population and intensified to 24.8% among individuals experiencing psychological stress, with even higher figures observed in cancer patients. In China, a study of 6213 cancer patients disclosed that 23.4% suffered from depression, 17.7% from anxiety, 9.3% from post-traumatic stress disorder, and 13.5% from hostility. In Italy, among 77 lymphoma patients undergoing chemotherapy, 36% reported anxiety, 31% depression, and 36% post-traumatic stress disorder. A comparative study by Lou et al. among cancer patients undergoing active treatment, those with a history of cancer, and healthy controls highlighted the distress of active cancer treatment recipients concerning COVID-19 protocols. Additionally, Ng et al. conducted telephone surveys on cancer survivors and healthy controls, revealing that 1 in 8 individuals reported clinical anxiety in both cohorts, while depression was more prevalent among cancer survivors. These findings underscore the profound psychological impact of the pandemic on the mental health of cancer patients [57, 58, 61–63].

The indirect impact of the COVID-19 pandemic on children ailing from cancer and their families is also a cause of concern. A study on cancer patients in Milan (children and young adolescents) revealed that cancer patients were worried and felt at risk of severe complications, with parental concern becoming an additional burden. The over-exerted healthcare system extends limited support to a vulnerable population, leading to lesser access to hospitals, increased incidence of intravenous treatments at home, and palliative care. A team of researchers utilized a validated artificial intelligence framework to conduct a real-time analysis of two datasets during lockdown, consisting of 2,469,822 tweets and 21,800 discussions by cancer patients. This approach was implemented through the Patient-Reported Information Multidimensional Exploration (PRIME). The most common emotional concerns revolved around delayed diagnosis, canceled, or missed treatments, and weakened immunity. Negative sentiment prevailed as the dominant emotion and required robust emotional support [57, 58, 60, 62, 64].

Infectious Diseases

COVID-19 comorbidity of patients with other infectious diseases: The societal and psychological impact of COVID-19 has disproportionately affected marginalized and vulnerable populations, including individuals with pre-existing health conditions, those struggling with substance abuse, and people of lower socioeconomic status. Groups such as individuals affected by HIV/AIDS, the homeless, migrants, residents of underdeveloped nations, African American and indigenous communities, as well as low-income, poorly educated, and undocumented individuals, have faced heightened challenges. Limited access to primary healthcare has left many in these high-risk groups especially vulnerable to the pandemic's effects, placing them at the greatest risk from strained healthcare systems.

Additionally, the high mortality rate has led to complex forms of bereavement, deeply impacting the emotional fabric of communities. Discrimination and stigma have also been significant issues for patients with illnesses like cancer, prompting some to hide their diagnosis from employers or avoid COVID-19 testing and treatment out of fear of job loss or further marginalization. These compounded social and psychological stresses underscore the urgent need for targeted support and inclusive healthcare policies [65–67]. The doctors and healthcare workers (HCWs), the support system of society in these times of distress, who have worked tirelessly in coping with the deteriorating health conditions across the world due to a raging pandemic, have also been impacted by COVID-19. HCWs suffer from excessive workloads with erratic timings due to the over-exerted hospitals and health centers, accentuated by the extended duration of separation from family, inconvenient personal protection equipment usage, pressures of ethical dilemmas, physical isolation, and periods of quarantine. There is also the problem of balancing the risk of admitting patients to the hospital and the risk of contracting SARS-Cov-2 infection, leading to an emotional impact on staff, patients, and their respective families. They have also faced critical decisive moments where they have to judge between treating COVID patients or continuing with the routine oncological care for cancer patients because resources are limited, and governmental policies all across the world have increasingly shifted focus toward decelerating the pace of the spread of COVID-19 at the cost of delaying the treatment of other diseases, including cancer. In a population-based survey conducted in Shanghai, it was found that social support, dietary/nutritional support, and psychological counseling were the main needs of cancer patients. Vaccinated cancer patients showed reduced psychological anxiety and were less likely to need psychological counseling [54, 55, 63, 67–70].

Therefore, the need of the hour is proper analysis of the psychological impact of COVID-19 and appropriate referrals to psychiatric and psycho-oncology therapists. Beyond traditional approaches, innovative initiatives have surfaced globally. In Canada, clinicians are leveraging technological interventions for psychological support to cancer patients, utilizing methods like text messaging and toll-free dedicated numbers. These initiatives not only offer cost-effective mental health interventions but also adapt to the challenges imposed by the pandemic. Virtual meetings and support group systems have concurrently emerged as valuable tools, playing a crucial role in alleviating the psychological impact of the pandemic on cancer patients. The importance of empathy has emerged due to the difficulty in picking up social cues in the absence of physical interactions between clinicians and patients. Therefore, virtual encounters need to take into account the importance of being humane and empathetic attitude while dealing with patients during such times [71].

Global Health Landscape in the COVID-19 Pandemic

The global health landscape has been profoundly shaken by the COVID-19 pandemic. By mid-October 2021, over 240 million people had contracted the virus, resulting in nearly five million deaths. Beyond the immediate toll, millions of survivors grapple with persistent symptoms, disrupting their return to normal life. Mental distress has surged, impacting individuals across diverse backgrounds. Notably, the virus has disproportionately affected those from disadvantaged socioeconomic backgrounds, heightening the risk of infection and mortality. Healthcare disruptions have rippled through various sectors, affecting cancer screenings, nonurgent surgeries, and emergency care. The introduction of vaccinations in 2021 marked a turning point, significantly reducing the risk of severe illness and death. However, challenges persist, including vaccination hesitancy in certain populations and concerns about waning vaccine effectiveness. As the world navigates this ongoing crisis, the resilience of healthcare systems and collective efforts to overcome these hurdles remain pivotal [72–76]. The COVID-19 pandemic has had a major impact on OECD countries, with over 110 million reported infections and 2.1 million deaths by mid-October 2021. This accounts for nearly half of all cases and deaths worldwide. The actual number of infections is likely even higher [14, 77–79]. The number of cases and deaths has varied from country to country, with some regions experiencing multiple peaks. In Europe, there were peaks in late 2020 and early 2021, with some countries seeing a resurgence in cases in July 2021. The United States and Canada saw a similar pattern, but with a spike in cases and deaths in August and September 2021. In Latin America, the situation varied, with peaks in infections and deaths occurring at different times in different countries. Asia-Pacific OECD countries generally had lower incidence and death rates, but some of them experienced an increase in cases in the third quarter of 2021. These variations in the pandemic can be attributed to differences in containment and mitigation strategies, healthcare system capabilities, and the timing of their implementation. Case fatality rates decreased over time due in part to increased case detection, vaccination campaigns, improved disease management, and stronger healthcare system capacity. Other factors, such

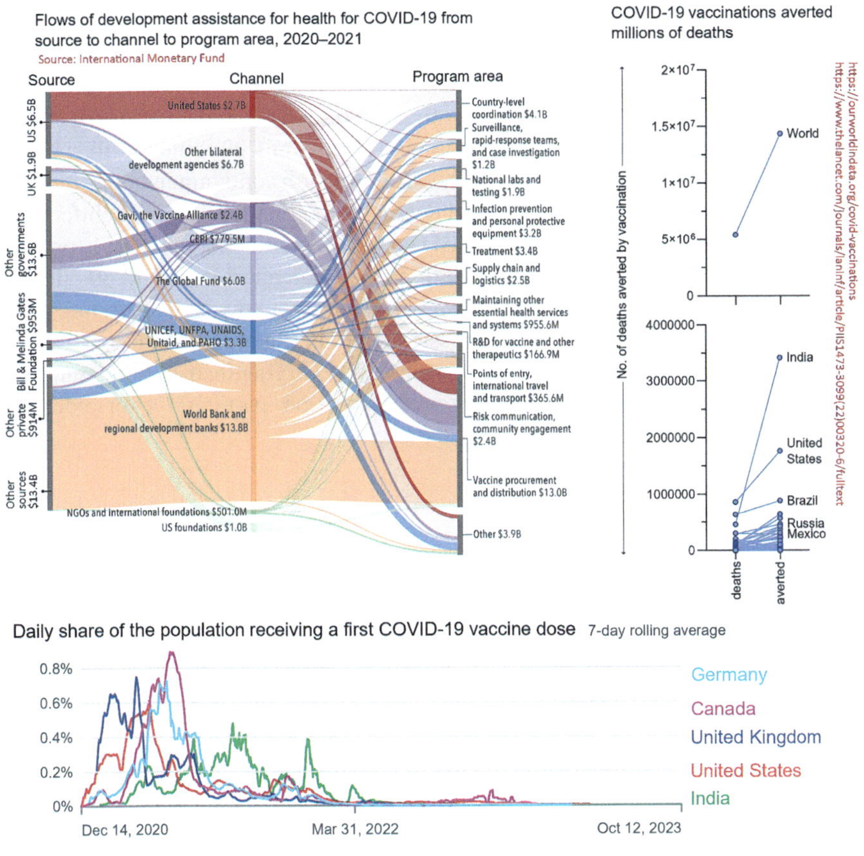

Fig. 26.5 Challenges and consequences in vaccination. The figure depicts the challenges faced in the health sector during COVID-19 pandemic and vaccination by the government and other private sources

as geography, demographics, and the prevalence of risk factors such as obesity, also contributed to differences in infection and mortality rates [80, 81] (Fig. 26.5). The emergence of "variants of concern," particularly the Delta variant, has played a significant role in the evolution of the pandemic. The Delta variant, initially detected in October 2020, swiftly asserted dominance in most OECD countries by mid-2021. It is more transmissible and leads to more severe infections, with a higher risk of hospitalization and death, particularly among the unvaccinated population. The introduction of COVID-19 vaccines in 2021 marked a pivotal advancement in the worldwide campaign to combat the pandemic. These vaccines, approved in OECD nations, have proven to be highly effective in significantly lowering the chances of symptomatic infection, hospitalization, and mortality. For instance, two doses of mRNA vaccines like Pfizer-BioNTech and Moderna exhibit real-world efficacy of over 85%. Furthermore, these vaccines provide even stronger protection against severe illness and fatality. While their efficacy is somewhat reduced in preventing symptomatic infection with the Delta variant, they remain highly efficient in reducing hospitalization and mortality. Factors influencing the speed of vaccination rollout include regulatory processes, vaccine procurement and distribution, infrastructure, healthcare workforce capacity, and vaccine hesitancy. Early vaccination campaigns in countries like Israel, the United Kingdom, and the United States, along with containment measures, led to a significant spread of infection and deaths in the first half of 2021. Infections increased again in these and other OECD countries starting in June 2021, primarily among the unvaccinated, as vaccination rates plateaued at around 60–70% of the population. Data showed that unvaccinated individuals had significantly higher rates of infection, hospital admissions, and ICU admissions. Booster doses have been introduced in 15 OECD countries, with varying distribution strategies, and initial evidence suggests they enhance protection, particularly for older age groups [82–84].

Conclusion and Future Perspectives

The rapid development of vaccines has significantly helped reduce infections and mortality rates from COVID-19. However, the pandemic continues to leave a profound impact on physical, mental, social, and economic well-being. Unequal vaccine access, inconsistent public awareness, and limited healthcare capacities have contributed to the spread and mutation of the virus into new variants. Long-term health effects, including persistent comorbidities and psychological trauma, remain pressing concerns worldwide. Addressing the therapeutic challenges posed by SARS-CoV-2, particularly among individuals with respiratory and cardiovascular conditions, is essential as these comorbidities have led to widespread loss of life and increased health vulnerabilities. Future strategies must focus on stabilizing socioeconomic disruptions caused by the pandemic. Key initiatives should include promoting safe and hygienic practices, prioritizing care for high-risk groups, and integrating psychological support for healthcare workers. Embracing telemedicine, remote work, and a flexible, resilient work culture will be vital in adapting to the altered landscape. Ensuring access to essential supplies, prioritizing mental health, and fostering a positive, proactive approach will be essential to building a more resilient society—a way forward toward a peaceful coexistence (Fig. 26.6).

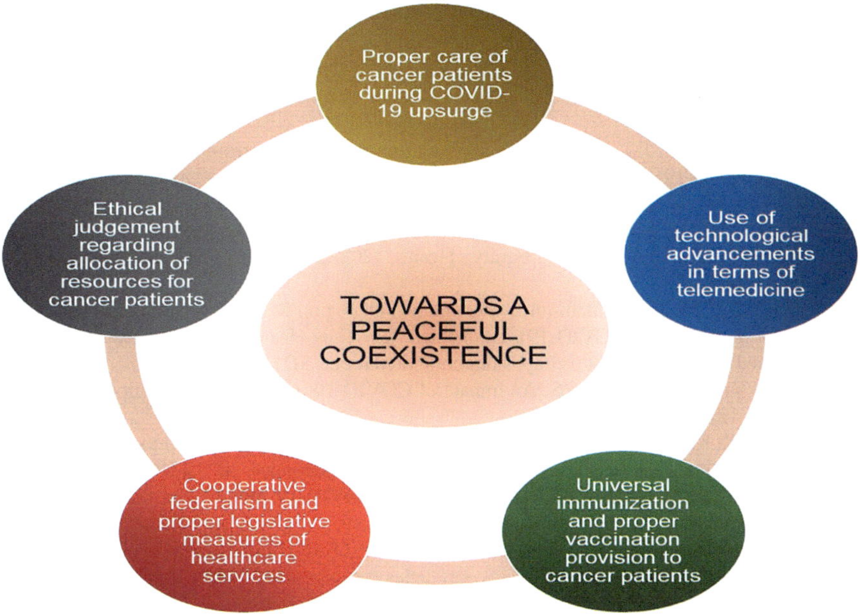

Fig. 26.6 Way forward toward a peaceful coexistence. The diagram highlights the steps that will lead to COVID-19 care, technological advancements, vaccination, legislative measures, and cancer comorbidities in terms of improved therapeutic outcomes and increased attention to handling the scenario

Acknowledgments The authors are deeply indebted to the researchers and publishers, whose scientific literature helped in this review. We are also thankful to the site BioRender.com as a few pictorial motifs in figures were partially created with BioRender.com tools.

Authors' Contributions All authors conceived the idea and wrote and edited the manuscript. All authors have read, agreed, and approved the final draft of the manuscript.

Funding This work is jointly supported by the Department of Science and Technology (NanoMission: DST/NM/NT/2018/105(G); SERB: EMR/2017/000992) and Focused Basic Research (FBR): #MLP-142, HCT, CSIR, Govt. of India.

Conflicts of Interest There is no competing and conflict of interests.

References

1. Han HJ, Nwagwu C, Anyim O, Ekweremadu C, Kim S. COVID-19 and cancer: from basic mechanisms to vaccine development using nanotechnology. Int Immunopharmacol. 2021;90:107247.
2. Wu Z, McGoogan JM. Characteristics of and important lessons from the coronavirus disease 2019 (COVID-19) outbreak in China: summary of a report of 72 314 cases from the Chinese Center for Disease Control and Prevention. JAMA. 2020;323(13):1239–42.
3. Allan M, Lièvre M, Laurenson-Schafer H, et al. The World Health Organization COVID-19 surveillance database. Int J Equity Health. 2022;21(S3):167. https://doi.org/10.1186/s12939-022-01767-5.
4. Rajnik M, Cascella M, Cuomo A, Dulebohn SC, Di Napoli R. Features, evaluation, and treatment of coronavirus (COVID-19). Uniformed Serv Univ Health Sci. Published online 2021. Accessed 27 Nov 2023. https://apps.dtic.mil/sti/citations/AD1127230.
5. Chavda VP, Ping FF, Chen ZS. An impact of COVID-19 on cancer care: an update. Vaccine. 2022;10(12):2072.
6. Ghosh MK, Kumar S, Ganguly KK, et al. COVID-19 and cancer: insights into their association and influence on genetic and epigenetic landscape. Epigenomics. 2023;15(4):227–48.
7. Kumar S, Basu M, Ghosh P, Ansari A, Ghosh MK. COVID-19: clinical status of vaccine development to date. Br J Clin Pharmacol. 2023;89:114–49.
8. Kumar S, Basu M, Ghosh P, Pal U, Ghosh MK. COVID-19 therapeutics: clinical application of repurposed drugs and futuristic strategies for target-based drug discovery. Genes Dis. 2023;10(4):1402–28.
9. Long B, Brady WJ, Koyfman A, Gottlieb M. Cardiovascular complications in COVID-19. Am J Emerg Med. 2020;38(7):1504–7.
10. Sykes DL, Holdsworth L, Jawad N, Gunasekera P, Morice AH, Crooks MG. Post-COVID-19 symptom burden: what is long-COVID and how should we manage it? Lung. 2021;199(2):113–9. https://doi.org/10.1007/s00408-021-00423-z.
11. Hosseinzadeh P, Zareipour M, Baljani E, Moradali MR. Social consequences of the COVID-19 pandemic. A systematic review. Investig Educ Enferm. 2022;40(1). Accessed 19 Nov 2023. http://www.scielo.org.co/scielo.php?pid=S0120-53072022000100010&script=sci_arttext.
12. Singh K, Kondal D, Mohan S, et al. Health, psychosocial, and economic impacts of the COVID-19 pandemic on people with chronic conditions in India: a mixed methods study. BMC Public Health. 2021;21(1):685. https://doi.org/10.1186/s12889-021-10708-w.
13. Saladino V, Algeri D, Auriemma V. The psychological and social impact of Covid-19: new perspectives of well-being. Front Psychol. 2020;11:2550.
14. Unraveling COVID-19 diagnostics: a roadmap for future pandemic—nature cell and science. Accessed 15 Nov 2024. https://cellnatsci.com/2958-695X/article/10-61474-ncs-2024-00026.
15. Muftah M. Impact of social media on learning English language during the COVID-19 pandemic. PSU Res Rev. 2024;8(1):211–26.
16. Khan M, Khan N, Begum S, Qureshi MI. Digital future beyond pandemic outbreak: systematic review of the impact of COVID-19 outbreak on digital psychology. Foresight. 2024;26(1):1–17.
17. Marler EK, Bruce MJ, Abaoud A, et al. The impact of COVID-19 on university students' academic motivation, social connection, and psychological well-being. Scholarsh Teach Learn Psychol. 2024;10(3):320.
18. Sanyaolu A, Okorie C, Marinkovic A, et al. Comorbidity and its impact on patients with COVID-19. SN Compr Clin Med. 2020;2(8):1069–76. https://doi.org/10.1007/s42399-020-00363-4.
19. Gasmi A, Peana M, Pivina L, et al. Interrelations between COVID-19 and other disorders. Clin Immunol. 2021;224:108651. https://doi.org/10.1016/j.clim.2020.108651.

20. Adir Y, Saliba W, Beurnier A, Humbert M. Asthma and COVID-19: an update. Eur Respir Rev. 2021;30(162). https://doi.org/10.1183/16000617.0152-2021.
21. Higham A, Mathioudakis A, Vestbo J, Singh D. COVID-19 and COPD: a narrative review of the basic science and clinical outcomes. Eur Respir Rev. 2020;29(158). https://doi.org/10.1183/16000617.0199-2020.
22. Ejaz H, Alsrhani A, Zafar A, et al. COVID-19 and comorbidities: deleterious impact on infected patients. J Infect Public Health. 2020;13(12):1833–9. https://doi.org/10.1016/j.jiph.2020.07.014.
23. Leung JM, Niikura M, Yang CWT, Sin DD. COVID-19 and COPD. Eur Respir J. 2020;56(2). https://doi.org/10.1183/13993003.02108-2020.
24. Sreenath K, Batra P, Vinayaraj EV, et al. Coinfections with other respiratory pathogens among patients with COVID-19. Microbiol Spectr. 2021;9(1):e0016321. https://doi.org/10.1128/Spectrum.00163-21.
25. Lodha R, Puranik M, Kattal N, Kabra SK. Social and economic impact of childhood asthma. Indian Pediatr. 2003;40(9):874–9.
26. Carr TF, Fajt ML, Kraft M, et al. Treating asthma in the time of COVID. J Allergy Clin Immunol. 2023;151(4):809–17. https://doi.org/10.1016/j.jaci.2022.12.800.
27. Nocon A, Booth T. The social impact of asthma. Fam Pract. 1991;8(1):37–41. https://doi.org/10.1093/fampra/8.1.37.
28. Patterson GE, McIntyre KM, Clough HE, Rushton J. Societal impacts of pandemics: comparing COVID-19 with history to focus our response. Front Public Health. 2021;9:630449. https://doi.org/10.3389/fpubh.2021.630449.
29. Poureslami I, FitzGerald JM, Tregobov N, Goldstein RS, Lougheed MD, Gupta S. Health literacy in asthma and chronic obstructive pulmonary disease (COPD) care: a narrative review and future directions. Respir Res. 2022;23(1):361. https://doi.org/10.1186/s12931-022-02290-5.
30. Cravo A, Attar D, Freeman D, Holmes S, Ip L, Singh SJ. The importance of self-management in the context of personalized care in COPD. Int J Chron Obstruct Pulmon Dis. 2022;17:231–43. https://doi.org/10.2147/COPD.S343108.
31. Franssen FME, Smid DE, Deeg DJH, et al. The physical, mental, and social impact of COPD in a population-based sample: results from the longitudinal aging study Amsterdam. NPJ Prim Care Respir Med. 2018;28(1):30. https://doi.org/10.1038/s41533-018-0097-3.
32. Yohannes AM. COPD patients in a COVID-19 society: depression and anxiety. Expert Rev Respir Med. 2021;15(1):5–7. https://doi.org/10.1080/17476348.2020.1787835.
33. Lee SC, Son KJ, Han CH, Park SC, Jung JY. Impact of COPD on COVID-19 prognosis: a nationwide population-based study in South Korea. Sci Rep. 2021;11(1):3735. https://doi.org/10.1038/s41598-021-83226-9.
34. Lenferink A, van der Palen J, Effing T. The role of social support in improving chronic obstructive pulmonary disease self-management. Expert Rev Respir Med. 2018;12(8):623–6. https://doi.org/10.1080/17476348.2018.1489723.
35. Azevedo RB, Botelho BG, de Hollanda JVG, et al. Covid-19 and the cardiovascular system: a comprehensive review. J Hum Hypertens. 2021;35(1):4–11. https://doi.org/10.1038/s41371-020-0387-4.
36. Bansal M. Cardiovascular disease and COVID-19. Diabetes Metab Syndr. 2020;14(3):247–50. https://doi.org/10.1016/j.dsx.2020.03.013.
37. Chen N, Zhou M, Dong X, et al. Epidemiological and clinical characteristics of 99 cases of 2019 novel coronavirus pneumonia in Wuhan, China: a descriptive study. Lancet. 2020;395(10223):507–13. https://doi.org/10.1016/S0140-6736(20)30211-7.
38. Guan WJ, Ni ZY, Hu Y, et al. Clinical characteristics of coronavirus disease 2019 in China. N Engl J Med. 2020;382(18):1708–20. https://doi.org/10.1056/NEJMoa2002032.

39. Mofijur M, Fattah IMR, Alam MA, et al. Impact of COVID-19 on the social, economic, environmental and energy domains: lessons learnt from a global pandemic. Sustain Prod Consum. 2021;26:343–59. https://doi.org/10.1016/j.spc.2020.10.016.
40. Tobler DL, Pruzansky AJ, Naderi S, Ambrosy AP, Slade JJ. Long-term cardiovascular effects of COVID-19: emerging data relevant to the cardiovascular clinician. Curr Atheroscler Rep. 2022;24(7):563–70. https://doi.org/10.1007/s11883-022-01032-8.
41. Khawaja SA, Mohan P, Jabbour R, et al. COVID-19 and its impact on the cardiovascular system. Open Heart. 2021;8(1). https://doi.org/10.1136/openhrt-2020-001472.
42. Marchi J, Johansson N, Sarkadi A, Warner G. The impact of the COVID-19 pandemic and societal infection control measures on children and adolescents' mental health: a scoping review. Front Psych. 2021;12:711791. https://doi.org/10.3389/fpsyt.2021.711791.
43. Schnitzler L, Janssen LMM, Evers SMAA, et al. The broader societal impacts of COVID-19 and the growing importance of capturing these in health economic analyses. Int J Technol Assess Health Care. 2021;37:e43. https://doi.org/10.1017/S0266462321000155.
44. Naylor-Wardle J, Rowland B, Kunadian V. Socioeconomic status and cardiovascular health in the COVID-19 pandemic. Heart. 2021;107(5):358–65. https://doi.org/10.1136/heartjnl-2020-318425.
45. Lewek J, Jatczak-Pawlik I, Maciejewski M, Jankowski P, Banach M. COVID-19 and cardiovascular complications - preliminary results of the LATE-COVID study. Arch Med Sci. 2021;17(3):818–22. https://doi.org/10.5114/aoms/134211.
46. Sabatino J, De Rosa S, Di Salvo G, Indolfi C. Impact of cardiovascular risk profile on COVID-19 outcome. A meta-analysis. PloS One. 2020;15(8):e0237131. https://doi.org/10.1371/journal.pone.0237131.
47. Kreatsoulas C, Anand SS. The impact of social determinants on cardiovascular disease. Can J Cardiol. 2010;26(Suppl C):8C–13C. https://doi.org/10.1016/s0828-282x(10)71075-8.
48. Liu C, Zhao Y, Okwan-Duodu D, Basho R, Cui X. COVID-19 in cancer patients: risk, clinical features, and management. Cancer Biol Med. 2020;17(3):519–27. https://doi.org/10.20892/j.issn.2095-3941.2020.0289.
49. Dimou A. Areas of uncertainty in SARS-CoV-2 vaccination for cancer patients. Vaccine. 2022;10(12). https://doi.org/10.3390/vaccines10122117.
50. Saini G, Aneja R. Cancer as a prospective sequela of long COVID-19. BioEssays News Rev Mol Cell Dev Biol. 2021;43(6):e2000331. https://doi.org/10.1002/bies.202000331.
51. Mohanty A, Agnihotri S, Mehta A, Rawal S. COVID-19 and cancer: sailing through the tides. Pathol Res Pract. 2021;221:153417. https://doi.org/10.1016/j.prp.2021.153417.
52. Gosain R, Abdou Y, Singh A, Rana N, Puzanov I, Ernstoff MS. COVID-19 and cancer: a comprehensive review. Curr Oncol Rep. 2020;22(5):53. https://doi.org/10.1007/s11912-020-00934-7.
53. Bakouny Z, Hawley JE, Choueiri TK, et al. COVID-19 and cancer: current challenges and perspectives. Cancer Cell. 2020;38(5):629–46. https://doi.org/10.1016/j.ccell.2020.09.018.
54. Al-Quteimat OM, Amer AM. The impact of the COVID-19 pandemic on cancer patients. Am J Clin Oncol. 2020;43(6):452–5. https://doi.org/10.1097/COC.0000000000000712.
55. De Guzman R, Malik M. Dual challenge of cancer and COVID-19: impact on health care and socioeconomic systems in Asia Pacific. JCO Glob Oncol. 2020;6:906–12. https://doi.org/10.1200/GO.20.00227.
56. Raymond E, Thieblemont C, Alran S, Faivre S. Impact of the COVID-19 outbreak on the management of patients with cancer. Target Oncol. 2020;15(3):249–59. https://doi.org/10.1007/s11523-020-00721-1.
57. Mohammed AH, Blebil A, Dujaili J, Rasool-Hassan BA. The risk and impact of COVID-19 pandemic on immunosuppressed patients: cancer, HIV, and solid organ transplant recipients. AIDS Rev. 2020;22(3):151–7. https://doi.org/10.24875/AIDSRev.20000052.
58. Moraliyage H, De Silva D, Ranasinghe W, et al. Cancer in lockdown: impact of the COVID-19 pandemic on patients with cancer. Oncologist. 2021;26(2):e342–4. https://doi.org/10.1002/onco.13604.

59. Patt D, Gordan L, Diaz M, et al. Impact of COVID-19 on cancer care: how the pandemic is delaying cancer diagnosis and treatment for American seniors. JCO Clin Cancer Inform. 2020;4:1059–71. https://doi.org/10.1200/CCI.20.00134.
60. Figueroa JD, Gray E, Pashayan N, et al. The impact of the Covid-19 pandemic on breast cancer early detection and screening. Prev Med. 2021;151:106585. https://doi.org/10.1016/j.ypmed.2021.106585.
61. Ahmad A, Mueller C, Tsamakis K. Covid-19 pandemic: a public and global mental health opportunity for social transformation? BMJ. 2020;369:m1383. https://doi.org/10.1136/bmj.m1383.
62. Onder G, Rezza G, Brusaferro S. Case-fatality rate and characteristics of patients dying in relation to COVID-19 in Italy. JAMA. 2020;323(18):1775–6. https://doi.org/10.1001/jama.2020.4683.
63. Tsamakis K, Gavriatopoulou M, Schizas D, et al. Oncology during the COVID-19 pandemic: challenges, dilemmas and the psychosocial impact on cancer patients. Oncol Lett. 2020;20(1):441–7. https://doi.org/10.3892/ol.2020.11599.
64. Hall PA, Sheeran P, Fong GT, et al. Biobehavioral aspects of the COVID-19 pandemic: a review. Psychosom Med. 2021;83(4):309–21. https://doi.org/10.1097/PSY.0000000000000932.
65. Shreffler J, Petrey J, Huecker M. The impact of COVID-19 on healthcare worker wellness: a scoping review. West J Emerg Med. 2020;21(5):1059–66. https://doi.org/10.5811/westjem.2020.7.48684.
66. Mehta S, Machado F, Kwizera A, et al. COVID-19: a heavy toll on health-care workers. Lancet Respir Med. 2021;9(3):226–8. https://doi.org/10.1016/S2213-2600(21)00068-0.
67. Shim M, Cappella JN, Han JY. How does insightful and emotional disclosure bring potential health benefits?: study based on online support groups for women with breast cancer. J Commun. 2011;61(3):432–64. https://doi.org/10.1111/j.1460-2466.2011.01555.x.
68. Teijaro JR, Farber DL. COVID-19 vaccines: modes of immune activation and future challenges. Nat Rev Immunol. 2021;21(4):195–7. https://doi.org/10.1038/s41577-021-00526-x.
69. Desai A, Gainor JF, Hegde A, et al. COVID-19 vaccine guidance for patients with cancer participating in oncology clinical trials. Nat Rev Clin Oncol. 2021;18(5):313–9. https://doi.org/10.1038/s41571-021-00487-z.
70. Del Vecchio BG, Calabrese E, Biancone L, Monteleone G, Paoluzi OA. The impact of COVID-19 pandemic in the colorectal cancer prevention. Int J Color Dis. 2020;35(10):1951–4. https://doi.org/10.1007/s00384-020-03635-6.
71. Attai DJ, Cowher MS, Al-Hamadani M, Schoger JM, Staley AC, Landercasper J. Twitter social media is an effective tool for breast cancer patient education and support: patient-reported outcomes by survey. J Med Internet Res. 2015;17(7):e188. https://doi.org/10.2196/jmir.4721.
72. Cypress BS. COVID-19: the economic impact of a pandemic on the healthcare delivery system in the United States. Nurs Forum. 2022;57(2):323–7. https://doi.org/10.1111/nuf.12677.
73. Ghosh MK, Kumar S, Begam S, Ghosh S, Basu M. GBM immunotherapy: exploring molecular and clinical frontiers. Life Sci. 2024;356:123018. https://doi.org/10.1016/j.lfs.2024.123018.
74. Kumar S, Basu M, Ghosh MK. Chaperone-assisted E3 ligase CHIP: a double agent in cancer. Genes Dis. 2021;9(6). https://doi.org/10.1016/j.gendis.2021.08.003.
75. Kumar S, Chatterjee M, Ghosh P, Ganguly KK, Basu M, Ghosh MK. Targeting PD-1/PD-L1 in cancer immunotherapy: an effective strategy for treatment of triple-negative breast cancer (TNBC) patients. Genes Dis. 2023;10(4):1318–50.
76. Kumar S, Ansari A, Basu M, Ghosh S, Begam S, Ghosh MK. Carbon nanotubes in cancer diagnosis and treatment: current trends and future perspectives. Adv Ther. 2025;8:2400283. https://doi.org/10.1002/adtp.202400283.
77. Kaye AD, Okeagu CN, Pham AD, et al. Economic impact of COVID-19 pandemic on healthcare facilities and systems: international perspectives. Best Pract Res Clin Anaesthesiol. 2021;35(3):293–306.

78. Kumar S, Basu M, Ghosh MK. E3 ubiquitin ligases and deubiquitinases in colorectal cancer: emerging molecular insights and therapeutic opportunities. Biochim Biophys Acta BBA-Mol Cell Res. 2024;1871:119827.
79. Sarkar S, Kumar S, Saha G, Basu M, Ghosh MK. Glioma nanotherapy: unleashing the synergy of dual-loaded DIM and TMZ. Int J Pharm. 2024;665:124697.
80. Lam ICH, Zhang R, Man KKC, et al. Persistence in risk and effect of COVID-19 vaccination on long-term health consequences after SARS-CoV-2 infection. Nat Commun. 2024;15(1):1716.
81. Kisa S, Kisa A. A comprehensive analysis of COVID-19 misinformation, public health impacts, and communication strategies: scoping review. J Med Internet Res. 2024;26:e56931.
82. Hungerford D, Cunliffe NA. Real world effectiveness of covid-19 vaccines. BMJ. 2021;374. Accessed 17 Nov 2023. https://www.bmj.com/content/374/bmj.n2034.full.
83. Zheng C, Shao W, Chen X, Zhang B, Wang G, Zhang W. Real-world effectiveness of COVID-19 vaccines: a literature review and meta-analysis. Int J Infect Dis. 2022;114:252–60.
84. Duho KCT, Taylor A, Amamoo RS, Amamoo RS. COVID-19 vaccination, supply chain and concerns on vaccine hesitancy in Africa. World Financ Rev Sept Ed. 2021:36–40. https://ssrn.com/abstract=3969368.

GPSR Compliance

The European Union's (EU) General Product Safety Regulation (GPSR) is a set of rules that requires consumer products to be safe and our obligations to ensure this.

If you have any concerns about our products, you can contact us on ProductSafety@springernature.com

In case Publisher is established outside the EU, the EU authorized representative is:

Springer Nature Customer Service Center GmbH
Europaplatz 3
69115 Heidelberg, Germany

Batch number: 10755222

Printed by Printforce, the Netherlands